Chemoinformatics and Advanced Machine Learning Perspectives:
Complex Computational Methods and Collaborative Techniques

Huma Lodhi
Imperial College, UK

Yoshihiro Yamanishi
Mines ParisTech—Institut Curie—Inserm U900, France

T0320488

Medical Information Science
REFERENCE

MEDICAL INFORMATION SCIENCE REFERENCE

Hershey · New York

Director of Editorial Content: Kristin Klinger
Director of Book Publications: Julia Mosemann
Acquisitions Editor: Lindsay Johnston
Development Editor: Christine Bufton
Publishing Assistant: Deanna Jo Zombro
Typesetter: Deanna Jo Zombro
Production Editor: Jamie Snavely
Cover Design: Lisa Tosheff

Published in the United States of America by
    Medical Information Science Reference (an imprint of IGI Global)
    701 E. Chocolate Avenue
    Hershey PA 17033
    Tel: 717-533-8845
    Fax: 717-533-8661
    E-mail: cust@igi-global.com
    Web site: http://www.igi-global.com

Library of Congress Cataloging-in-Publication Data

Chemoinformatics and advanced machine learning perspectives : complex computational methods and collaborative techniques / Huma Lodhi and Yoshihiro Yamanashi, editors.
    p. cm.
  Summary: "This book is a timely compendium of key elements that are crucial for the study of machine learning in chemoinformatics, giving an overview of current research in machine learning and their applications to chemoinformatics tasks"--Provided by publisher. Includes bibliographical references and index. ISBN 978-1-61520-911-8 (hardcover) -- ISBN 978-1-61520-912-5 (ebook) 1. Cheminformatics. 2. Machine learning. I. Lodhi, Huma M. II. Yamanashi, Yoshihiro, 1976-
  QD39.3.E46C45 2010
  542'.85--dc22
                        2009054321

British Cataloguing in Publication Data
A Cataloguing in Publication record for this book is available from the British Library.

All work contributed to this book is new, previously-unpublished material. The views expressed in this book are those of the authors, but not necessarily of the publisher.

List of Reviewers

Andras Szilagyi, *University of Kansas, USA*
Roman Rosipal, *Medical University of Vienna, Austria*
Jurgen Bajorath, *University of Bonn, Germany*
Masahiro Hattori, *Tokyo University of Technology, Japan*
Huma Lodhi, *Brunel University, UK*
Viviana Consonni, *Milano-Bicocca University Environmental, Italy*
Nicos Angelopoulos, *Edinburgh University, UK*
Peter Willett, *University of Sheffield, UK*
Yoshihiro Yamanishi, *Mines ParisTech-Institut Curie-Inserm U900, France*
Koji Tsuda, *National Institute of Advanced Industrial Science and Technology (AIST), Japan*
Michael Schmuker, *Freie Universit ät Berlin, Germany*
Rahul Singh, *San Francisco State University, USA*
Jean-Philippe Vert, *Mines ParisTech-Institut Curie-Inserm U900, France*
Kimito Funatsu, *The University of Tokyo, Japan*
Hisashi Kashima, *University of Tokyo, Japan*
Shuxing Zhang, *The University of Texas at M.D. Anderson Cancer Center, USA*
Holger Froehlich, *Cellzome AG, Germany*

Table of Contents

Section 1
Similarity Design in Chemical Space

Section 2
Graph-Based Approaches in Chemoinformatics

Section 3
Statistical and Bayesian Approaches for Virtual Screenin

Detailed Table of Contents

Section 1
Similarity Design in Chemical Space

Hisashi Kashima, University of Tokyo, Japan
Hiroto Saigo, Kyushu Institute of Technology, Japan
Masahiro Hattori, Tokyo University of Technology, Japan
Koji Tsuda, AIST Computational Biology Research Center, Japan

We review graph kernels which is one of the state-of-the-art approaches using machine learning techniques for computational predictive modeling in chemoinformatics. We introduce a random walk graph kernel that defines a similarity between arbitrary two labeled graphs based on label sequences generated by random walks on the graphs. We introduce two applications of the graph kernels, the prediction of properties of chemical compounds and prediction of missing enzymes in metabolic networks. In the latter application, we propose to use the random walk graph kernel to compare arbitrary two chemical reactions, and apply it to plant secondary metabolism.

Holger Fröhlich, Bonn-Aachen International Center for IT (B-IT), Germany

Prediction models for absorption, distribution, metabolic and excretion properties of chemical compounds play a crucial rule in the drug discovery process. Often such models are derived via machine learning techniques. Kernel based learning algorithms, like the well known support vector machine (SVM) have gained a growing interest during the last years for this purpose. One of the key concepts of SVMs is a kernel function, which can be thought of as a special similarity measure. In this Chapter we describe optimal assignment kernels for multi-labeled molecular graphs. The optimal assignment kernel is based on the idea of a maximal weighted bipartite matching of the atoms of a pair of molecules. At the same time the physico-chemical properties of each single atom are considered as well as the neighborhood in the molecular graph. Later on our similarity measure is extended to deal with reduced graph representations, in which certain structural elements, like rings, donors or acceptors, are condensed in one single node of the graph. Comparisons of the optimal assignment kernel with other graph kernels as well as with classical descriptor based models show a significant improvement in prediction accuracy.

Chapter 3

Jean-Philippe Vert, Mines ParisTech, Institut Curie and INSERM U900, France

We review an approach, proposed recently by Mahé, Ralaivola, Stoven, and Vert (2006), for ligand-based virtual screening with support vector machines using a kernel based on the 3D structure of the molecules. The kernel detects putative 3-point pharmacophores, and generalizes previous approaches based on 3-point pharmacophore fingerprints. It overcomes the categorization issue associated with the discretization step usually required for the construction of fingerprints, and leads to promising results on several benchmark datasets.

Chapter 4

Martin Whittle, University of Sheffield, UK
Valerie J. Gillet, University of Sheffield, UK
Peter Willett, University of Sheffield, UK

This chapter analyses the use of similarity fusion in similarity searching of chemical databases. The ranked retrieval of molecules from a database can be modelled using both analytical and simulation approaches describing the similarities between an active reference structure and both the active and the non-active molecules in the database. The simulation model described here has the advantage that it can handle unmatched molecules, i.e., those that occur in one of the ranked similarity lists that are to be fused but that do not occur in the other. Our analyses provide insights into why the results of similarity fusion are often inconsistent when used for virtual screening.

Chapter 5

Viviana Consonni, University of Milano-Bicocca, Italy
Roberto Todeschini, University of Milano-Bicocca, Italy

The objective of this chapter is to investigate the chemical information encompassed by autocorrelation descriptors and elucidate their role in QSAR and drug design. After a short introduction to molecular descriptors from a historical point of view, the chapter will focus on reviewing the different types of autocorrelation descriptors proposed in the literature so far. Then, some methodological topics related to multivariate data analysis will be overviewed paying particular attention to analysis of similarity/ diversity of chemical spaces and feature selection for multiple linear regressions. The last part of the chapter will deal with application of autocorrelation descriptors to study similarity relationships of a set of flavonoids and establish QSARs for predicting affinity constants, Ki, to the GABAA benzodiazepine receptor site, BzR.

Section 2
Graph-Based Approaches in Chemoinformatics

In standard QSAR (Quantitative Structure Activity Relationship) approaches, chemical compounds are represented as a set of physicochemical property descriptors, which are then used as numerical features for classification or regression. However, standard descriptors such as structural keys and fingerprints are not comprehensive enough in many cases. Since chemical compounds are naturally represented as attributed graphs, graph mining techniques allow us to create subgraph patterns (i.e., structural motifs) that can be used as additional descriptors. In this chapter, we present theoretically motivated QSAR algorithms that can automatically identify informative subgraph patterns. A graph mining subroutine is embedded in the mother algorithm and it is called repeatedly to collect patterns progressively. We present three variations that build on support vector machines (SVM), partial least squares regression (PLS) and least angle regression (LARS). In comparison to graph kernels, our methods are more interpretable, thereby allows chemists to identify salient subgraph features to improve the druglikeliness of lead compounds.

#### Chapter 7

To annotate the biological function of a protein molecule, it is essential to have information on its 3D structure. Many successful methods for function prediction are based on determining structurally conserved regions because the functional residues are proved to be more conservative than others in protein evolution. Since the 3D conformation of a protein can be represented by a contact map graph, graph matching algorithms are often employed to identify the conserved residues in weakly homologous protein pairs. However, the general graph matching algorithm is computationally expensive because graph similarity searching is essentially a NP-hard problem. Parallel implementations of the graph matching are often exploited to speed up the process. In this chapter, we review theoretical and computational approaches of graph theory and the recently developed graph matching algorithms for protein function prediction.

Section 3
Statistical and Bayesian Approaches for Virtual Screenin

Chapter 8

Kiyoshi Hasegawa, Chugai Pharmaceutical Company, Japan
Kimito Funatsu, University of Tokyo, Japan

In quantitative structure-activity/property relationships (QSAR and QSPR), multivariate statistical methods are commonly used for analysis. Partial least squares (PLS) is of particular interest because it can analyze data with strongly collinear, noisy and numerous X variables, and also simultaneously model several response variables Y. Furthermore, PLS can provide us several prediction regions and diagnostic plots as statistical measures. PLS has evolved or changed for copying with sever demands from complex data X and Y structure. In this review article, we picked up four advanced PLS techniques and outlined their algorithms with representative examples. Especially, we made efforts to describe how to disclose the embedded inner relations in data and how to use their information for molecular design.

Chapter 9

Roman Rosipal, Medical University of Vienna, Austria & Pacific Development and Technology,
    LLC, USA

In many areas of research and industrial situations, including many data analytic problems in chemistry, a strong nonlinear relation between different sets of data may exist. While linear models may be a good simple approximation to these problems, when nonlinearity is severe they often perform unacceptably. The nonlinear partial least squares (PLS) method was developed in the area of chemical data analysis. A specific feature of PLS is that relations between sets of observed variables are modeled by means of latent variables usually not directly observed and measured. Since its introduction, two methodologically different concepts of fitting existing nonlinear relationships initiated development of a series of different nonlinear PLS models. General principles of the two concepts and representative models are reviewed in this chapter. The aim of the chapter is two-fold i) to clearly summarize achieved results and thus ii) to motivate development of new computationally efficient nonlinear PLS models with better performance and good interpretability.

Chapter 10

Martin Vogt, Rheinische Friedrich-Wilhelms-Universität, Germany
Jürgen Bajorath, Rheinische Friedrich-Wilhelms-Universität, Germany

Computational screening of in silico-formatted compound libraries, often termed virtual screening (VS), has become a standard approach in early-phase drug discovery. In analogy to experimental high-throughput screening (HTS), VS is mostly applied for hit identification, although other applications such as database filtering are also pursued. Contemporary VS approaches utilize target structure and/or ligand information as a starting point. A characteristic feature of current ligand-based VS approaches is that many of these methods differ substantially in the complexity of the underlying algorithms and also of the molecular representations that are utilized. In recent years, probabilistic VS methods have become increasingly popular in the field and are currently among the most widely used ligand-based approaches. In this contribution, we will introduce and discuss selected methodologies that are based on Bayesian principles.

Chapter 11

Nicos Angelopoulos, Edinburgh University, UK

Andreas Hadjiprocopis, Higher Technical Institute, Cyprus

Malcolm D. Walkinshaw, Edinburgh University, UK

In high throughput screening a large number of molecules are tested against a single target protein to determine binding affinity of each molecule to the target. The objective of such tests within the pharmaceutical industry is to identify potential drug-like lead molecules. Current technology allows for thousands of molecules to be tested inexpensively. The analysis of linking such biological data with molecular properties is thus becoming a major goal in both academic and pharmaceutical research. This chapter details how screening data can be augmented with high-dimensional descriptor data and how machine learning techniques can be utilised to build predictive models. The pyruvate kinase protein is used as a model target throughout the chapter. Binding affinity data from a public repository provide binding information on a large set of screened molecules. We consider three machine learning paradigms: Bayesian model averaging, Neural Networks, and Support Vector Machines. We apply algorithms from the three paradigms to three subsets of the data and comment on the relative merits of each. We also used the learnt models to classify the molecules in a large in-house molecular database that holds commercially available chemical structures from a large number of suppliers. We discuss the degree of agreement in compounds selected and ranked for three algorithms. Details of the technical challenges in such large scale classification and the ability of each paradigm to cope with these are put forward. The application of machine learning techniques to binding data augmented by high-dimensional can provide a powerful tool in compound testing. The emphasis of this work is on making very few assumptions or technical choices with regard to the machine learning techniques. This is to facilitate application of such techniques by non-experts.

Section 4
Machine Learning Approaches for Drug Discovery, Toxicology, and Biological Systems

Chapter 12

Shuxing Zhang, The University of Texas at M.D. Anderson Cancer Center, USA

Machine learning techniques have been widely used in drug discovery and development, particularly in the areas of cheminformatics, bioinformatics and other types of pharmaceutical research. It has been demonstrated they are suitable for large high dimensional data, and the models built with these methods can be used for robust external predictions. However, various problems and challenges still exist, and new approaches are in great need. In this Chapter, we will review the current development of machine learning techniques, and especially focus on several machine learning techniques we developed as well as their application to model building, lead discovery via virtual screening, integration with molecular docking, and prediction of off-target properties. We will suggest some potential different avenues to unify different disciplines, such as cheminformatics, bioinformatics and systems biology, for the purpose of developing integrated in silico drug discovery and development approaches.

Chapter 13

Rahul Singh, San Francisco State University, USA

The problem of modeling and predicting complex structure-property relationships, such as the absorption, distribution, metabolism, and excretion of putative drug molecules is a fundamental one in contemporary drug discovery. An accurate model can not only be used to predict the behavior of a molecule and understand how structural variations may influence molecular property, but also to identify regions of molecular space that hold promise in context of a specific investigation. However, a variety of factors contribute to the difficulty of constructing robust structure activity models for such complex properties. These include conceptual issues related to how well the true bio-chemical property is accounted for by formulation of the specific learning strategy, algorithmic issues associated with determining the proper molecular descriptors, access to small quantities of data, possibly on tens of molecules only, due to the high cost and complexity of the experimental process, and the complex nature of bio-chemical phenomena underlying the data. This chapter attempts to address this problem from the rudiments: we first identify and discuss the salient computational issues that span (and complicate) structure-property modeling formulations and present a brief review of the state-of-the-art. We then consider a specific problem: that of modeling intestinal drug absorption, where many of the aforementioned factors play a role. In addressing them, our solution uses a novel characterization of molecular space based on the notion of surface-based molecular similarity. This is followed by identifying a statistically relevant set of molecular descriptors, which along with an appropriate machine learning technique, is used to build the structure-property model. We propose simultaneous use of both ratio and ordinal error-measures for model construction and validation. The applicability of the approach is demonstrated in a real world case study.

Chapter 14

Huma Lodhi, Imperial College London, UK

Predicting mutagenicity is a complex and challenging problem in chemoinformatics. Ames test is a biological method to assess mutagenicity of molecules. The dynamic growth in the repositories of molecules establishes a need to develop and apply effective and efficient computational techniques to solving chemoinformatics problems such as identification and classification of mutagens. Machine learning methods provide effective solutions to chemoinformatics problems. In this chapter we review learning techniques that have been developed and applied to the problem of identification and classification of mutagens.

Chapter 15

Michael Schmuker, Freie Universität Berlin, Germany
Gisbert Schneider, Johann-Wolfgang-Goethe Universität, Germany

The purpose of the olfactory system is to encode and classify odorants. Hence, its circuits have likely evolved to cope with this task in an efficient, quasi-optimal manner. In this chapter we present a three-step approach that emulate neurocomputational principles of the olfactory system to encode, transform and classify chemical data. In the first step, the original chemical stimulus space is encoded by virtual receptors. In the second step, the signals from these receptors are decorrelated by correlation-dependent lateral inhibition. The third step mimics olfactory scent perception by a machine learning classifier. We observed that the accuracy of scent prediction is significantly improved by decorrelation in the second stage. Moreover, we found that although the data transformation we propose is suited for dimensionality reduction, it is more robust against overdetermined data than principal component scores. We successfully used our method to predict bioactivity of drug-like compounds, demonstrating that it can provide an effective means to connect chemical space with biological activity.

Section 5
Machine Learning Approaches for Chemical Genomics

Chapter 16

Yoshihiro Yamanishi, Mines ParisTech – Institut Curie – INSERM U900, France
Hisashi Kashima, IBM Tokyo Research Laboratory, Japan

In silico prediction of compound-protein interactions from heterogeneous biological data is critical in the process of drug development. In this chapter we review several supervised machine learning methods to predict unknown compound-protein interactions from chemical structure and genomic sequence information simultaneously. We review several kernel-based algorithms from two different viewpoints: binary classification and dimension reduction. In the results, we demonstrate the usefulness of the methods on the prediction of drug-target interactions and ligand-protein interactions from chemical structure data and genomic sequence data.

Chapter 17

Masahiro Hattori, Tokyo University of Technology, Japan
Masaaki Kotera, Kyoto University, Japan

Chemical genomics is one of the cutting-edge research areas in the post-genomic era, which requires a sophisticated integration of heterogeneous information, i.e., genomic and chemical information. Enzymes play key roles for dynamic behavior of living organisms, linking information in the chemical space and genomic space. In this chapter, we report our recent efforts in this area, including the development of a similarity measure between two chemical compounds, a prediction system of a plausible enzyme for a given substrate and product pair, and two different approaches to predict the fate of a given compound in a metabolic pathway. General problems and possible future directions are also discussed, in hope to attract more activities from many researchers in this research area.

Preface

Chemoinformatics endeavors to study and solve complex chemical problems by using computational tools and methods. It involves storing, and analyzing data, and drawing inferences from chemical information. Recent advances in high throughput technologies and generation of large amount of data have generated huge interest to design, analyze and apply novel computational and more specifically learning methodologies to solving chemical problems. The advances in machine learning for chemoinformatics establish a need for a comprehensive text on the subject. The book addresses the need by presenting in-depth description of novel learning algorithms and approaches for foundational topics ranging from virtual screening to chemical genomics.

The book is designed for multidisciplinary audiences. The intended audiences are researchers, scientists and experts in the fields ranging from chemistry, biology, to machine learning. It is stimulating, clear and accessible to its readers. It provides useful and efficient tools to experts in industry including pharmaceutical, agrochemical and biotechnology companies. It is aimed at fostering collaborations between experts from chemistry, biology and machine learning that is crucial for the advances in chemoinformatics.

The chapters of the book are organized into five sections. The first section presents methods for computing similarity in chemical spaces. Kernel methods are well known class of machine learning algorithms that give state-of-the performance. Chapters one and two present novel kernel functions for the problems in chemoinformatics. Graph kernels that can be viewed as effective approaches to compute similarity and predict the properties of chemical compounds are described in detail in the first chapter. The next chapter explains kernels, namely optimal assignment kernels, for predicting absorption, distribution, metabolic and excretion (ADME) properties of compounds. Other useful kernels, pharmacophore kernels, are presented in chapter three. The use of data fusion in chemoinformatics is the topic of chapter four. The last chapter of the section one clearly introduces molecular features like autocorrelation descriptors and their effectiveness in building quantitative structure activity relational models.

The complexity of chemical problems has led the researchers to develop techniques that are based on the integration of different areas. Section two of the book presents methodologies that combine graph mining and machine learning techniques, for example, chapter six introduces a framework for integrating graph mining algorithm with support vector machines (SVMs), partial least squares (PLS) regression and least angle regression (LARS) in order to extract informative or discriminative chemical fragments. Chapter seven describes graph matching algorithms to compare 3D structures of protein molecules to predict the biological functions.

Section three is based on explaining important elements of chemoinformatics. It gives an in-depth description of statistical and Bayesian techniques. Partial least square methods are well-known for their effectiveness to constructing accurate models in chemoinformatics. Chapter eight and nine not only explain PLS techniques but also motivate the further development and enhancement of these methods. Virtual screening is a very useful tool in drug design and development. A number of Bayesian approaches are presented in chapter ten and the efficacy of their applications to virtual screening is validated. The next chapter further explains the use of Bayesian and non-Bayesian learning methods for large scale virtual screening.

Integration of ideas from different fields like chemoinformatics, bioinformatics and systems biology is important for the development of methods to solving the challenging chemical problems. The research presented in section four is aimed to enhance the design of theses techniques. In chapter twelve an overview of a number of learning methods ranging from classical to modern approaches is given. The chapter also establishes efficacy for the tasks in chemoinformatics. Chapter thirteen highlights the key issues that need to be addressed to construct structure activity relationships models and introduces a useful tool for modeling human intestinal absorption. Mutagenicity is an unfavorable characteristic of drugs that can cause adverse effects. In chapter fourteen an overview of inductive logic programming (ILP) techniques, propositional methods within ILP, ensemble methods, probabilistic, and kernel methods is given to detect and identify mutagenic compounds. Chapter fifteen describes an exciting research avenue that is classification of odorants by using machine learning.

The final section of the book is based on introducing machine learning for chemical genomics that requires useful computational approaches to investigate the relationship between chemical space of possible compounds and genomic space of possible genes or proteins. Chapter sixteen addresses the issue by presenting a number of machine learning approaches for predicting drug-target and ligand-protein interactions from the integration of chemical and genomic data on a large scale. Chapter seventeen presents research that integrates genomics and metabolomics to analyze enzymatic reactions on metabolic pathways, hence introducing tools that can solve challenges ranging from environmental issues to health problems.

As described in preceding paragraphs the book presents cutting edge tools and strategies to solving problems in chemoinformatics. It explains key elements of the filed. Authors have highlighted many future research directions that will foster multi-disciplinary collaborations and hence will lead to significant development and progress in the field of chemoinoformatics.

Huma Lodhi and Yoshihiro Yamanishi
November 2009

Acknowlegment

We want to thank IGI Global for the help in the processing of the book.

Huma Lodhi and Yoshihiro Yamanishi
November 2009

Section 1
Similarity Design in Chemical Space

Chapter 1
Graph Kernels for Chemoinformatics

Hisashi Kashima
University of Tokyo, Japan

Hiroto Saigo
Kyushu Institute of Technology, Japan

Masahiro Hattori
Tokyo University of Technology, Japan

Koji Tsuda
AIST Computational Biology Research Center, Japan

ABSTRACT

The authors review graph kernels which is one of the state-of-the-art approaches using machine learning techniques for computational predictive modeling in chemoinformatics. The authors introduce a random walk graph kernel that defines a similarity between arbitrary two labeled graphs based on label sequences generated by random walks on the graphs. They introduce two applications of the graph kernels, the prediction of properties of chemical compounds and prediction of missing enzymes in metabolic networks. In the latter application, the authors propose to use the random walk graph kernel to compare arbitrary two chemical reactions, and apply it to plant secondary metabolism.

INTRODUCTION

In chemoinformatics and bioinformatics, it is effective to automatically predict the properties of chemical compounds and proteins with computer-aided methods, since this can substantially reduce the costs of research and development by screening out unlikely compounds and proteins from the candidates for 'wet' experiment. Data-driven predictive modeling is one of the main research topics in chemoinformatics and bioinformatics. A chemical compound can be represented as a graph (Figure 1) by considering the atomic species (such as C, Cl, and H) as the vertex labels, and the bond types (such as s (single bond) and d (double bond)) as the edge labels. Similarly, chemical reactions can be analyzed as graphs where the chemical compounds and their relationships during the reaction are considered as vertices and edges,

DOI: 10.4018/978-1-61520-911-8.ch001

Figure 1. A chemical compound can be represented as an undirected graph

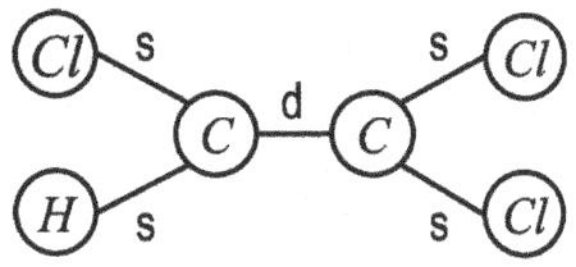

respectively (Figure 3). Prediction based on such graph representations by using machine learning techniques will help discover new chemical and biological knowledge and lead in designing drugs. However, it is not obvious how to apply machine learning methods to graph data, since ordinary machine learning methods generally assume that the data is readily represented as vector structures called feature vectors.

Recently, *kernel methods* (Bishop, 2006, Shawe-Taylor & Cristianini, 2004) have attracted considerable attention in the field of machine learning and data mining, since they can handle (non-vectorial) structural data as if it was represented as vectorial data through the use of kernel functions. In the framework of the kernel methods, various kernel functions have been proposed for different types of structured data, such as sequences (Lodhi, Saunders, Shawe-Taylor, Cristianini, & Watkins, 2002, Leslie, Eskin, & Noble, 2002, Leslie, Eskin, Weston, & Noble, 2003), trees (Collins & Duffy, 2002, Kashima & Koyanagi, 2002), and graphs (Kashima, Tsuda, & Inokuchi, 2003, Gärtner, Flach, & Wrobel, 2003). They have been successfully applied to tasks in various areas including bioinformatics and chemoinformatics (Schölkopf, Tsuda, & Vert, 2004).

In the first part of this chapter, we introduce two approaches to graph data analysis, the pattern-based methods and the kernel methods, and then give a brief introduction to the kernel methods. We introduce one of the state-of-the-art graph learning methods called the *random walk graph kernel* (Kashima et al., 2003, Gärtner et al., 2003) and its extensions. The random walk graph kernel uses random walks on two given labeled graphs

to generate label sequences, and the two graphs are compared based on the label sequences. Although the number of possible label sequences is infinite, we introduce an efficient algorithm for computing the kernel function without an explicit enumeration of the label sequences.

The second part of this chapter is devoted to applications of the random walk graph kernel. First, we introduce an application to predict the properties of chemical compounds. We use two data sets, one for predicting the carcinogenicity of chemical compounds, and the other for predicting whether or not each of the given compounds has mutagenicity. The results show that the graph kernel is comparable to a pattern-based method.

Next, we introduce a new application of the graph kernel, which is the prediction of missing enzymes in a plant's secondary metabolism. The secondary metabolic pathways in a plant are important for finding candidate enzymes that may be pharmacologically relevant. However, there are many enzymes whose functions are still undiscovered especially in organism-specific metabolic pathways. We propose representing chemical reactions as *reaction graphs*, and using the random walk kernel on the reaction graphs to automatically assign EC numbers to unknown enzymatic reactions in a metabolic network. The reaction graph kernel compares pairs of chemical reactions, since the targets of classification are themselves chemical reactions. In our experiment using the KEGG/REACTION database, our reaction graph kernel in combination with the kernel nearest neighbor method showed 83% accuracy in classifying the 4,610 reactions into 124 classes. In addition, we predicted missing enzymatic reactions in the plant's secondary metabolism in the KEGG database. As expected, our reaction graph kernel method shows higher coverage than e-zyme, an existing rule-based system. We discuss the biochemical validity of some of the individual predictions.

BACKGROUND

In this section, we introduce two different approaches to graph data analysis, then give a brief introduction to the kernel methods.

Two Approaches to Graph Data Analysis

There are two widely used approaches to analyzing graph-structured data from the machine learning and data mining perspectives. One approach involves *pattern-based methods* (Washio & Motoda, 2003), which use explicit vectorization of graphs, and the other involves *kernel methods* (Bishop, 2006, Shawe-Taylor & Cristianini, 2004), which use implicit vectorization of the data.

In the pattern-based methods, each graph is explicitly vectorized by using structural patterns. These methods first enumerate structural patterns such as paths and subgraphs using pattern mining algorithms, and then each graph is represented as a binary feature vector of the presence or absence of each pattern in the graph. The number of possible pattern is enormous and hence naive enumeration of the patterns is prohibitive, and so efficient branch-and-bound heuristics for finding the patterns are studied in this field of data mining. One advantage of the pattern-based methods is that they can identify structural patterns used for modeling, since they construct the feature vectors explicitly.

The other approach is the kernel methods approach that we focus on in this chapter. In contrast to the pattern-based methods, they do not construct feature vectors explicitly so as to avoid the computational difficulty caused by the enormous number of structural patterns. One of the most famous kernel methods is the *support vector machine* (SVM) (Vapnik, 2000). Since Vapnik's seminal work, many linear supervised and unsupervised learning algorithms have been kernelized (Bishop, 2006, Shawe-Taylor & Cristianini, 2004), including ridge regression, perceptrons, *k*-nearest neighbours, Fisher discriminant analysis, principal component analysis (PCA), k-means clustering, and independent component analysis (ICA). These kernelized algorithms have been shown to perform very well in many real-world problems.

The key technique is the use of similarity functions called *kernel functions*. The kernel methods have favorable characteristics through their usage of the kernel functions. It is particularly notable that the complexity of the kernel methods is independent of the dimensionality of the feature space induced by a kernel function. Recalling that the feature vectors for graph-structured data are of high dimensionality, this is well suited to graphs. This implies that we can implement efficient graph learners if we can design an appropriate similarity metric between paired graphs that can be computed efficiently.

Introduction to Kernel Methods

In this subsection, we briefly review kernel methods, and explain why the kernel methods can efficiently solve graph learning problems.

In order to illustrate the nonlinearization procedures for linear algorithms through kernels, we give the perceptron algorithm for binary classification as an example. Let us consider the linear parametric model

$$f(x;w) \equiv sign(\langle w,x \rangle)$$

where $x \in \mathbb{R}^d$ is an input vector called a feature vector, $w \in \mathbb{R}^d$ is a parameter vector, and $\langle \cdot, \cdot \rangle$ denotes the inner product. The *sign* function returns +1 if the argument is positive, and returns -1 otherwise, which means that the function f defined by the parameter vector **w** classifies a given feature vector into one of the two classes, +1 or -1.

Given input-output training data $\{(x_i, y_i) | x_i \in \mathbb{R}^d, y_i \in \{+1, -1\}\}_{i=1}^{\ell}$, the perceptron algorithm determines the parameter w so that the number of mistakes is minimized. The perceptron algorithm processes the training examples one by one in a sequential manner. At each step, the perceptron algorithm picks one example, say, the *i*-th example in the training data. The perceptron algorithm uses the current parameter **w** to make a prediction for the example. If the prediction is correct, i.e. *sign*($\langle w, x_i \rangle$)=y_i, the perceptron algorithm does nothing. If the prediction is wrong, i.e. *sign*($\langle w, x_i \rangle$)≠y_i, the perceptron algorithm updates the parameter by using the following update rule.

$$w \leftarrow w + y_i x_i$$

After the update, the prediction for the example by using the updated parameter becomes $\langle w, x_i \rangle$ +$y_i \| x_i \|^2$, which means that the "confidence of the positive class" of the prediction increases by $\| \mathbf{x}_i \|^2$ if the example is a positive example with y_i=+1, or it decreases (or the "confidence of the negative class" increases) by $\| x_i \|^2$ if the example is a negative example with y_i=-1.

Now we kernelize the perceptron algorithm. Equation (2) implies that the parameter vector can be represented as a linear combination of the feature vectors as

$$w \equiv \sum_{i=1}^{\ell} \alpha_i x_i.$$

Then Equation (1) can be expressed using the inner product as

$$f(x; \alpha) = \sum_{i=1}^{\ell} \alpha_i \langle x_i, x \rangle.$$

Now let us define the *kernel function* as

$$k(x, x') \equiv \langle x, x' \rangle$$

Then the parametric model is expressed as

$$f(\alpha; x) = \sum_{i=1}^{\ell} \alpha_i k(x_i, x)$$

With the change of the parameter, the update rule (2) is also modified as

$$\alpha_i \leftarrow \begin{cases} \alpha_i + 1 & \text{if } y_i = +1, \\ \alpha_i - 1 & \text{otherwise}(y_i = -1). \end{cases}$$

The new parameter vector α≡α$_1$,α$_2$,...,α$_\ell$) is an ℓ-dimensional vector, while *w* is a *d*-dimensional vector. Therefore, now the dependency of the computational complexity of each step on the number ℓ of training samples is more significant than the input dimensionality *d*. Note that in this kernel formulation, the data samples are accessed only *through* the kernel functions, and the input vectors are not directly handled. Conversely, if there exists a kernel function which corresponds to the inner product between **x** and **x'**, then the use of the kernel formulation would be beneficial. In the context of graph learning, while the original formulation of Equation (1) suffers from the high dimensionality of the feature vectors of graphs, the kernel formulation of Equation (4) can offer efficient graph learning, if we can devise an appropriate and efficient similarity metric between the two graphs. It is worthwhile to note that, not all similarity metrics can be used as kernel functions. In order to be a valid kernel, the existence of the corresponding inner product must be guaranteed, which is satisfied if the similarity metric is positive semidefinite (Schölkopf & Smola, 2002). Such a kernel function is called a *Mercer kernel* or a *reproducing kernel* (Mercer, 1909, Aronszajn, 1950).

Note that the kernel representation of Equation (4) is based on the fact that the parameter vector **w** is expressed by a linear combination of training samples (see Equation (3)). Generally, the existence of such a kernel formulation is justified by the *representer theorem* (Wahba, 1987).

GRAPH KERNELS

In this section, we introduce graph kernels that define similarity metrics between two labeled graphs. First we introduce the random walk graph kernel proposed by Kashima et al. (2003), which is almost identical to the graph kernel proposed by Gärtner et al. (2003). The random walk kernel has several different variants, and we review other graph kernels and recent extensions of the random walk graph kernel.

Random Walk Graph Kernel

The key idea behind the random walk graph kernel is to use random walks on the given graphs to generate label sequences, and each graph is represented as a bag of label sequences from the random walks. The similarity of two graphs are defined as the number of common label sequences weighted by the probability of the corresponding walks (or more precisely, the probability of common label sequences being generated). The random walk graph kernel is a valid kernel, since it is interpreted as an inner product in the feature space spanned by the label sequences.

Let us assume that we want to define a similarity metric between two labeled graphs $G_1=(V_1,E_1,L_1(V_1))$ and $G_2=(V_2,E_2,L_2(V_2))$, where V_1 and V_2 are sets of vertices, E_1 and E_2 are sets of edges, and L_1 and L_2 are sets of labels of the vertices. We assume that the edges are not labeled, but we can convert labeled edges to labeled vertices, which doubles the number of vertices.

We consider a joint random walk over the two graphs G_1 and G_2 to define our graph kernel. First, we define a random walk over one graph. Let $\mathbf{u}_1(t)$ be a $|V_1|$-dimensional vector representing the probability distribution of the position of the random walk over the vertices in G_1 at time t. The random walk starts with an initial distribution $\mathbf{u}_1(0)$. One possible choice of $\mathbf{u}_1(0)$ is the uniform distribution over V_1. At each time step t, the random walk terminates with probability $1-\lambda_1$,

where $0<\lambda_1<1$. The random walk proceeds with probability λ_1, and moves to the next vertex by using a transition matrix $\mathbf{T}_1$. The (i,j)-th element of $\mathbf{T}_1$ indicates the probability of a transition from the j-th vertex to the i-th vertex in G_1. One possible choice of T_1 is the normalized adjacency matrix of G_1. The dynamics of the random walks over G_1 are given as

$$\mathbf{u}_1(t)=\lambda_1\mathbf{T}_1\mathbf{u}_1(t\text{-}1)$$

For example, when a random walk stops at time t, the probability distribution over V_1 is represented as $(1\text{-}\lambda_1)(\lambda_1\mathbf{T}_1)^t\mathbf{u}_1(0)$ A random walk over G_2 is defined by using $\mathbf{u}_2(t)$, λ_2, and $\mathbf{T}_2$ defined accordingly.

Since we want to compute the probability of two label sequences produced by the random walks matching, we consider the joint random walk using $\mathbf{T}_1$ and $\mathbf{T}_2$ over G_1 and G_2, respectively. Specifically, the joint distribution of the two random walks is given as $\mathbf{U}(t)\equiv\mathbf{u}_1(t)\otimes\mathbf{u}_2(t)^\mathrm{T}$, where $\otimes$ indicates the Kronecker product. Noting that the two random walks are independent of each other, the dynamics for the joint random walk is given as

$$\mathbf{U}(t) = \left(\lambda_1\mathbf{T}_1\right)\mathbf{U}(t-1)\left(\lambda_2\mathbf{T}_2^\top\right).$$

Let $\mathbf{M}$ be a $|V_1|\times|V_2|$ matrix, whose (i_1,i_2)-th elements is 1 if the i_1-th node in V_1 and the i_2-th node in V_2 have an identical label, and is 0 otherwise. The dynamics of the "label matching" joint random walks are represented as

$$\mathbf{V}(t) = \mathbf{M} * \left(\left(\lambda_1\mathbf{T}_1\right)\mathbf{V}(t-1)\left(\lambda_2\mathbf{T}_2^\top\right)\right),$$

where $*$ is the Hadamard (element-wise) product, and $\mathbf{V}(0)\equiv\mathbf{M}*\mathbf{U}(0)$.

Then the matching probability (which is the graph kernel) is given as

$$K(G_1, G_2) \equiv (1 - \lambda_1)(1 - \lambda_2) \sum_{i_1, i_2} \sum_{t=0}^{\infty} [\mathbf{V}]_{i_1, i_2}(t),$$

where the (i_1, i_2)-th element of $\mathbf{V}$ is denoted by $[\mathbf{V}]_{i_1, i_2}$. Now our goal is reduced to computing the infinite sum $\overline{\mathbf{V}} \equiv \sum_{t=0}^{\infty} \mathbf{V}(t)$. From Equation (5), we have the relation

$$\overline{\mathbf{V}} = \mathbf{M} * \left(\left(\lambda_1 \mathbf{T}_1 \right) \overline{\mathbf{V}} \left(\lambda_2 \mathbf{T}_2^{\top} \right) \right) + \mathbf{V}(0),$$

so we can use the fixed point iteration

$$\overline{\mathbf{V}} \leftarrow \mathbf{M} * \left(\left(\lambda_1 \mathbf{T}_1 \right) \overline{\mathbf{V}} \left(\lambda_2 \mathbf{T}_2^{\top} \right) \right) + \mathbf{V}(0),$$

used by Vishwanathan, Borgwardt, and Schraudolph (2007) to update the current solution starting from $\overline{\mathbf{V}} \leftarrow \mathbf{V}(0)$. The computational complexity of each update is $O(|E_1||V_2| + |E_2||V_1| + |V_1||V_2|)$, where the first term and the second term are for applying $\mathbf{T}_1$ and $\mathbf{T}_2$ to $\overline{\mathbf{V}}$, respectively. The third term is for the application of $\mathbf{M}$, so it can be replaced by the number of non-zero elements in $\mathbf{M}$. Therefore, $\overline{\mathbf{V}}$ can be updated very efficiently if the graphs and $\mathbf{M}$ are sparse. The iteration is continued until convergence, but usually a few dozen steps are sufficient.

Note that the values of $\mathbf{M}$ are not limited to zero and one, and can also take any values between zero and one according to the similarities between the labels. They can be regarded as "label matching probabilities". As we will see in the following sections, each vertex can itself be a labeled graph, so each graph can be represented as a graph of graphs. In such a case, a label matching probability matrix $\mathbf{M}$.is replaced by a pre-calculated graph kernel similarity matrix.

Other Graph Kernels

Several extensions have recently been proposed for the random walk kernel.

Mahé, Ueda, Akutsu, Perret, and Vert (2004) offer two modifications to the random walk kernel. The first one is to prevent tottering. One of the drawbacks of the random walk kernel is that it can "totter" around the same position. Modifying the random walk probability prevents the random walk from going back to the vertex just passed. The second modification is enriching the vertex labels by using the Morgan index. The Morgan index process starts by annotating each vertex with the degree of the vertex (as the number of edges attached to the vertex), and continues to replace the score by the sum of the scores of the neighboring vertices several times. The Morgan index basically tries to incorporate the local structure around each vertex into the score of the vertex.

Ramon and Gärtner (2003) and Mahé and Vert (2008) created another extension to the random walk kernel by using tree-structured patterns. In the tree-pattern graph kernel, the random walk is replaced by a branching random walk. The branching walk generates tree-structured branching sequences that are more expressive than the linear sequences generated by the original random walk kernel.

Borgwardt and Kriegel (2005) proposed an efficient alternative called the shortest path kernel. The shortest path kernel transforms given graphs into shortest path graphs, where an edge between two vertices is weighted by the shortest path distance between the two vertices, and the kernel is computed as the random walk kernel of length one.

Horváth, Gärtner, and Wrobel (2004) proposed a cyclic-pattern kernel based on a different design principle from the random walk kernel. The cyclic-pattern kernel decomposes graphs into sets of cycles and trees, and the kernel between two graphs is defined as the intersection of the two corresponding pattern sets. Fröhlich, Wegner, Sieker, and Zell (2006) also proposed a decomposition approach specialized for chemoinformatics. They decompose chemical compounds into specific substructures such as aromatic cycles and carbonyl groups, and use the weighted bipartite graph matching to define a similarity metric between two bag-of-substructure representations of the graphs.

Application to Classification of Chemical Compounds

In this section, we apply the graph kernel to prediction of properties of chemical compounds (Ralaivola, Swamidass, Saigo, & Baldi, 2005, Pierre, Ueda, Akutsu, Perret, & Vert, 2005). We used two datasets, the PTC dataset (Helma, King, Kramer, & Srinivasan, 2001) and the Mutag dataset (Srinivasan, Muggleton, King, & Sternberg, 1996).

The PTC dataset is the results of pharmaceutical experiments. A total of 417 compounds were given to four types of test animals: Male Mouse (MM), Female Mouse (FM), Male Rat (MR), and Female Rat (FR). According to their carcinogenicity, each compound is assigned one of the following labels: {EE,IS,E,CE,SE,P,NE,N} where CE, SE and P indicate 'relatively active', and NE and N indicate 'relatively inactive', and EE, IS and E indicate 'cannot be decided'. In order to simplify the problem, we relabeled CE, SE and P as 'positive', and NE and N as 'negative'. The task is to predict whether the label of a given compound is positive or negative for each type of test animal. Thus we reduced the task to four two-class classification problems.

In the Mutag dataset, the task is a two-class classification problem to predict whether or not each of the 188 compounds has mutagenicity. The data is summarized in Table 1. Since the edges of the graphs are also labeled, we converted them to vertex-labeled graphs.

We defined the probability distributions for random walks as follows. We set the initial distributions to be uniform distributions over the vertices. Also, normalized adjacency matrices

were used as the transition probabilities. The random walk probability $\lambda \equiv \lambda_1 \equiv \lambda_2$ was varied from 0.1 to 0.9. In our observation, 20-30 iterations were sufficient for convergence with the fixed point iteration method. For the classification learning algorithm, we used the voted kernel perceptron (Freund & Shapire, 1999), whose performance is known to be comparable to an SVM.

We compared the graph kernel with the pattern-based method, where the feature vectors were constructed by using the frequent paths (Kramer & De Raedt, 2001), and the voted perceptron was used. There are other methods which enumerate more complicated substructures such as subgraphs (Inokuchi, Washio, & Motoda, 2000), but we focused on the path patterns whose features are similar to ours. The minimum support parameter was set to 0.5%, 1%, 3%, 5%, 10% and 20% of the number of compounds.

Table 2 and Table 3 show the classification accuracies in the five two-class problems measured by leave-one-out cross-validation. The pattern-based method was better in MR, FR and Mutag, but the graph kernel was better in MM and FM. Therefore, no general tendencies were found upon which to conclude which method is better. Thus we can say that the performances are comparable.

Table 1. Several statistics of the two datasets

	MM	FM	MR	FR	Mu-tag
number of positive examples	129	143	152	121	125
number of negative examples	207	206	192	230	63
max. number of vertices	109	109	109	109	40
avg. number of vertices	25.0	25.2	25.6	26.1	31.4
max. degree	4	4	4	4	4
number of vertex labels	21	19	19	20	8
number of edge labels	4	4	4	4	4

Table 2. Classification accuracies (%) for the pattern-based method. The 'MinSup' shows the ratio of the minimum support parameter

MinSup	MM	FM	MR	FR	Mutag
0.5%	60.1	57.6	61.3	**66.7**	88.3
1%	**61.0**	**61.0**	**62.8**	63.2	87.8
3%	58.3	55.9	60.2	63.2	**89.9**
5%	60.7	55.6	57.3	63.0	86.2
10%	58.9	58.7	57.8	60.1	84.6
20%	**61.0**	55.3	56.1	61.3	83.5

Table 3. Classification accuracies (%) for our graph kernel. The parameter λ is the parameter for the random walks, which controls the effect of the length of the label sequences

λ	MM	FM	MR	FR	Mutag
0.9	62.2	59.3	57.0	62.1	84.3
0.8	62.2	61.0	57.0	62.4	83.5
0.7	64.0	61.3	56.7	62.1	**85.1**
0.6	**64.3**	61.9	56.1	63.0	**85.1**
0.5	64.0	61.3	56.1	64.4	83.5
0.4	62.8	61.9	54.4	65.8	83.0
0.3	63.1	62.5	54.1	63.2	81.9
0.2	63.4	**63.4**	54.9	64.1	79.8
0.1	62.8	61.6	**58.4**	**66.1**	78.7

For the PTC dataset, the accuracies by the two approaches are not significantly better than random predictions although we can still see some improvements. This is probably because the graph representation of molecular structures does not have a sufficient expressive power for this particular task. Chemical compounds are actually three-dimensional objects and also have other complex features. Additional properties such as spatial conformations, atom masses and geometric bond lengths might be introduced for improving the result.

APPLICATIONS TO PREDICTING EC NUMBERS OF UNKNOWN ENZYMATIC REACTIONS IN THE SECONDARY METABOLISM

In this section, we extend the random walk kernel to handle "graphs of graphs" for comparing two chemical reactions, and apply it to discovering missing enzymes in a plant's secondary metabolism.

Chemical Reaction Prediction Problem

A metabolic network represents the transition or transformation of chemical compounds, where enzymes are represented as edges, and chemical compounds are represented as vertices. With the recent developments of pathway databases such as KEGG PATHWAY (Kanehisa et al., 2006) and EcoCyc (Keseler et al., 2005), much more information about chemical compounds and the roles of enzymes in biological systems has become available. In particular, many secondary metabolites found in plants are known to have roles in the defenses against pathogens, and have been attracting increasing attention from scientists for more than a decade (Bennett & Wallsgrove, 1994). However the organism-specific metabolic networks are not complete, and there are many "missing enzymes" (Yamanishi et al., 2007). Yamanishi et al. have proposed a way to find these missing enzymes by combining different sources of information, such as protein sequences, gene expression patterns, and the locations of genes in the chromosomes. However, their method assumes that the EC (Enzyme Commission) numbers are correctly assigned to the metabolic pathways, even though this is often not true in reality. For example, Figure 2 is a part of a terpenoid biosynthesis pathway, but there are many enzymes whose EC numbers are not yet assigned (denoted as "? " in the boxes in the figure).

Figure 2. A part of a terpenoid biosynthesis pathway extracted from KEGG/PATHWAY

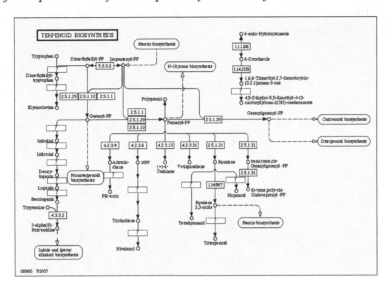

The EC number represents a hierarchical categorization of enzymes with respect to the enzymatic reactions they catalyze. The process of assigning EC numbers is done manually by the Joint Commission on Biological Nomenclature (JCBN) of the International Union of Biochemistry and Molecular Biology (IUBMB) and the International Union of Pure and Applied Chemistry (IUPAC), however, this assignment process is so slow that many enzymes are still unannotated. Fulfilling such missing EC numbers on a pathway can be casted as a multi-class classification problem given substrates and products as an input under an assumption that any enzymatic reaction can be classified into one of existing enzymatic classes. Recently, Kotera, Okuno, Hattori, and Goto (2004) proposed an automatic EC number assignment system for metabolic reactions using the RC numbers. However, this system is based on the detection of maximum common subgraphs between chemical compounds, so the shift of a large chemical group is not correctly detected (Arita, 2003). Also, their e-zyme system is a rule-based method and does not allow approximate matching, which results in poor coverage. In many cases, e-zyme rejects a query because none of the rules matches (Yamanishi et al., 2007). In addition, e-zyme does not produce any reasonable similarity metric, which becomes problematic, especially when the e-zyme system returns a long list of candidates.

An example of metabolic chemical reaction is represented by

$$Loganin + NADPH + H^+ + Oxygen <=> Secologanin + NADP^+ + H_2O$$

Given such a chemical reaction, one important task is to predict the EC number of the enzyme catalyzing the reaction. In this case, the enzyme is *secologanin synthase* (EC 1.3.3.9), which turns a substrate (*Loganin*) into a product (*Secologanin*) with *NADPH* as a cofactor. Publicly available databases such as KEGG offer a large number of enzyme-reactant pairs. Given a chemical reaction without enzyme information as a query, we would like to look up the entries in the database whose reactions are similar to the query reaction. Therefore, a reasonable similarity metric is key to solving this problem.

Figure 3. The reaction graph for the reaction Loganin+NADPH+H⁺+Oxygen<=>Secologanin+NAD P⁺+H₂O, which is catalyzed by secologanin synthase (EC 1.3.3.9). Edges except for 'main', 'cofactor' are all 'group' edges in this reaction

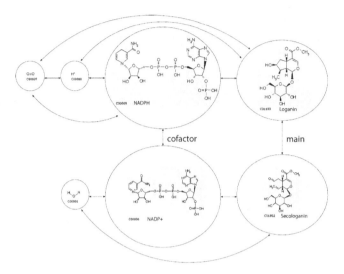

Reaction Graph Kernel

As a canonical representation of chemical reactions, we propose representing a metabolic reactions as *reaction graphs*. A reaction graph is a "graph of graphs", because each vertex contains a graph representing a chemical compound. A reaction graph consists of vertices, which are compounds in a reaction, and edges which denote the relationships between compounds. The edge labels are chosen from 'main', 'leave', 'cofactor', 'transferase', 'ligase', and 'group'. The example is the reaction graph of Equation (6) (Figure 3).

To evaluate the similarity of two reaction graphs, we use the random walk kernel (Kashima et al., 2003) in a recursive way. We first compute all of the pairwise similarities of the vertices (the chemical compounds) using the random walk kernel. Then the compound-wise similarities are used as the label matching probabilities M for the upper-level graph kernel. For the edge kernel representing reaction similarities in the reaction-wise kernel and the atom similarities and the bond similarities in the compound-wise kernel, we simply use the Dirac kernel which returns 1 if the two labels in comparison are identical, and 0 otherwise.

Leave-One-Out Prediction of Missing Enzymes

In order to evaluate the reaction graph kernel similarity, we collected metabolic reactions from KEGG/REACTION. Following the pre-process used by Kotera et al. (2004), we did not use reactions which (i) do not have EC numbers, (ii) include chemical compounds whose structures are not available, (iv) have classes 97 and 99, (v) have only one reaction in the same subsubclass. This pre-processing found in 4,610 reactions in 6 classes, 50 subclasses, and 124 subsubclasses.

In this experiment, we withheld one reaction from the database, and predicted its EC number using all of the other reaction-enzyme pairs. Since the numbers of class, subclass and subsubclass are large, we used the nearest neighbor approach based on the reaction graph kernels. For the calculation of the reaction graph kernels, we used Chemcpp[1] with the "non-tottering" option (Pierre et al., 2005). The random walk parameter of the lower-level and upper-level graph kernels were selected from {099,0.9,08}, respectively, and 0.9 was used for both kernels, since it performed best in the experiments.

In reality, it is not often the case that the whole reaction graph of the query is known, so So we considered three settings, RPAIR, main-pair, and full-edge. In RPAIR, only reactant pairs are known, where the reactant pair information can be obtained from KEGG/RPAIR database (Figure 4 left). In main-pair, only the main-pair is known (Figure 4 right).

The leave-one-out accuracy is reported in Table 4. Clearly, predictions up to the second digit (subclass) and to the third digit (subsubclass) are more difficult. We did not test up to the fourth digit, since the last digit is often used just as a serial number (Kotera et al., 2004). We observed that additional edges in the reaction graphs help improve the classification performance. Notice that RPAIR corresponds to the same setting used in e-zyme, but the use of full-edge turned out to be strongly advantageous in discriminating small changes in similar lower class reactions. According to (Kotera et al., 2004), the e-zyme system has similar precision, as long as they provide an answer. However, its coverage is much lower than our method, as shown in the next section. In this experiment, our coverage is 100% as we did not reject any queries. The accuracy should become even higher if some queries are rejected.

Table 4. Leave-one-out cross validation accuracy

setting	class (%)	subclass (%)	subsubclass (%)
full-edge	94.8	86.0	82.5
RPAIR	92.3	81.4	78.1
main-pair	77.8	69.8	66.2

Predicting Unannotated Reactions in Plant Secondary Metabolism

In order to further evaluate the proposed method, we performed a blind test, where we tested only those reactions whose EC numbers are not yet assigned in the secondary metabolism of plants. First we collected metabolic reactions from the KEGG "Biosynthesis of Secondary Metabolites-Reference pathway" data. From the resulting 60 reactions, we removed 16 reactions which are either non-enzymatic reactions or multi-step reactions whose systems are too complicated, based on an expert's judgment. Then we tested the e-zyme and reaction graph kernels on the remaining 44 reactions.

E-zyme returned answers for only 14 queries (reactions). This is because e-zyme is a rule-based method, and can only match very similar reactions. Reaction graph kernels allow approximate matching, and returned answers for all the 44 reactions. The performance on a blind test is reported both for e-zyme and reaction graph kernels in Table 5. To compare reaction graph kernels with e-zyme, which does not have a reasonable similarity metric,

Figure 4. RPAIR graph (left) and main-pair graph (right)

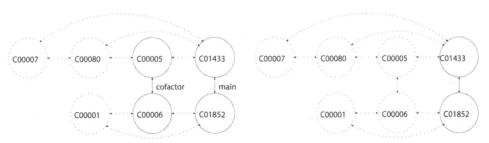

Table 5. Number of correct predictions in top k candidates for 44 unknown reactions

	TOP 1	TOP 3	TOP 5
e-zyme	7	16	21
RGK	11	30	44

we sorted the candidate EC numbers for e-zyme by the number of the reactions matching the query (majority voting). Reaction graph kernels could detect twice as many correct reactions as e-zyme by looking at the top 5 candidates.

A list of newly annotated reactions is presented in Figure 5, together with prediction results of e-zyme and reaction graph kernels (RGK). For reaction graph kernels, the Z-score ($z = \dfrac{x - \mu}{\sigma}$, where x is a raw score, and μ and σ are the mean and the standard deviation of the candidate scores) is calculated so that one can find the candidate with a saliently higher score than others. The biochemical grounds for the manual assignment of the EC numbers are presented in the "Comments"

column. As can be seen in Figure 5, some reactions progress in multiple steps and have several correct EC numbers. However, neither reaction graph kernels nor e-zyme consider such situations, which remains for future research.

FUTURE RESEARCH DIRECTION

There are several possible research directions for graph kernels. Scalability is an important issue for handling large chemical compounds with many thousands of vertices or chemical reactions with many constituents. Recent advances in graph kernels provide several simple and efficient alternatives (Vishwanathan & Smola, 2003). In order to realize large scale virtual screening, further acceleration is desired. Interpretability is another problem. Basically, the models learned by the kernel methods are difficult to interpret. Bakir, Alexander, and Tsuda (2004) proposed solving a pre-image problem to reconstruct a graph from its feature space representation, while Akutsu and

Figure 5. A list of newly annotated reactions in plant secondary metabolism. "-" in the e-zyme column means that no answer was available for that query. CXXXX is a KEGG compound ID. Correctly assigned EC numbers are highlighted in bold fonts

Reaction	manual assignment	RGK (Z score)	e-zyme	Comments
C 02362 = > C 06625	[2.6.3 + 4.1.1]	4.1.3 (3.5)	-	After eliminating a carbon dioxide (4.1.1).
		4.1.3 (3.5)		
		2.8.3 (3.5)		
C 01221 = > C 01138	[1.1.1]	3.1.3 (4.5)	1.1.1	
		2.7.7 (4.1)	1.1.3	
		2.7.1 (3.5)	1.4.3	
C 01456 + C 00729 = > C 02046	[3.1.1]	3.1.1 (4.3)	6.1.1	Structures of C02046 and C01479 are almost the same
		1.14.11 (3.3)	3.1.1	
		1.14.11 (3.2)	3.2.1	at a carboxylic-ester bond, like R03563.
C 06175 = > C 01735	[2.1.1 + 1.1.1]	1.1.1 (5.3)	-	After removing a methyl group of C06175 (2.1.1),
		1.1.1 (5.3)		C01735 will be produced by oxidation of CH-OH group (1.1.1).
		1.1.1 (5.1)		
C 01494 = > C 01752	[1.14.- + 2.4.1 + 5.2.1 + 3.2.1]	1.2.1 (5.7)		
		1.2.1 (5.7)		
		1.14.13 (5.2)		
C 02646 = > C 12206	[1.14.-]	1.14.11 (4.6)	1.14.13	
		1.3.1 (4.5)	1.14.16	
		1.3.1 (4.4)	1.14.15	
C 00806 + C 00235 = > C 06067	[2.5.1]	2.5.1 (4.4)	-	2-Isopentenyl group of C00235 will be transferred to benzene ring of C00806.
		2.5.1 (3.2)		
		2.5.1 (2.8)		
C 03206 = > C 06570	[1.1.1]	1.14.13 (6.7)	-	This reaction is an oxidation of CH-OH using NAD+ as cofactor.
		1.1.1 (3.2)		
		1.1.1 (3.2)		
C 02105 = > C 06167	[1.5.-]	1.21.3 (6.6)	-	The chemical structures of C02105 and C05178 are almost the same
		1.5.1 (6.4)		
		1.14.21 (6.2)		
C 06173 = > C 06175	[2.1.1]	1.1.1 (6.1)	2.1.1	
		1.14.21 (6.0)	1.14.13	
		1.1.1 (5.7)		
C 04712 = > C 15666	[1.11.1]	1.4.3 (4.7)		
		2.6.1 (4.1)		
		3.1.1 (3.8)		
C 06173 = > C 06172	[2.1.1 + 1.1.1]	1.1.1 (5.9)	-	The reason is the same with the above reaction.
		1.14.13 (5.7)		
		1.1.1 (5.6)		
C 00137 = > C 01214	[1.1.1 + 2.6.1]	3.2.1 (3.9)	5.1.1	After oxidation reaction of CH-OH to aldehyde with NAD+ (1.1.1),
		1.1.1 (3.9)	5.1.3	an amino group will be transferred like R00576 (2.6.1).
		1.1.1 (3.8)	5.2.1	
C 06568 = > C 06569	[1.14.20]	1.14.11 (4.0)	-	This is one of oxidoreductases acting on C06568 and 2-oxoglutarate.
		2.1.3 (3.4)		
		1.14.20 (3.4)		

Fukagawa (2005) proposed reconstructing graphs from the path feature frequencies.

In our approach, the EC number is used for classification of the chemical reactions, while in Borgwardt et al. (2005), the EC numbers are used for the classification of the enzymatic proteins. Little is known about the dual problem, which is how the EC number of a specific chemical reaction corresponds with a specific class of enzymatic proteins. This relationship can play a key role in bridging chemical information and genomic information. Recently, with the progress of ongoing plant and microbial metabolomics projects, more and more plant genomes and plant pathways are becoming available. This data in combination with the graph kernel would help uncover unknown enzymatic reactions in the secondary metabolite pathways of plants. It would be of particular interest to integrate genomic data with chemical data (Yamanishi et al., 2007). Handling of a multi step reactions is another research direction. In a KEGG pathway, an unclear pathway is represented as a single step reaction as if a reaction between a substrate and a product had proceeded in one step. However, it is often the case that a series of valid chemical reactions is found in the pathway of a closely related organism. Therefore, it would be a desirable feature for a reaction graph kernel that takes into account this kind of prior knowledge.

CONCLUSION

In this chapter, the authors reviewed the use of graph kernels, which is one of the state-of-the-art approaches for computational predictive modeling using machine learning techniques in chemoinformatics. The authors introduced a random walk graph kernel which defines a similarity metric between two graphs based on label sequences generated by random walks on the graphs.

In the second part of this chapter, they introduced two applications of the graph kernels, prediction of the properties of chemical compounds and prediction of missing EC numbers in metabolic networks. The authors proposed using graph kernels on reaction graphs which compare pairs of chemical reactions, and used this method for plant secondary metabolism.

REFERENCES

Akutsu, T., & Fukagawa, D. (2005). Inferring a graph from path frequency. In *Proceedings of the sixteenth annual symposium on combinatorial pattern matching* (p. 371-382). New York: Springer.

Arita, M. (2003). In silico atomic tracing by substrate-product relationships in escherichia coli intermediary metabolism. *Genome Research, 13,* 2455–2466. doi:10.1101/gr.1212003

Aronszajn, N. (1950). Theory of reproducing kernels. *Transactions of the American Mathematical Society, 68,* 337–404.

Bakir, G., Alexander, Z., & Tsuda, K. (2004). Learning to find graph pre-images. In *Dagm-symposium* (pp. 253–261).

Bennett, R. N., & Wallsgrove, R. M. (1994). Secondary metabolites in plant defence mechanisms. *The New Phytologist, 127*(4), 617–633. doi:10.1111/j.1469-8137.1994.tb02968.x

Bishop, C. (2006). *Pattern recognition and machine learning.* New York: Springer.

Borgwardt, K. M., & Kriegel, H.-P. (2005). Shortest-path kernels on graphs. In *Proceedings of the international conference on data mining* (Vol. 0, pp. 74-81).

Borgwardt, K. M., Ong, C. S., Schönauer, S., Vishwanathan, S. V. N., Smola, A. J., & Kriegel, H. P. (2005). Protein function prediction via graph kernels. *Bioinformatics (Oxford, England), 21*(suppl 1), i47–i56. doi:10.1093/bioinformatics/bti1007

Collins, M., & Duffy, N. (2002). Convolution kernels for natural language. [Cambridge, MA: MIT Press.]. *Advances in Neural Information Processing Systems, 14*, 625–632.

Freund, Y., & Shapire, R. (1999). Large margin classification using the perceptron algorithm. *Machine Learning, 37*(3), 277–296. doi:10.1023/A:1007662407062

Fröhlich, H., Wegner, J., Sieker, F., & Zell, A. (2006). Kernel functions for attributed molecular graphs - a new similarity based approach to adme prediction in classification and regression. *QSAR & Combinatorial Science, 25*(4), 317–326. doi:10.1002/qsar.200510135

Gärtner, T., Flach, P., & Wrobel, S. (2003). On graph kernels: Hardness results and efficient alternatives. In *Proceedings of the sixteenth annual conference on computational learning theory* (pp. 129–143).

Helma, C., King, R., Kramer, S., & Srinivasan, A. (2001). The Predictive Toxicology Challenge 2000-2001. *Bioinformatics (Oxford, England), 17*(1), 107–108. doi:10.1093/bioinformatics/17.1.107

Horváth, T., Gärtner, T., & Wrobel, S. (2004). Cyclic pattern kernels for predictive graph mining. In *Proceedings of the tenth acm sigkdd international conference on knowledge discovery and data minig* (p. 158-167).New York: ACM Press.

Inokuchi, A., Washio, T., & Motoda, H. (2000). An Apriori-based algorithm for mining frequent substructures from graph data. In *Proceedings of the fourth european conference on machine learning and principles and practice of knowledge discovery in databases* (pp. 13-23).

Kanehisa, M., Goto, S., Hattori, M., Aoki-Kinoshita, K., Itoh, M., & Kawashima, S. (2006). From genomics to chemical genomics: new developments in KEGG. *Nucleic Acids Research, 34*, D354–D357. doi:10.1093/nar/gkj102

Kashima, H., & Koyanagi, T. (2002). Kernels for semi-structured date. In *Proceedings of the nineteenth international conference on machine learning* (pp. 291–298). San Francisco, CA: Morgan Kaufmann.

Kashima, H., Tsuda, K., & Inokuchi, A. (2003). Marginalized kernels between labeled graphs. In *Proceedings of the twentieth international conference on machine learning* (pp. 321–328). San Francisco, CA: Morgan Kaufmann.

Keseler, I., Collado-Vides, J., Gama-Castro, S., Ingraham, J., Paley, S., & Paulsen, I. (2005). EcoCyc: a comprehensive database resource for escherichia coli. *Nucleic Acids Research, 33*, D334–D337. doi:10.1093/nar/gki108

Kotera, M., Okuno, Y., Hattori, M., & Goto, S. (2004). Computational assignment of the ec numbers for genomic-scale analysis of enzymatic reactions. *Journal of the American Chemical Society, 126*, 16487–16498. doi:10.1021/ja0466457

Kramer, S., & De Raedt, L. (2001). Feature construction with version spaces for biochemical application. In *Proceedings of the eighteenth international conference on machine learning* (pp. 258–265).

Leslie, C., Eskin, E., & Noble, W. (2002). The spectrum kernel: A string kernel for SVM protein classification. In R. B. Altman, A. K. Dunker, L. Hunter, K. Lauerdale, & T. E. Klein (Eds.), *Proceedings of the pacific symposium on biocomputing* (pp. 566–575). Hackensack, NJ: World Scientific.

Leslie, C., Eskin, E., Weston, J., & Noble, W. (2003). Mismatch string kernels for svm protein classification . In Becker, S., Thrun, S., & Obermayer, K. (Eds.), *Advances in neural information processing systems 15* (pp. 467–476). Cambridge, MA: MIT Press.

Lodhi, H., Saunders, C., Shawe-Taylor, J., Cristianini, N., & Watkins, C. (2002). Text classification using string kernels. *Journal of Machine Learning Research, 2*, 419–444. doi:10.1162/153244302760200687

Mahé, P., Ueda, N., Akutsu, T., Perret, J.-L., & Vert, J.-P. (2004). Extensions of marginalized graph kernels. In R. Greiner & D. Schuurmans (Eds.), *Proceedings of the Twenty-First International Conference on Machine Learning* (p. 552-559). New York: ACM Press.

Mahé, P., & Vert, J.-P. (2008). *Graph kernels based on tree patterns for molecules.* Machine Learning.

Mercer, J. (1909). Functions of positive and negative type, and their connection with the theory of integral equations. *Proceedings of the Royal Society of London. Series A, Containing Papers of a Mathematical and Physical Character, 83*(559), 69–70. doi:10.1098/rspa.1909.0075

Pierre, M., Ueda, N., Akutsu, T., Perret, J.-L., & Vert, J.-P. (2005). Graph kernels for molecular structure-activity relationship analysis with support vector machines. *Journal of Chemical Information and Modeling*, 939–951.

Ralaivola, L., Swamidass, S. J., Saigo, H., & Baldi, P. (2005). Graph kernels for chemical informatics. *Neural Networks, 18*(8), 1093–1110. doi:10.1016/j.neunet.2005.07.009

Ramon, J., & Gärtner, T. (2003). Expressivity versus efficiency of graph kernels. In *Proceedings of the first international workshop on mining graphs, trees and sequences.*

Schölkopf, B., & Smola, A. J. (2002). *Learning with kernels: Support vector machines, regularization, optimization, and beyond.* Cambridge, MA: MIT Press.

Schölkopf, B., Tsuda, K., & Vert, J.-P. (Eds.). (2004). *Kernel methods in computational biology.* Cambridge, MA: The MIT Press.

Shawe-Taylor, J., & Cristianini, N. (2004). *Kernel methods for pattern analysis.* New York: Cambridge University Press.

Srinivasan, A., Muggleton, S., King, R. D., & Sternberg, M. (1996). Theories for mutagenicity: a study of first-order and feature based induction. *Artificial Intelligence, 85*(1-2), 277–299. doi:10.1016/0004-3702(95)00122-0

Vapnik, V. (2000). *The nature of statistical learning theory.* New York: Springer.

Vishwanathan, S., & Smola, A. (2003). Fast kernels for string and tree matching . In Becker, S., Thrun, S., & Obermayer, K. (Eds.), *Advances in neural information processing systems 15* (pp. 569–576). Cambridge, MA: MIT Press.

Vishwanathan, S. V. N., Borgwardt, K., & Schraudolph, N. (2007). Fast computation of graph kernels. [Cambridge, MA: MIT Press.]. *Advances in Neural Information Processing Systems, 19*, 1449–1456.

Wahba, G. (1987). Spline models for observational data . In *Cbms-nsf regional conference series in applied mathematics (Vol. 59).* SIAM.

Washio, T., & Motoda, H. (2003). State of the art of graph-based data mining. *SIGKDD Explorations, 5*(1), 59–68. doi:10.1145/959242.959249

Yamanishi, Y., Mihara, H., Osaki, M., Muramatsu, H., Esaki, N., & Sato, T. (2007). Prediction of missing enzyme genes in a bacterial metabolic network. *The FEBS Journal*, 2262–2273. doi:10.1111/j.1742-4658.2007.05763.x

ENDNOTE

[1]   available from http://chemcpp.sourceforge.net/

Chapter 2
Optimal Assignment Kernels for ADME *in Silico* Prediction

Holger Fröhlich
Bonn-Aachen International Center for IT (B-IT) [1], Germany

ABSTRACT

Prediction models for absorption, distribution, metabolic and excretion properties of chemical compounds play a crucial rule in the drug discovery process. Often such models are derived via machine learning techniques. Kernel based learning algorithms, like the well known support vector machine (SVM) have gained a growing interest during the last years for this purpose. One of the key concepts of SVMs is a kernel function, which can be thought of as a special similarity measure. In this Chapter the author describes optimal assignment kernels for multi-labeled molecular graphs. The optimal assignment kernel is based on the idea of a maximal weighted bipartite matching of the atoms of a pair of molecules. At the same time the physico-chemical properties of each single atom are considered as well as the neighborhood in the molecular graph. Later on our similarity measure is extended to deal with reduced graph representations, in which certain structural elements, like rings, donors or acceptors, are condensed in one single node of the graph. Comparisons of the optimal assignment kernel with other graph kernels as well as with classical descriptor based models show a significant improvement in prediction accuracy.

INTRODUCTION

ADME *in Silico* Prediction

The development of a new drug is often compared with finding a needle in a haystack (Kubinyi, 2004). Therefore, rational approaches for drug

design began to develop about 30 years ago with the aim to significantly reduce the amount of *in vivo* animal experiments. With the dramatic increase of computer performance during the last years there has been an increasing interest in *virtual screening* methods (HJ. Böhm & Schneider, 2000). The goal is to filter out a significant amount of "uninteresting" chemicals, that cannot be used as potential drugs, *in silico* in an early stage of

DOI: 10.4018/978-1-61520-911-8.ch002

the drug discovery process. Thereby, especially the so-called *ADME* (*A*bsorption, *D*istribution, *M*etabolism, *E*xcretion) properties of a compound are of great interest (Kubinyi, 2002, 2003, 2004, Waterbeemd & Gifford, 2003): As most drugs are given orally for reasons of convenience, the compound is dissolved in the gastro-intestinal tract. It then has to be absorbed through the gut wall and pass the liver to get into the blood circulation. The percentage of the compound dose reaching the circulation is called the *bioavailability*. From there, the potential drug will have to be distributed to various tissues and organs in the body. The extend of distribution will depend on the structural and physico-chemical properties of the compound. For some drugs it will be further necessary to enter the central nervous system by crossing the blood-brain barrier. Finally, the chemical has to bind to its molecular target, for example, a receptor or ion channel, and exert its desired action.

The body will eventually try to eliminate a drug. Hence, for many drugs this requires metabolism or *biotransformation*. This takes place partly in the gut wall during absorption, but primarily in the liver. Traditionally, a distinction is made between phase I and phase II metabolism, although these do not necessarily occur sequentially. In phase I metabolism, a molecule is functionalized, for example, through oxidation, reduction or hydrolysis. In phase II metabolism, the functionalized compound is further transformed in so-called conjugation reactions, e.g. glucuronidation, sulfation or conjugation with glutathione.

The clearance of a drug from the body mainly takes place via the liver (hepatic clearance or metabolism, and biliary excretion) and the kidney (renal excretion). The *half-life* ($t_{1/2}$) of a compound is the time taken for its concentration in the blood plasma to be reduced by 50%. It is a function of the clearance and volume of distribution, and determines how often a drug needs to be administered.

QSPR *(*Quantitative Structure Property Relationship*)* methods try to predict *in silico* various ADME, but also physico-chemical properties, which have an important impact on a drug's pharmacokinetic and metabolic fate in the body. Among others, today models for forecasting oral absorption, bioavailability, degree of blood-brain barrier penetration, clearance and volume of distribution are available. Additionally, there are methods for predicting physico-chemical properties, such as e.g. lipophilicity and water solubility (Waterbeemd & Gifford, 2003). Similarly, QSAR (Quantitative Structure Activity Relationship) methods are used to forecast the biological activity/inactivity of an untested ligand for a target protein (Kubinyi, 2002, 2003, 2004).

The basic assumption behind all QSAR/QSPR approaches is that the molecular properties in question can be derived from certain aspects of the molecular structure only. This implies that structurally similar compounds have similar biological or physico-chemical properties as well. In practice this supposition is often fulfilled, but there are also counter examples (Kubinyi, 2002, 2003).

Often, *ADME models* are derived via machine learning methods. Hence, one needs an abstract representation of a chemical compound in the computer. Classically, this is done by a large amount of *descriptors* (= features in machine learning language), which represent global molecular properties, like the polar surface area (Waterbeemd & Gifford, 2003), the distribution of certain physico-chemical properties, like the Radial Distribution Function (RDF) descriptor, the frequency of the occurrence of certain atomic patterns (fingerprints), invariances or characteristics of the molecular graph (topological indices) or others (Todeschini & Consonni, 2000). In conclusion, for each chemical compound one can calculate hundreds or even thousands of descriptors, which are of potential interest. The bottom line is that each molecule, which by itself is a complex three dimensional and dynamic object, is described in a simplified manner by a vector representation, which allows the easy use of classical machine learning algorithms.

Representation of Molecules as Multi-Labeled Graphs

A difficulty coming along with a high dimensional vector representation of molecules is that descriptors irrelevant for the prediction problem at hand may degrade the performance of the ADME model. Hence, determining the subset of relevant descriptors is of high importance (Xue et al., 2004, Byvatov & Schneider, 2004, Agrafiotis & Xu, 2003, Godden, Furr, Xue, Stahura, & Bajorath, 2003, J. Wegner, Fröhlich, & Zell, 2003b). This is, however, often a computationally demanding and difficult problem. On the other hand, embedding of the data in a lower dimensional space via projection techniques, like Principal Component Analysis (Hotelling, 1933), often leads to results, which are hard to interpret.

Hence, an appealing idea is to directly work on a representation of chemical compounds as *multi-labeled graphs* without explicitly calculating any descriptor information. An advantage of this method is that the problem of descriptor selection becomes almost irrelevant, because all computations are carried out directly on the molecular structures represented as labeled graphs. Atoms in a molecule are represented as nodes in the graph and bonds as edges between nodes. For the prediction of ADME properties it is crucial to take into account physico-chemical features of atoms and bonds. These features can be represented as labels of the nodes and edges, respectively. It is also possible to encode structural aspects into the labels, like the membership of an atom to a ring, to a donor, an acceptor, etc. The multi-labeled graph representation can give us a detailed description of the topology of a molecule without making any a-priori assumptions on the relevance of certain chemical descriptors for the whole molecule. It is clear that thereby a crucial point is to capture the characteristics of each single atom and bond by its physico-chemical properties – e.g. electrotopological state (Todeschini & Consonni, 2000), partial charge (Gast-eiger & Marsili, 1978) – which are encoded in the labels (see Evaluation Section for more detail).

Multi-labeled graphs offer a detailed representation of the molecular 2D structure, which is well suited for ADME *in silico* prediction. However, a principle question arising in this context is, how different graph structures should be compared. One possibility is a conversion of the molecular graph into a tree. This approach was proposed by Rarey and Dixon (1998). Their *feature tree* representation integrates the multi-label information of atoms and also allows for the efficient calculation of a similarity measure between molecules. In contrast, many graph mining methods (Washio & Motoda, 2003) ignore the multi-label information of the graph structure and are therefore not well suited for ADME prediction. Examples of graph mining methods in the literature are *maximum common substructure* approaches (Raymond, Gardiner, Willett, & Rascal, 2002), the *pattern-discovery* (PD) algorithm (Raedt & Kramer, 2001) or *atom-pair descriptors* (Carhart, Smith, & Venkataraghavan, 1985). Although these approaches are not preferable for ADME models, they have been successfully applied in other domains of chemoinformatics, like toxicity prediction (Raedt & Kramer, 2001).

Kernel Functions for Multi-Labeled Graphs

Kernel based learning algorithms, like support vector machines (SVMs), have become increasingly popular during the last years (Boser, Guyon, & Vapnik, 1992, Cortes & Vapnik, 1995, Schölkopf & Smola, 2002, Shawe-Taylor & Cristianini, 2004). One of the central ideas of these algorithms is the use of a positive semidefinite kernel function representing a dot product in some high dimensional Hilbert space. Kernel functions can be naturally interpreted as similarity measures. Apart from vectorial data, in the literature kernel functions between complex structured objects, such as strings, graphs or trees have been proposed

(Vishwanathan & Smola, 2004, Leslie, Kuang, & Eskin, 2004, Kashima, Tsuda, & Inokuchi, 2003).

In Kashima et al. (2003) the authors propose a kernel function between multi-labeled graphs, which they call *marginalized graph kernel*: Its idea is to compute the expected match of all pairs of random walk label sequences up to infinite length. An efficient computation can be carried out in a time complexity proportional to the product of the size of both graphs by solving a system of linear simultaneous equations. Kashima et al. show that also the geometric and the exponential graph kernel (Gärtner, Flach, & Wrobel, 2003) can be seen as special variants of the marginalized graph kernel (Kashima, Tsuda, & Inokuchi, 2004).

The approach we would like to present here is most similar to the marginalized graph kernel, but has a clearer interpretation from a chemists point of view. It better reflects a chemist's intuition on the similarity of molecular multi-labeled graphs. Rather than comparing label sequences, the main idea of our approach is that the similarity of two molecular graphs mainly depends on the matching of certain substructures, e.g. aromatic rings, carbonyl groups, etc., and the way they are connected. I.e. two molecules are more similar the better structural elements from both graphs fit together and the more these structural elements are connected in a similar way. Thereby the physico-chemical properties of each single atom and bond in both structures are considered.

On an atomic level this leads to the idea of computing a maximum weighted bipartite matching (*optimal assignment*) of atoms in one structure to those in another one, including for each atom information on the neighborhood and on its physico-chemical properties. The maximum weighted bipartite matching is a classical problem from graph theory. It can be computed efficiently in cubic time complexity (Kuhn, 1955) and allows an easy interpretation from the chemistry side.

A natural extension of our approach is to represent each molecule not on an atomic level,

but in form of a *reduced graph*. Thereby certain structural motifs, like rings, donors, acceptors, are collapsed into one node of the molecular graph, whereas remaining atoms are removed. This allows to concentrate on important structural features, where the definition of what an important structural feature actually is, is induced by the problem at hand and may be given by expert knowledge. In the literature this procedure is known as *pharmacophore mapping* (Martin, 1998).

The work presented in this Chapter is a summary of our original published papers (Fröhlich, Wegner, & Zell, 2005, Fröhlich, Wegner, Sieker, & Zell, 2005, 2006). Other authors in the meanwhile have proposed modifications and extensions of the optimal assignment kernel (Rupp, Proschak, & Schneider, 2007, Fechner, Jahn, Hinselmann, & Zell, 2009), which are, however, not further described here.

The Chapter is organized as follows: In the next Section we begin by defining so called "optimal assignment kernels" as a general class of symmetric similarity measures and show how they can be used in combination with SVMs. Given this result we can introduce our optimal assignment kernel for chemical graphs and show how it can be computed efficiently. Next we investigate the extension of the optimal assignment kernel by means of the reduced graph representation. In Section "Evaluation" we show experimental results of our method compared to the marginalized graph kernel and to classical descriptor-based ADME models on four datasets and show that in several cases we can significantly outperform marginalized graph kernels as well as descriptor-based models with and without performing automatic descriptor selection by means of Recursive Feature Elimination (RFE) (Guyon, Weston, Barnhill, & Vapnik, 2002). We also demonstrate the good performance of the reduced graph representation. Finally, we draw conclusions from our work and point out directions of future research.

OPTIMAL ASSIGNMENT KERNELS

Definition

Let X be some domain of structured objects (e.g. graphs). Let us denote the parts of some object x (e.g. the nodes of a graph) by $x[1],..,x[|x|]$, i.e. x consists of $|x|$ parts, while another object y consists of $|y|$ parts. Let X' denote the domain of all parts, i.e. $x[i] \in X'$ for $1 \leq i \leq |x|$. Further let π be some permutation of either an $|x|$-subset of natural numbers $\{1,..,|y|\}$ or an $|y|$-subset of $\{1,..,|x|\}$ (this will be clear from context).

Definition 1 *(Optimal Assignment Kernels)*

Let $k_1: X' \times X' \to \mathbb{R}$ be some non-negative and symmetric similarity measure (not necessarily a positive semidefinite kernel). Then $k_A: X \times X \to \mathbb{R}$ with

$$k_A(x,y) := \begin{cases} \max_\pi \sum_{i=1}^{|x|} k_1(x[i], y[\pi(i)]) & \text{if } |y| \geq |x| \\ \max_\pi \sum_{j=1}^{|y|} k_1(x[\pi(j)], y[j]) & \text{otherwise} \end{cases}$$

is called an *optimal assignment kernel*.

This definition captures the idea of a maximal weighted bipartite matching (optimal assignment) of the parts of two objects. Each part of the smaller of both structures is assigned to exactly one part of the other structure such that the overall similarity score between both structures is maximized (Figure 2).

Unfortunately, the optimal assignment kernel does not always lead to a positive semidefinite gram matrix (Fröhlich, 2006, Vert, 2008). As it is highly desirable to have a positive semidefinite kernel matrix for SVM training, a way out is given by shifting the spectrum of the $n \times n$ kernel matrix on the training data. This can be achieved by transforming the original kernel matrix $\mathbf{K}$ as

$$\mathbf{K} \leftarrow \mathbf{K} - \lambda_{min}\mathbf{I} \tag{1}$$

where λ_{min} is the smallest negative eigenvalue of $\mathbf{K}$, if there is any (and zero otherwise). To see this consider the eigenvector decomposition of $\mathbf{K}$:

$$\mathbf{K} - \lambda_{min}\mathbf{I} = \mathbf{V}\Lambda\mathbf{V}^T - \lambda_{min}\mathbf{I} = \mathbf{V}\Lambda\mathbf{V}^T - \lambda_{min}\mathbf{V}\mathbf{V}^T = \mathbf{V}(\Lambda - \lambda_{min}\mathbf{I})\mathbf{V}^T \tag{2}$$

where $\mathbf{V}$ is an orthogonal matrix with columns being the eigenvectors and Λ a diagonal matrix containing the eigenvalues.

Optimal Assignment Kernels for Chemical Compounds

We now turn to the construction of an optimal assignment kernel for molecular graphs. Let us assume we have two molecules M and M', which have atoms $a_1,..,a_n$ and $a'_1,..,a'_m$. Let us further assume we have some non-negative kernel k_{nei}, which compares a pair of atoms $(a_h, a'_{h'})$ from both molecules, including information on their neighborhoods, membership to certain substructures (like aromatic systems, donors, acceptors, and so on) and other physico-chemical properties (e.g. mass, partial charge and others). Then the similarity between M and M' can be calculated using the optimal assignment kernel

$$k_A(M,M') = \begin{cases} \max_\pi \sum_{h=1}^{n} k_{nei}(a_h, a'_{\pi(h)}) & \text{if } m \geq n \\ \max_\pi \sum_{h'=1}^{m} k_{nei}(a_{\pi(h')}, a'_{h'}) & \text{otherwise} \end{cases} \tag{3}$$

That means we assign each atom of the smaller of both molecules to exactly one atom of the bigger molecule such that the overall similarity score is maximized. The *optimal assignment* is a classical problem from graph theory and can be computed efficiently in $O(\max(n,m)^3)$ (Kuhn, 1955). Although this seems to be a drawback compared to the quadratic time complexity of marginalized graph kernels, we have to point out, that marginalized graph kernels have to be iteratively computed until convergence, and thus in practice, depending on the size of the molecules, there might be no real difference in computation time.

In order to prevent larger molecules automatically to achieve a higher kernel value than smaller ones, we further normalize our kernel (Schölkopf & Smola, 2002), i.e.

$$k_A(M,M') \leftarrow \frac{k_A(M,M')}{\sqrt{k_A(M,M)k_A(M',M')}} \quad (4)$$

Finally, given the results from the last Section, before SVM training we transform the kernel matrix on our training data to assure its positive semi-definiteness.

We now have to define the kernel k_{nei}. For this purpose let us suppose we have non-negative kernels k_{atom} and k_{bond}, which compare the atom and bond labels, respectively. A natural choice for both kernels is the RBF-kernel. The set of labels associated with each atom or bond can be interpreted as feature vectors. A complete list of the atom and bond features used for our kernel can be found in Table 1.

As the individual features for an atom or bond can live on different numerical scales, it is beneficial to normalize them, e.g. to mean 0 and standard deviation 1. Let us denote by $a \rightarrow n_i(a)$ the bond connecting atom a with its ith neighbor $n_i(a)$. Let us further denote by $|a|$ the number of neighbors of atom a. We now define a kernel R_0, which compares all direct neighbors of atoms

(a,a') as the optimal assignment kernel between all neighbors of a and a' and the bonds leading to them, i.e.

$$R_0(a,a') = \frac{1}{|a|}\max_\pi \sum_{i=1}^{|a|} k_{atom}(n_i(a), n_{\pi(i)}(a')) \cdot k_{bond}(a \rightarrow n_i(a), a' \rightarrow n_{\pi(i)}(a'))$$
$$(5)$$

where we assumed $|a| \geq |a'|$ for the sake of simplicity of notation. As an example consider the C-atom 3 in the left and the C-atom 5 in the right structure of Figure 1: If our only atom and bond features were the element type and bond order, respectively, and k_{atom} and k_{bond} would simply count a match by 1 and a mismatch by 0, our kernel $R_0(a_3, a'_5)$ would tell us that 2 of 3 possible neighbors of atom 3 in the left structure match with the neighbors of atom 5 in the right structure. It is worth mentioning that the computation of R_0 can be done in $O(1)$ time complexity as for chemical compounds $|a|$ and $|a'|$ can be upper bounded by a small constant (e.g. 4).

Of course it would be beneficial not to consider the match of direct neighbors only, but also that of indirect neighbors and atoms having a larger topological distance. For this purpose we can evaluate R_0 not at (a,a') only, but also at all pairs of neighbors, indirect neighbors and so on, up to some topological distance L. In our example that would mean we also evaluate $R_0(a_2, a'_2)$, $R_0(a_4, a'_2)$, $R_0(a_7, a'_2)$, and so on. If we consider the mean of all these values and add them to $k_{atom}(a,a') + R_0(a,a')$, this leads to the following definition of the kernel k_{nei}:

Table 1. Atom and bond features chosen for the optimal assignment kernel

features	nominal and ordinal	real valued
atom	in donor, in acceptor, in donor or acceptor (M. Böhm & Klebe, 2002), in terminal carbon, in aromatic system (Bonchev & Rouvray, 1990), negative/positive, in ring (Figueras, 1996), in conjugated environment, free electrons, implicit valence, heavy valence, hybridization, is chiral, is axial	electrotopological state, mass, graph potentials, electron-affinity, van-der-Waals volume, electrogeometrical state, electronegativity (Pauling), intrinsic state (Todeschini & Consonni, 2000), Gasteiger/Marsili partial charge (Gast-eiger & Marsili, 1978)
bond	order, in aromatic system(Bonchev & Rouvray, 1990), in ring (Figueras, 1996), is rotor, in carbonyl/amide/ primary amide/ester group	geometric length

Figure 1. Intuition of the optimal assignment kernel. (a) Matching regions of two molecular graphs; (b) Direct and indirect neighbors of atom 3 in the left and atom 5 in the right molecule

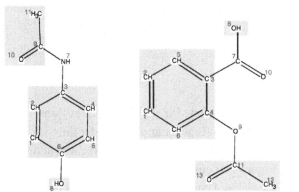

$$k_{nei}(a,a') = k_{atom}(a,a') + R_0(a,a') + \sum_{\ell=1}^{L} \gamma(\ell)R_\ell(a,a')$$
(6)

Here R_ℓ denotes the mean of all R_0 evaluated at neighbors of topological distance ℓ, and $\gamma(\ell)$ is a decay parameter, which reduces the influence of neighbors that are further away and depends on the topological distance ℓ to (a,a'). It makes sense to set $\gamma(\ell)=p(\ell)p'(\ell)$, where $p(\ell),p'(\ell)$ are the probabilities for molecules M,M' that neighbors with topological distance ℓ are considered.

A key observation is that R_ℓ can be computed efficiently from $R_{\ell-1}$ via the recursive relationship

$$R_\ell(a,a') = \frac{1}{|a||a'|}\sum_{i,j} R_{\ell-1}(n_i(a),n_j(a'))$$
(7)

I.e. we can compute k_{nei} by iteratively revisiting all direct neighbors of (a,a') only. In case that L is set to a constant we thus still have a $O(1)$ time complexity for the calculation of k_{nei}. In case that $L\to\infty$, we can prove the following theorem:

Theorem 2 *Let* $\gamma(\ell) = (\hat{p}_1\hat{p}_2)^\ell$ *with* $\hat{p}_1,\hat{p}_2 \in (0,1)$. *If there exists a* $C\in\mathbb{R}^+$, *such that* $k_{atom}(a,a')\leq C$ *for all* a,a' *and* $k_{bond}(a\to n_i(a),a'\to n_j(a'))\leq C$ *for*

all $a\to n_i(a),a'\to n_j(a')$, *then* (6) *converges for* $L\to\infty$.

Figure 2. Possible assignments of atoms from molecule 2 to those of molecule 1. The kernel function k measures the similarity of a pair of atoms (a_i, a') including information on structural and physico-chemical properties. The goal is to find the matching, which assigns each atom from molecule 2 to exactly one atom from molecule 1, such that the overall similarity score, i.e. the sum of edge weights in the bipartite graph, is maximized

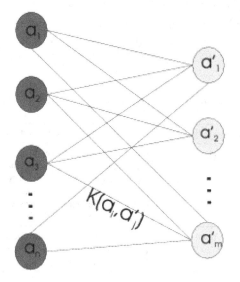

It is

$$R_0(a, a') \leq \frac{\min(|a|, |a'|)}{\max(|a|, |a'|)} C^2 \leq C^2$$

and thus

$$R_1(a, a') \leq \frac{1}{|a||a'|} \sum_{i=1}^{|a|} \sum_{j=1}^{|a'|} C^2 = C^2$$

Hence, also $R_\ell(a,a') \leq C^2$ for $\ell = 1,..,L$. Therefore we have

$$(6) \leq C + C^2 + (\hat{p}_1 \hat{p}_2)^1 C^2 + \cdots + (\hat{p}_1 \hat{p}_2)^L C^2 = C + C^2 + C^2 \sum_{\ell=1}^{L} (\hat{p}_1 \hat{p}_2)^\ell$$

which converges for $L \to \infty$.

Note, that the boundedness of k_{atom} and k_{bond} is especially fulfilled, if we take the RBF-kernel for both.

To briefly summarize, our method works as follows: We first compute the similarity of all atom and bond features using the kernels k_{atom} and k_{bond}. Having these results, we can compute the match of direct neighbors R_ℓ for each pair of atoms from both molecules by means of (5). From R_0 we can compute $R_1,..,R_L$ by iteratively revisiting all direct neighbors of each pair of atoms and computing the recursive update Formula (7). Having k_{atom} and $R_1,..,R_L$ directly gives us k_{nei}, the final similarity score for each pair of atoms, which includes information on memberships to substructures as well as physico-chemical properties. With k_{nei} we finally compute the optimal assignment kernel between two molecules M and M' using (3) and (4). Thereby, in our present implementation (Equation: 3) is calculated by the *Hungarian method* (Kuhn, 1955). Alternatively one can use the algorithm described in Mehlhorn and Näher (1999). We prefer the Hungarian method because it is more commonly used for solving the optimal assignment problem. In a last step, before SVM training, the kernel matrix is made positive semi-definite, as explained in the last Section.

Reduced Graph Representation

The main intuition of our method lies in the matching of substructures from both molecules. In the last Section we achieved this by using structural, neighborhood and other characteristic information for each single atom and bond, and computing the optimal assignment kernel between atoms of both molecules. A natural extension of this idea is to collapse structural features, like rings, donors, acceptors and others, into a single node of the graph representation of a molecule. Atoms not matching a-priori defined types of structural features can be removed (Martin, 1998). This allows us to concentrate on important structural elements of a molecule, where the definition of what an important structural element actually is, depends on the ADME problem at hand and can be given by expert knowledge in form of so-called **SMARTS**[2,3] patterns. The high relevance of such a pharmacophore mapping for ADME models is also reported in Chen, Rusinko, Tropsha, and Young (1999) and Oprea, Zamora, and Ungell (2002). If atoms match more than one SMARTS pattern, a structural feature consists of the smallest substructure that cannot be further divided into subgroups with regard to all patterns. That means in our reduced graph we may get a substructure node describing a ring only and another one describing both, a ring and an acceptor.

There are two principle problems arising in the implemention of the reduced graph: First, if certain atoms are removed from the molecular graph, then we may obtain nodes, which are disconnected from the rest of the graph. We reconnect them by new edges again such that these new edges preserve the neighborhood information, i.e. if before we had $a \to b$ and $b \to c$ and atom b is removed, we introduce an edge $a \to c$. The new edge contains information on the topological and geometrical distance of the substructures connected by it.

Thereby the topological distance between two substructures is calculated as the minimal topological distance between the atoms belonging to them, whereas the geometrical distance is computed between the centers of gravity in order to conserve information on the 3D structure of the substructures (Figure 3).

Second, we define how the feature vectors for each single atom and bond included in a substructure can be transferred to the whole substructure. This is solved by recursively applying our method from the last Section, if two substructures have to be compared.

A principle advantage of the reduced graph representation lies in the fact that complete substructures and their neighbor substructures can be compared at once. This allows us to concentrate on relevant structural aspects of the molecular graph. By means of SMARTS patterns in principle it is possible to define arbitrary structural features to be condensed in one node of the reduced molecular graph. That means in some sense one can change the "resolution" at which one looks at the molecule. This way one achieves an even higher

flexibility as offered for instance by the feature tree approach (Rarey & Dixon, 1998), because rather than considering the average over atom and bond features contained in a substructure, substructure nodes are compared on an atomic level and hence less structural information is lost. From a computational side a reduced graph representation may be useful for larger molecules, because the effort for computing the optimal assignment is reduced.

EVALUATION

Datasets

Human Intestinal Absorption (HIA)

The HIA (Human Intestinal Absorption) dataset consists of 164 structures from different sources. The dataset is a collection of Wessel, Jurs, Tolan, and Muskal (1998) (82 structures), Gohlke et al. (2001) (49 structures), Palm, Stenburg, Luthman, and Artursson (1997) (8 structures), Balon, Riebesehl, and Müller (1999) (11 structures), Kansy, Senner, and Gubernator (1998) (6 structures), Yazdanian, Glynn, Wright, and Hawi (1998) (6 structures) and Yee (1997) (2 structures). The molecules are divided into two classes "high oral bioavailability" (106 structures) and "low oral bioavailability" (58 structures) based on a histogram binning (J. Wegner, Fröhlich, & Zell, 2003a). Multiple occurrences of one molecule were removed. After the usual elimination of the hydrogen atoms, the maximal molecule size was 57 and the average size 25 atoms.

Bioavailability (Yoshida)

The Yoshida dataset (Yoshida & Topliss, 2000) has 265 molecules that we divided into two classes "high bioavailability" (bioavailability >= 50%, 159 structures) and "low bioavailability" (bioavailability < 50%, 106 structures). The maximal

Figure 3. Example of a conversion of a molecule into its reduced graph representation with edge labels containing topological distances

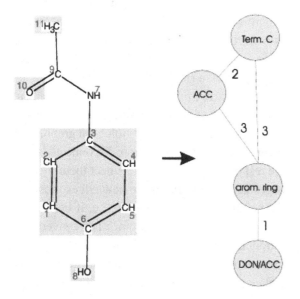

molecule size was 36 and the average size 20 atoms, after removing hydrogen.

Blood Brain Barrier Crossing (BBB)

The BBB (Blood Brain Barrier) dataset (Hou & Xu, 2003) consists of 109 structures having a maximal molecule size of 33 and an average size of 16 atoms after removing hydrogen. The target is to predict the logBB value, which describes up to which degree a drug can cross the blood-brain-barrier.

Aqueous Solubility (Huuskonen)

Finally, we investigated the Huuskonen dataset (Huuskonen, 2000), which has 1264 molecules with a maximal size of 47 and an average size of 13 atoms after removing hydrogen. The goal is to predict the aqueous solubility of a structure measured by the logS value.

Each dataset consists of energy-minimized structures using the MOE all-atom-pair force field method (Halgren, 1998), and was tested for duplicate molecules.

Descriptor Calculation for Comparison

To compare our approach with a classical descriptor based model for each dataset we calculated all descriptors available in the open source software *JOELib*[4] (J. K. Wegner, 2006). This way for the HIA dataset we obtained 6603, for the Yoshida dataset 5857, for the BBB dataset 5456, and for the Huuskonen dataset 6172 descriptors. The difference is due to the removal of constant descriptors. Missing values in descriptors were replaced by mean values. JOELib contains various 0D - 3D descriptors[5]. Besides others, these include the Radial Distribution Function descriptor, the Moreau-Broto autocorrelation, the Global Topological Charge Index and Burden\primes Modified Eigenvalues (Todeschini & Consonni, 2000). The descriptors are based on the following

atom properties: atom mass (tabulated), valence (calculated, based on graph connectivity), conjugated environment (calculated, SMARTS based), van-der-Waals volume (tabulated), electron affinity (tabulated), electro-negativity (tabulated, Pauling), graph potentials (calculated, graph theoretical), Gasteiger-Marsili partial charges (calculated, iterative — Gast-eiger and Marsili, 1978), intrinsic state (calculated), electrotopological state (calculated), electrogeometrical state (calculated — Todeschini and Consonni, 2000).

Results

Example Matching of Two Molecules

Before turning to the evaluation results, in Figure 4 we show an optimal assignment calculated by our method for the two example molecules, which were taken from the HIA dataset. As one can see, the optimal assignment indeed nicely matches the ring atoms and the atoms of the carbonyl groups and thus implements the intuition explained in the introduction.

Comparison to Descriptor-Based Models and the Marginalized Graph Kernel

Let us now turn to the evaluation of our method. We compared the optimal assignment kernel to the marginalized graph kernel (MG) using the same atom and bond features, to a full descriptor model (DESC) and a Recursive Feature Elimination based descriptor selection model (DESCSEL). For the OA and the MG kernel we considered the kernels k_{atom} and k_{bond} as a product of two RBF kernels for the real valued and the nominal atom/bond features. The width of these RBF kernels was set such that $\exp(-D/(2\sigma^2))=0.1$ (D = number of atom/bond features). Thereby all atom and bond features were approximately normalized to mean 0 and standard deviation 1 by considering the means and standard deviations of the atom

25

Figure 4. Two molecules from the HIA dataset and the optimal assignment computed by our method

acetaminophen   acetylsalicylic acid

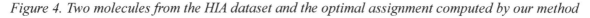

and bond properties over all our datasets. Furthermore, we explicitly set k_{bond} to 0, if one bond was in an aromatic system and the other not, or if both bonds had a different bond order. This corresponds to the multiplication with a δ-kernel. The probabilities to reach neighbors with topological distance ℓ was set to $p(\ell) = p'(\ell) = 1 - \dfrac{1}{L}$ with $L=3$ (compare Equation 6). This allows us to consider the neighborhood of a whole 6-ring of an atom. We used a SVM on the HIA and Yoshida classification datasets and a Support Vector Regression (ϵ-SVR) on the BBB and Huuskonen regression problems. The prediction strength was evaluated by means of 10-fold cross-validation. Thereby on the classification problems we ensured that the ratio of examples from both classes in the actual training set was always the same (stratified cross-validation). On each actual training fold a model selection for the necessary parameters was performed by evaluating each candidate parameter set by an extra level of 10-fold cross-validation employing the parameter optimization procedure described in Fröhlich and Zell (2005). For the MG kernel the model selection included testing termination probabilities p_t=0.1,0.3,0.5,0.7. The soft-margin parameter C was chosen from the interval $[2^{-2},2^{14}]$. On the regression datasets (BBB, Huuskonen) the width of the ϵ-tube was chosen from the interval $[2^{-8},2^{-1}]$. For the descrip-

tor based models we also tuned the width of the RBF kernel in the range $\sigma'/4,..,4\sigma'$ where σ' was set with the same heuristic mentioned above. All descriptor values (also the target values logBB, logS in case of the regression datasets) were normalized to mean 0 and standard deviation 1 on each training fold, and the calculated scaling parameters were then applied to normalize the descriptor values in the actual test set.

Figure 5 - 6 show the results we obtained. Using our OA kernel we outperformed the DESC model statistically significant on the BBB and the Huuskonen dataset ($p<0.1\%$). Thereby statistical significance was tested by a two-tailed paired t-test at significance level 10%. On the BBB dataset we also achieved significantly better results than the DESCSEL model ($p = 6.49\%$). At the same time our OA kernel circumvented the high computational burden of computing thousands of descriptors followed by a descriptor selection algorithm.

Compared to the MG kernel our OA kernel utilizing the *Hungarian method* for the assignment calculation was significantly better on the Yoshida dataset ($p = 6.57\%$), the BBB dataset ($p = 4.65\%$) and the Huuskonen dataset ($p < 1\%$). At the same time using our JAVA implementation on an AMD single core Opteron 64 Bit 2.4GHz machine the computation time for the OA kernel was always **below** that for the MG kernel (Table 3).

Figure 5. OA/OARG-Kernel vs. marginalized graph kernel and descriptor-based models: 10-fold cross-validation results (class loss in %) for the HIA dataset (legt) and the Yoshida dataset (right)

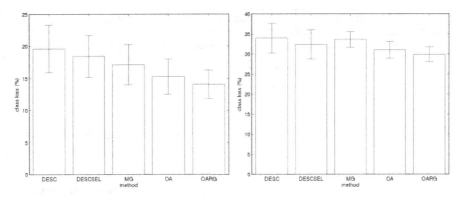

Reduced Graph Representation

We also investigated the effect of the reduced graph representation (OARG kernel — Figures 5 and 6). Thereby in the reduced graph representation only direct neighbors were considered to compute (i.e. $L' = 1$), whereas for the comparison of nodes representing structural elements we used $L = 3$ atomic neighbors as before. We considered the following pharmacophore features defined by SMARTS patterns: ring, donor, donor/acceptor, terminal carbon, positive charge, negative charge. The precise definitions of these patterns can be found in Table 2. Molecules, which did not contain any of these features and hence lead to an empty graph, were represented by the pattern "*", i.e. as one node. This only effected two compounds in the BBB dataset: N_2 and C_2HF_3BrCl.

Except for the Huuskonen dataset this choice of pharmacophoric features lead to similar cross-validation results of the OARG kernel than the original OA kernel. Obviously, on the Huuskonen dataset the defined SMARTS patterns did not capture the underlying chemistry in the right way. However, the improvement to the MG kernel was significant on the HIA dataset ($p = 9.79\%$), the Yoshida dataset ($p = 3.19\%$) and the BBB dataset ($p = 0.71\%$). Likewise, the OARG kernel significantly outperformed the DESC and the DESCSEL model on the BBB dataset ($p = 7.11\%$).

This demonstrates that on these datasets the reduced graph representation, although using less structural information than the original OA kernel, covers well the relevant chemical and biological aspects of the molecules.

The computation times for the OARG kernel are slightly higher than those for the OA kernel, because of the effort needed to match the SMARTS patterns and to construct the reduced graph, but the differences are not significant (Table 3).

Table 2. SMARTS patterns used for defining the reduced graph representation

Name	SMARTS
ring	[R]
H-bond donor	[(NH2]-a), ND1H3, ND2H2, ND3H1, ND2H1, nD1H3, nD2H2, nD3H1, nD2H1, (Cl-*), (Br-*), (I-*)]
H-bond acceptor	[(N#C-[C,c]), OD1X1, oD1X1, OD2X2, oD2X2, ND3X3, nD3X2, ND2X2, nD2X2, ND1X1, nD1X1]
H-bond donor/acceptor	[([NH2]-A), ([OH]-*)]
terminal carbon	[CH3, CD1H2, CD1H1, cH3, cD1H2, cD1H1]
positive charge	[+,++,+++]
negative charge	[-,--,---]

Table 3. Average computation times (ms ± standard deviation) for one kernel function evaluation for the different kernels

Method	HIA	Yoshida	BBB	Huuskonen
MG	5.66±6.34	4.4±3.74	3.14±10.12	1.87±1.81
OA	**4.90±7.22**	**3.33±3.63**	**2.81±9.50**	**1.58±1.47**
OARG	5.92±9.00	4.23±8.30	3.38±13.68	1.83±1.86

CONCLUSION

The author introduced a new similarity measure for chemical compounds based on a representation of molecules as multi-labeled graphs, which is well suited for ADME *in silico* prediction. The basic idea of the *optimal assignment kernel* is to compute an optimal assignment of the atoms of one molecule to those of another one, including information on neighborhood, membership to certain structural elements and other characteristics. The optimal assignment kernel can be computed efficiently by means of existing algorithms, such as the *Hungarian method* (Kuhn, 1955) or the algorithm described in Mehlhorn and Näher (1999). The author showed how the inclusion of neighborhood information for each single atom can be done efficiently via a recursive update equation, even if not only direct neighbors are considered. Comparisons to the marginalized graph kernel by Kashima et al. (2003) in several cases showed a significant reduction of the prediction error on our ADME datasets. At the same time the author's results were at least as good as a classical descriptor model with state-of-the-art SVM descriptor selection. Thereby it is important to point out that in contrast to such a model there was no data dependent adaption of the kernel. The author thinks that this is a special benefit of the kernel approach as it guarantees a unified, highly flexible, easy and fast way to obtain reliable ADME models. However, for the future it might also make sense to perform some form of feature selection among a given set of atomic properties. This can in principle be solved by using e.g. the RFE algorithm. Compared to marginalized graph kernels all in all the author sees the main advantage of his method that it better reflects a chemist\ primes intuition on the similarity of molecules.

As an extension of this approach the author introduced a reduced graph representation, in which certain structural elements are collapsed

Figure 6. OA/OARG-Kernel vs. marginalized graph kernel and descriptor-based models: 10-fold cross-validation results (mean squared error) for the BBB (left) and the Huuskonen (right) dataset

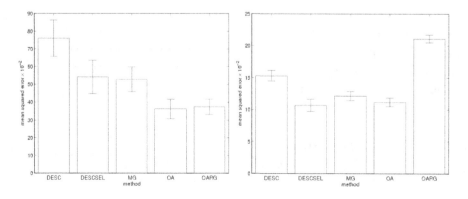

into a single node of the molecular graph and hence allow to view molecules at different user-specified levels of resolution. The main benefit lies in the fact that expert knowledge on important structural features can be included.

For the future, the optimal assignment kernel could also give the opportunity to deduce *pharmacophores*. Especially for this purpose the reduced graph representation is advantageous as well. Another open issue related to the reduced graph representation is how an optimal reduced graph could be found by looking at the data. This task certainly involves the necessity to look at the physico-chemical properties of the individual atoms. However, SMARTS patterns currently do not allow the specification of queries including physico-chemical properties. Therefore, more sophisticated ways of defining reduced graphs are required. A promising approach pointing in this direction was proposed recently (Proschak, Wegner, Schüller, Schneider, & Fechner, 2007). The combination of the Molecular Query Language with our approach would be an interesting field of research for the future.

Finally, it should be mentioned that the whole method presented here relies on a mulit-labeled graph representation of molecule, which is a simplification and abstraction. Hence, there is still the need for a kernel computation taking into account other molecular aspects, such as 3D and confirmational information. An interesting extension of the optimal assignment kernel in that direction has been published recently by Fechner et al. (2009). The more we incorporate human knowledge into the calculation of a similarity measure, the more realistic results and thus better prediction models we should get in the future.

The JAVA implementation of the optimal assignment kernel as a plugin for the open source chemoinformatics library JOELib can be retrieved from the author upon request.

ACKNOWLEDGMENT

I would like to thank all people from the group of Prof. Zell at the Center for Bioinformatics Tübingen (ZBIT), Tübingen, Germany. This includes especially Jörg Wegner, who contributed his expert knowledge on chemoinformatics and was co-author of several of my papers. Special thanks go also to Klaus Beyreuther for IT support.

REFERENCES

Agrafiotis, D., & Xu, H. (2003). A Geodisc Framework for Analyzing Molecular Similarities. *Journal of Chemical Information and Computer Sciences*, *43*, 475–484. doi:10.1021/ci025631m

Balon, K., Riebesehl, B., & Müller, B. (1999). Drug liposome partitioning as a tool for the prediction of human passive intestinal absorption. *Pharmaceutical Research*, *16*, 882–888. doi:10.1023/A:1018882221008

Böhm, H. J., & Schneider, G. (2000). *Virtual screening for bioactive molecules*. Weinheim: Wiley-VCH.

Böhm, M., & Klebe, G. (2002). Development of a new hydrogen-bond descriptor and their application to comparative mean field analysis. *Journal of Medicinal Chemistry*, *45*, 1585–1597. doi:10.1021/jm011039x

Bonchev, D., & Rouvray, D. H. (Eds.). (1990). *Chemical Graph Theory: Introduction and Fundamentals* (*Vol. 1*). London, UK: Gordon and Breach Science Publishers.

Boser, B., Guyon, M., & Vapnik, V. (1992). A training algorithm for optimal margin classifiers. In D. Haussler (Ed.), *Proc. 5th ann. acm workshop on comp. learning theory*. Pittsburgh, PA: ACM Press.

Byvatov, E., & Schneider, G. (2004). Svm-based feature selection for characterization of focused compound collections. *Journal of Chemical Information and Computer Sciences, 44*(3), 993–999. doi:10.1021/ci0342876

Carhart, R., Smith, D., & Venkataraghavan, R. (1985). Atom pairs as molecular features in structure-activity studies: Definition and applications. *Journal of Chemical Information and Computer Sciences, 25*, 64–73. doi:10.1021/ci00046a002

Chen, X., Rusinko, A., Tropsha, A., & Young, S. S. (1999). Automated Pharmacophore Identification for Large Chemical Data Sets. *Journal of Chemical Information and Computer Sciences, 39*, 887–896. doi:10.1021/ci990327n

Cortes, C., & Vapnik, V. (1995). Support vector networks. *Machine Learning, 20*, 273–297. doi:10.1007/BF00994018

Fechner, N., Jahn, A., Hinselmann, G., & Zell, A. (2009, Feb). Atomic local neighborhood flexibility incorporation into a structured similarity measure for qsar. *J Chem Inf Model*. Retrieved from http://dx.doi.org/10.1021/ci800329r

Figueras, J. (1996). Ring Perception Using Breadth–First Search. *Journal of Chemical Information and Computer Sciences, 36*, 986–991. doi:10.1021/ci960013p

Fröhlich, H. (2006). *Kernel methods in chemo- and bioinformatics*. Berlin: Logos-Verlag. (PhD-Thesis)

Fröhlich, H., Wegner, J., Sieker, F., & Zell, A. (2005). Optimal assignment kernels for attributed molecular graphs . In Raedt, L. D., & Wrobel, S. (Eds.), *Proc. int. conf. machine learning* (pp. 225–232). ACM Press.

Fröhlich, H., Wegner, J., Sieker, F., & Zell, A. (2006). Kernel functions for attributed molecular graphs – a new similarity based approach to adme prediction in classification and regression. *QSAR & Combinatorial Science, 25*(4), 317–326. doi:10.1002/qsar.200510135

Fröhlich, H., Wegner, J., & Zell, A. (2005). Assignment kernels for chemical compounds. In *Proc. int. joint conf. neural networks* (pp. 913 - 918).

Fröhlich, H., & Zell, A. (2005). Efficient parameter selection for support vector machines in classification and regression via model-based global optimization. In *Proc. int. joint conf. neural networks* (pp. 1431 - 1438).

Gärtner, T., Flach, P., & Wrobel, S. (2003). On graph kernels: Hardness results and efficient alternatives. In *Proc. 16th ann. conf. comp. learning theory and 7th ann. workshop on kernel machines*.

Gasteiger, J., & Marsili, M. (1978). A New Model for Calculating Atomic Charges in Molecules. *Tetrahedron Letters, 34*, 3181–3184. doi:10.1016/S0040-4039(01)94977-9

Godden, J., Furr, J., Xue, L., Stahura, F., & Bajorath, J. (2003). Recursive median partitioning for virtual screening of large databases. *Journal of Chemical Information and Computer Sciences, 43*, 182–188. doi:10.1021/ci0203848

Gohlke, H., Dullweber, F., Kamm, W., März, J., Kissel, T., & Klebe, G. (2001). Prediction of human intestinal absorption using a combined simmulated annealing/backpropagation neural network approach . In Hültje, H. D., & Sippl, W. (Eds.), *Rational approaches drug des* (pp. 261–270). Barcelona: Prous Science Press.

Guyon, I., Weston, J., Barnhill, S., & Vapnik, V. (2002). Gene Selection for Cancer Classification using Support Vector Machines. *Machine Learning, 46*, 389–422. doi:10.1023/A:1012487302797

Halgren, T. A. (1998). Merck molecular force field. I–V. MMFF94 Basics and Parameters. *Journal of Computational Chemistry*, *17*, 490–641. doi:10.1002/(SICI)1096-987X(199604)17:5/6<490::AID-JCC1>3.0.CO;2-P

Hotelling, H. (1933). Analysis of a complex of statistical variables into principal components. *J. Educat. Psychol.*, *24*, 417 - 441 &498 - 520.

Hou, T., & Xu, X. (2003). Adme evaluation in drug discovery. 3. modelling blood-brain barrier partitioning using simple molecular descriptors. *Journal of Chemical Information and Computer Sciences*, *43*(6), 2137–2152. doi:10.1021/ci034134i

Huuskonen, J. (2000). Estimation of Aqueous Solubility for Diverse Set of Organic Compounds Based on Molecular Topology. *Journal of Chemical Information and Computer Sciences*, *40*, 773–777. doi:10.1021/ci9901338

Kansy, M., Senner, F., & Gubernator, K. (1998). Physicochemical high throughput screening: Parallel artificial membrane permeation assay in the description of passive absorption processes. *Journal of Medicinal Chemistry*, *41*, 1007–1010. doi:10.1021/jm970530e

Kashima, H., Tsuda, K., & Inokuchi, A. (2003). Marginalized kernels between labeled graphs. In *Proc. 20th int. conf. on machine learning*.

Kashima, H., Tsuda, K., & Inokuchi, A. (2004). Kernels for graphs . In Schölkopf, B., Tsuda, K., & Vert, J. P. (Eds.), *Kernel methods in computational biology* (pp. 155–170). Cambridge, MA: MIT Press.

Kubinyi, H. (2002). From Narcosis to Hyperspace: The History of QSAR. *Quant. Struct. Act. Relat.*, *21*, 348–356. doi:10.1002/1521-3838(200210)21:4<348::AID-QSAR348>3.0.CO;2-D

Kubinyi, H. (2003). Drug research: myths, hype and reality. *Nature Reviews. Drug Discovery*, *2*, 665–668. Retrieved from http://home.t-online.de/home/kubinyi/nrdd-pub-08-03.pdf. doi:10.1038/nrd1156

Kubinyi, H. (2004). Changing paradigms in drug discovery. In M. H. et al. (Ed.), *Proc. int. beilstein workshop* (pp. 51 - 72). Berlin: Logos-Verlag.

Kuhn, H. (1955). The hungarian method for the assignment problem. *Naval Res. Logist. Quart.*, *2*, 83–97. doi:10.1002/nav.3800020109

Leslie, C., Kuang, R., & Eskin, E. (2004). Inexact matching string kernels for protein classification . In Schölkopf, B., Tsuda, K., & Vert, J. P. (Eds.), *Kernel methods in computational biology* (pp. 95–112). Cambridge, MA: MIT Press.

Martin, Y. (1998). Pharmacophore mapping . In Martin, Y., & Willett, P. (Eds.), *Designing bioactive molecules* (pp. 121–148). Oxford University Press.

Mehlhorn, K., & Näher, S. (1999). *The LEDA Platform of Combinatorial and Geometric Computing*. Cambridge University Press.

Oprea, T. I., Zamora, I., & Ungell, A. L. (2002). Pharmacokinetically based mapping device for chemical space navigation. *Journal of Combinatorial Chemistry*, *4*, 258–266. doi:10.1021/cc010093w

Palm, K., Stenburg, P., Luthman, K., & Artursson, P. (1997). Polar molecular surface properties predict the intestinal absorption of drugs in humans. *Pharmaceutical Research*, *14*, 586–571. doi:10.1023/A:1012188625088

Proschak, E., Wegner, J. K., Schüller, A., Schneider, G., & Fechner, U. (2007). Molecular query language (mql)–a context-free grammar for substructure matching. *Journal of Chemical Information and Modeling*, *47*(2), 295–301. Retrieved from http://dx.doi.org/10.1021/ci600305h. doi:10.1021/ci600305h

Raedt, L. D., & Kramer, S. (2001). Feature construction with version spaces for biochemical application. In *Proc. 18th int. conf. on machine learning* (pp. 258 - 265).

Rarey, M., & Dixon, S. (1998). Feature trees: A new molecular similarity measure based on tree-matching. *Journal of Computer-Aided Molecular Design, 12*, 471–490. doi:10.1023/A:1008068904628

Raymond, J., Gardiner, E., Willett, P., & Rascal, P. (2002). Calculation of graph similarity using maximum common edge subgraphs. *The Computer Journal, 45*(6), 631–644. doi:10.1093/comjnl/45.6.631

Rupp, M., Proschak, E., & Schneider, G. (2007). Kernel approach to molecular similarity based on iterative graph similarity. *Journal of Chemical Information and Modeling, 47*(6), 2280–2286. Retrieved from http://dx.doi.org/10.1021/ci700274r. doi:10.1021/ci700274r

Schölkopf, B., & Smola, A. J. (2002). *Learning with Kernels*. Cambridge, MA: MIT Press.

Shawe-Taylor, J., & Cristianini, N. (2004). *Kernel methods for pattern analysis*. Cambridge, UK: Cambridge University Press.

Todeschini, R., & Consonni, V. (Eds.). (2000). *Handbook of Molecular Descriptors*. Weinheim: Wiley–VCH.

van de Waterbeemd, H., & Gifford, E. (2003). ADMET *In Silico* Modelling: Towards Prediction Paradise? *Nature Reviews. Drug Discovery, 2*, 192–204. doi:10.1038/nrd1032

Vert, J. P. (2008). *The optimal assignment kernel is not positive definite*. Retrieved from http://www.citebase.org/abstract?id=oai:arXiv.org:0801.4061

Vishwanathan, S., & Smola, A. (2004). Fast Kernels for String and Tree Matching . In Schölkopf, B., Tsuda, K., & Vert, J. P. (Eds.), *Kernel methods in computational biology* (pp. 113–130). Cambridge, MA: MIT Press.

Washio, T., & Motoda, H. (2003). State of the art of graph-based data mining. *SIGKDD Explorations Special Issue on Multi-Relational Data Mining, 5*.

Wegner, J., Fröhlich, H., & Zell, A. (2003a). Feature selection for Descriptor based Classification Models: Part II - Human Intestinal Absorption (HIA). *Journal of Chemical Information and Computer Sciences, 44*, 931–939. doi:10.1021/ci034233w

Wegner, J., Fröhlich, H., & Zell, A. (2003b). Feature Selection for Descriptor based Classificiation Models: Part I - Theory and GA-SEC Algorithm. *Journal of Chemical Information and Computer Sciences, 44*, 921–930. doi:10.1021/ci0342324

Wegner, J. K. (2006). *Data Mining und Graph Mining auf molekularen Graphen - Cheminformatik und molekulare Kodierungen fï¿œr ADME/Tox & QSAR-Analysen*. Unpublished doctoral dissertation, Eberhard-Karls Universität Tübingen.

Wessel, M. D., Jurs, P. C., Tolan, J. W., & Muskal, S. M. (1998). Prediction of Human Intestinal Absorption of Drug Compounds from Molecular Structure. *Journal of Chemical Information and Computer Sciences, 38*, 726–735. doi:10.1021/ci980029a

Xue, Y., Li, Z. R., Yap, C. W., Sun, L. Z., & Chen, X. (2004). Effect of molecular descriptor feature selection in support vector machine classification of pharmacokinetic and toxicological properties of chemical agents. *Journal of Chemical Information and Computer Sciences, 44*(5), 1630–1638. doi:10.1021/ci049869h

Yazdanian, M., Glynn, S., Wright, J., & Hawi, A. (1998). Correlating partitioning and caco-2 cell permeability of structurally diverse small molecular weight compounds. *Pharmaceutical Research, 15*, 1490–1494. doi:10.1023/A:1011930411574

Yee, S. (1997). In vitro permeability across caco-2 cells (colonic) can predict in vivo (small intestinal) absorption in man - fact or myth. *Pharmaceutical Research, 14,* 763–766. doi:10.1023/A:1012102522787

Yoshida, F., & Topliss, J. (2000). QSAR model for drug human oral bioavailability. *Journal of Medicinal Chemistry, 43,* 2575–2585. doi:10.1021/jm0000564

ENDNOTES

[1]   The work presented here was done while the author was employed at the Center for Bioinformatics Tübingen (ZBIT), Tübingen, Germany

[2]   Daylight Chemical Information Systems Inc., http://www.daylight.com

[3]   SMARTS, similar to regular expressions, is a language for describing patterns on molecular graphs.

[4]   http://sourceforge.net/projects/joelib

[5]   A complete list of all actually implemented descriptors can be found under http://www.ra.cs.uni-tuebingen.de/software/joelib/tutorial/descriptors/descriptors.html.

APPENDIX

Preventing the "Tottering"

If we evaluate R_0 (Equation 5) at all neighbors of a certain topological distance ℓ, we also revisit atoms and bonds that we have considered at topological distance ℓ-1. To prevent this "tottering", we can make our decay factor γ dependent not just on the topological distance, but also on the path of visited atoms and bonds. Thereby we have to explicitly forbid paths of the form $a \to n_i(a) \to a$. This can be achieved by setting

$$\gamma'(\ell, a, a', n_i(a), n_j(a')) = \begin{cases} 0 & \exists k : n_k(n_i(a)) = a \lor \exists t : n_t(n_j(a')) = a' \\ \gamma(\ell) & \text{otherwise} \end{cases} \tag{11}$$

This requires the following changes in our computation for k_{nei}:

$$k_{nei}(a, a') = k_{atom}(a, a') + \frac{1}{|a||a'|} \sum_{i,j} r_0(a, a', n_i(a), n_j(a')) \tag{12}$$

$$+ \sum_{\ell=1}^{L} \left(\frac{1}{|a||a'|} \sum_{i,j} \gamma'(\ell, a, a', n_i(a), n_j(a')) \cdot r_\ell(a, a', n_i(a), n_j(a')) \right) \tag{13}$$

$$r_\ell(a, a', n_i(a), n_j(a')) = \frac{1}{|n_i(a)||n_j(a')|} \cdot$$

$$\sum_{k,t} r_{\ell-1}(n_i(a), n_j(a'), n_k(n_i(a)), n_l(n_j(a')))$$

$$r_0(a, a', n_i(a), n_j(a')) = k_{atom}(n_i(a), n_j(a')) k_{bond}(a \to n_i(a), a' \to n_j)a'))$$

That means we can compute k_{nei} by iteratively revisiting the direct neighbors and indirect neighbors of (a, a'). In contrast to (6) in (8) we do not use an optimal assignment kernel for the direct neighbors of (a, a'), but compute the average match here. Experimentally the modified kernel never leads to an improvement of the prediction error (Table 4).

Table 4. Effect of the OA kernel with prevention of the "tottering" (OAopt) compared to original OA kernel

Method	HIA	Yoshida	BBB	Huuskonen
OA	15.26±2.74	30.98±2.07	36.24±5.48	11.17±0.6
OAopt	15.92±2.64	32.46±2.63	37.79±6.11	11.71±0.53

Chapter 3
3D Ligand–Based Virtual Screening with Support Vector Machines

Jean-Philippe Vert
Mines ParisTech, Institut Curie and INSERM U900, France

ABSTRACT

The author reviews an approach, proposed recently by Mahé, Ralaivola, Stoven, and Vert (2006), for ligand-based virtual screening with support vector machines using a kernel based on the 3D structure of the molecules. The kernel detects putative 3-point pharmacophores, and generalizes previous approaches based on 3-point pharmacophore fingerprints. It overcomes the categorization issue associated with the discretization step usually required for the construction of fingerprints, and leads to promising results on several benchmark datasets.

INTRODUCTION

Computational models play an important role in early-stage drug discovery, in particular for lead identification and optimization. Starting from a list of molecules with experimentally determined binding affinity to a particular therapeutic target, as typically obtained by high-throughput screening (HTS), the goal of lead optimization is to find additional molecules with good binding affinity. The resulting leads are then further optimized, in particular to improve their pharmacokinetical and toxicological profiles, eventually leading to new candidate drugs.

Lead identification and optimization are usually performed by screening large databanks of small molecules, e.g., created by combinatorial chemistry, to find active molecules. Since experimental screening remains costly and time-consuming when large banks are concerned, and given the immensity of the space of small molecules which may be synthesized, *in silico* screening provides an interesting complementary approach to identify active molecules. An *in silico* screening is based on a model which can predict the activity of candidate molecules from their structure. Two general classes of models are often used. First, if the 3D structure of the target is known, then *docking* models predict whether

DOI: 10.4018/978-1-61520-911-8.ch003

a small molecule can inhibit it by simulating its 3D structure and estimating the binding affinity of the protein-ligand complex. Docking involves difficult optimization problems to find the optimal 3D conformation of the molecule, binding configuration, and estimating its free energy. It is therefore rarely used on very large chemical databanks and is furthermore limited by the need to know in advance the 3D structure of the target. A second common approach, sometimes used in parallel or as an alternative to docking, is *ligand-based* virtual screening. In that case an initial set of molecules with known binding affinity is used to build a predictive model that relates the structure of a molecule to its activity. The model can then be used to screen candidate molecules by ranking them in terms of their predicted activity by the model. This approach, often referred to as *quantitative structure-activity relationship* (QSAR), does not require the structure of the target, and usually results in computationally fast tools to predict the activity of candidate molecules. In this chapter we focus on this later, ligand-based virtual screening approach.

Ligand-based approaches usually involve statistical and machine learning procedure. Indeed, they can be formulated as the problem of estimating the attribute (activity) of patterns (molecules) given a set of patterns with known attributes. When the activity is considered as a real-valued attribute (e.g., free binding energy), we recognize a problem of *regression*, while when the attribute is categorical (e.g., active vs. not active) we are confronted with a problem of *supervised binary classification*, or *pattern recognition*. We note that, apart from ligand-based virtual screening, machine learning has been used to solve other problems in chemoinformatics such as toxicity or ADME in silico prediction, by simply modifying the definition of the attribute to be predicted. Both regression and pattern recognition have been much studied in statistics and machine learning, and a number of algorithms are available to attack them, ranging from linear models such as least-square regression,

partial least-square (PLS) or linear discriminant analysis (LDA) to nonlinear methods such as neural networks, nearest neighbor or decision trees. A particularity of the ligand-based problem is that the patterns are molecules, while most algorithms for regression and pattern recognition work with vectors. Hence, in order to apply these algorithms to our problems, the molecules must first be converted to vectors of numerical features. This vectorization of the molecules turns out to be an important but difficult step. The problem of constructing numerical features, usually referred to as *descriptors* in chemoinformatics, remains one of the most debated and challenging issue in chemoinformatics. Indeed, a molecule is not easily and unambiguously described by a small set of numerical descriptors, and many descriptors have been proposed to describe various properties or features of the molecules (Todeschini & Consonni, 2002). Common descriptors include general properties of the molecules, such as the molecular weight, 2D descriptors with encode information about the 2D structure of the molecules, such as topological descriptors, hydrophobicity or substructure fingerprints, or 3D descriptors which capture geometric aspects of the molecule seen as a shape in the 3D space, such as quantum mechanical descriptors or shape indices (Figure 1).

The profusion of possible descriptors raises theoretical and practical issues. First, having many descriptors makes the dimension of the regression or pattern recognition problem large, which is often equivalent to making it hard to solve with a limited number of training examples. Second, computing and storing in memory many descriptors for large banks of molecules requires time and computational resources that may become limiting factors. Therefore, a common strategy in chemoinformatics is to select a limited number of "well-chosen" descriptors among the variety of possibilities. The criteria that should guide this selection for a given problem remain, however, largely debated.

Figure 1. Many descriptors have been proposed to characterize the 2D (left) or 3D (right) structures of molecules, here a drug (Sarmazenil, $C_{15}H_{14}ClN_3O_3$) from the benzodiazepine family

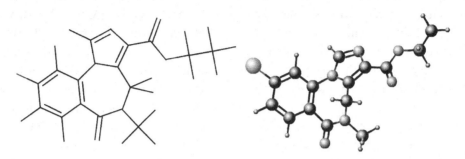

In recent years, a new strategy has been investigated to overcome the issue of choosing a limited number of descriptors. This strategy is based on a particular property of a class of algorithms for regression and pattern recognition, collectively known as *kernel methods* (Schölkopf & Smola, 2002, Shawe-Taylor & Cristianini, 2004, Schölkopf, Tsuda, & Vert, 2004). While the most famous kernel method is the *support vector machine* (SVM), other methods such as *kernel perceptron* or *kernel PLS* are also gaining popularity in chemoinformatics since they generalize methods well-known in the field. The particular property of these algorithms is that they are not restricted to work only on vectorial data. Instead, they can work with any sort of data as soon as a distance or a similarity measure, which must fulfill some technical conditions, is defined on the set of patterns. In order to use these algorithms for ligand-based screening, it therefore suffices to define a measure of distance or similarity between molecules, and not necessarily to encode molecules using descriptors. Interestingly, any representation of molecules as vectors can be turned into a measure of similarity, e.g., by taking the inner product between vectors. All kernel methods can therefore be used with all existing descriptors. However, the converse is not true: one can imagine many measures of similarity between molecules which cannot be associated to finite-dimensional representations of the molecules. Hence kernel methods generalize and offer more possibilities

for ligand-based virtual screening than methods based on a representation of molecules as vectors of descriptors.

In this chapter, we illustrate this idea on the problem of using information about the 3D structure of the molecules for ligand-based virtual screening with SVM. An interesting information that can be extracted from the 3D structure is the presence (or absence) of specific arrangements of triplets of atoms, often called pharmacophores, which may be responsible for the activity of the molecule. Encoding this information into a vector of descriptors raises issues due to the need to discretize continuous-valued distances into a finite number of bins. When kernel methods are used, we show that the discretization step is not required, and can be advantageously replaced by continuous measures of similarities between putative pharmacophores.

KERNELS AND KERNEL METHODS

Kernel methods form a rich family of algorithms for data analysis and machine learning. For example, kernel principal component analysis (KPCA) can be used to project the data to a low-dimensional subspace to visualize them, kernel ridge regression or kernel perceptron can be used to solve regression problems, and SVM can be used for pattern recognition problem. They share in common the use of positive definite kernels to manipulate data. More precisely, let us denote by X the space of patterns to be analyzed, i.e., the space of molecules in our case. A positive definite kernel on X is a similarity function $K:X \times X \to \mathbb{R}$ which fulfills two conditions: (i) it is symmetric, in the sense that $K(x,x')=K(x,x')$ for any two patterns x and x' in X and (ii) it is positive definite, in the sense that for any set of patterns $x_1,...,x_n$ the square $n \times n$ matrix of pairwise similarities $K=(K(x_i,x_j))_{1 \le i,j \le n}$ is positive semidefinite, i.e., has only nonnegative eigenvalues. Many common similarity measures fulfill these conditions. For examples, is the patterns are vectors, then the inner product $x^T x'$ and the Gaussian kernel $\exp(-\|x-x'\|^2/2\sigma^2)$ are both positive definite kernels. Kernel methods can be used to analyze data as soon as a positive definite kernel is defined on the space of data. For example, in the case of regression of pattern recognition, SVM estimate a function $f:X \to \mathbb{R}$ which has the form

$$f(x) = \sum_{i=1}^{n} \alpha_i K(x_i, x),$$

where $x_1,...,x_n$ are training patterns, K is a positive definite kernel, and $\alpha_1,...,\alpha_n$ are weights estimated during the learning step to optimize some objective function.

Kernel methods have recently gained a lot of popularity in many fields, in particular chemoinformatics, both because of their good performance in practice and the flexibility they offer in the choice of the positive definite kernel (Ivanciuc, 2007). A large body of research has focused on the use of kernel methods with classical descriptors, but several groups have also started to investigate the possibility to design more general positive definite kernels for molecules. Kernels for 2D structures have by far attracted the most attention, triggered by the seminal work of Kashima, Tsuda, and Inokuchi (2003) and Gärtner (2002) who showed how to define a kernels encoding the information of an infinite number of descriptors counting the number of linear fragments in the molecular graphs. These kernels were subsequently extended to encode more informative

linear fragments (Mahé, Ueda, Akutsu, Perret, & Vert, 2005), or nonlinear fragments (Ramon & Gärtner, 2003, Mahé & Vert, 2009). Alternatively Fröhlich, Wegner, Sieker, and Zell (2005) proposed to directly define a similarity measure between 2D structures of molecules by optimally matching the vertices of the molecular graphs, and obtain good performance although the similarity function is not positive definite (Vert, 2008).

While the 2D structure of molecules contains a lot of relevant information to predict its activity, the 3D configuration of atoms in space is also likely to play an important role in the activity of a molecule. Indeed, the binding of a molecule to a particular site of a protein target often involves geometric complementarity between the molecule and the target, to allow the formation of weak bounds and of a stable complex. Not surprisingly, ligand-based virtual screening approaches sometimes use the 3D structure of molecules, requiring methods to represent and compare 3D structures. The comparison of 3D structures can for example rely on optimal alignments in the 3D space (Lemmen & Lengauer, 2000), or on the comparison of features extracted from the structures (Xue & Bajorath, 2000). Features of particular importance in this context are subsets of two to four atoms together with their relative spatial organization, also called *pharmacophores*. Discovering pharmacophores common to a set of known inhibitors to a drug target can be a powerful approach to screening other candidate molecules containing the pharmacophores, as well as a first step towards the understanding of the biological phenomenon involved (Holliday & Willett, 1997, Finn, Muggleton, Page, & Srinivasan, 1998). Alternatively, pharmacophore fingerprints, that is, bitstrings representing a molecule by the pharmacophores it contains, has emerged as a potential approach to apply statistical learning methods for SAR, although sometimes with mixed results (Matter & Pötter, 1999, Brown & Martin, 1997, Bajorath, 2001). In spite of the importance of 3D descriptors in ligand-based virtual screening, relatively little attention has been paid in the development of positive definite kernels for 3D structures, which would pave the way to the use of kernel methods. Notable exceptions include the work of Swamidass et al. (2005), who propose to compare the histograms of pairwise distances between atom classes in 3D, and the work of Mahé et al. (2006), reviewed in the next section, who compare 3D structures through the candidate 3-point pharmacophores they contain.

THE PHARMACOPHORE KERNEL

Rational drug design largely relies on the paradigm of site-ligand shape and functional group complementarity in order to explain the affinity of a ligand for its macromolecular receptor. Complementarity is here understood as the ability to form stabilizing interactions between a ligand and a site, e.g., hydrophobic contacts, hydrogen bonds, and salt bridges. From the ligand point of view, this suggests that the ability to bind a target is driven by the presence of particular geometric configuration of pharmacophoric features such as hydrophobic or aromatic, hydrogen-bond acceptors and donors, and positively or negatively charged ions. Such a configuration responsible for the activity of the ligand is called a *pharmacophore*. As we are interested in descriptors of 3D structures which can help predict the activity of molecules, this suggests to consider putative pharmacophores as relevant features.

More precisely, we define a *putative three-point pharmacophore* as any three-dimensional arrangement of three pharmacophoric features, like acceptor atom/acceptor atom/hydrophobic point, which may be responsible for the biological activity of a molecule. The triangle formed by the three features is completely characterized by the three distances between all pairs of features. Formally, we therefore define a putative pharmacophore p as a triplet of features, together with the 3 pairwise distances between the 3 features in the 3D space.

Denoting by F the finite set of features considered, a putative pharmacophore can therefore be represented as a 6-tuple $p=(f_1,f_2,f_3,d_1,d_2,d_3)$, where $f_1,f_2,f_3 \in F$ are the three features and $d_1,d_2,d_3 \in \mathbb{R}$ are the three pairwise distances. Let us denote by P the space of possible pharmacophores. Given a molecule M represented as a list of m pharmacophoric features with their 3D coordinates, we can extract $n_m=m(m-1)(m-2)$ putative pharmacophores by taking all triplets of features, together with their three pairwise distances in the 3D space. In other words, we represent a molecule M as a list $\Phi(M)$ of putative pharmacophores $p_1,\dots,p_{n_m} \in \mathcal{P}$.

The space of putative pharmacophores is continuous by nature. Indeed, besides the discrete triplet of features involved in the triplet, the set of possible pairwise distances spans a continuous subset of $\mathbb{R}^3$. One strategy to simplify the description $\Phi(M)$ of a molecule and make it amenable to computation is then to discretize the continuous space P into a finite number of regions, and just record how many putative pharmacophores in $\Phi(M)$ fall in each region. This would result in a fingerprint for each molecule, indexed by the regions defined in the space of pharmacophores, which could then be used for 3D ligand-based virtual screening (Matter & Pötter, 1999, Brown & Martin, 1997, Bajorath, 2001). For example, Pickett, Mason, and McLay (1996) discretize each distance in the range $2-24\mathring{A}$ using six distance ranges, to obtain a global discretization of the pharmacophore space P into 5,916 regions. More formally, let us denote by $Q=(q_1,\dots,q_L)$ the finite set of regions used to discretize P, and by $d:P \rightarrow Q$ the discretization function that maps any putative pharmacophore $p \in P$ to one region $d(p) \in Q$. For a molecule M described by a set of putative pharmacophores $\Phi(M)$, the 3D fingerprint representing M is the vector $\Psi(M)=(\Psi_q(M))_{q \in Q}$ such that $\Psi_q(M)$ is the number of putative pharmacophores $p \in \Phi(M)$ which are mapped to the region $q \in Q$.

Although convenient in practice, this discretization has the potential disadvantage that some information is lost between the list of putative pharmacophores in *P*, and its discretized version in the fingerprint. This loss of information and the resulting problem is illustrated in Figure 2. For example, two putative pharmacophores may be very close to each other in the continuous space *P*, but may fall on two different regions just because the region boundary goes between the two points in *P*. Alternatively, two putative pharmacophores may belong to the same region and be counted together in the fingerprint, although one may be responsible for the activity of a molecules and not the other; this would be the case if the biological boundary between active and non-active putative pharmacophores in *P* splits a region in two.

Mahé et al. (2006) proposed to remedy both issues related to the discretization of *P* in the construction of 3D pharmacophore fingerprints. They noticed that if a kernel method is to be used to analyze the 3D pharmacophore fingerprint,

then what matters is not the fingerprints themselves, but the positive definite kernel they induce. Considering the inner product between fingerprint as a natural kernel $K(M,M')$ between two 3D structures of molecules, they noticed that the kernel can be expressed as follows:

$$
\begin{aligned}
K(M,M') \quad &= \Psi(M)^{\top}\Psi(M') \\
&= \sum_{q \in \mathcal{Q}} \Psi_q(M)\Psi_q(M') \\
&= \sum_{q \in \mathcal{Q}} \left(\sum_{p \in \Phi(M)} \mathbf{1}\big(d(p) = q\big) \right)\left(\sum_{p \in \Phi(M)} \mathbf{1}\big(d(p) = q\big) \right) \\
&= \sum_{p \in \Phi(M)} \sum_{p' \in \Phi(M')} \left(\sum_{q \in \mathcal{Q}} \mathbf{1}\big(d(p) = q\big)\mathbf{1}\big(d(p') = q\big) \right) \\
&= \sum_{p \in \Phi(M)} \sum_{p' \in \Phi(M')} \mathbf{1}\big(d(p) = d(p')\big).
\end{aligned}
\tag{1}
$$

If we denote by $K_p(p,p')$ the similarity function between putative pharmacophores induced by the discretization step as follows:

Figure 2. Illustration of the issues associated to the discretization of a continuous space with a partition. Although x_1 is closer to x_3 than to x_2 in the continuous space, the order is inverted after discretization since x_1 and x_2 fall in the same region, but not x_3. The regions should be as large as possible to limit the risk of having neighboring points, such as x_1 and x_3 to fall on two different regions; however they should not be too large otherwise distant points, such as x_1 and x_2, would fall in the same region and be considered the same after discretization

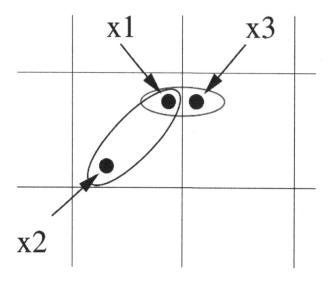

$$K_P(p, p') = \begin{cases} 1 & \text{if } d(p) = d(p'), \\ 0 & \text{otherwise}, \end{cases} \qquad (2)$$

then the previous computation can be simplified as:

$$K(M, M') = \sum_{p \in \Phi(M)} \sum_{p' \in \Phi(M')} K_P(p, p').$$

In other words, the resulting kernel for molecules in a particular example of a convolution kernel (Haussler, 1999): it decomposes as the sum, over all pairs of putative pharmacophores in both molecules, of a basic kernel K_p between individual pharmacophores. Interestingly, we observe that the discretization process influences the kernel $K(M,M')$ only through the definition of the basic kernel K_p through (2): the similarity between two putative pharmacophores is 1 if they fall in the same region of the discretized space, 0 otherwise. In order to overcome the limitations of categorization errors, Mahé et al. (2006) propose to replace this binary-valued kernel by any other kernel with takes into account the geometry of the space of pharmacophores, without enforcing categorization of the space. For example, they suggest to use the following Gaussian kernel between two pharmacophores $p=(f_1,f_2,f_3,d_1,d_2,d_3)$ and $p'=(f_1,f_2,f_3,d_1,d_2,d_3)$ as follows:

$$K_P(p,p') = \begin{cases} \exp\left(-\gamma((d_1 - d_{1'})^2 - (d_2 - d_{2'})^2 - (d_3 - d_{3'})^2)\right) & \text{if } f_1 = f_{1'}, f_2 = f_{2'} \text{ and } f_3 = f_{3'}, \\ 0 & \text{otherwise}, \end{cases}$$
$$(3)$$

The key difference between (2) and (3) is that the latter takes real values, which monotonically decrease to 0 when the shapes of the triangles become more and more different, while the former can take only 0/1 values and is not monotonically decreasing with the difference between triangles (measured as the Euclidean distance between the vector of pairwise distances). Hence the similarity defined by (3) can be thought of as a smooth form of the similarity defined by (2).

The number of pharmacophores in a molecule M with m pharmacophoric features being of the order of $O(m^3)$, the complexity of computing the kernel between two molecules with respectively m and m' features is $O(m^3 m'^3)$. Mahé et al. (2006) discuss various strategies to speed up the computation, including the use of pharmacophore fingerprints based on discretization of the pharmacophore space, which can reduce the complexity of the kernel computation to $O(m^3 + m'^3)$. A free and publicly available implementation of those variants is available in the ChemCPP software[1].

EXPERIMENTS

In order to validate the relevance of using 3D descriptors for virtual screening, and of using a real-valued 3D pharmacophore kernel instead of its binary counterpart obtained by discretizing the pharmacophore space, Mahé et al. (2006) tested the approach on four datasets previously analyzed by Sutherland, O'Brien, and Weaver (2003):

- the *BZR* dataset, a set of 405 ligands for the benzodiazepine receptor,
- the *COX* dataset, a set of 467 cyclooxygenase-2 inhibitors,
- the *DHFR* dataset, a set of 756 inhibitors of dihydrofolate reductase,
- the *ER* dataset, a set of 1009 estrogen receptor ligands.

For all datasets, a 3D conformation of each molecule is given, together with a quantitative measure of its binding affinity to its target. The activity of a molecule is then transformed into a binary value to separate active vs inactive compounds. Training and validation sets are defined in advance in order to allow a fair comparison of different methods, as summarized in Table 1.

The pharmacophoric features used to define the putative pharmacophores are simply each atom's type (carbon, oxygen, nitrogen, etc...),

Table 1. Number of positive and negative examples in the training and validation sets of each benchmark

	Train		Validation	
	Pos	**Neg**	**Pos**	**Neg**
BZR	94	87	63	62
COX	87	91	61	64
DHFR	84	149	42	118
ER	110	156	70	110

together with their partial charge which can be positive, negative or neutral. Hence each atom of a molecule is considered a pharmacophoric feature in the set $\{C^+, C^0, C^-, O^+, O^0, O^-, \dots\}$, and if a molecule has a atoms then we consider $a(a-1)(a-2)$ putative pharmacophores. The partial charges account for the contribution of each atom to the total charge of the molecule, and were computed with the QuacPAC software developed by OpenEye[2]. It is important to note that, contrary to the physico-chemical properties of atoms, partial charges depend on the molecule and describe the spatial distribution of charges. Although the partial charges take continuous values, we simply kept their signs to define the pharmacophoric features, as basic indicators of charges in the description of pharmacophores. The distance between two such pharmacophoric features is computed as the Euclidean distance between the 3D coordinates of the corresponding atoms. The QSAR model is built with a SVM, whose regularization parameter is optimized over a grid by cross-validation on the training set. The main objective of the experiment is to compare the continuous version of the 3D pharmacophore kernel (3) with its discrete counterpart (2). Each kernel has a parameter which is also optimized by cross-validation on the training sets: γ in (3), and the number of bins to discretize the distances in (2). Finally, a state-of-the-art Tanimoto kernel based on the 2D structures of molecules is also tested as a baseline method, to test the potential gain obtained by including 3D information.

Table 2 shows the accuracy obtained with the different kernels on all datasets. In all cases, the continuous 3D pharmacophore kernel outperforms the other ones. A slight but consistent gain of 1 to 5% over the 3D fingerprint kernel illustrate the loss in performance due to categorization of pharmacophores into non-overlapping regions. Moreover both 3D kernels outperform the 2D kernel on all datasets, with a large margin on BZR and COX. In terms of computation time, it took of the order of 4 minutes to compute the 405 by 405 matrix of kernel values for the BZR database on a desktop computer (Pentium 4, 3.6GHz, 1GB). The discretized version with a split of the distance range into 24 bins ran 35 times faster. On larger databases, it may therefore provide an interesting alternative if model accuracy has to be balanced for computational speed.

DISCUSSION

In summary, the 3D pharmacophore kernel proposed by Mahé et al. (2006) relies on two observations: first, that the discretization process is not required if positive definite kernels are used, and second, that discretization induces a particular convolution kernel which can advantageously be replaced by another kernel which better takes into account the natural metric in the continuous space of pharmacophores. Obviously this observation may have direct applications in any field where discretization is carried out to summarize a distribution over a continuous-valued space by a finite histogram of counts.

Table 2. Accuracy of various kernels on the four benchmark datasets

Kernel	BZR	COX	DHFR	ER
2D (Tanimoto)	71.2	63.0	76.9	77.1
3D fingerprint	75.4	67.0	76.9	78.6
3D not discretized	**76.4**	**69.8**	**81.9**	**79.8**

An interesting generalization of this work would be to investigate the use of higher-order pharmacophores, e.g., involving four pharmacophoric features. When discretization is performed, then the precision of discretization degrades quickly when the number of features in the pharmacophores increases. For example, when f pharmacophoric features are used, and each distance is discretized into d bins, then the number of regions in the space of 3-point pharmacophores is $f^3 d^3$, since they must code three features and three distances. Taking f=6 features with only d=5 bins for distances already leads to 27, 000 descriptors (which

may be reduced by considering symmetries and redundancies). 4-point pharmacophores require the coding of four features and six distances, resulting in $f^4 d^6$ descriptors. Keeping f=6 features and d=5 bins would result in more than 20 million descriptors, which is much too large for usual applications. Decreasing the bin numbers to d=2 even leads to more than $80k$ descriptors. Hence, in this case, replacing the discretization by a direct measure of similarity in the space of putative pharmacophore may lead to effective methods for ligand-based virtual screening based on 4-point pharmacophoric descriptors.

An important point that needs to be addressed by kernel method is that, besides good performance, chemoinformaticians and chemists often want to understand the features important in the model in order to understand the molecular basis of the inhibition and be able to design new molecules with increased activity. The extraction of features or rules from a kernel method using a smooth kernel like a Gaussian kernel remains, however, a debated topic.

Finally, it should be pointed out that the 3D structure of a molecule is often difficult to define. It is more common to talk about sets of conformers, which correspond to the different conformations a molecule can take in 3D, and which can be generated from the 2D structure by various softwares. In that case kernel methods may also offer interesting opportunities to work with sets of conformers, e.g., using multi-instance kernels (Gärtner, Flach, Kowalczyk, & Smola, 2002).

REFERENCES

Bajorath, J. (2001). Selected concepts and investigations in compound classification, molecular descriptor analysis, and virtual screening. *Journal of Chemical Information and Computer Sciences*, *41*(2), 233–245. doi:10.1021/ci0001482

Brown, R. D., & Martin, Y. C. (1997). The information content of 2D and 3D structural descriptors relevant to ligand-receptor binding. *Journal of Chemical Information and Computer Sciences*, *37*, 1–9. doi:10.1021/ci960373c

Finn, P., Muggleton, S., Page, D., & Srinivasan, A. (1998). Pharmacophore discovery using the inductive logic programming language Progol. *Machine Learning*, *30*, 241–270. doi:10.1023/A:1007460424845

Fröhlich, H., Wegner, J. K., Sieker, F., & Zell, A. (2005). Optimal assignment kernels for attributed molecular graphs. In *Proceedings of the 22nd international conference on machine learning* (pp. 225 - 232). New York: ACM Press.

Gärtner, T. (2002). Exponential and Geometric Kernels for Graphs. *In NIPS Workshop on Unreal Data: Principles of Modeling Nonvectorial Data*.

Gärtner, T., Flach, P., Kowalczyk, A., & Smola, A. (2002). Multi-Instance Kernels. In C. Sammut & A. Hoffmann (Eds.), *Proceedings of the Nineteenth International Conference on Machine Learning* (pp. 179-186). San Francisco: Morgan Kaufmann.

Haussler, D. (1999). *Convolution Kernels on Discrete Structures* (Tech. Rep. No. UCSC-CRL-99-10). UC Santa Cruz.

Holliday, J. D., & Willett, P. (1997). Using a genetic algorithm to identify common structural features in sets of ligands. *Journal of Molecular Graphics & Modelling, 15*(4), 221–232. doi:10.1016/S1093-3263(97)00080-6

Ivanciuc, O. (2007). Applications of support vector machines in chemistry. In K. B. Lipkowitz & T. R. Cundari (Eds.), *Reviews in computational chemistry* (Vol. 23, pp. 291–400). Weiheim: Wiley-VCH.

Kashima, H., Tsuda, K., & Inokuchi, A. (2003). Marginalized Kernels between Labeled Graphs. In T. Faucett & N. Mishra (Eds.), *Proceedings of the Twentieth International Conference on Machine Learning* (pp. 321-328). New York: AAAI Press.

Lemmen, C., & Lengauer, T. (2000). Computational methods for the structural alignment of molecules. *Journal of Computer-Aided Molecular Design, 14*(3), 215–232. doi:10.1023/A:1008194019144

Mahé, P., Ralaivola, L., Stoven, V., & Vert, J.-P. (2006). The pharmacophore kernel for virtual screening with support vector machines. *Journal of Chemical Information and Modeling, 46*(5), 2003–2014. doi:10.1021/ci060138m

Mahé, P., Ueda, N., Akutsu, T., Perret, J.-L., & Vert, J.-P. (2005). Graph kernels for molecular structure-activity relationship analysis with support vector machines. *Journal of Chemical Information and Modeling, 45*(4), 939–951. doi:10.1021/ci050039t

Mahé, P., & Vert, J. P. (2009). Graph kernels based on tree patterns for molecules. *Machine Learning, 75*(1), 3–35. doi:10.1007/s10994-008-5086-2

Matter, H., & Pötter, T. (1999). Comparing 3D pharmacophore triplets and 2D fingerprints for selecting diverse compound subsets. *Journal of Chemical Information and Computer Sciences, 39*(6), 1211–1225. doi:10.1021/ci980185h

Pickett, S. D., Mason, J. S., & McLay, I. M. (1996). Diversity profiling and design using 3D pharmacophores: Pharmacophores-Derived Queries (PQD). *Journal of Chemical Information and Computer Sciences, 36*(6), 1214–1223. doi:10.1021/ci960039g

Ramon, J., & Gärtner, T. (2003). Expressivity versus efficiency of graph kernels. In T. Washio & L. De Raedt (Eds.), *Proceedings of the First International Workshop on Mining Graphs, Trees and Sequences* (pp. 65-74).

Schölkopf, B., & Smola, A. J. (2002). *Learning with Kernels: Support Vector Machines, Regularization, Optimization, and Beyond.* Cambridge, MA: MIT Press.

Schölkopf, B., Tsuda, K., & Vert, J.-P. (2004). *Kernel Methods in Computational Biology.* Cambridge, MA: MIT Press.

Shawe-Taylor, J., & Cristianini, N. (2004). *Kernel Methods for Pattern Analysis.* New York: Cambridge University Press.

Sutherland, J. J., O'Brien, L. A., & Weaver, D. F. (2003). Spline-fitting with a genetic algorithm: a method for developing classification structure-activity relationships. *Journal of Chemical Information and Computer Sciences, 43*(6), 1906–1915. doi:10.1021/ci034143r

Swamidass, S. J., Chen, J., Bruand, J., Phung, P., Ralaivola, L., & Baldi, P. (2005). Kernels for small molecules and the prediction of mutagenicity, toxicity and anti-cancer activity. *Bioinformatics (Oxford, England), 21*(Suppl. 1), i359–i368. doi:10.1093/bioinformatics/bti1055

Todeschini, R., & Consonni, V. (2002). *Handbook of molecular descriptors.* New York: Wiley-VCH.

Vert, J.-P. (2008). *The optimal assignment kernel is not positive definite* (Tech. Rep. No. 0801.4061). Arxiv.

Xue, L., & Bajorath, J. (2000). Molecular descriptors in chemoinformatics, computational combinatorial chemistry, and virtual screening. *Combinatorial Chemistry & High Throughput Screening*, 3(5), 363–372.

ENDNOTES

[1] http://chemcpp.sourceforge.net

[2] http://www.eyesopen.com/products/applications/quacpac.html

Chapter 4

A Simulation Study of the Use of Similarity Fusion for Virtual Screening

Martin Whittle
University of Sheffield, UK

Valerie J. Gillet
University of Sheffield, UK

Peter Willett
University of Sheffield, UK

ABSTRACT

This chapter analyses the use of similarity fusion in similarity searching of chemical databases. The ranked retrieval of molecules from a database can be modelled using both analytical and simulation approaches describing the similarities between an active reference structure and both the active and the non-active molecules in the database. The simulation model described here has the advantage that it can handle unmatched molecules, i.e., those that occur in one of the ranked similarity lists that are to be fused but that do not occur in the other. Our analyses provide insights into why the results of similarity fusion are often inconsistent when used for virtual screening.

INTRODUCTION

The discovery of novel bioactive molecules in the agrochemical and pharmaceutical industries is both costly and time-consuming, and there is hence much interest in techniques that can increase the cost-effectiveness of the discovery process. One such technique is virtual screening: the use of computational methods to rank a database of chemical molecules in order of decreasing probability of bioactivity. Attention can then be focused on those molecules at the top of the ranking since these are most likely to exhibit the activity of interest and are hence prime candidates for acquisition (or synthesis) and detailed biological screening (Alvarez & Shoichet, 2005; Bajorath, 2002; Eckert & Bajorath, 2007; Lengauer, Lemmen, Rarey, & Zimmermann, 2004; Oprea & Matter, 2004.

Many different approaches to virtual screening have been described in the literature, including:

DOI: 10.4018/978-1-61520-911-8.ch004

similarity searching (as discussed further below); pharmacophore mapping (where an attempt is made to identify the substructural features common to a set of known bioactive molecules); machine learning methods such as neural networks, support vector machines or decision trees (which can be used to classify an unknown molecule as a drug or a non-drug) and docking (which involves determining the degree of complementarity of a potential ligand to the binding site of the biological target). In this chapter, we focus on similarity searching, which is probably the simplest of the available techniques and which involves ranking a database of molecules in order of decreasing similarity to a known bioactive reference structure, (Eckert & Bajorath, 2007; Willett, 2006a). Given a ranked database, a set of molecules is retrieved by setting a threshold and then retrieving those molecules that come above this threshold in the ranking, e.g., the top-ranked 1000 molecules. The resulting set of retrieved molecules can then be used to determine the effectiveness of the similarity search, using one of the evaluation criteria that have been described in the literature (Edgar, Holliday, & Willett, 2000; Jain & Nicholls, 2008; Truchon & Bayly, 2007).

The rationale for similarity searching is the Similar Property Principle (Johnson & Maggiora, 1990), which states that molecules that are structurally similar are also likely to have similar properties. The ranking in a similarity search is effected using a quantitative measure of structural similarity, and many different types of measure have been reported in the literature (Bender & Glen, 2004; Sheridan & Kearsley, 2002; Willett, 2009). There have been several comparative studies that seek to assess the relative merits of different measures when used under the same conditions (Brown & Martin, 1996; Glen & Adams, 2006; Maldonado, Doucet, Petitjean, & Fan, 2006; Willett, 2006a; Willett, 2009). However, while it has been possible to identify measures of wide applicability it has not been possible to identify any single approach that will

always result in optimal performance (Sheridan, 2007). This has led to the idea of using not one but multiple similarity measures, combining the resulting rankings using a technique known as *similarity fusion* (Willett, 2006b).

The basic fusion approach involves computing a similarity score (approximating, directly or indirectly, the probability of activity) for each molecule in a database using several different similarity measures. These sets of scores are then combined in some way to give a new, fused score that will provide a better ranking of the database than will any single similarity measure. Given a known reference structure that is to be searched against a database using m different similarity measures, a pseudo-code description of similarity fusion can hence be summarized as follows:

- For each of the m similarity measures
  - Compute the similarity to the reference structure for each database molecule
- Use a fusion rule to combine the set of m similarities for each database molecule to give a new fused similarity
- Rank the database in decreasing order of the fused similarities, and apply a cut-off to retrieve some number of the top-ranked molecules.

BACKGROUND

The input to a fusion procedure can be either a set of similarity scores (such as those obtained using 2D fingerprints and the Tanimoto coefficient) or a set of ranks (obtained by sorting the similarity scores into decreasing order). In what follows, we have used linear range scaling to define the score

$$S^*\left(j\right) = \frac{S\left(j\right) - S_{\min}}{S_{\max} - S_{\min}} \qquad (1)$$

where $S(j)$ is the similarity between the reference structure and database structure j, S_{max} and S_{min} are the maximum and minimum values in the ranked list and $S^*(j)$ is the scaled similarity. More generally, if multiple similarity measures are being used then we denote the scaled similarity obtained using the k-th measure as $S_k^*(j)$ (with $1<=k<=m$, where m is the number of similarity lists that are to be combined when similarity fusion takes place).

In similarity fusion, the m lists of scaled similarity values $S_k^*(j)$ are combined using a fusion rule F to give a fused score $S_{FUS}(j)$ for each structure:

$$S_{FUS}(j) = F_{k=1}^m \left[S_k^*(j) \right] \qquad (2)$$

If structure j is not found in one of the lists (the *unmatched* case that is considered further below) then the scaled similarity is assumed to be zero. In our work we have used two fusion rules: SUM and MAX, defined by

$$S_{FUS}(j) = \sum_{k=1}^m \left[S_k^*(j) \right] \qquad (3)$$

and

$$S_{FUS}(j) = \max \left[S_1^*(j), S_2^*(j), S_3^*(j), \ldots\ldots S_m^*(j) \right] \qquad (4)$$

respectively. The reader should note that there are many other fusion rules available, such as those that have been used for fusing the outputs of search engines in textual information retrieval (Belkin, Kantor, Fox, & Shaw, 1995; Hsu & Taksa, 2005).

We have described previously an analytical model of similarity fusion based on bivariate distributions that seek to describe the relationships between pairs of sets of similarity scores, and tested this model using data from the *MDL Drug Data Report* (MDDR) database (available from Symyx Technologies at http://www.mdli.com/products/knowledge/drug_data_report/index.jsp). We describe this model briefly here to provide the context for the simulation described in the following section; full details are provided by Whittle, Gillet, Willett, & Loesel (2006a).

Our analytical approach is based on modelling the distribution of the similarities between the reference structure and each of the molecules in the database that is being searched. More specifically, we are interested in modelling this distribution for each of two similarity measures, x and y, when they are used on their own to generate a ranking of the database that is to be searched, and in modelling this distribution when similarity fusion is used to give a combined measure xy as a result of applying the chosen fusion rule. In this paper, we consider the combination of just two ranked lists, but the reader should note that both the analytical and simulation approaches can be applied to the combination of more than two lists.

The operation of the two fusion rules – MAX and SUM – is illustrated in Figure 1. Here, the two axes represent similarity values for the two similarity measures x and y: thus, the top right-hand corner (or bottom left-hand corner) represents a molecule getting a scaled similarity of unity (or of zero) with both measures x and y. The shaded area represents that part of the database that will be retrieved when fusion is applied to the two ranked lists at some particular cut-off similarity. For example, in Figure 1(a) the retrieved molecules for MAX fusion occupy the L-shaped area at the top-right part of the diagram. The regions extend to fill the boxes as the chosen cut-off similarity is increased. If, as one would hope, the individual similarity measures have been successful in clustering actives towards the upper parts of the x and y axes, then there is clear potential for the fused output to contain even more actives than would the individual, non-fused outputs.

How can we model this behaviour? When we apply a cut-off to a ranked database some of the retrieved molecules – hopefully a large number

Figure 1. Sketch of regions for retrieved values using: (a) MAX and (b) SUM fusion rules based on the combination of two similarity measures x and y. The shaded area in each case exemplifies the part of the joint probability distribution that will be retrieved by fusing the two similarity lists when the MAX or SUM rule is applied with appropriate cut-off values for the measures x and y

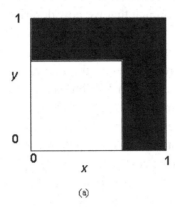
(a)

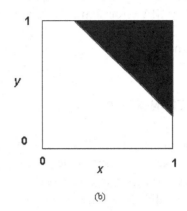
(b)

of them – will prove to be active on biological testing, with the remainder inactive. If a similarity measure is to perform well then the active similarities must in some way be biased towards the reference structure: this is a reflection of the well-known Similar Property Principle (Johnson & Maggiora, 1990) and can be modelled using a simple linear term. Two measures may also give a comparable rank ordering of similarity values leading to a non-zero correlation coefficient when the sets of values are compared: this can be modelled using a cross-term that reflects the degree of correlation.

The similarity joint probability densities $\phi(x,y)$ for the retrieved active molecules can be expressed as the sum of a uniform density ϕ_0, a pair of linear bias terms with coefficients ϕ_x, ϕ_y describing the individual measures x and y and a term with the coefficient ϕ_{xy} describing the degree of correlation.

$$\varphi\left(x,y\right) = \varphi_0 + \varphi_x\left(x - \tfrac{1}{2}\right) + \varphi_y\left(y - \tfrac{1}{2}\right) + \varphi_{xy}\left(x - \tfrac{1}{2}\right)\left(y - \tfrac{1}{2}\right) \tag{5}$$

for the range $0 \leq x \leq 1$, $0 \leq y \leq 1$. Here, $\phi(x,y)$ denotes a function, and the coefficients ϕx, ϕy, ϕxy are simple positive numbers. In each equation above, since x

and y range between 0 and 1, the terms $(x - \frac{1}{2})$ and $(y - \frac{1}{2})$ are multiplicative factors that vary linearly between $-\frac{1}{2}$ and $+\frac{1}{2}$. A similar function can be used to describe the similarity joint probability density for the retrieved non-active molecules:

$$\Phi\left(x,y\right) = \Phi_0 + \Phi_x\left(x - \tfrac{1}{2}\right) + \Phi_y\left(y - \tfrac{1}{2}\right) + \Phi_{xy}\left(x - \tfrac{1}{2}\right)\left(y - \tfrac{1}{2}\right) \tag{6}$$

These two equations provide the basic building blocks for modelling the retrieval of active and of non-active molecules as a result of similarity fusion. Specifically, they can be used to estimate the numbers of active and of non-active molecules that will be retrieved over each part of the joint similarity distribution for the two measures x and y and the comparable numbers when the two measures are used for separate similarity searches (Whittle, Gillet, Willett, & Loesel, 2006a). Different fusion rules will cover different parts of the joint probability distribution (as demonstrated in Figure 1), and it is hence possible to determine how different fusion rules will affect (either positively or negatively) screening performance, as estimated by some criterion of retrieval effectiveness. Whittle, Gillet, Willett, & Loesel (2006a) were able to show that both the SUM and MAX rules could, in principle at least, improve the effectiveness of conventional, non-fused similarity searching at the top of a ranking; they were also able to show that SUM would be expected to perform better than MAX for similarity fusion, as has been observed in practice (Ginn, Willett & Bradshaw, 2000).

THE SIMULATION APPROACH

A limitation of the analytical model is its assumption that the two ranked sets of molecules that are to be fused contain exactly the same molecules (albeit probably in a different order in each list). This will be the case if an entire database is ranked since every molecule will have a score that can serve to locate its place in the ranking when the similarities are ordered in decreasing value. However, the whole point of virtual screening is to enable attention to be restricted to just some small fraction (e.g., the top-1% or the top-1000 molecules) of the database after ranking so that subsequent attention can be focused on the acquisition and testing of this small fraction. When two (or more) different similarity measures are used to rank the database, it is extremely unlikely that the sets of top-ranked molecules in the two cases will be the same (unless the two similarity measures are monotonic). Thus while both lists may contain, e.g., 1000 molecules, some (or indeed many) of the molecules appearing in the first ranked list will not appear in the second list and vice versa. The extent of these *unmatched* molecules can be quantified by means of the ratio of the number of unique structures occurring in both of the lists to the number of unique structures appearing in either list. Whittle, Gillet, Willett, & Loesel (2006b) quote values for this match ratio, M_R ($0 <= M_R <= 1$), ranging between 0.069 (i.e., lists with very few common structures) and 0.926 (i.e., lists with a very large degree of overlap) for searches of the MDDR database involving pairs of similarity coefficients and lists containing 1000

top-ranked structures. There can thus be very many unmatched molecules in some cases, and the analytical model cannot be applied to such data; this is not, however, the case with the simulation approach that we now describe.

The approach we have taken to simulating similarity fusion is straightforward and comprises three elements: the generation of suitable data with known distributions; a fusion algorithm; and a means of analysing the results. For simplicity, we assume that we are combining just two ranked lists, but the simulation can easily be extended to the combination of multiple lists if required.

The basic approach uses random variables x_i, y_i with values in the range $0 - 1$ representing, e.g., the similarity values obtained in two searches using different similarity measures. Random variables can be chosen from any given distribution using a rejection method (Press, Teukolsky, Vetterling, & Flannery, 1994) that we have adapted for bivariate distributions as follows. A pair of random numbers, x_i, y_i, is first chosen on the range $0 - 1$ from a flat distribution. A third random number, s ($0 < s < 1$), is then chosen and compared with the desired distribution $f(x,y)$. If $s \leq f(x_i, y_i)$ then the values x_i and y_i are accepted but if $s > f(x_i, y_i)$ then they are discarded and the process is repeated until acceptable values have been obtained. For our simulations, $f(x,y)$ is just a scaled version of the bivariate probability density $\phi(x,y)$ used in the analytical model, with a peak value of 1 to minimise the number of rejections needed. In the simulations described here, 10^5 values are taken in total with some fraction, λ, of these randomly assigned as active, and the remaining $1 - \lambda$ randomly assigned as non-active. In general the distribution parameters for the values assigned as active and non-active are different: if they were not then fusion could not result in performance enhancement.

The SUM fusion rule is implemented by calculating values of a new quantity, z_i, from:

$$z_i = \frac{1}{2}(x_i + y_i) \tag{7}$$

and the MAX fusion rule is implemented by calculating z_i from

$$z_i = \max\{x_i, y_i\} \tag{8}$$

The reader should note that the normalising factor of ½ for the SUM rule is designed to maintain the range of resultant values between 0 and 1 and hence to facilitate comparison with the MAX rule: defined in this way the SUM rule is equivalent to the mean of the two results.

Values x_i, y_i, z_i are then independently ranked with the highest values top, and the number of actives – i.e. those values with index i that have been flagged – counted in each case up to a given cut-off rank position. Knowing the total number of pairs flagged as active these values are then turned into recall values, where the recall is the fraction of the active molecules that are retrieved above the cut-off. Comparison of recall values obtained using the fused results (z_i) and the single measures (x_i, y_i) then leads directly to values of what we shall refer to as the "enhancement factor", ΔR, i.e., the ratio of the recalls with and without fusion.

Simulations were carried out using appropriately correlated and biased distributions with the objective of mutually validating the model and theoretical low-order expansion predictions. Values of x_i and y_i representing the actives were generated using the rejection method with probabilities given by

$$p(x,y) = \frac{\varphi_0 + \varphi_{xy}\left(x - \frac{1}{2}\right)\left(y - \frac{1}{2}\right)}{\varphi_0 + \varphi_{xy}\left(x_{max} - \frac{1}{2}\right)\left(y_{max} - \frac{1}{2}\right)} \tag{9}$$

for the correlated distributions and

$$p(x,y) = \frac{\varphi_0 + \varphi_x\left(x - \frac{1}{2}\right) + \varphi_y\left(y - \frac{1}{2}\right)}{\varphi_0 + \varphi_x\left(x_{max} - \frac{1}{2}\right) + \varphi_y\left(y_{max} - \frac{1}{2}\right)} \tag{10}$$

for the biased distributions (as in the analytical model described above). For these simulations $x_{max} = y_{max} = 1.0$ represent the maximum allowed values of x and y, with the denominator in each case designed to ensure a maximum probability of 1.0. Results were obtained with $\phi 0$ set at 0.5, $\phi xy_{=0}$.2 and $\phi x = \phi_y = 0._1$. Identical methods were used to generate values representing the non-actives but in this case flat, random distributions were chosen, i.e., for these the densities were set to $\Phi x = \Phi y_= \Phi x_y = 0_{,0}; \Phi 0 = 0._5$. Thus, the biased distributions for the active molecules represent the fact that are likely to be clustered towards the top of a ranked list (as suggested by the Similar Property Principle) while the flat distributions for the non-active molecules represent the fact they are unlikely to be clustered but are, instead, spread equally throughout the ranking. Random deviates x_i, y_i ar_e chosen on the range $0 - 1$ and represent, for example, the similarity values obtained by two experiments: currently 100000 values are taken in total, a number that is comparable with the number of molecules in the MDDR database. The fraction of actives, λ, was set to a value of 0.5. The results obtained using these parameters were quite noisy and were hence averaged over 100 cycles to obtain acceptable results.

The accuracy of the simulation is demonstrated in Figure 2 (a) and (b), which compare the results from the analytical model theory with those of the simulation obtained with the same parameters over the full range of normalised rank N* $(0 <= N* <= 1)$, i.e., the integer rank position of each molecule has been normalised into the range 0-1. In Figure 2, the results from the analytical model and from the simulation are shown as points and as lines, respectively (note that ΔR can be either positive or negative depending on whether the

Figure 2. Comparison of the enhancement factor ΔR plotted against the normalised rank N^* for the analytical model (full lines) and for the simulation (symbols) using (a) the SUM rule; (b) the MAX rule. In each case the analytical and simulation results were obtained using the same parameters. All results: $\Phi_0 = \phi 0_{-} 0.5$, $\Phi x_{y=} \Phi x_{-} \Phi y_{-} 0.0$. For actives with: $\circ$ correlated densities $\phi xy_{=0} . 2$; $\square$ ramp biased densities $\phi x = \phi_{y} = 0 ._{1}$

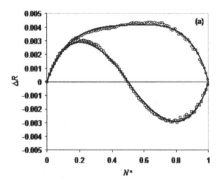 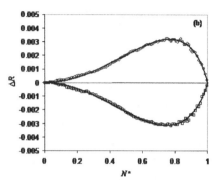

fusion has had a positive or a negative effect on retrieval). The agreement is clearly excellent and can be demonstrated for other values of the input parameters.

We have noted that one of the advantages of the simulation approach is that it can encompass unmatched values, whereas this is not possible in the analytical model, and the simulation was hence extended to include unmatched values. To do this, proportions of the retrieved active and retrieved non-active values are randomly set to zero in one list or the other when the values are generated. Notably, when generated in this way, the distribution of non-zero unmatched values in each list is determined by the distribution parameters (excepting correlation) entered for the matched values; thus if the xi values are biased towards high similarity by a positive value of ϕx then the unmatched values in this list will retain this bias. To isolate the influence of this effect some comparative results were obtained in which the non-zero part of the unmatched pair was randomised at generation time to give a flat distribution.

The lists of values generated as described above were then fused using the SUM and MAX rules and the results are shown in Figure 3. For this set of results a smaller fraction of retrieved actives were assumed by putting $\lambda = 0.1$. In the figure, the y axis represents the change in recall,

ΔR and the x axis represents the normalised rank position N^*, with parameter settings as follows. All results: $\phi 0_{=} 0.1, \phi xy_{=0} . 0$; $\Phi 0 =_{0} .9$, $\Phi xy_{=\phi} x =_{\phi} y =_{0} .0$. In each figure, the main series shows the effect of changing the active–active biased density for partially matched active–active values and completely unmatched active–non-active values: $M_R^{act} = 0.1$, $M_R^{nact} = 0.0$. For active–active values with ramp biased densities: $\phi x = \phi y = 0\,0, \Diamond$; $\phi x = \phi y\_0.0_{1}, \Delta$; $\phi x = \phi y = 0\,02, \square$; $\phi x = \phi y = 0. 0_{3}, +; \phi x = \phi y = 0.04, \times; \phi x = \phi y = 0.05, \blacklozenge$ Th_{e}se are compared with the result $\phi x = \phi y = 0.05$ for $_{-} -_{,}$ in which the unmatched values have been chosen from flat distributions and, $\blacksquare$, the result for fully matched values $M_R^{act} = 1.0$, $M_R^{nact} = 1.0$ and $\phi x = \phi y = 0.05$.

$Th_{e} cu_{r}$ves in Figure 3 can be understood in terms of the fusion regions sketched in Figure 1. The sharp changes in gradient of the curves reflect the discontinuous nature of the unmatched distributions. In the main series shown in these figures the active–active values are partially matched, while the active–non-active values are fully unmatched. Since the unmatched active-non-active values lie on the axes of the boxes shown in Figure 1, they are not initially accessed by the SUM-rule region, and SUM-fusion is therefore relatively successful compared with single value

Figure 3. Simulated fusion enhancement, ΔR plotted against the normalised rank N*: using the SUM (a) and MAX (b) fusion rules. These diagrams show the effect of changes in the value of the active–active match ratio for completely unmatched active-non-active values (see text)

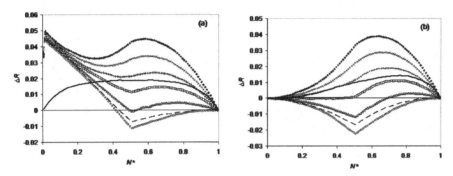

recall, which does. At $N^* \sim 0.01$, the diagonal is reached after retrieving only active values and the SUM-rule region begins to access the non-active values located on the axes. The fused recall therefore begins to fall and in some cases becomes negative. The cusp at $N^* \sim 0.5$ reflects the point at which the single measure regions have extended across the box and reached the far side, at which point they begin to collect the active–non-active values on that axes.

The result for fully matched values (the black lines in Figure 3) is shown for comparison on the same normalised rank scale and is exceeded over the whole range by the results which include a relatively high bias in the distributions. That this is due to the bias in the unmatched active–active values and not some kind of coupling between matched and unmatched contributions is seen by comparison with the result of using randomised unmatched values with matched values taken from a distribution with $\phi x = \phi y = 0.05$. This curve lies only just above that for the flat distribution $\phi x = \phi y = 0.0$, the difference being caused by the small fraction of matched active–active values which retain the bias. Results for the MAX-rule, Figure 3(b), do not show the correspondingly high values at low N^* because the "arms" of the MAX-rule region (see Figure 1(a)) do initially access unmatched values lying on the axes. Area

for area, the MAX-rule region accesses slightly more of these than the single value recall regions and the MAX rule is therefore disadvantageous in these circumstances. However, the negative fusion enhancement seen over the whole range for $\phi x = \phi y = 0.0$, is turned positive by sufficiently biased unmatched active–active values. Nevertheless, the enhancement never quite reaches the highest values seen using the SUM-rule.

For this system, the inclusion of relatively more unmatched active–non-active than active–active values leads to significantly increased fusion enhancement at low N^* when using the SUM rule. However, as lower ranks (high N^*) are accessed, this advantage is lost unless the distributions of unmatched active–active values in both lists are biased towards high similarity. In this case, sufficiently strong bias can lead to significant enhancement over the whole available range of rank.

The results of changing values of active–non-active match ratio are shown in Figure 4. Here, the similarity distributions are all flat and the active–active values are completely matched, i.e., $M_R^{act} = 1.0$, which leads to an order-of-magnitude increase in the response compared with Figure 3 where $M_R^{act} = 0.1$. The parameter used here are as follows. All results are for flat distributions: $\phi 0 = 0.1; \Phi 0 = 0.9, \phi x_y = \phi x = \phi y_ 0.0; \Phi x_{y=} \Phi x_ \Phi y_ 0.0$.

Figure 4. Simulated fusion enhancement, ΔR plotted against the normalised rank N* using the SUM (a) and MAX (b) rules. These diagrams show the effect of changes in the value of the active–non-active match ratio for completely matched active-active values (see text)

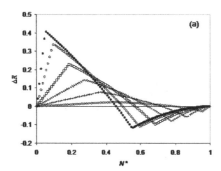 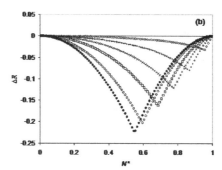

In each case, the active–active values are completely matched; $M_R^{act} = 1.0$, and the series shows the effect of changing the proportion of unmatched active–non-active values. For active–non-active values with match ratios: $M_R^{nact} = 0.9$, ◊; $M_R^{nact} = 0.7$, +; $M_R^{nact} = 0.5$, ×; $M_R^{nact} = 0.3$, □; $M_R^{nact} = 0.1$, ○; $M_R^{nact} = 0.0$, ●.

Figure 4 shows how increasing the unmatched proportion of active–non-active values (decreasing M_R^{nact}) directly increases the effect of fusion. The initial peak in the SUM-rule also moves to lower values of N* as the density of matched active–non-active values decreases. In terms of the SUM-rule retrieval region (see Figure 1(b)) this allows more actives to be accumulated at lower cost in rank before the box-diagonal is reached. Over this region only matched values are collected. The comparable results for the MAX fusion rule are entirely negative and emphasise the importance of biased distributions for any enhancement in this case.

DISCUSSION

The availability of many different types of similarity measure has spurred interest in the use of fusion methods to combine the rankings that result when different measures are used, with the hope that this combination will result in enhanced screening effectiveness. The experiments that have been carried out (Willett, 2006b) show that this hope is often realised in practice. However, fusion does not always result in enhancement, and even when it does the level is very variable. Thus, while fusion is generally regarded as beneficial at a purely practical level, it has not been possible to rationalise the results to date. This lack of predictive power spurred our interest in the development of analytical and simulation tools that could be used to model what happens when similarity rankings are fused.

The analytical model focuses on the effects of differential correlation, bias and commonality between active–active and active–non-active values as obtained by the two similarity measures that are being combined. The model shows that enhancement can occur when there is a suitable combination of differences between the active–active and active–non-active multivariate distributions and the geometrical difference between the regions of the multivariate distributions that the fusion rules access. The analytical model has shown fair agreement with experimental similarity data from the MDDR database (Whittle, Gillet, Willett, & Loesel, 2006b). However, it requires the entire ranked lists that are to be fused, so that it is possible to match the position of each and every database structure in one of the rankings with the corresponding position in the other ranking, and this information is often not available in an operational virtual screening context, where attention inevitably focuses on just that small fraction of top-ranked molecules. It is for this reason that we developed the simulation approach described in this chapter. The approach gives results comparable to the analytical model where the latter can be applied, but also allows statements about the effectiveness of fusion where this is not the case and where there are multiple unmatched similarity values.

Taken together, the two approaches show that the three different contributions that we have studied lead to very different behaviour when operated on by the MAX or SUM fusion rules. When the values representing two different similarity measures between an active reference structure and an active database structure are correlated more than those of the corresponding active–non-active similarity values, then data fusion using the SUM rule can lead to enhanced recall at the top end of the fused ranking. The extent of this enhancement lessens, however, as one moves down the ranking and less similar database structures are retrieved. There is thus potential benefit in using the SUM rule for similarity fusion; this is not the case with the MAX rule, where our models suggest no enhancement at any point in the ranking, a finding that is in agreement with experimental studies (Ginn, Willett, & Bradshaw, 2000). However, if there is reduced correlation of the active–active similarities relative to the active–non-active values (or anti-correlation of the active–active similarities if the active–non-active values are uncorrelated) then the use of the MAX rule may be worthwhile.

It seems not unreasonable to expect, and has been noted in previous experimental studies (Willett, 2006b), that performance enhancement is most likely to be obtained by fusing two good rankings that involve different sets of structures. This empirical rule is supported by our results using biased distributions since we find that the best fusion results are obtained in our model when both active-active distributions are equally biased towards high similarity. This is true for the SUM rule and, but to a lesser extent, for the MAX rule. The simulation results show clearly that the inclusion of unmatched similarity values can make a marked and positive contribution to the effectiveness of fusion; however, the nature of this contribution depends on the rank position that is chosen. At, or near to, the top of a ranking, higher active–active overlap than active–non-active overlap is a factor that leads to positive fusion enhancement for the SUM rule. However, this criterion is insufficient lower down the ranking unless the active–active values are also biased towards high similarity. The MAX rule is again found to be much less well suited to fusion than is the SUM rule.

If only matched values need to be considered, then positive fusion results should be obtained, at least for some cut-off rank positions, if two conditions hold for the bivariate distribution of active-active values: the distribution differs from the active–non-active distribution; and the distribution better matches the appropriate retrieval region for the chosen fusion rule (MAX or SUM in our analyses) than do either of the single measure distributions. The situation when unmatched values need to be considered is more complex, as we have seen above.

CONCLUSION

The increasing costs of drug discovery have spurred interest in the use of virtual screening to prioritize molecules for synthesis and testing. There are many different methods available for virtual screening of chemical databases, and this has led to interest in the use of similarity fusion to combine the outputs resulting from different methods. Most of the work to date has been experimental in nature (Willett, 2006b), with little attempt having been made to provide a theoretical analysis of, and rationalisation for, the observed behaviour. The work reported here seeks to rectify this omission using a simulation approach that complements our previous studies based on a purely analytic description of similarity fusion (Whittle, Gillet, Willett, & Loesel, 2006a,b): the analytical model was designed as a tool to further the understanding of similarity fusion whilst simulation extends the model towards the real-world virtual screening scenario, where the ranked output lists contain only a small fraction of the database that is being searched.

Our studies have enabled us to understand the origin of some common observations relating to similarity fusion; however, this has involved the use of highly idealized distributions and even these result in very complex behaviour, especially when unmatched lists are involved. The observed complexity arises from the involvement of subtle interactions between multiple factors that, taken together, make it very difficult to predict with any degree of certainty the effect of data fusion on the effectiveness of a similarity searching system. For example, assume that we wish to fuse the ranked lists resulting from similarity searches based on the Tanimoto coefficient and the cosine coefficient: then predicting the effect of this fusion involves consideration of the distributions of retrieved–active and retrieved-non-active molecules in both lists for both the matched and the unmatched molecules, a total of eight different distributions. If such distributional information is available, then the tools described here and in our previous papers can be used to predict the effect of using the SUM or MAX fusion rules, and it would be possible to extend our analyses to encompass some of the other fusion rules that have been suggested in the literature (Ginn, Willett & Bradshaw, 2000); indeed, one of the starting points for our work was the wish to design novel types of fusion rule that could make better use of the available information than the rather simple rules that have been used to date. In the event, perhaps the main

conclusion that we draw from our studies is that while it is now possible to rationalise the results of fusion experiments retrospectively, it is still very difficult to do this predictively.

In conclusion, we note that whilst the analysis reported here and in our previous papers has been carried out in the context of similarity-based virtual screening, we believe that our methods may be applicable in related contexts that produce a ranking of a set of molecules based on some quantitative scoring function. Thus, fusion techniques are widely used in virtual screening systems based on protein-ligand docking (Feher, 2006) and analogous techniques are also starting to be used in virtual screening systems based on different types of machine learning (Givehchi & Schneider, 2005; Jorissen & Gilson, 2005). Our methods may also be applicable to the multiple ranking systems used in text retrieval (Belkin, Kantor, Fox & Shaw, 1995; Hsu & Taksa, 2005).

REFERENCES

Alvarez, J., & Shoichet, B. (Eds.). (2005). *Virtual Screening in Drug Discovery*. Boca Raton, FL: CRC Press.

Bajorath, J. (2002). Integration of virtual and high-throughput screening. *Nature Reviews. Drug Discovery*, *1*, 882–894. doi:10.1038/nrd941

Belkin, N. J., Kantor, P., Fox, E. A., & Shaw, J. B. (1995). Combining the evidence of multiple query representations for information retrieval. *Information Processing & Management*, *31*, 431–448. doi:10.1016/0306-4573(94)00057-A

Bender, A., & Glen, R. C. (2004). Molecular similarity: a key technique in molecular informatics. *Organic & Biomolecular Chemistry*, *2*, 3204–3218. doi:10.1039/b409813g

Brown, R. D., & Martin, Y. C. (1996). Use of structure-activity data to compare structure-based clustering methods and descriptors for use in compound selection. *Journal of Chemical Information and Computer Sciences, 36,* 572–584. doi:10.1021/ci9501047

Eckert, H., & Bajorath, J. (2007). Molecular similarity analysis in virtual screening: foundations, limitation and novel approaches. *Drug Discovery Today, 12,* 225–233. doi:10.1016/j.drudis.2007.01.011

Edgar, S. J., Holliday, J. D., & Willett, P. (2000). Effectiveness of retrieval in similarity searches of chemical databases: A review of performance measures. *Journal of Molecular Graphics & Modelling, 18*(4-5), 343–357. doi:10.1016/S1093-3263(00)00061-9

Feher, M. (2006). Consensus scoring for protein-ligand interactions. *Drug Discovery Today, 11,* 421–428. doi:10.1016/j.drudis.2006.03.009

Ginn, C. M. R., Willett, P., & Bradshaw, J. (2000). Combination of molecular similarity measures using data fusion. *Perspectives in Drug Discovery and Design, 20,* 1–16. doi:10.1023/A:1008752200506

Givehchi, A., & Schneider, G. (2005). Multispace classification for predicting GPCR ligands. *Molecular Diversity, 9,* 371–383. doi:10.1007/s11030-005-6293-4

Glen, R. C., & Adams, S. E. (2006). Similarity metrics and descriptor spaces - which combinations to choose? *QSAR & Combinatorial Science, 25,* 1133–1142. doi:10.1002/qsar.200610097

Hsu, D. F., & Taksa, I. (2005). Comparing rank and score combination methods for data fusion in information retrieval. *Information Retrieval, 8,* 449–480. doi:10.1007/s10791-005-6994-4

Jain, A. N., & Nicholls, A. (2008). Recommendations for evaluation of computational methods. *Journal of Computer-Aided Molecular Design, 22,* 133–139. doi:10.1007/s10822-008-9196-5

Johnson, M. A., & Maggiora, G. M. (Eds.). (1990). *Concepts and Applications of Molecular Similarity.* New York: John Wiley.

Jorissen, R. N., & Gilson, M. K. (2005). Virtual screening of molecular databases using a support vector machine. *Journal of Chemical Information and Computer Sciences, 45,* 549–561. doi:10.1021/ci049641u

Lengauer, T., Lemmen, C., Rarey, M., & Zimmermann, M. (2004). Novel technologies for virtual screening. *Drug Discovery Today, 9,* 27–34. doi:10.1016/S1359-6446(04)02939-3

Maldonado, A. G., Doucet, J. P., Petitjean, M., & Fan, B.-T. (2006). Molecular similarity and diversity in chemoinformatics: from theory to applications. *Molecular Diversity, 10,* 39–79. doi:10.1007/s11030-006-8697-1

Oprea, T. I., & Matter, H. (2004). Integrating virtual screening in lead discovery. *Current Opinion in Chemical Biology, 8,* 349–358. doi:10.1016/j.cbpa.2004.06.008

Press, W. H., Teukolsky, S. A., Vetterling, W. T., & Flannery, B. P. (1994). *Numerical Recipies in C* (2nd ed.). Cambridge, UK: Cambridge University Press.

Sheridan, R. P. (2007). Chemical similarity searches: when is complexity justified? *Expert Opinion on Drug Discovery, 2,* 423–430. doi:10.1517/17460441.2.4.423

Sheridan, R. P., & Kearsley, S. K. (2002). Why do we need so many chemical similarity search methods? *Drug Discovery Today, 7,* 903–911. doi:10.1016/S1359-6446(02)02411-X

Truchon, J.-F., & Bayly, C. I. (2007). Evaluating virtual screening methods: good and bad metrics for the "early recognition" problem. *Journal of Chemical Information and Modeling, 47*, 488–508. doi:10.1021/ci600426e

Whittle, M., Gillet, V. J., Willett, P., & Loesel, J. (2006a). Analysis of data fusion methods in virtual screening: theoretical model. *Journal of Chemical Information and Modeling, 46*, 2193–2205. doi:10.1021/ci049615w

Whittle, M., Gillet, V. J., Willett, P., & Loesel, J. (2006b). Analysis of data fusion methods in virtual screening: similarity and group fusion. *Journal of Chemical Information and Modeling, 46*, 2206–2219. doi:10.1021/ci0496144

Willett, P. (2006a). Similarity-based virtual screening using 2D fingerprints. *Drug Discovery Today, 11*, 1046–1053. doi:10.1016/j.drudis.2006.10.005

Willett, P. (2006b). Data fusion in ligand-based virtual screening. *QSAR & Combinatorial Science, 25*, 1143–1152. doi:10.1002/qsar.200610084

Willett, P. (2009). Similarity methods in chemoinformatics. *Annual Review of Information Science & Technology, 43*, 3–71.

ADDITIONAL READING

Baber, J. C., Shirley, W. A., Gao, Y., & Feher, M. (2006). The use of consensus scoring in ligand-based virtual screening. *Journal of Chemical Information and Modeling, 46*, 277–288. doi:10.1021/ci050296y

Ginn, C. M. R., Turner, D. B., Willett, P., Ferguson, A. M., & Heritage, T. W. (1997). Similarity searching in files of three-dimensional chemical structures: evaluation of the EVA descriptor and combination of rankings using data fusion. *Journal of Chemical Information and Computer Sciences, 37*, 23–37. doi:10.1021/ci960466u

Goodman, I. R., Mahler, R. P. S., & Nguyen, H. T. (1997). *Mathematics of Data Fusion*. Norwell, MA: Kluwer.

Hall, D. L. (1992). *Mathematical Techniques in Multisensor Data Fusion*. Northwood, MA: Artech House.

Kearsley, S. K., Sallamack, S., Fluder, E. M., Andose, J. D., Mosley, R. T., & Sheridan, R. P. (1996). Chemical similarity using physicochemical property descriptors. *Journal of Chemical Information and Computer Sciences, 36*, 118–127. doi:10.1021/ci950274j

Klein, L. A. (1999). *Sensor and Data Fusion Concepts and Applications* (2nd ed.). Bellingham: SPIE Optical Engineering Press.

Kogej, T., Engkvist, O., Blomberg, N., & Muresan, S. (2006). Multifingerprint based similarity searches for targeted class compound selection. *Journal of Chemical Information and Modeling, 46*, 1201–1213. doi:10.1021/ci0504723

Nikolova, N., & Jaworska, J. (2003). Approaches to measure chemical similarity - a review. *Quantitative Structure-Activity Relationships and Combinatorial Science, 22*, 1006–1026.

Raymond, J. W., Jalaie, M., & Bradley, P. P. (2004). Conditional probability: a new fusion method for merging disparate virtual screening results. *Journal of Chemical Information and Computer Sciences, 44*, 601–609. doi:10.1021/ci034234o

Salim, N., Holliday, J. D., & Willett, P. (2003). Combination of fingerprint-based similarity coefficients using data fusion. *Journal of Chemical Information and Computer Sciences, 43*, 435–442. doi:10.1021/ci025596j

Sheridan, R. P., Miller, M. D., Underwood, D. J., & Kearsley, S. K. (1996). Chemical similarity using geometric atom pair descriptors. *Journal of Chemical Information and Computer Sciences, 36*, 128–136. doi:10.1021/ci950275b

Williams, C. (2006). Reverse fingerprinting, similarity searching by group fusion and fingerprint bit importance. *Molecular Diversity, 10,* 311–332. doi:10.1007/s11030-006-9039-z

Zhang, Q., & Muegge, I. (2006). Scaffold hopping through virtual screening using 2D and 3D similarity descriptors: ranking, voting, and consensus scoring. *Journal of Medicinal Chemistry, 49,* 1536–1548. doi:10.1021/jm050468i

Chapter 5
Structure–Activity Relationships by Autocorrelation Descriptors and Genetic Algorithms

Viviana Consonni
University of Milano-Bicocca, Italy

Roberto Todeschini
University of Milano-Bicocca, Italy

ABSTRACT

Quantitative Structure-Activity Relationships (QSARs) are models relating variation of molecule properties, such as biological activities, to variation of some structural features of chemical compounds. Three main topics take part of the QSAR/QSPR approach to the scientific research: the representation of molecular structure, the definition of molecular descriptors and the chemoinformatics tools. Molecular descriptors are numerical indices encoding some information related to the molecular structure. They can be both experimental physico-chemical properties of molecules and theoretical indices calculated by mathematical formulas or computational algorithms. In the last few decades, much interest has been addressed to studying how to encompass and convert the information encoded in the molecular structure into one or more numbers used to establish quantitative relationships between structures and properties, biological activities or other experimental properties. Autocorrelation descriptors are a class of molecular descriptors based on the statistical concept of spatial autocorrelation applied to the molecular structure. The objective of this chapter is to investigate the chemical information encompassed by autocorrelation descriptors and elucidate their role in QSAR and drug design. After a short introduction to molecular descriptors from a historical point of view, the chapter will focus on reviewing the different types of autocorrelation descriptors proposed in the literature so far. Then, some methodological topics related to multivariate data analysis will be overviewed paying particular attention to analysis of similarity/ diversity of chemical spaces and feature selection for multiple linear regressions. The last part of the chapter will deal with application of autocorrelation descriptors to study similarity relationships of a set of flavonoids and establish QSARs for predicting affinity constants, Ki, to the $GABA_A$ benzodiazepine receptor site, BzR.

DOI: 10.4018/978-1-61520-911-8.ch005

INTRODUCTION

Quantitative Structure–Activity Relationships (QSARs) are the final result of the process which starts with a suitable description of molecular structures and ends with some inference, hypothesis, prediction on the behaviour of molecules in environmental, biological, and physico-chemical systems in analysis. They are models playing a relevant role in chemistry, pharmaceutical sciences, environmental protection policy, toxicology, ecotoxicology, health research and quality control.

QSARs are based on the assumption that the structure of a molecule (for example, its geometric, steric and electronic properties) must contain the features responsible for its physical, chemical, and biological properties and on the ability to capture these features into one or more numerical descriptors. According to the *congenericity principle*, similar compounds have similar activities and activity changes gently in the chemical space.

By QSAR models, the biological activity (or property, reactivity, etc.) of a new designed or untested chemical can be inferred from the molecular structure of similar compounds whose activities (properties, reactivity, etc.) have already been assessed.

It has been nearly 45 years since the QSAR modelling firstly was used into the practice of agrochemistry, drug design, toxicology, industrial and environmental chemistry. Its growing power in the following years may be mainly attributed to the rapid and extensive development in methodologies and computational techniques that have allowed to delineate and refine the many variables and approaches used to model molecular properties (Martin, 1979; Kubinyi, 1993; Hansch & Leo, 1995; van de Waterbeemd, Testa, & Folkers, 1997; Devillers, 1998; Kubinyi, Folkers, & Martin, 1998; Martin, 1998; Charton & Charton, 2002; Gasteiger, 2003; Oprea, 2004). Furthermore, the interest in QSARs is more and more growing because nowadays these tools are not only used for research purposes but also to produce data on chemicals in the interest of time and cost effectiveness.

The development of *QSAR/QSPR models* is a quite complex process (Figure 1). Important stages of this process are a) selection of the set of molecules the modelling procedure is applied to, and the set of molecular descriptors which will define the model chemical space; b) selection of the training set for the model estimation and the test set for model validation; c) application of the validated model(s) to design new molecules with desirable properties and /or predict the properties of future molecules.

In recent years, "*The use of information technology and management has become a critical part of the drug discovery process. Chemoinformatics is the mixing of those information resources to transform data into information and information into knowledge for the intended purpose of making better decisions faster in the area of drug lead identification and organization.*" (Brown, 1998). In fact, *chemoinformatics* encompasses the design, creation, organization, management, retrieval, analysis, dissemination, visualization, and use of chemical information (Gasteiger, 2003; Oprea, 2003); molecular descriptors play a fundamental role in all these processes being the basic tool to transform chemical information into a numerical code necessary to apply informatics procedures.

Molecular descriptors are formally mathematical representations of a molecule obtained by a well specified algorithm applied to a defined molecular representation or a well specified experimental procedure: *the molecular descriptor is the final result of a logic and mathematical procedure which transforms chemical information encoded within a symbolic representation of a molecule into a useful number or the result of some standardized experiment* (Todeschini & Consonni, 2000, p. 303).

Figure 1. Role of models in QSAR/QSPR framework

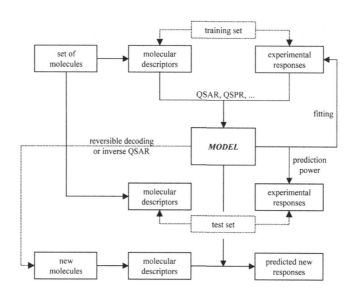

Each molecular descriptor takes into account a small part of the whole chemical information contained into the real molecule and, as a consequence, the number of descriptors is continuously increasing with the increasing request of deeper investigations on chemical and biological systems.

Different descriptors are different ways or perspectives to view a molecule, taking into account the various features of its chemical structure.

Evidence of the interest of the scientific community in the molecular descriptors is provided by the huge number of descriptors proposed up today: more than 5000 of descriptors derived from different theories and approaches are actually defined and computable by using dedicated software tools (Todeschini & Consonni, 2009).

Moreover, a growing interest in QSAR has been generated by the New Chemicals Policy of the European Commission (REACH: Registration, Evaluation and Authorisation of Chemicals) that explicitly states that at chemical registration level the registrant should include information from alternative sources (e.g. from QSARs, etc.) which may assist in identifying the presence or absence of hazardous properties of the substance and which can, in certain cases, replace the results of animal tests. Obviously, for the purposes of the REACH legislation, it is essential to use QSAR models that produce reliable estimates, and, to this regard, model validation and evaluation of applicability domain to avoid data extrapolation have become of primary concern in the QSAR development process, as well as reliable and effective modelling strategies for dealing with the huge chemical information now available.

This book chapter aims at introducing and discussing the role of *molecular autocorrelation descriptors* in QSAR and drug design. These are descriptors of molecules based on the concepts of spatial autocorrelation, which is a measure of the degree to which a spatial phenomenon is correlated to itself in space, or, in other words, the degree to which the observed value of a variable at one locality is independent of values of the variable at neighbouring localities. The spatial pattern of a property distribution is defined by the arrangement of individual entities in space and the spatial relationships among them.

BACKGROUND

Molecular descriptors can be considered as the most important realization of the idea of Crum-Brown. His M.D. thesis at the University of Edinburgh (1861), entitled "On the Theory of Chemical Combination", shows that he was a pioneer of mathematical chemistry science. In it, he developed a system of graphical representation of compounds which is basically identical to that used today. His formulae were the first that showed clearly both valency and linking of atoms in organic compounds.

Later, Crum-Brown and Fraser (1867; 1868) proposed the existence of a correlation between biological activity of different alkaloids and their molecular constitution. More specifically, the physiological action of a substance in a certain biological system (Φ) was defined as a function (f) of its chemical constitution (C): $\Phi = f$(C). Thus, an alteration in chemical constitution, ΔC, would be reflected by an effect on biological activity, $\Delta\Phi$. This equation can be considered the first general formulation of a quantitative structure-activity relationship.

The seminal work of Hammett (1938), based on his equation (1935; 1937), gave rise to the "σ–ρ" culture in the delineation of substituent effects on organic reactions, whose aim was the search for linear free energy relationships (LFERs): steric, electronic and hydrophobic constants were derived for several substituents and used in an additive model to estimate the biological activity of congeneric series of compounds.

The first theoretical QSAR/QSPR approaches, that relate biological activities and physicochemical properties to theoretical numerical indices derived from the molecular structure, date back to the end of 1940s. The *Wiener index* (Wiener, 1947) and the *Platt number* (Platt, 1947), proposed in 1947 to model the boiling point of hydrocarbons, were the first theoretical molecular descriptors derived from concepts of the graph theory and known with the name of *topological*

indices (TIs). These are numerical quantifiers of molecular topology that are mathematically derived in a direct and unambiguous manner from the graph representation of a molecule, usually a H-depleted molecular graph (Kier & Hall, 1976; Bonchev & Trinajstic, 1977; Balaban, Motoc, Bonchev, & Mekenyan, 1983; Rouvray, 1983; Basak, Magnuson, & Veith, 1987; Trinajstic, 1992; Randic, 1993; Diudea & Gutman, 1998; Ivanciuc & Balaban, 1999). They can be sensitive to one or more structural features of the molecule such as size, shape, symmetry, branching and cyclicity and can also encode chemical information concerning atom type and bond multiplicity. In fact, topological indices were proposed divided into two categories: *topostructural* and *topochemical indices* (Basak, Gute, & Grunwald, 1997). Topostructural indices encode only information about adjacency and distances between atoms in the molecular structure; topochemical indices quantify information about topology but also specific chemical properties of atoms such as their chemical identity and hybridisation state.

In the mid-1960s, the pioneering works of Hansch (1962), Fujita (1964), Free and Wilson (1964), who proposed models of additive substituent contributions to biological activities, led to explosive development in QSAR analysis.

The use of quantum-chemical descriptors in QSAR/QSPR modelling dates back to early 1970s (Kier, 1971), although they actually were conceived several years before to encode information about relevant properties of molecules in the framework of quantum-chemistry.

As a natural extension of the topological representation of a molecule, the geometrical aspects of molecular structures were taken into account since the mid-1980s, leading to the development of the 3D-QSAR. Geometrical descriptors, also called *3D-molecular descriptors*, are defined in several different ways but always derived from the three-dimensional structure of the molecule, more specifically from x, y, z Cartesian coordinates of the molecule atoms (Ivanciuc, 2001; Todeschini

& Consonni, 2003). Generally, geometrical descriptors are calculated either on some optimized molecular geometry obtained by the methods of the computational chemistry or from crystallographic coordinates.

In the late 1980s, a new strategy for describing molecules was proposed, based on molecular interaction fields. The focus of this approach is to identify and characterize quantitatively the interactions between the molecule and the receptor's active site. Molecules are placed in a 3D lattice constituted by several thousands of evenly spaced grid points and a probe (such as a water molecule, methyl group, hydrogen, etc.) is used to map a surface of molecules on the basis of the molecule interactions with the probe. The formulation of a lattice model to compare molecules by aligning them in 3D space and extracting chemical information from molecular interaction fields was first proposed by Goodford (1985) in the GRID method and then by Cramer, Patterson, Bunce (1988) in the Comparative Molecular Field Analysis (CoMFA).

Finally, an increasing interest of the scientific community has been showing in recent years for virtual screening and design of chemical libraries, for which several similarity/diversity approaches, cell-based methods and scoring functions have been proposed mainly based on substructure descriptors such as molecular fingerprints (Gasteiger, 2003; Oprea, 2004). Substructure descriptors are counts of occurrences of predefined structural features (functional groups, augmented atoms, pharmacophore point pairs, atom pairs and triangles, surface triangles, etc.) in molecules or binary variables specifying their presence/absence (Crowe, Lynch, & Town, 1970; Adamson, Lynch, & Town, 1971). They are string representations of chemical structures usually designed to enhance the efficiency of chemical database screening and analysis. Each bin or set of bins of the string is associated with a structural feature or pattern. Characterization of molecules by these substructure descriptors is evident to chemists and directly related to similarity/diversity of chemi-

cal structures (Varmuza, Demuth, Karlovits, & Scsibrany, 2005).

AUTOCORRELATION DESCRIPTORS OF MOLECULES

Spatial autocorrelation coefficients are frequently used in molecular modelling and QSAR to account for spatial distribution of molecular properties.

The simplest descriptor P for a molecular property is obtained by summing the (squared) atomic property values. Mathematically:

$$P = \sum_{i=1}^{A} p_i^2 \tag{1}$$

where A is the number of atoms in a molecule and P the global property which depends on the kind of molecule atoms and not on the molecular structure; p_i is the property of the ith atom.

An extension of this global property descriptor that combines chemical information given by property values in specified molecule regions and structural information are the spatial autocorrelation descriptors. These are based on a conceptual dissection of the molecular structure and the application of an autocorrelation function to molecular properties measured in different molecular regions.

Autocorrelation functions AC_k for ordered discrete sequences of n values $f(x_i)$ are based on summation of the products of the ith value and the $(i + k)$th value as:

$$AC_l = \sum_{i=1}^{n-l} f\left(x_i\right) \cdot f\left(x_{i+l}\right) \tag{2}$$

where $f(x)$ is any function of the variable x and k is the *lag* representing an interval of x, σ^2 is the variance of the function values and μ their mean. The lag assumes values between 1 and K, where the maximum value K can be $n - 1$; however, in several applications, K is chosen equal to a small number (e.g. K £ 8). A lag value of zero corresponds to the sum of the squared centred values of the function. *Autocovariances* are calculated in the same way but omitting standardization by σ^2. The function $f(x)$ is usually a time-dependent function such as a time-dependent electrical signal, or a spatial-dependent function such as the population density in space. Then, autocorrelation measures the strength of a relationship between observations as a function of the time or space separation between them (Moreau & Turpin, 1996).

Spatial autocorrelation indicates the extent to which the occurrence of one feature is influenced by similar features in the adjacent area. This exists when there is systematic spatial variation in the values of a given variable. This variation can exist in two forms: positive and negative spatial autocorrelation. In the case of positive autocorrelation, the value of a variable at a given location tends to be similar to the values of that variable in nearby locations. This means that if the value of some variable is low in a given location, the presence of positive spatial autocorrelation indicates that nearby values are also low. Conversely, negative spatial autocorrelation is characterized by dissimilar values in nearby locations (e.g. a small value of a variable may be surrounded by large values in nearby locations).

Autocorrelation descriptors of chemical compounds are calculated by using various molecular properties that can be represented at the atomic level or molecular surface level or else. These molecular descriptors are invariant to translation and rotation, since a property of the autocorrelation function is that it does not change when the origin of the x variable is shifted.

Based on the same principles as the autocorrelation descriptors, but calculated contemporarily on two different properties $f(x)$ and $g(x)$, *crosscorrelation descriptors* are calculated to measure the strength of relationships between the two considered properties. For any two ordered sequences comprised of a number of discrete values, the crosscorrelation is calculated by summing the products of the ith value of the first sequence and the $(i + k)$th value of the second sequence, as:

$$AC_l = \sum_{i=1}^{n-l} f\left(x_i\right) \cdot f\left(x_{i+l}\right) \tag{3}$$

where n is the lowest cardinality of the two sets and k the considered *lag* representing an interval of x. *Crosscovariance descriptors* are calculated in the same way but omitting standardization by the two standard deviations $\sigma_{f(x)}$ and $\sigma_{g(x)}$.

The most common spatial molecular autocorrelation descriptors are obtained taking the molecule atoms as the set of discrete points in space and an atomic property as the function evaluated at those points.

Common atomic properties used to calculate autocorrelation descriptors are atomic masses, van der Waals volumes, atomic electronegativities, atomic polarizabilities, covalent radii, etc. Alternatively, quantities, which are local vertex invariants derived from the molecular graph, such as the topological vertex degrees (i.e., the number of adjacent vertices) and the intrinsic states or E-state indices of Kier and Hall (1990) can be used as the atomic properties.

The diagram of autocorrelation values against the lag (e.g. from 0 to K) is called *autocorrelogram*; this can be viewed as a molecular profile, useful to describe a chemical compound in similarity/diversity analysis studies. Moreover, autocorrelation descriptors have been demonstrated useful in QSAR studies as they are unique for a given molecular structure, do not require any molecule alignment being invariant to roto-translation, encode information on structural fragments and therefore seem to be particularly suitable for describing differences in congeneric series of molecules.

A typical disadvantage of all the autocorrelation descriptors might be that the original information on the molecular structure or surface cannot be reconstructed. In drug design, to overcome this drawback *Maximum Auto-CrossCovariance descriptors* (or *MACC descriptors*) were proposed; these are autocorrelation and crosscorrelation descriptors calculated taking into account only the maximum product of molecular properties for each lag k:

$$ACC_l = \sum_{i=1}^{n-l} f\left(x_i\right) \cdot f\left(y_{i+l}\right) \qquad (4)$$

where f and g represent two molecular properties; three different maximal values can be retained for the three possible combinations $(+, +)$, $(-,-)$, and $(-, +)$. In some applications, similar molecular descriptors are calculated omitting property centring as follows:

$$ACC_l = \sum_{i=1}^{n-l} f\left(x_i\right) \cdot f\left(y_{i+l}\right) \qquad (5)$$

Topological Autocorrelation Descriptors

Topological autocorrelation descriptors are spatial autocorrelations calculated on a molecular graph; in this case, the lag k coincides with the *topological distance* between any pair of vertices (i.e., the number of edges along the shortest path between two vertices).

Moreau-Broto Autocorrelation Descriptors

Moreau (1980) and Broto (1984) are the first who applied an autocorrelation function to the molecular graph to measure the distribution of atomic properties on the molecule topology. They called *Autocorrelation of a Topological Structure* (*ATS*) the final vector comprised of autocorrelation functions calculated as:

$$ATS_d = \sum_{i=1}^{A} \sum_{j=1}^{A} \delta_{ij} \cdot \left(w_i \cdot w_j\right)_d = \mathbf{w}^{\mathbf{T}} \cdot {}^{\mathbf{m}}\mathbf{B} \cdot \mathbf{w} \qquad (6)$$

where w is any atomic property, A is the number of atoms in a molecule, k is the lag, and d_{ij} is the topological distance between ith and jth atoms; $\delta(d_{ij}; k)$ is a Dirac-delta function equal to 1 if $d_{ij} = k$, zero otherwise. The autocorrelation ATS_0 defined for path of length zero is calculated as:

$$ATS_0 = \sum_{i=1}^{A} w_i^2 \qquad (7)$$

that is, the sum of the squares of the atomic properties. Atomic properties w should be centred by subtracting the average property value in the molecule in order to obtain proper autocorrelation values. Hollas (2002) demonstrated that, only if properties are centred, all autocorrelation descriptors are uncorrelated thus resulting more suitable for subsequent statistical analysis.

For each atomic property w, the set of the autocorrelation terms defined for all existing topological distances in the graph is the *ATS* descriptor defined as:

$$\{ATS_0, ATS_1, ATS_2, ..., ATS_D\}_w \qquad (8)$$

where D is the topological diameter, that is, the maximum distance in the graph.

Average spatial autocorrelation descriptors are obtained dividing each term by the corresponding number of contributions, thus avoiding any dependence on molecular size:

$$\overline{ATS_d} = \frac{1}{\Delta} \cdot \sum_{i=1}^{A} \sum_{j=1}^{A} \delta_{ij} \cdot \left(w_i \cdot w_j\right)_d \qquad (9)$$

where Δ_k is the sum of the Dirac-delta, that is, the total number of vertex pairs at distance equal to k (Wagener, Sadowski, & Gasteiger, 1995).

An example of calculation of Moreau-Broto autocorrelation descriptors is reported for 4-hydroxy-2-butanone. The H-depleted molecular graph is:

Atomic masses are used as the weighting scheme for molecule atoms: $w_1 = w_2 = w_3 = w_4 = 12$; $w_5 = w_6 = 16$. Then, autocorrelation terms for lag k from 0 to 4 are:

$$ATS_0 = w_1^2 + w_2^2 + w_3^2 + w_4^2 + w_5^2 + w_6^2 = 12^2 + 12^2 + 12^2 + 12^2 + 16^2 + 16^2 = 1088$$

$$ATS_1 = w_1 \cdot w_2 + w_2 \cdot w_3 + w_3 \cdot w_4 + w_4 \cdot w_5 + w_2 \cdot w_6 = 12 \cdot 12 + 12 \cdot 12 + 12 \cdot 12 + 12 \cdot 16 + 12 \cdot 16 = 816$$

$$ATS_2 = w_1 \cdot w_3 + w_1 \cdot w_6 + w_2 \cdot w_4 + w_3 \cdot w_5 + w_2 \cdot w_6 = 12 \cdot 12 + 12 \cdot 16 + 12 \cdot 12 + 12 \cdot 16 + 12 \cdot 16 = 864$$

$$ATS_3 = w_1 \cdot w_4 + w_2 \cdot w_5 + w_4 \cdot w_6 = 12 \cdot 12 + 12 \cdot 16 + 12 \cdot 16 = 528$$

$$ATS_4 = w_1 \cdot w_5 + w_6 \cdot w_5 = 12 \cdot 16 + 16 \cdot 16 = 448$$

Moran and Geary Autocorrelation Descriptors

Moran and Geary coefficients are autocorrelation functions mainly applied in ecological studies to describe the spatial distribution of environmental features; they provide an indication of the type and degree of spatial autocorrelation present in a data set. They are applied to the molecular structure in the same way as the Moreau-Broto function; however, unlike Moreau-Broto function, Moran and Geary functions give real autocorrelation accounting explicitly for the mean and variance of each property.

The *Moran coefficient*, applied to a molecular graph, is calculated as (Moran, 1950):

$$I(d) = \frac{\frac{1}{\Delta} \cdot \sum_{i=1}^{A} \sum_{j=1}^{A} \delta_{ij} \cdot \left(w_i - \overline{w}\right) \cdot \left(w_j - \overline{w}\right)}{\frac{1}{A} \cdot \sum_{i=1}^{A} \left(w_i - \overline{w}\right)^2}$$

(10)

where w_i is any atomic property, $\overline{w}$ is its average value on the molecule, A is the number of atoms, k is the considered lag and d_{ij} the topological distance between ith and jth atoms; $\delta(d_{ij}; k)$ is the Dirac-delta equal to 1 if $d_{ij} = k$, zero otherwise. Δ_k is the number of vertex pairs at distance equal to k.

Moran coefficient is in the range from approximately -1 to $+1$. Positive values represent positive spatial autocorrelation, that is, similar values are spatially clustered, whereas negative values negative spatial autocorrelation, that is, neighbouring values are dissimilar. A zero value indicates no spatial autocorrelation.

The *Geary coefficient*, denoted by c_k, is defined as (Geary, 1954):

$$c(d) = \frac{\frac{1}{2\Delta} \cdot \sum_{i=1}^{A} \sum_{j=1}^{A} \delta_{ij} \cdot \left(w_i - w_j\right)^2}{\frac{1}{(A-1)} \cdot \sum_{i=1}^{A} \left(w_i - \overline{w}\right)^2}$$

(11)

where w_i is any atomic property and $\overline{w}$ its average value on the molecule; k is the considered lag and d_{ij} the topological distance between ith and jth atoms; $\delta(d_{ij}; k)$ is the Dirac-delta and Δ_k the number of vertex pairs at distance equal to k.

Geary coefficient is a distance-type function varying from zero to infinite. Strong autocorrelation produces low values of this index; moreover, positive autocorrelation translates in values between 0 and 1 whereas negative autocorrelation produces values larger than 1; therefore, the Geary's c value of one suggests that no spatial autocorrelation is present.

Molecular Electronegativity Edge Vector

This is a modification of the Moreau-Broto autocorrelation function defined using reciprocal topological distances in conjunction with the Pauling atom electronegativities χ^{PA} (Li, Fu, Wang, & Liu, 2001). The autocorrelation value for each kth lag is calculated as:

$$VMEE_k = \sum_{i=1}^{A-1} \sum_{j=i+1}^{A} \frac{\chi_i^{PA} \cdot \chi_j^{PA}}{d_{ij}} \cdot \delta\left(d_{ij}; k\right) \quad (12)$$

where d_{ij} is the topological distance between the ith and jth atoms. This autocorrelation vector was proposed for modelling biological activities of dipeptides.

ACC Transforms

Standing for *Auto-CrossCovariance transforms*, ACC transforms are autocovariances and crosscovariances calculated from sequential data with the aim of transforming them into uniform-length descriptors suitable for QSAR modelling. ACC transforms were originally proposed to describe peptide sequences (Wold, Jonsson, Sjöström, Sandberg, & Rännar, 1993; Sjöström, Rännar, & Wieslander, 1995). In order to calculate ACC transforms, each amino acid position in the peptide sequence was defined in terms of a number of amino acid properties; in particular, three orthogonal z-scores, derived from a Principal Component Analysis (PCA) of 29 physico-chemical properties of the 20 coded amino acids, were originally used to describe each amino acid. Then, for each peptide sequence, auto- and crosscovariances with lags k = 1, 2,..., K, were calculated as:

$$ACC_k\left(j, j\right) = \sum_{i=1}^{n-k} \frac{z_i\left(j\right) \cdot z_{i+k}\left(j\right)}{n-k}$$
$$ACC_k\left(j, m\right) = \sum_{i=1}^{n-k} \frac{z_i\left(j\right) \cdot z_{i+k}\left(m\right)}{n-k} \quad (13)$$

where j and m indicate two different z-scores, n is the number of amino acids in the sequence, and index i refers to amino acid position in the sequence; z-score values, being derived from PCA, are used directly because they are already mean centred.

TMACC descriptors, standing for *Topological MAximum CrossCorrelation descriptors*, are a variant of the ACC transforms for molecular graphs (Melville & Hirts, 2007). These are crosscovariances calculated taking into account the topological distance d_{ij} between the atoms i and j and four basic atomic properties: 1) partial charges, accounting for electrostatic properties (Gasteiger & Marsili, 1980); 2) molar refractivity parameters, accounting for steric properties and polarizabilities (Wildman & Crippen, 1999); 3) logP values, accounting for hydrophobicity (Wildman & Crippen, 1999); 4) logS values, accounting for solubility and solvation phenomena (Hou, Xia, Zhang, & Xu, 2004).

The general formula for the calculation of *TMACC* descriptors is:

$$TMACC\left(w_1, w_2; k\right) = \frac{1}{\Delta_k} \cdot \sum_{i=1}^{A} \sum_{j=1}^{A} w_{1i} \cdot w_{2j} \cdot \delta\left(d_{ij}; k\right) \quad (14)$$

where w_1 and w_2 are two different atomic properties, A is the number of atoms in the molecule, k is the lag and d_{ij} the topological distance between ith and jth atoms; Δ_k is the number of atom pairs located at topological distance k, and $\delta(d_{ij}; k)$ is the Dirac-delta. If only one property is considered, that is, $w_1 = w_2$, autocovariances are obtained. Because all the selected properties, except for molar refractivity, can assume both positive and negative values, these are treated as different properties and crosscovariance terms are also calculated between positive and negative values of each property. Therefore, 7 autocovariance terms and 12 crosscovariance terms constitute the final *TMACC* vector.

Estrada Generalized Topological Indices

Variable topological indices are local and graph invariants containing adjustable parameters whose values are optimised in order to improve the statistical quality of a given regression model. Sometimes also called flexible descriptors or optimal descriptors, their flexibility in modelling is useful to obtain good models; however, due to the increased number of parameters needing to be optimised, they require more intensive validation procedures to generate predictive models.

These molecular descriptors are called "variable" because their values are not fixed for a molecule but change depending on the training set and the property to be modelled.

The *Estrada Generalized Topological Index* (GTI) is a general strategy to search for optimised quantitative-structure property relationship models based on variable topological indices (Estrada, 2001; Estrada & Matamala, 2007). The main objective of this approach is to obtain the best optimised molecular descriptors for each property under study. The family of GTI descriptors is comprised of autocorrelation functions defined by the following general form:

$$GTI = \sum_{k=1}^{D} C_k\left(x_0, p_0\right) \cdot \eta^{(k)} \qquad (15)$$

where the summation goes over the different topological distances in the graph, D being the topological diameter, that is, the maximum topological distance in the graph, and accounts for the contributions $\eta^{(k)}$ of pairs of vertices located at the same topological distance k. Each contribution $\eta^{(k)}$ is scaled by two real parameters x_0 and p_0 through the $C_k(x_0, p_0)$ coefficient defined as:

$$C_k\left(x_0, p_0\right) = k^{p_0} \cdot x_0^{p_0(k-1)} \qquad (16)$$

By definition, the C_k coefficient is equal to one for any pair of adjacent vertices ($k = 1$), regardless of the parameter values.

The term $\eta^{(k)}$ defines the contribution of all those interactions due to the pairs of vertices at distance k in the graph as:

$$\eta^{(k)} = \frac{1}{2} \cdot \sum_{i=1}^{A} \sum_{j=1}^{A} \langle i, j \rangle \cdot \delta\left(d_{ij}; k\right) \qquad (17)$$

where A is the number of vertices in the graph, $\delta(d_{ij}; k)$ is a Dirac-delta function equal to one if the topological distance d_{ij} is equal to k, and zero otherwise. The term $\langle i,j \rangle$ is the 'geodesic-bracket' term encoding information about the molecular shape on the basis of a connectivity-like formula as:

$$\langle i, j \rangle = \frac{1}{2} \cdot \left(u_i \cdot v_j + v_i \cdot u_j\right) \qquad (18)$$

where u and v are two functions of the variable parameters x and p and can be considered as generalized vertex degrees defined as:

$$u_i\left(x_1, p_1, \mathbf{w}\right) = \left[w_i + \delta_i + \sum_{k=2}^{D} k \cdot x_1^{k-1} \cdot {}^k f_i\right]^{p_1} \qquad (19)$$

$$v_i\left(x_2, p_2, \mathbf{s}\right) = \left[s_i + \delta_i + \sum_{k=2}^{D} k \cdot x_2^{k-1} \cdot {}^k f_i\right]^{p_2} \qquad (20)$$

where δ_i is the simple vertex degree of the ith vertex, that is, the number of adjacent vertices, and ${}^k f_i$ is its vertex distance count, that is, the number of vertices at distance k from the ith vertex. The scalars $x_0, x_1, x_2, p_0, p_1, p_2, \mathbf{w}$ and $\mathbf{s}$ define a ($2A + 6$)-dimensional real space of parameters; $\mathbf{w}$ and $\mathbf{s}$ are two A-dimensional vectors collecting atomic properties. The first six parameters $x_0, x_1, x_2, p_0,$

p_1, and p_2 are free parameters to be optimised, whereas the parameters **w** and **s** are predefined quantities used to distinguish among the different atom-types. For each combination of the possible values of these parameters a different topological index is obtained for a molecule. It has to be noted that several of the well known topological indices can be calculated by the GTI formula by settling specific combinations of the parameters; for instance, for **w** = (0, 0,...., 0) and **s** = (0, 0,...., 0), the index GTI reduces to the *Wiener index* (Wiener, 1947) when $x_0 = 1$, x_1 = any, x_2 = any, p_0 =1, $p_1 = 0$, $p_2 = 0$, while GTI coincides with the *Randić connectivity index* (Randic, 1975) when $x_0 = 0$, $x_1 = 0$, $x_2 = 0$, $p_0 =1$, $p_1 = -1/2$, $p_2 = -1/2$.

Geometrical Autocorrelation Descriptors

Autocorrelation descriptors can also be calculated from 3D-spatial molecular geometry. In this case, the distribution of a molecular property can be evaluated by a mathematical function $f(x, y, z)$, x, y, and z being the spatial coordinates, either defined for each point of molecular space or molecular surface (i.e., a continuous property such as electronic density or molecular interaction energy) or only for points occupied by atoms (i.e., atomic properties) (Broto, Moreau, & Vandycke, 1984b; Broto & Devillers, 1990; Wagener et al., 1995).

As calculation of geometrical descriptors requires the knowledge of the spatial (x, y, z) atomic coordinates, geometrical autocorrelation descriptors usually provide more information and discrimination ability than the topological counterpart, also being suitable for distinguishing of similar molecular structures and molecule conformations. However, despite their high information content, geometrical descriptors usually show some drawbacks. They require geometry optimization and therefore the overhead to calculate them. Moreover, for flexible molecules, several molecule conformations can be available: on one hand, new information is available and can

be exploited, but, on the other hand, the problem complexity can significantly increase.

3D-TDB Autocorrelation Descriptors

3D-TDB descriptors, standing for *3D-Topological Distance Based descriptors*, are a variant of the average Moreau-Broto autocorrelations encoding also information about the 3D spatial separation between two atoms (Klein, Kaiser, & Ecker, 2004). *TDB-steric descriptors*, denoted by S, are defined for each lag k as:

$$S_k = \frac{1}{\Delta_k} \cdot \sum_{i=1}^{A-1} \sum_{j=i+1}^{A} \left(R_i^{\text{cov}} \cdot r_{ij} \cdot R_j^{\text{cov}} \right) \cdot \delta \left(d_{ij}; k \right)$$

(21)

where Δ_k is the number of atom pairs located at a topological distance k, r_{ij} is the geometric distance between the ith and jth atoms, and R^{cov} is the atomic covalent radius accounting for steric properties of atoms.

In a similar way, *TDB-electronic descriptors*, denoted by X, are defined as:

$$X_k = \frac{1}{\Delta_k} \cdot \sum_{i=1}^{A-1} \sum_{j=i+1}^{A} \left(\chi_i \cdot r_{ij} \cdot \chi_j \right) \cdot \delta \left(d_{ij}; k \right)$$

(22)

where χ is the sigma orbital electronegativity accounting for electronic properties of atoms.

Surface Autocorrelation Vector

The autocorrelation of molecular surface properties is a general approach for the description of property measures on the molecular surface by using uniform-length descriptors which are comprised of the same number of elements regardless of the size of the molecule (Sadowski, Wagener, & Gasteiger, 1995; Wagener et al., 1995). This approach is an extension of Moreau-Broto autocorrelation function to 3D molecular geometry.

Since geometrical distances r_{ij} can have any real positive value, some ordered distance intervals need to be specified each defined by a lower and upper value of r_{ij}. All distances falling in the same interval are considered identical.

To generate 3D autocorrelation descriptors of molecular surface properties, first, a number of points are randomly distributed on the molecular surface with a user-defined density and in an orderly manner to ensure a continuous surface. Then, the *Surface Autocorrelation Vector* (SAV) is derived by calculating for each lag k the sum of the products of the property values at two surface points located at a distance falling into the kth distance interval. This value is then normalized by the number Δ_k of the geometrical distances r_{ij} in the interval:

$$A\left(k\right) = \frac{1}{\Delta_k} \cdot \sum_{i=1}^{N-1} \sum_{j=i+1}^{N} w_i \cdot w_j \cdot \delta\left(r_{ij}; k\right) \qquad (23)$$

where N is the number of surface points, and k represents a distance interval defined by a lower and upper bound.

It was demonstrated that to obtain the best Surface Autocorrelation Vectors for QSAR modelling, the van der Waals surface is better than other molecular surfaces. Then, surface should have no fewer than five grid points per $Å^2$ and a distance interval no greater than 1 Å should be used in the distance binning scheme. Autocorrelation values calculated for a number of distance intervals constitute a unique fingerprint of the molecule, thus resulting suitable for similarity/diversity analysis of molecules.

3D – ACC Transforms

ACC transforms, originally proposed to encode chemical information of peptide sequences, were successively used to describe Molecular Interaction Fields (MIFs) typical of CoMFA analysis, which are 3D fields of molecule – probe interaction energies measured at several thousands of evenly spaced grid points. In this case, the distance between grid points along each coordinate axis, along the diagonal or along any intermediate direction is used as the *lag* to calculate auto- and crosscovariances (Clementi et al., 1993). The crosscovariance terms are calculated by the products of the interaction energy values for steric and electrostatic fields in grid points at distances equal to the lag. Different kinds of interactions, namely positive-positive, negative-negative, and positive-negative, are usually kept separated, thus resulting in 10 ACC terms for each lag. The major drawback of ACC transforms derived from molecular interaction fields is that their values depend on molecule orientation along the axes.

GETAWAY Descriptors

The GETAWAY (GEometry, Topology, and Atom-Weights AssemblY) descriptors are derived from the Molecular Influence Matrix (**H**), which is a representation of the molecular structure, defined as (Consonni, Todeschini, & Pavan, 2002):

$$\mathbf{H} = \mathbf{M} \times (\mathbf{M}^T\mathbf{M})^{-1} \times \mathbf{M}^T \qquad (24)$$

where **M** is the molecular matrix, that is, a rectangular matrix ($A \times 3$) whose rows represent the molecule atoms and the columns the atom Cartesian coordinates (x, y, z) with respect to any rectangular coordinate system with axes X, Y, Z. Atomic coordinates are assumed to be calculated with respect to the geometrical centre of the molecule in order to obtain translational invariance. The molecular influence matrix is a square symmetric $A \times A$ matrix, where A represents the number of atoms, and shows rotational invariance with respect to the molecule coordinates, thus resulting independent of molecule alignment rules. The diagonal elements h_{ii} of this matrix range from 0 to 1 and encode atomic information related to the "influence" of each molecule atom in determining the whole shape of the molecule; in effect, mantle atoms always have higher h_{ii} values than atoms near the molecule centre.

Combining the elements of the molecular influence matrix **H** with those of the geometry matrix **G**, which encodes spatial relationships between pairs of atoms in terms of inter-atomic Euclidean distances, another symmetric $A \times A$ molecular matrix, called *influence/distance matrix* and denoted by **R**, was derived as the following:

$$[\text{R}]_{ij} \equiv \left[\frac{\sqrt{h_{ii} \cdot h_{jj}}}{r_{ij}} \right]_{ij} \qquad i \neq j \qquad (25)$$

where h_i and h_j are the leverages of the atoms i and j, and r_{ij} is their geometric distance. Each off-diagonal element i-j is calculated by the ratio of the geometric mean of the corresponding ith and jth diagonal elements of the matrix **H** over the interatomic distance r_{ij} provided by the geometry matrix **G**.

Most of the GETAWAY descriptors are autocorrelation vectors obtained by double-weighting the molecule atoms in such a way as to account for chemical information together with 3D information encoded by the elements of the molecular influence matrix **H** or influence/distance matrix **R**.

HATS indices are defined by analogy with the Moreau-Broto autocorrelation descriptors *ATS*, weighting each atom of the molecule by physico-chemical properties combined with the diagonal elements of the molecular influence matrix **H**, thus also accounting for the 3D features of the molecules:

$$HATS_k(w) = \sum_{i=1}^{A-1} \sum_{j>i} (w_i \cdot h_{ii}) \cdot (w_j \cdot h_{jj}) \cdot \delta(k; d_{ij}) \qquad k = 0, 1, 2, ..., d \qquad (26)$$

where w is an atomic weighting and $\delta(d_{ij}; k)$ a Dirac-delta function equal to one when the topological distance d_{ij} between atoms i and j is equal to k, and zero otherwise.

H indices are filtered autocorrelation descriptors defined as:

$$H_k(w) = \sum_{i=1}^{A-1} \sum_{j>i} h_{ij} \cdot w_i \cdot w_j \cdot \delta(k; d_{ij}; h_{ij}) \qquad k = 0, 1, 2, ..., d \qquad (27)$$

where h_{ij} are the off-diagonal elements of the molecular influence matrix **H** and the Dirac-delta function $\delta(d_{ij}; h_{ij}; k)$ is here defined as:

$$\delta(k; d_{ij}; h_{ij}) = \begin{cases} 1 & if \quad d_{ij} = k \quad and \quad h_{ij} > 0 \\ 0 & if \quad d_{ij} \neq k \quad or \quad h_{ij} \leq 0 \end{cases} \qquad (28)$$

While the *HATS* indices make use of the diagonal elements of the matrix **H**, the *H* indices exploit the off-diagonal elements, which can be either positive or negative. In order to emphasize interactions between spatially near atoms, only off-diagonal positive h values are used. In effect, for a given *lag* (i.e., topological distance) the product of the atom properties is multiplied by the corresponding h_{ij} value and only those contributions with a positive h_{ij} value are considered. This means that, for a given atom i, only those atoms j at topological distance d_{ij} with a positive h_{ij} value are considered because they may have the chance to interact with the ith atom.

R indices are defined in the same way as the *H* indices, by using the off-diagonal elements of the influence/distance matrix **R** instead of the elements of the matrix **H**:

$$R_k(w) = \sum_{i=1}^{A-1} \sum_{j>i} \frac{\sqrt{h_{ii} \cdot h_{jj}}}{r_{ij}} \cdot w_i \cdot w_j \cdot \delta(k; d_{ij}) \qquad k = 1, 2, ..., d \qquad (29)$$

In this case, no filtering is applied, because geometrical distances r_{ij} act as a smoothing function.

The *maximal R indices* were proposed in order to take into account local aspects of the molecule and allow reversible decoding; only the maximum property product between atom pairs at a given topological distance (*lag*) is retained:

$$R_k^+ \left(w \right) = \max \left(\sum_{i=1}^{A} \sum_{j \neq i} \frac{\sqrt{h_{ii} \cdot h_{jj}}}{r_{ij}} \cdot w_i \cdot w_j \cdot \delta \left(k; d_{ij} \right) \right) \qquad k = 1, 2, ..., d$$

(30)

The atomic weighting schemes applied for GETAWAY descriptor calculation usually are atomic mass (*m*), atomic polarizability (*p*), Sanderson atomic electronegativity (*e*), and atomic van der Waals volume (*v*); they can also be derived from the unweighted molecular structures.

Atom-Type Autocorrelation Descriptors

A special case of autocorrelation function is the *Atom-Type AutoCorrelation* (*ATAC*), which is calculated by summing property values only of atoms of given types. The simplest atom-type autocorrelation is given by:

$$A T A C_k \left(s, t \right) = \sum_{i=1}^{A} \sum_{j=1}^{A} \delta \left(i; s \right) \cdot \delta \left(j; t \right) \cdot \delta \left(d_{ij}; k \right)$$

(31)

where *s* and *t* denote two different atom-types, *A* is the number of molecule atoms; $\delta(i; s)$ is a Dirac-delta function equal to 1 if the atom *i* is of type *s*, and zero otherwise; analogously, $\delta(j; t)$ is equal to 1 if the atom *j* is of type *t*, and zero otherwise; $\delta(d_{ij}; k)$ is equal to one if the inter-atomic distance d_{ij} is equal to the lag *k*, and zero otherwise.

This descriptor is defined for each pair of atom-types and simply encodes the occurrence numbers of the given atom-type pair at different distance values. It can be normalized by using two different procedures: the first one consists in dividing each $ATAC_k$ value by the total number of atom pairs at distance *k* regardless of their types; the second one consists in dividing each $ATAC_k$ value by a constant, which can be equal

to the total number of atoms in the molecule or, alternatively, to the total number of (*s,t*) atom-type pairs in the molecule.

Note that if atom-types *s* and *t* coincide, that is *s* = *t*, then the atom-type autocorrelation is calculated as:

$$A T A C_k \left(s, s \right) = \frac{1}{2} \cdot \sum_{i=1}^{A} \sum_{j=1}^{A} \delta \left(i; s \right) \cdot \delta \left(j; s \right) \cdot \delta \left(d_{ij}; k \right)$$

(32)

where the fraction ½ is used to avoid counting twice a given pair of atom-types. *TDB-atom type descriptors*, denoted by *I*, are examples of atom-type autocorrelations proposed by Klein (2004) and calculated only for pairs of atoms of the same type.

Atom-types can be defined in different ways; they can be defined in terms of the simple chemical identity or account also for atom connectivity, hybridisation states and pharmacophoric features.

Atom-type autocorrelations have been used to derive some vectors of substructure descriptors such as Atom Pairs (APs) (Carhart, Smith, & Venkataraghavan, 1985) and CATS descriptors (Schneider, Neidhart, Giller, & Schmid, 1999; Fechner, Franke, Renner, Schneider, & Schneider, 2003). To calculate Atom Pairs, the atom-type is defined by the element itself, number of heavy-atom connections and number of π electron pairs on each atom.

CATS descriptors are holographic vectors where each bin encodes the number of times a pair of pharmacophore point types occurs in the molecule. The five defined potential pharmacophore points are: hydrogen-bond donor (D), hydrogen-bond acceptor (A), positively charged or ionisable (P), negatively charged or ionisable (N), and lipophilic (L). If an atom does not belong to any of the five PPP types it is not considered. Moreover, an atom is allowed to be assigned to one or two PPP types.

For each molecule, the number of occurrences of all 15 possible pharmacophore point pairs (DD, DA, DP, DN, DL, AA, AP, AN, AL, PP, PN, PL, NN, NL, LL) is determined and then associated with the topological distance between the two considered points. Topological distances of 0 – 9 bonds are considered leading to a final 150-dimensional autocorrelation vector.

Distance-counting descriptors (or *SE-vectors*) are a particular implementation of Atom Pairs proposed by Clerc & Terkovics (1990). These are holographic vectors encoding information on the occurrence frequency of any combination of two atom-types and a distance relationship between them. All the paths and not only the shortest one between any pair of atom-types are considered in the original proposal. Based on the shortest path (i.e., topological distance), revised SE-vectors were proposed by Baumann (2002) and called *SESP-Top vectors* and *SESP-Geo vectors*.

GA-BASED MULTIPLE LINEAR REGRESSION

In multivariate data analysis a very important problem to be faced is the data multicollinearity, which greatly influences results of several statistical and *chemometrics* methods, like regression analysis. In such methods it is essential to estimate how much data variability is related to systematic and potentially useful information rather than random noise and chance correlation. Random noise or chance correlation are always potentially present, making models both unstable and unreliable and giving undesired and sudden bias in data exploration; if noise and chance correlation are dominant in the data, then the results of any multivariate data analysis will be strongly affected by errors and, consequently, become difficult to interpret. Especially in multivariate regression analysis, data correlation greatly influences results due to well known troubles like model instability, overfitting, and errors in the regression coefficients.

Moreover, in QSAR studies there are thousands of descriptors available for describing molecules and there often is no *a priori* knowledge about which molecular features are more responsible for a specific property, then there is a need to search for the subsets of the most appropriate descriptors.

In the last few years, a great attention of the scientific community has been paid to the techniques devoted to the variable selection. Two main approaches can be used for extracting not redundant but relevant variables from the pool of available variables: the *variable reduction* and the *variable selection*.

Variable reduction refers to the procedure that aims at selecting a subset of variables able to preserve as much information of the original data as possible, but eliminating redundancy, noise and linearly or near-linearly dependent variables. When the variables to be used in a model are chosen on the basis of general principles and not accounting for a specific goal (i.e., some experimental property to model), the term variable reduction should be more properly used than variable selection. Variable reduction methods can be useful for pre-treatment of large data sets since they allow discarding variables that contain redundant or minimal information. Reducing the number of variables often allows one to better investigate data structure and obtain more stable results from multivariate modelling methods.

Variable subset selection techniques (VSS), unlike variable reduction techniques, take into account the specific property to be modelled. For instance, in QSAR analysis, these techniques aim at finding the subset of molecular descriptors which lead to the best predictive model for the studied property.

The exhaustive search, sometimes called All Possible Models (APM), can be applied in all but the simplest cases, this procedure being highly demanding when molecules are characterized by a number of molecular descriptors.

Several different variable selection techniques are nowadays available: besides the classical Stepwise Regression (SWR), proposed by Efroymson (1960) and based on alternating forward selection and backward elimination, more powerful machine learning techniques were devised and are largely used for variable selection purposes: Genetic Algorithms (GAs) (Leardi, 1994; Luke, 1994), Simulated Annealing (SA) (Zheng & Tropsha, 2000), Tabu Search (TS) (Baumann, Albert, & von Korff, 2002), and Evolutionary Programming (EP) (Kubinyi, 1994).

Several modifications of the original Partial Least Squares (PLS) regression method were also proposed with the aim to perform variable selection and, among them, Iterative Variable Selection for PLS (IVS-PLS) (Lindgren, Geladi, Rännar, & Wold, 1994) and Uninformative Variable Elimination by PLS (UVE-PLS) (Centner et al., 1996) are the most popular.

The general approach to the variable subset selection is shown in Figure 2. The first step (A) is the definition of the algorithm performing the selection of one or more variables within the whole set of candidate variables. This step can be performed by selecting the variables by a random strategy or by using a genetic strategy (based on repeated reproduction and mutation steps), or other approaches. Then, from each subset of variables a model is calculated.

The second step (B) is the evaluation of the quality associated to each model by using proper optimization functions (often called fitness functions). In this phase, both the method for estimating the models and the fitness function to be optimized have to be previously defined.

The most popular regression methods used in the model estimation are Ordinary Least Squares (OLS or Multiple Linear Regression, MLR), Partial Least Squares (PLS), Back-Propagation Artificial Neural Network (BP-ANN), and k-Nearest Neighbour (k-NN) estimator.

In regression studies, the most popular fitness function is the model predictive ability (Q^2) based on leave-one-out or leave-more-out, even if the leave-one-out procedure is the most common during the model selection phase.

The acceptability of a final regression model (step C) should not be evaluated simply looking at its predictive ability but considering also additional rules. For instance, models whose differences between R^2 and Q^2 (obtained by the leave-one-out procedure) are too large (Sutter, Peterson, & Jurs, 1997) should be rejected because a significant decreasing in the predictive ability can be expected in their practical use on new chemicals. Therefore, in order to prevent the acceptability of not real predictive models and/or chance correlated models, severe optimization functions need to be used as the AIC index (Akaike, 1974), the LOF function (Friedman, 1988), the FIT function (Kubinyi, 1996) and the RQK functions (Todeschini, Consonni, Mauri, & Pavan, 2004), these last including more than one rule for the model acceptability.

The iterative step (D) depends on the chosen variable selection technique. During the iterative procedure, the conditions for the stop are checked and the accepted models are properly managed.

The simplest VSS techniques preserve only the best model at each step, providing a final unique model at the end of the optimization pro-

Figure 2. General scheme of the variable selection approach

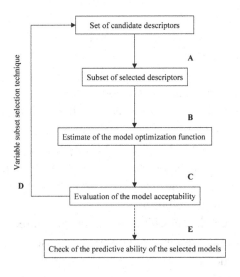

cedure. Other strategies (typically, for example, those based on the genetic algorithms) provide a population of accepted models or more than one population of models.

The final selected model(s) is further processed (E) to check its effective predictive ability and the eventual presence of chance correlation. To this end, strong validation procedures are applied such as bootstrap or using an external data set; chance correlation can be evaluated by the Y-randomization test (Lindgren, Hansen, Karcher, Sjöström, & Eriksson, 1996; Topliss & Edwards, 1979).

Genetic Algorithms (GAs) are an evolutionary method widely used for complex optimisation problems in several fields such as robotics, chemistry and QSAR (Goldberg, 1989). A specific application of GAs is variable subset selection (GA-VSS) (Leardi, Boggia, & Terrile, 1992). GAs perform the selection of the most relevant variables by considering populations of models generated through a reproduction process and optimised according to a defined *objective function* related to model quality.

The GA procedure is based on the evolution of a *population* of models, which are ordered according to some objective function. In genetic algorithm terminology, each population individual is called *chromosome* and is a binary vector, where each bin (a *gene*) corresponds to a variable (1 if included in the model, 0 otherwise). Then, each chromosome represents a model given by a subset of variables.

Once the objective function to optimise is defined, the model population size P (e.g., P = 100) and the maximum number L of allowed variables in a model (e.g., L = 5) have to be defined; the minimum number of allowed variables is usually assumed equal to one. Moreover, the user must define both the *crossover probability* and the *mutation probability*.

Genetic algorithm evolution is a three-fold process: random initialisation, crossover, and mutation.

The random initialisation is the initial phase where random models with a number of variables between 1 and L are generated. The value of the selected objective function of each model is calculated in a process called *evaluation*. The models are then ordered with respect to the selected objective function, that is, the model quality.

In the crossover stage, from the actual population, pairs of models are selected randomly or with a probability function of their quality. Then, from each pair of selected models (*parents*), a new model is generated, preserving the common characteristics of the parents (i.e., variables excluded in both models remain excluded, variables included in both models remain included) and mixing the opposite characteristics according to the crossover probability. If the generated son coincides with one of the individuals already present in the actual population, it is rejected; otherwise, it is evaluated. If the objective function value is better than the worst value in the population, the model is included in the population, in the place corresponding to its rank; otherwise, it is no longer considered. This procedure is repeated for several pairs.

After a number of crossover iterations, the population proceeds through the mutation process. This means that for each individual of the population every gene is randomly changed into its opposite or left unchanged. Mutated individuals are evaluated and included in the population if their quality is acceptable. This process is controlled by mutation probability which is commonly set at low values, thus allowing only a few mutations and new individuals not too far away from the generating individual.

Unlike the classical genetic algorithm, crossover and mutation steps can be kept disjoint. Population crossover and mutation are alternatively repeated until a stop condition is encountered (e.g., a user-defined maximum number of iterations) or the process is ended arbitrarily.

An important characteristic of the GA-VSS method is that it provides not a single model but a population of acceptable models; this characteristic, sometimes considered a disadvantage, makes the evaluation of variable relationships with response from different points of view possible.

SIMILARITY ANALYSIS OF CHEMICAL SPACES

Since molecules can be described in different ways depending on the selected molecular representation and algorithms for generating molecular descriptors, methods for comparison of the different viewpoints are required. In QSAR and drug design field, different results can derive if the same collection of molecules is analysed in different chemical spaces. Thus, information about differences between chemical spaces should be acquired before attempting analysis of molecule relationships in different chemical spaces.

The *chemical space* is defined as the *p*-dimensional space constituted by a set of *p* molecular descriptors selected to represent the studied compounds; chemical space design is generally recognized as a crucial step for the successful application of QSPR/QSAR methods (Oprea, Zamora, & Ungell, 2002; Dutta, Guha, Jurs, & Chen, 2006; Eckert, Vogt, & Bajorath, 2006; Landon & Schaus, 2006).

In different chemical spaces, the same molecules can behave differently with the consequence that different relationships may arise. Then, comparison among chemical spaces can enhance knowledge about a pool of molecules and selection of chemical spaces that are really diverse can help to find the most diverse molecules when the objective is molecule library design or to select the most similar molecules to a lead compound when the objective is to find out new drug candidates.

Let A and B be two different chemical spaces, that is, two sets of molecular descriptors calculated for the same *n* molecules. The simplest way to measure the diversity between the two chemical spaces disregards the actual descriptor values and simply consists in computing the number of diverse descriptors defining the two spaces, that is, the Hamming distance:

$$d_H = b + c \qquad (33)$$

where b is the number of molecular descriptors defining the space A but not the space B, and c the number of molecular descriptors defining the space B but not the space A. Hamming distance usually has an upward bias since it overestimates the actual distance between two chemical spaces, due to the fact that variable correlation is not accounted for.

In order to overcome this drawback, a distance measure between two sets of variables was recently proposed with the name *Canonical Measure of Distance*, or simply *CMD index* (Todeschini, Ballabio, Consonni, Manganaro, & Mauri, 2009); it is defined as:

$$CMD_{AB} = p_A + p_B - 2 \cdot \sum_{j=1}^{M} \sqrt{\lambda_j} \qquad 0 \le CMD_{AB} \le \left(p_A + p_B \right)$$

$$(34)$$

where A and B are the two sets of variables being compared, p_A and p_B are the number of variables in set A and B, respectively; λ are the eigenvalues of the symmetrical cross-correlation matrix and M is the number of non-vanishing eigenvalues.

The cross-correlation matrix contains the pairwise correlation coefficients between variables of the two sets; it is an unsymmetrical matrix $\mathbf{C_{AB}}$ of size $(p_A \times p_B)$ or $\mathbf{C_{BA}}$ of size $(p_B \times p_A)$. The symmetrical cross-correlation matrix is derived by the following inner product:

$$\mathbf{Q_A} = \mathbf{C_{AB}} \times \mathbf{C_{BA}} \text{ or } \mathbf{Q_B} = \mathbf{C_{BA}} \times \mathbf{C_{AB}} \qquad (35)$$

where $\mathbf{Q_A}$ and $\mathbf{Q_B}$ are two different square symmetrical matrices, one of size $p_A \times p_A$ and the other of size $p_B \times p_B$. Although these symmetrical matrices are different, their M non-zero eigenvalues coincide, M being the minimum rank between $\mathbf{Q_A}$ and $\mathbf{Q_B}$.

The *Canonical Measure of Correlation* or *CMC index* was also derived from the non-vanishing eigenvalues λ of the symmetrical cross-correlation matrices as the following:

$$CMC_{AB} = \frac{\sum_{j=1}^{M} \sqrt{\lambda_j}}{\sqrt{p_A \cdot p_B}} \qquad 0 \leq CMC_{AB} \leq 1$$

(36)

where the numerator measures the inter-set common variance and the denominator is its theoretical maximum value. This index is related to the multidimensional correlational structure between two sets of variables and thus is a suitable index for measuring the degree of similarity between two chemical spaces. If no correlation exists between any pair of molecular descriptors from the two spaces, then $CMC = 0$ and CMD index reduces to the Hamming distance.

A CASE STUDY: THE FLAVONOID DATA SET

Some naturally occurring flavonoids, and synthetic flavone derivatives, have been demonstrated to have high binding affinity to the GABA$_A$ benzodiazepine receptor site (BzR) (Huang et al., 2001; Hong & Hopfinger, 2003). Therefore, flavonoids may provide important leads for the development of new psychoactive drug candidates with the same action as the benzodiazepines (BZDs), which are anxiolytic, anticonvulsant, muscle relaxant, and sedative-hypnotic, but also have a series of unwanted side effects.

38 flavonoids, taken from the literature (Huang et al., 2001), were used as the training set for QSAR modelling; moreover additional four compounds were used for model validation. The chemical structures of the 42 flavonoids and their BzR site binding affinities (–logKi) are reported in

Table 1. Ki is the affinity constant to the GABA$_A$ benzodiazepine receptor site.

The three-dimensional structure of each flavonoid was built in the neutral form by means of the software HyperChem 8.0 (2007); the minimum energy conformation of each structure was searched for by quantum mechanical method without any geometric constraint.

Most of the molecular descriptors used to analyse flavonoids were calculated by the software DRAGON 5.5 (2007). DRAGON provides more than 3000 descriptors ranging from the simplest counts of structural features to more complex geometrical descriptors. CATS, VMEE, and TDB descriptors were calculated by specific routines written by the Authors.

The first part of this study aims at comparing the chemical information encoded by the different types of autocorrelation descriptors to find out the types that are the most diverse and hence provide the most diverse information. Obviously, the final results will depend on the specific data set used for comparison and in order to draw more general conclusion this study should be extended to other data sets.

For all the 42 available flavonoids the following autocorrelation descriptors were calculated: Moreau-Broto autocorrelations (ATS), Moran coefficients (MATS), Gaery coefficients (GATS), GETAWAY descriptors (HATS, H, R, R+), binary Atom Pairs (bAP), and frequency Atom Pairs (fAP), CATS descriptors, molecular electronegativity edge vector (VMEE), TDB-steric descriptors (TDB-S), and TDB-electronic descriptors (TDB-X). Each series of autocorrelations, except for Atom Pairs and CATS, was calculated for lags from 1 to 8. Only two atomic properties were considered: atomic mass (m) and atomic Sanderson electronegativity (e). Pauling electronegativity was only used to calculate VMEE. H and HATS indices were also derived from unweighted molecular structures (u). Binary Atom Pairs only account for the presence (1) or absence (0) of a given atom-type pair at each

Table 1. Chemical structures and BzR site binding affinities (–logKi, Ki in uM) of the analyzed Flavonoids

ID	I	II	III	IV	V	VI	VII	VIII	IX	X	–LogKi
1	H	H	H	H	H	H	H	H	H	H	6.00
2	H	F	H	H	H	OH	H	H	H	H	5.60
3	H	Cl	H	H	H	OH	H	H	H	H	6.07
4	H	Br	H	H	H	OH	H	H	H	H	6.22
5	H	F	H	H	H	NO$_2$	H	H	H	H	6.74
6	H	Cl	H	H	H	NO$_2$	H	H	H	H	8.10
7	H	Cl	H	H	H	H	OCH$_3$	H	H	H	5.90
8	H	Br	H	H	H	H	OCH$_3$	H	H	H	5.68
9	H	Br	H	H	NO$_2$	H	H	H	H	H	6.68
10	H	NO$_2$	H	H	H	H	Br	H	H	H	7.60
11	H	Cl	H	H	F	H	H	H	H	H	6.38
12	H	Br	H	H	F	H	H	H	H	H	6.42
13	H	H	H	H	H	F	H	H	H	H	5.45
14	H	F	H	H	H	F	H	H	H	H	6.04
15	H	Cl	H	H	H	F	H	H	H	H	6.93
16	H	Br	H	H	H	F	H	H	H	H	7.38
17	H	H	H	H	H	H	F	H	H	H	5.44
18	H	F	H	H	H	H	F	H	H	H	5.60
19	H	Cl	H	H	H	H	F	H	H	H	6.74
20	H	Br	H	H	H	H	F	H	H	H	6.94
21	H	H	H	H	H	Cl	H	H	H	H	6.21
22	H	F	H	H	H	Cl	H	H	H	H	6.70
23	H	Cl	H	H	H	Cl	H	H	H	H	7.64
24	H	Br	H	H	H	Cl	H	H	H	H	7.77
25	H	H	H	H	H	Br	H	H	H	H	6.38
26	H	F	H	H	H	Br	H	H	H	H	6.63
27	H	Cl	H	H	H	Br	H	H	H	H	7.64
28	H	Br	H	H	H	Br	H	H	H	H	7.72
29	H	Br	H	H	H	H	H	H	H	H	7.15
30	H	Br	H	H	H	H	NO$_2$	H	H	H	6.70

continues on following page

Table 1. continued

ID	I	II	III	IV	V	VI	VII	VIII	IX	X	–LogKi
31	H	NO_2	H	H	H	NO_2	H	H	H	H	7.92
32	H	Br	H	H	H	NO_2	H	H	H	H	9.00
33	OH	Br	OH	H	H	H	H	H	Br	H	6.15
34	OH	H	OH	H	H	H	H	H	H	H	5.52
35	OH	H	OH	H	H	H	OH	H	H	H	5.52
36	OH	H	OH	H	Cl	H	H	H	H	H	5.10
37	OH	H	OH	H	F	H	H	H	H	H	5.10
38	OH	OCH_3	OH	H	H	H	OH	H	H	H	6.00
39	$C_6H_9O_7$	OH	OH	H	H	H	H	H	H	H	4.11
40	OH	OH	OH	H	H	H	H	H	H	H	5.25
41	OH	OH	OH	H	H	H	OH	H	H	H	4.92
42	OH	H	OH	H	H	H	H	H	OCH_3	H	5.69

topological distance, whereas frequency Atom Pairs count the number of occurrences. Atom-types were defined on the basis of the chemical element and topological distances range from 1 to 10. CATS were calculated for hydrogen-bond donor (D), hydrogen-bond acceptor (A), positively charged or ionisable (P), negatively charged or ionisable (N), and lipophilic (L) atom-types and for topological distances from 0 to 9.

22 types of autocorrelation descriptors were thus considered in this study. Constant and near-constant variables were omitted.

To quantitatively measure the extent to which the chemical spaces defined by the considered descriptors differ, the Canonical Measure of Distance (CMD) and Canonical Measure of Correlation (CMC) were calculated for each pair of autocorrelation types. While CMD measures the dissimilarity between two chemical spaces, CMC measures their similarity. The complement to 1 of CMC, that is CMC* = 1 – CMC, was also calculated because, unlike CMD, this is a dissimilarity measure independent of the number of descriptors defining the chemical space. Then, the multidimensional scaling was applied on the final square (22 × 22) dissimilarity matrix comprised of the CMC* values calculated for all the possible pairs of autocorrelation types.

The MultiDimensional Scaling (MDS) is a multivariate analysis technique that enables the identification of a subspace of the original *p*-dimensional space into which the points can be projected and in which the inter-object dissimilarities are approximated as well as possible by the corresponding inter-point distances (Krzanowski, 1988). The final result of such a technique is a geometrical model of the objects in analysis, which allows a visual investigation of the relationships between the objects.

The MDS scatter plot of the first two dimensions is shown in Figure 3. Each vector of autocorrelations is represented by a point; near points indicate similar types of descriptors.

The first consideration is that the two weighting schemes for atoms (i.e., atomic mass *m* and electronegativity *e*) provide different autocorrelation descriptors, which are located in two distinct regions of the MDS space. Moreover, frequency Atom Pairs, as expected, are clearly separated from all the other types of descriptors, meaning that the chemical space they define is quite different from the others. Also CATS descriptors are not much correlated with all the other vectors of autocorrelations, having the maximal correlation with frequency Atom Pairs (fAP). The Moreau-Broto autocorrelation vector based on atomic

Figure 3. Multidimensional scaling plot of the first two dimensions calculated by using the dissimilarity matrix of 22 different vectors of autocorrelation descriptors for the flavonoid data set

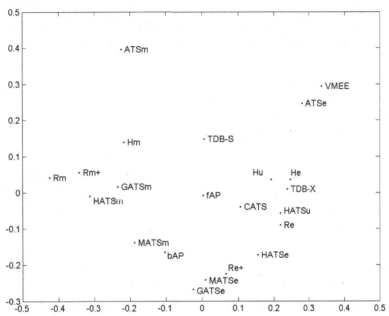

masses (ATSm) is on average not much correlated with others, the only relevant correlations (i.e., CMC > 0.7) being those with GETAWAY indices Hm and Rm. Moran and Geary coefficients are quite correlated with each other, whereas they are not much correlated with Moreau-Broto autocorrelations. Binary Atom Pairs (bAP) have a large correlation (i.e., CMC > 0.9) with frequency Atom Pairs (fAP) and a relevant correlation also with all the GETAWAY descriptors. The molecular electronegativity edge vector (VMEE) based on Pauling atomic electronegativity is, as expected, largely correlated with Moreau-Broto autocorrelations ATSe based on Sanderson atomic electronegativity and a similar equation. TDB-steric (TDB-S) and TDB-electronic (TDB-X) descriptors are significantly correlated with each other and also with HATSm and HATSe, respectively. This result is not surprising considering the fact that both TDB and HATS descriptors account for geometrical information of a molecular structure and atomic steric properties (i.e., mass or covalent radius) in one case and electronic properties (i.e., atomic electronegativity) in the latter.

Large correlation is observed among GETAWAY descriptors; in particular, H indices derived from unweighted molecular structures (Hu) have a large correlation (i.e., CMC > 0.9) with H indices based on atomic electronegativity (He), and a nearly large correlation (i.e., CMC > 0.7) with HATSu, HATSe, Re, Rm+, bAP, and CATS. Furthermore, indices HATSm, Rm, and Rm+ define very similar chemical spaces. Finally, maximal R indices (Rm+ and Re+) seem to reproduce most of the information encoded by HATS indices (HATSm and HATSe), meaning that information given by geometrical inter-atomic distances is not here necessary for describing relationships among the congeneric and almost planar flavonoids.

Diversity between chemical spaces was further investigated by the aid of Principal Component Analysis (PCA). Principal Component Analysis is a multivariate statistical technique widely used for approximating relevant information in data matrices (Jolliffe, 1986; Krzanowski, 1988); it is the most common method for explorative data analysis, which aims at reducing variables and, at the same time, preserve the multidimensional correlation structure of the data. The score plots of the first two principal components (PCs) allowed visual inspection of the chemical spaces defined by the different vectors of autocorrelations. Figures 4 and 5 show the first two PC loading and score plots of ATSm, GATSe, HATSu, and CATS descriptors, which were selected as the most diverse in the previous analysis.

Moreau-Broto autocorrelations ATSm suffer from a large internal correlation as it can be seen in the loading plot of the first two PCs (Figure 4a); all the ATS calculated at different lags explain the same information and this is a consequence of the fact that they are based on non-centred atomic weightings. The first two PCs explain almost 93% of the total data variability, thus the score plot of Figure 4b is a good approximation of the real chemical space, where almost all compounds are similar to each other whereas three

outliers are clearly separated: the unsubstituted parent structure (i.e., 1), compound 33 that has two bromine and two hydroxyl substituents, and compound 39 that has a very bulk substituent (i.e., 5-carboxyglucosyloxy).

Geary coefficients (GATSe) calculated at increasing lags represent different sources of information as it can be seen in the loading plot of Figure 4c. Also in this case the score plot of the first two PCs (Figure 4d) can be considered a good approximation of the real chemical space, explaining almost 71% of the total variance. Relationships among flavonoids are completely different from those in the ATS chemical space, since here there is a well separated cluster comprised of compounds 5, 6, 9, 10, 30, 31, and 32, which are all those flavonoids with at least one nitro group. Moreover, other two small clusters are present in the right: one is comprised of compounds 1, 29, 25, and 28 that have hydrogen or bromine substituents and the other one is comprised of compounds 21, 24, and 27 that also have chlorine.

Figure 4. Loading (a, c) and score (b, d) plots of the first two PCs for the flavonoid data set, calculated by using in (a) and (b) Moreau-Broto autocorrelations based on atomic masses (ATSm), and in (c) and (d) Geary coefficients based on atomic electronegativities (GATSe)

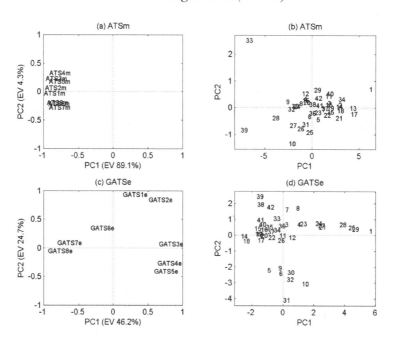

Figure 5. Loading (a, c) and score (b, d) plots of the first two PCs for the flavonoid data set, calculated by using in (a) and (b) HATS indices from unweighted molecular structures, and in (c) and (d) CATS descriptors

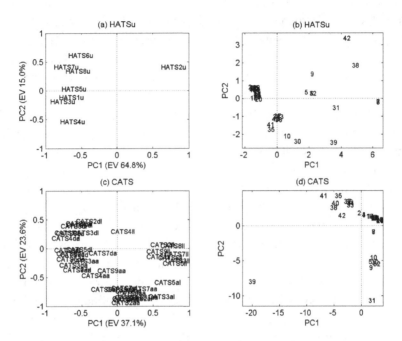

HATSu indices are based only on geometrical structural information. Most of them are largely correlated with each other, the only relevant different being HATS2u (Figure 5a). The score plot of Figure 5b shows that there are a number of compounds in the right, which are those characterized by bulky substituents such as $-NO_2$ and $-OCH_3$, while in the lower left there is a compact cluster comprised of compounds 33–37, 40, and 41, which have more than one hydroxyl group.

The first two PCs of CATS descriptors explain about 61% of the total information; the first PC is mainly dedicated to explain the difference between lipophilic substituents and the other ones, while the second PC explains the difference between hydrogen-bond donors and hydrogen-bond acceptors (Figure 5c). The score plot of Figure 5d shows an apparent outlier that is compound 39 with a large number of oxygen atoms. Furthermore, compound 31, having two nitro groups, results isolated in the lower right. Three well defined clusters appear in the upper right: one cluster includes compounds 5, 6, 9, 10, 30, and 32, which have one nitro group; the cluster in the upper right is comprised of compounds 33–38, 40, 41, and 42, which have a number of hydroxyl groups; the last cluster includes the remaining compounds mainly having halogens as the substituents.

In the second part of this study the modelling ability of autocorrelation descriptors was investigated toward the BzR site binding affinity of the considered flavonoids. The first 38 flavonoids were used for model training, whereas the last four compounds for model validation as they had been used as test set in previous studies (Huang, Song, & Wang, 1997; Hong & Hopfinger, 2003). The following molecular descriptors were calculated: Moreau-Broto autocorrelations (ATS), Geary (GATS) and Moran (MATS) coefficients, GETAWAY indices (H, HATS, R, and R+), binary (bAP) and frequency (fAP) Atom Pairs, CATS descriptors, molecular electronegativity edge vector (VMEE), and TDB descriptors. These descriptors were calculated both from the unweighted molecular structures (*u*) and by using different atomic weightings: masses (*m*), electronegativity (*e*), polarizability (*p*), and van der Waals volume (*v*). Moreover, for each series of autocorrelations, the descriptor for lag zero, which is the sum of the squared atomic properties, was also calculated.

All regression models were calculated by using genetic algorithms for feature selection implemented in MobyDigs 1.1 (2003); this is a software application that extends the genetic algorithm strategy based on the evolution of a single population of models to a more complex genetic strategy based on the evolution of more than one population. Model populations evolve independently of one another and, after a number of iterations, they can be merged according to different criteria, thus generating a new population with different evolutionary capabilities.

The preliminary search for the best models based on populations comprised of the single series of autocorrelations did not provide acceptable results and this is due to the fact that autocorrelation descriptors mainly encode information on local features of molecules and thus are not able to model global molecular properties alone. Consequently, we decided to add to each series also some molecular properties such as logP, molar refractivity, degree of unsaturation, molecular weight, and total counts of atom- and bond-types. Models containing 1 to 4 descriptors were searched for by maximizing the leave-one-out cross-validated determination coefficient (i.e. Q^2).

Once the best models had been selected, bootstrapping (5000 iterations) and Y-scrambling test (300 response randomization) were applied for further validation. *Bootstrap validation* (Efron, 1982; Efron, 1987; Cramer III, Bunce, Patterson, & Frank, 1988) is a technique by which the training set is repeatedly – thousands of times – built by including the same initial number (*n*) of objects, but selected from the raw data with replacement. Objects not selected for model training are instead used as the test set for model

validation. This procedure allows building models from equal-sized training sets and checking the model predictive ability by non-equal size test sets. The *Y-scrambling* or *permutation test* (Lindgren et al., 1996; Topliss & Edwards, 1979) is a statistical tool able to detect the presence of chance correlation in a model and/or the lacking of model robustness; it consists of repeating the model calculation procedure with randomized responses and subsequent probability assessment of the resultant statistics (i.e. R^2 and Q^2). Only models passing Y-scrambling test were retained.

To check multicollinearity of model variables, which can lead to errors in regression coefficients, the *multivariate K correlation index* (*Kx*) was calculated for each regression. This index represents the measure of the overall correlation among the independent variables of a model (Todeschini, 1997); it is defined in terms of the distribution of the eigenvalues obtained by diagonalization of the data set correlation matrix. The *K* correlation index is a measure of redundancy, taking the value of 1 when all of the variables are correlated and 0 when they are orthogonal; only models with *Kx* smaller than 0.5 were retained.

The final selected models for each type of autocorrelation descriptors are reported in the following. For all the models, the training set objects are 38 (*n* = 38) and the objects left out for external validation are 4 (n_{ext} = 4). R^2 indicates the model determination coefficient measuring the model fitting ability, Q^2_{LOO} and Q^2_{BOOT} the squared explained variance determined by leave-one-out cross-validation and bootstrapping, respectively; *Kx* is the overall correlation of the model descriptors.

Model 1

$$-\log(\text{Ki}) = 4.5635 + 0.1369 \cdot \text{ALOGP2} + 5.9338 \cdot \text{MATS1p}$$
$$R^2 = 0.70 \qquad Q^2_{LOO} = 0.63 \qquad Q^2_{BOOT} = 0.56 \qquad Kx = 0.05$$

$$(37)$$

Model 2

$$-\log(\text{Ki}) = -31.1986 + 0.1471 \cdot \text{ALOGP2} + 8.4643 \cdot \text{Ui} + 2.5097 \cdot \text{GATS2e}$$
$$R^2 = 0.71 \qquad Q^2_{LOO} = 0.63 \qquad Q^2_{BOOT} = 0.57 \qquad Kx = 0.44$$
$$(38)$$

Model 3

$$-\log(\text{Ki}) = -20.1999 + 0.1365 \cdot \text{ALOGP2} + 6.2708 \cdot \text{Ui} + 3.3172 \cdot \text{H8m}$$
$$R^2 = 0.74 \qquad Q^2_{LOO} = 0.66 \qquad Q^2_{BOOT} = 0.59 \qquad Kx = 0.22$$
$$(39)$$

Model 4

$$-\log(\text{Ki}) = 1.3895 + 2.1358 \cdot \text{HATS0e} + 20.0245 \cdot \text{HATS2p}$$
$$R^2 = 0.70 \qquad Q^2_{LOO} = 0.61 \qquad Q^2_{BOOT} = 0.56 \qquad Kx = 0.11$$
$$(40)$$

Model 5

$$-\log(\text{Ki}) = -25.9994 + 0.1590 \cdot \text{ALOGP2} + 7.2166 \cdot \text{Ui} + 2.6730 \cdot \text{R6e}$$
$$R^2 = 0.72 \qquad Q^2_{LOO} = 0.63 \qquad Q^2_{BOOT} = 0.53 \qquad Kx = 0.15$$
$$(41)$$

Model 6

$$-\log(\text{Ki}) = 4.3440 + 38.7628 \cdot \text{R2v}^+$$
$$R^2 = 0.38 \qquad Q^2_{LOO} = 0.32 \qquad Q^2_{BOOT} = 0.27 \qquad Kx = 0$$
$$(42)$$

Model 7

$$-\log(\text{Ki}) = 4.2127 + 0.1524 \cdot \text{ALOGP2} + 1.7919 \cdot \text{B05[N-O]}$$
$$R^2 = 0.75 \qquad Q^2_{LOO} = 0.68 \qquad Q^2_{BOOT} = 0.63 \qquad Kx = 0.15$$
$$(43)$$

Model 8

$$-\log(\text{Ki}) = 4.1457 + 0.1558 \cdot \text{ALOGP2} + 0.6487 \cdot \text{F04[C-N]}$$
$$R^2 = 0.72 \qquad Q^2_{LOO} = 0.65 \qquad Q^2_{BOOT} = 0.59 \qquad Kx = 0.18$$
$$(44)$$

Model 9

$$-\log(\text{Ki}) = 4.5004 - 0.6937 \cdot \text{CATS5\_DL} + 0.4006 \cdot \text{CATS5\_AL}$$
$$R^2 = 0.71 \qquad Q^2_{LOO} = 0.66 \qquad Q^2_{BOOT} = 0.63 \qquad Kx = 0.04$$
$$(45)$$

Model 10

$$-\log(\text{Ki}) = -4.0522 + 39.0066 \cdot \text{TDB-S6}$$
$$R^2 = 0.46 \qquad Q^2_{LOO} = 0.40 \qquad Q^2_{BOOT} = 0.35 \qquad Kx = 0$$
$$(46)$$

The molecular descriptors selected in the best models are: ALOGP2, which is the squared logP calculated by the atom contribution method of Ghose and Crippen (1998); Moran coefficient al lag 1 based on atomic polarizability (MATS1p); an unsaturation index accounting for multiple bonds (Ui); Geary coefficient at lag 2 based on atomic electronegativity (GATS2e); GETAWAY index H at lag 8 based on atomic mass (H8m); GETAWAY indices HATS at lag 0 based on atomic electronegativity (HATS0e) and at lag 2 based on atomic polarizability (HATS2p); GETAWAY index R at lag 6 based on atomic electronegativity (R6e); GETAWAY maximal index R at lag 2 based on atomic volume (R2v$^+$); presence of atom-type pair N–O at topological distance 5 (B05[N-O]); number of occurrences of atom-pair C–N at topological distance 4 (F04[C-N]); number of occurrences of hydrogen-bond donor – lipophilic pair at topological distance 5 (CATS5_DL); number of occurrences of hydrogen-bond acceptor – lipophilic pair at topological distance 5 (CATS5_AL); TDB-steric descriptor at lag 6 (TDB-S6).

Note that no acceptable models were found by using Moreau-Broto autocorrelations (ATS) and molecular electronegativity edge vector (VMEE); all the models based on these descriptors had problems of chance correlation and low predictive ability. Moreover, also TDB descriptors showed both scarce fitting and predictive ability as one can see from model 10 (Equation 46).

The general finding is that autocorrelation descriptors alone are not very effective in modelling BzR site binding affinities of flavonoids and also with the help of other global molecular descriptors they do not provide very satisfactory model statistics; however, their models are quite general to be able to well predict the binding affinity of the four external test compounds. Observed and calculated binding affinities for test compounds by all the ten selected models are listed in Table 2.

It is noteworthy that ALOGP2, that is, the squared Ghose-Crippen logP, and the unsaturation index Ui are additional descriptors present in several models. The model obtained using ALOGP2 alone gives the following statistics:

$$R^2 = 0.31 \qquad Q^2_{LOO} = 0.25 \qquad Q^2_{BOOT} = 0.19 \qquad Kx = 0$$

and the model obtained by using both the descriptors ALOGP2 and Ui the following:

$$R^2 = 0.69 \qquad Q^2_{LOO} = 0.63 \qquad Q^2_{BOOT} = 0.57 \qquad Kx = 0.16$$

Therefore, it can be concluded that for the studied data set addition of autocorrelation descriptors in models 2, 3, and 5 actually does not improve model quality.

FUTURE RESEARCH DIRECTIONS

The scientific community is showing more and more increasing interest in the QSAR field. Several chemoinformatics methods were specifically conceived trying to solve QSAR problems, answering the demand to know in deeper way chemical systems and their relationships with biological systems.

Several questions are still open and matter of debate, such as the problem of the validation strategies to obtain predictive models, the interpretability of complex molecular descriptors, the introduction of new modelling tools, the evaluation of the applicability domain of a model for new compounds, the assessment of the validity of the congenericity principle for the data under study, etc.

Moreover, the need to deal with biological systems described by peptide/protein or DNA sequences, to describe proteomic maps, or to give effective answers to ecological and wealth problems, further pushes toward new emerging scientific fields where mathematics, statistics, chemistry, and biology and their inter-relationships may produce new effective useful knowledge.

Finally, it is nowadays observed new trends in molecular descriptor generation, which lead to an ever-growing number of molecular descriptors proposed in the literature. Not all the proposed descriptors really explain new chemical information and are useful for QSAR and drug design; much effort should be addressed to investigate information encoded by these descriptors and their relationships with the more traditional ones.

Table 2. Binding affinities to the BzR site of the four test compounds calculated by means of the ten selected models (Equations 37–46)

						Models					
		1	**2**	**3**	**4**	**5**	**6**	**7**	**8**	**9**	**10**
Comp.	*Obs.*	**MATS**	**GATS**	**H**	**HATS**	**R**	**R+**	**bAP**	**fAP**	**CATS**	**TDB**
39	**4.11**	4.11	5.21	4.95	4.06	5.46	5.35	4.27	4.21	4.31	3.87
40	**5.25**	5.13	5.06	5.05	5.23	5.03	6.01	5.04	4.99	5.01	5.78
41	**4.92**	4.99	4.91	4.92	5.07	5.08	5.58	4.86	4.81	4.61	5.58
42	**5.69**	4.96	5.25	5.27	5.71	5.71	5.43	5.23	5.19	6.50	6.28
	rmse*	0.372	0.600	0.480	0.080	0.689	0.809	0.267	0.291	0.461	0.530

*root mean square error.

CONCLUSION

Autocorrelation descriptors constitute a class of molecular descriptors able to give information on logically organized fragments of the molecules. Several new kinds of molecular autocorrelation descriptors have been proposed in the last few years and, due to their usefulness in similarity/diversity analysis, in the screening of large chemical libraries and in QSAR modelling, the increasing interest for their properties and variants is apparent from the literature.

Techniques of multivariate data analysis were applied to investigate information encoded by different sets of molecular autocorrelation descriptors and get a deeper insight about similarity/diversity relationships among the chemical spaces they define. The most important finding is that Moreau-Broto autocorrelations that are the first ones proposed suffer from the drawback of being largely internally correlated thus resulting not very suitable for statistical analysis.

In general, the chemical spaces defined by autocorrelation descriptors are easily interpretable as these molecular descriptors encode local information such as substituent types, thus providing characterization of molecules evident to chemists.

The effectiveness of molecular autocorrelation descriptors in QSAR modelling was tested by regressing each single set of autocorrelations against the benzodiazepine receptor site binding affinity of some flavonoids. As they resulted not able to provide alone statistically significant regressions, some global molecular descriptors, such logP and the molecule degree of unsutaration, were introduced in models.

As it usually happens for indices able to only describe local features of molecules, searching for predictive QSARs often requires a combination of local descriptors, like autocorrelation descriptors, and global molecular descriptors, like, for example, molecular properties, topological and geometrical indices.

The last consideration is that the molecular electronegativity edge vector (VMEE), which was recently proposed, actually explains almost the same information as the Moreau-Broto autocorrelation vector and like this is not very effective in QSAR modelling.

REFERENCES

Adamson, G. W., Lynch, M. F., & Town, W. G. (1971). Analysis of Structural Characteristics of Chemical Compounds in a Large Computer-based File. Part II. Atom-Centred Fragments. *Journal of the Chemical Society*, (C), 3702–3706.

Akaike, H. (1974). A New Look at the Statistical Model Identification. *IEEE Transactions on Automatic Control*, *AC-19*, 716–723. doi:10.1109/TAC.1974.1100705

Balaban, A. T., Motoc, I., Bonchev, D., & Mekenyan, O. (1983). Topological Indices for Structure-Activity Correlations. In M.Charton & I. Motoc (Eds.), *Steric Effects in Drug Design (Topics in Current Chemistry, Vol. 114)* (pp. 21-55). Berlin, Germany: Springer-Verlag.

Basak, S. C., Gute, B. D., & Grunwald, G. D. (1997). Use of Topostructural, Topochemical, and Geometric Parameters in the Prediction of Vapor Pressure: A Hierarchical QSAR Approach. *Journal of Chemical Information and Computer Sciences*, *37*, 651–655. doi:10.1021/ci960176d

Basak, S. C., Magnuson, V. R., & Veith, G. D. (1987). Topological Indices: Their Nature, Mutual Relatedness, and Applications. In X.J.R.Avula, G. Leitmann, C. D. Jr. Mote, & E. Y. Rodin (Eds.), *Mathematical Modelling in Science and Technology* (pp. 300-305). Oxford, UK: Pergamon Press.

Baumann, K. (2002). An Alignment-Independent Versatile Structure Descriptor for QSAR and QSPR Based on the Distribution of Molecular Features. *Journal of Chemical Information and Computer Sciences, 42*, 26–35. doi:10.1021/ci990070t

Baumann, K., Albert, H., & von Korff, M. (2002). A systematic evaluation of the benefits and hazards of variable selection in latent variable regression. Part I. Search algorithm, theory and simulations. *Journal of Chemometrics, 16*, 339–350. doi:10.1002/cem.730

Bonchev, D., & Trinajstic, N. (1977). Information Theory, Distance Matrix, and Molecular Branching. *The Journal of Chemical Physics, 67*, 4517–4533. doi:10.1063/1.434593

Broto, P., & Devillers, J. (1990). Autocorrelation of Properties Distributed on Molecular Graphs . In Karcher, W., & Devillers, J. (Eds.), *Practical Applications of Quantitative Structure-Activity Relationships (QSAR) in Environmental Chemistry and Toxicology* (pp. 105–127). Dordrecht, The Netherlands: Kluwer.

Broto, P., Moreau, G., & Vandycke, C. (1984a). Molecular Structures: Perception, Autocorrelation Descriptor and SAR Studies. Autocorrelation Descriptor. *European Journal of Medicinal Chemistry, 19*, 66–70.

Broto, P., Moreau, G., & Vandycke, C. (1984b). Molecular Structures: Perception, Autocorrelation Descriptor and SAR Studies. Use of the Autocorrelation Descriptors in the QSAR Study of Two Non-Narcotic Analgesic Series. *European Journal of Medicinal Chemistry, 19*, 79–84.

Brown, F. K. (1998). Chemoinformatics: what is it and how does it impact drug discovery. *Annual Reports in Medicinal Chemistry, 33*, 375–384. doi:10.1016/S0065-7743(08)61100-8

Carhart, R. E., Smith, D. H., & Venkataraghavan, R. (1985). Atom Pairs as Molecular Features in Structure-Activity Studies: Definition and Applications. *Journal of Chemical Information and Computer Sciences, 25*, 64–73. doi:10.1021/ci00046a002

Centner, V., Massart, D. L., de Noord, O. E., De Jong, S., Vandeginste, B. G. M., & Sterna, C. (1996). Elimination of Uniformative Variables for Multivariate Calibration. *Analytical Chemistry, 68*, 3851–3858. doi:10.1021/ac960321m

Charton, M., & Charton, B. I. (2002). *Advances in Quantitative Structure-Property Relationships*. Amsterdam, The Netherlands: JAI Press.

Clementi, S., Cruciani, G., Riganelli, D., Valigi, R., Costantino, G., & Baroni, M. (1993). Autocorrelation as a Tool for a Congruent Description of Molecules in 3D QSAR Studies. *Pharmaceutical and Pharmacological Letters, 3*, 5–8.

Clerc, J. T., & Terkovics, A. L. (1990). Versatile Topological Structure Descriptor for Quantitative Structure/Property Studies. *Analytica Chimica Acta, 235*, 93–102. doi:10.1016/S0003-2670(00)82065-6

Consonni, V., Todeschini, R., & Pavan, M. (2002). Structure/Response Correlations and Similarity/Diversity Analysis by GETAWAY Descriptors. Part 1. Theory of the Novel 3D Molecular Descriptors. *Journal of Chemical Information and Computer Sciences, 42*, 682–692. doi:10.1021/ci015504a

Cramer, R. D. III, Bunce, J. D., Patterson, D. E., & Frank, I. E. (1988). Crossvalidation, Bootstrapping and Partial Least Squares Compared with Multiple Regression in Conventional QSAR Studies. *Quantitative Structure-Activity Relationships, 7*, 18–25. doi:10.1002/qsar.19880070105

Cramer, R. D. III, Patterson, D. E., & Bunce, J. D. (1988). Comparative Molecular Field Analysis (CoMFA). 1. Effect of Shape on Binding of Steroids to Carrier Proteins. *Journal of the American Chemical Society, 110*, 5959–5967. doi:10.1021/ja00226a005

Crowe, J. E., Lynch, M. F., & Town, W. G. (1970). Analysis of Structural Characteristics of Chemical Compounds in a Large Computer-based File. Part 1. Non-cyclic Fragments. *Journal of the Chemical Society*, (C), 990–997.

Crum-Brown, A. (1867). On an application of mathematics to chemistry. *Proceedings of the Royal Society of Edinburgh, VI*(73), 89–90.

Crum-Brown, A., & Fraser, T. R. (1868). On the connection between chemical constitution and physiological action. Part 1. On the physiological action of salts of the ammonium bases, derived from strychnia, brucia, thebia, codeia, morphia and nicotia. *Transactions of the Royal Society of Edinburgh, 25*, 151–203.

Devillers, J. (Ed.). (1998). *Comparative QSAR*. Washington, DC: Taylor & Francis.

Diudea, M. V., & Gutman, I. (1998). Wiener-Type Topological Indices. *Croatica Chemica Acta, 71*, 21–51.

DRAGON 5.5 (2007). Milano, Italy: Talete s.r.l.

Dutta, D., Guha, R., Jurs, P. C., & Chen, T. (2006). Scalable Partitioning and Exploration of Chemical Spaces Using Geometric Hashing. *Journal of Chemical Information and Modeling, 46*, 321–333. doi:10.1021/ci050403o

Eckert, H., Vogt, I., & Bajorath, J. (2006). Mapping Algorithms for Molecular Similarity Analysis and Ligand-Based Virtual Screening: Design of DynaMAD and Comparison with MAD and DMC. *Journal of Chemical Information and Modeling, 46*, 1623–1634. doi:10.1021/ci060083o

Efron, B. (1982). *The Jackknife, the Bootstrap and Other Resampling Planes*. Philadelphia, PA: Society for Industrial and Applied Mathematics.

Efron, B. (1987). Better Bootstrap Confidence Intervals. *Journal of the American Statistical Association, 82*, 171–200. doi:10.2307/2289144

Efroymson, M. A. (1960). Multiple Regression Analysis . In Ralston, A., & Wilf, H. S. (Eds.), *Mathematical Methods for Digital Computers*. New York: Wiley.

Estrada, E. (2001). Generalization of topological indices. *Chemical Physics Letters, 336*, 248–252. doi:10.1016/S0009-2614(01)00127-0

Estrada, E., & Matamala, A. R. (2007). Generalized Topological Indices. Modeling Gas-Phase Rate Coefficients of Atmospheric Relevance. *Journal of Chemical Information and Modeling, 47*, 794–804. doi:10.1021/ci600448b

Fechner, U., Franke, L., Renner, S., Schneider, P., & Schneider, G. (2003). Comparison of correlation vector methods for ligand-based similarity searching. *Journal of Computer-Aided Molecular Design, 17*, 687–698. doi:10.1023/B:JCAM.0000017375.61558.ad

Free, S. M., & Wilson, J. W. (1964). A Mathematical Contribution to Structure-Activity Studies. *Journal of Medicinal Chemistry, 7*, 395–399. doi:10.1021/jm00334a001

Friedman, J. H. (1988). *Multivariate Adaptive Regression Splines* (Tech. Rep. No. 102). Stanford, CA: Laboratory of Computational Statistics - Dept. of Statistics.

Fujita, T., Iwasa, J., & Hansch, C. (1964). A New Substituent Constant, p, Derived from Partition Coefficients. *Journal of the American Chemical Society, 86*, 5175–5180. doi:10.1021/ja01077a028

Gasteiger, J. (Ed.). (2003). *Handbook of Chemoinformatics. From Data to Knowledge in 4 Volumes*. Weinheim, Germany: Wiley-VCH.

Gasteiger, J., & Marsili, M. (1980). Iterative Partial Equalization of Orbital Electronegativity: A Rapid Access to Atomic Charges. *Tetrahedron, 36*, 3219–3228. doi:10.1016/0040-4020(80)80168-2

Geary, R. C. (1954). The Contiguity Ratio and Statistical Mapping. *Incorp. Statist., 5*, 115–145. doi:10.2307/2986645

Ghose, A. K., Viswanadhan, V. N., & Wendoloski, J. J. (1998). Prediction of Hydrophobic (Lipophilic) Properties of Small Organic Molecules Using Fragmental Methods: An Analysis of ALOGP and CLOGP Methods. *The Journal of Physical Chemistry A, 102*, 3762–3772. doi:10.1021/jp980230o

Goldberg, D. E. (1989). *Genetic Algorithms in Search, Optimization and Machine Learning.* Massachusetts, MA: Addison-Wesley.

Goodford, P. J. (1985). A Computational Procedure for Determining Energetically Favorable Binding Sites on Biologically Important Macromolecules. *Journal of Medicinal Chemistry, 28*, 849–857. doi:10.1021/jm00145a002

Hammett, L. P. (1935). Reaction Rates and Indicator Acidities. *Chemical Reviews, 17*, 67–79. doi:10.1021/cr60053a006

Hammett, L. P. (1937). The Effect of Structure upon the Reactions of Organic Compounds. Benzene Derivatives. *Journal of the American Chemical Society, 59*, 96–103. doi:10.1021/ja01280a022

Hammett, L. P. (1938). Linear Free Energy Relationships in Rate and Equilibrium Phenomena. *Transactions of the Faraday Society, 34*, 156–165. doi:10.1039/tf9383400156

Hansch, C., & Leo, A. (1995). *Exploring QSAR. Fundamentals and Applications in Chemistry and Biology.* Washington, DC: American Chemical Society.

Hansch, C., Maloney, P. P., Fujita, T., & Muir, R. M. (1962). Correlation of Biological Activity of Phenoxyacetic Acids with Hammett Substituent Constants and Partition Coefficients. *Nature, 194*, 178–180. doi:10.1038/194178b0

Hollas, B. (2002). Correlation Properties of the Autocorrelation Descriptor for Molecules. *MATCH Communications in Mathematical and in Computer Chemistry, 45*, 27–33.

Hong, X., & Hopfinger, A. J. (2003). 3D-Pharmacophores of Flavonoid Binding at the Benzodiazepine GABAA Receptor Site Using 4D-QSAR Analysis. *Journal of Chemical Information and Computer Sciences, 43*, 324–336. doi:10.1021/ci0200321

Hou, T.-J., Xia, K., Zhang, W., & Xu, X. (2004). ADME Evaluation in Drug Discovery. 4. Prediction of Aqueous Solubility Based on Atom Contribution Approach. *Journal of Chemical Information and Computer Sciences, 44*, 266–275. doi:10.1021/ci034184n

Huang, Q.-G., Song, W.-L., & Wang, L.-S. (1997). Quantitative Relationship Between the Physiochemical Characteristics as well as Genotoxicity of Organic Pollutants and Molecular Autocorrelation Topological Descriptors. *Chemosphere, 35*, 2849–2855. doi:10.1016/S0045-6535(97)00345-7

Huang, X., Liu, T., Gu, J., Luo, X., Ji, R., & Cao, Y. (2001). 3D-QSAR models of flavonoids binding at benzodiazepine site in GABA$_A$ receptors. *Journal of Medicinal Chemistry, 44*, 1883–1891. doi:10.1021/jm000557p

HyperChem 8.0 (2007). Hypercube, Inc.

Ivanciuc, O. (2001). 3D QSAR Models. In Diudea, M. V. (Ed.), *QSPR / QSAR Studies by Molecular Descriptors* (pp. 233–280). Huntington, NY: Nova Science.

Ivanciuc, O., & Balaban, A. T. (1999). The Graph Description of Chemical Structures. In Devillers, J., & Balaban, A. T. (Eds.), *Topological Indices and Related Descriptors in QSAR and QSPR* (pp. 59–167). Amsterdam, The Netherlands: Gordon & Breach Science Publishers.

Jolliffe, I. T. (1986). *Principal Component Analysis*. New York: Springer-Verlag.

Kier, L. B. (1971). *Molecular Orbital Theory in Drug Research*. New York: Academic Press.

Kier, L. B., & Hall, L. H. (1976). *Molecular Connectivity in Chemistry and Drug Research*. New York: Academic Press.

Kier, L. B., & Hall, L. H. (1990). An Electrotopological-State Index for Atoms in Molecules. *Pharmaceutical Research*, 7, 801–807. doi:10.1023/A:1015952613760

Klein, Ch. Th., Kaiser, D., & Ecker, G. (2004). Topological Distance Based 3D Descriptors for Use in QSAR and Diversity Analysis. *Journal of Chemical Information and Computer Sciences*, 44, 200–209. doi:10.1021/ci0256236

Krzanowski, W. J. (1988). *Principles of Multivariate Analysis*. New York: Oxford Univ. Press.

Kubinyi, H. (Ed.). (1993). *3D QSAR in Drug Design. Theory, Methods, and Applications*. Leiden, The Netherlands: ESCOM.

Kubinyi, H. (1994). Variable Selection in QSAR Studies. I. An Evolutionary Algorithm. *Quantitative Structure-Activity Relationships*, 13, 285–294.

Kubinyi, H. (1996). Evolutionary Variable Selection in Regression and PLS Analyses. *Journal of Chemometrics*, 10, 119–133. doi:10.1002/(SICI)1099-128X(199603)10:2<119::AID-CEM409>3.0.CO;2-4

Kubinyi, H., Folkers, G., & Martin, Y. C. (Eds.). (1998). *3D QSAR in Drug Design* (*Vol. 3*). Dordrecht, The Netherlands: Kluwer/ESCOM.

Landon, M. R., & Schaus, S. E. (2006). JEDA: Joint entropy diversity analysis. An information-theoretic method for choosing diverse and representative subsets from combinatorial libraries. *Molecular Diversity*, 10, 333–339. doi:10.1007/s11030-006-9042-4

Leardi, R. (1994). Application of Genetic Algorithms to Feature Selection Under Full Validation Conditions and to Outlier Detection. *Journal of Chemometrics*, 8, 65–79. doi:10.1002/cem.1180080107

Leardi, R., Boggia, R., & Terrile, M. (1992). Genetic Algorithms as a Strategy for Feature Selection. *Journal of Chemometrics*, 6, 267–281. doi:10.1002/cem.1180060506

Li, Z., Fu, B., Wang, Y., & Liu, S. (2001). On Structural Parametrization and Molecular Modeling of Peptide Analogues by Molecular Electronegativity Edge Vector (VMEE): Estimation and Prediction for Biological Activity of Dipeptides. *Journal of Chinese Chemical Society*, 48, 937–944.

Lindgren, F., Geladi, P., Rännar, S., & Wold, S. (1994). Interactive Variable Selection (IVS) for PLS. Part I: Theory and Algorithms. *Journal of Chemometrics*, 8, 349–363. doi:10.1002/cem.1180080505

Lindgren, F., Hansen, B., Karcher, W., Sjöström, M., & Eriksson, L. (1996). Model Validation by Permutation Tests: Applications to Variable Selection. *Journal of Chemometrics*, 10, 521–532. doi:10.1002/(SICI)1099-128X(199609)10:5/6<521::AID-CEM448>3.0.CO;2-J

Luke, B. T. (1994). Evolutionary Programming Applied to the Development of Quantitative Structure-Activity Relationships and Quantitative Structure-Property Relationships. *Journal of Chemical Information and Computer Sciences*, 34, 1279–1287. doi:10.1021/ci00022a009

Martin, Y. C. (1979). Advances in the Methodology of Quantitative Drug Design . In Ariëns, E. J. (Ed.), *Drug Design* (*Vol. VIII*, pp. 1–72). New York, NY: Academic Press.

Martin, Y. C. (1998). 3D QSAR: Current State Scope, and Limitations. In H.Kubinyi, G. Folkers, & Y. C. Martin (Eds.), *3D QSAR in Drug Design* (pp. 3-23). Dordrecht, The Netherlands: Kluwer/ESCOM.

Melville, J. L., & Hirts, J. D. (2007). TMACC: Interpretable Correlation Descriptors for Quantitative Structure-Activity Relationships. *Journal of Chemical Information and Modeling, 47,* 626–634. doi:10.1021/ci6004178

MobyDigs 1.0 (2003). Milano, Italy: Talete s.r.l.

Moran, P. A. P. (1950). Notes on Continuous Stochastic Phenomena. *Biometrika, 37,* 17–23.

Moreau, G., & Broto, P. (1980). The Autocorrelation of a Topological Structure: A New Molecular Descriptor. *New Journal of Chemistry, 4,* 359–360.

Moreau, G., & Turpin, C. (1996). Use of similarity analysis to reduce large molecular libraries to smaller sets of representative molecules. *Analusis, 24,* M17–M21.

Oprea, T. I. (2003). Chemoinformatics and the Quest for Leads in Drug Discovery . In Gasteiger, J. (Ed.), *Handbook of Chemoinformatics* (pp. 1509–1531). Weinheim, Germany: Wiley-VCH. doi:10.1002/9783527618279.ch44b

Oprea, T. I. (2004). 3D QSAR modeling in drug design . In Bultinck, P., De Winter, H., Langenaeker, W., & Tollenaere, J. P. (Eds.), *Computational Medicinal Chemistry for Drug Discovery* (pp. 571–616). New York: Marcel Dekker.

Oprea, T. I., Zamora, I., & Ungell, A.-L. (2002). Pharmacokinetically Based Mapping Device for Chemical Space Navigation. *Journal of Combinatorial Chemistry, 4,* 258–266. doi:10.1021/cc010093w

Platt, J. R. (1947). Influence of Neighbor Bonds on Additive Bond Properties in Paraffins. *The Journal of Chemical Physics, 15,* 419–420. doi:10.1063/1.1746554

Randic, M. (1975). On Characterization of Molecular Branching. *Journal of the American Chemical Society, 97,* 6609–6615. doi:10.1021/ja00856a001

Randic, M. (1993). Comparative Regression Analysis. Regressions Based on a Single Descriptor. *Croatica Chemica Acta, 66,* 289–312.

Rouvray, D. H. (1983). Should We Have Designs on Topological Indices? In King, R. B. (Ed.), *Chemical Applications of Topology and Graph Theory. Studies in Physical and Theoretical Chemistry* (pp. 159–177). Amsterdam, The Netherlands: Elsevier.

Sadowski, J., Wagener, M., & Gasteiger, J. (1995). Assessing Similarity and Diversity of Combinatorial Libraries by Spatial Autocorrelation Functions and Neural Networks. *Angewandte Chemie International Edition in English, 34,* 2674–2677. doi:10.1002/anie.199526741

Schneider, G., Neidhart, W., Giller, T., & Schmid, G. (1999). "Scaffold-Hopping" by Topological Pharmacophore Search: A Contribution to Virtual Screening. *Angewandte Chemie International Edition in English, 38,* 2894–2895. doi:10.1002/(SICI)1521-3773(19991004)38:19<2894::AID-ANIE2894>3.0.CO;2-F

Sjöström, M., Rännar, S., & Wieslander, Å. (1995). Polypeptide sequence property relationships in *Escherichia coli* based on auto cross covariances. *Chemometrics and Intelligent Laboratory Systems, 29,* 295–305. doi:10.1016/0169-7439(95)00059-1

Sutter, J. M., Peterson, T. A., & Jurs, P. C. (1997). Prediction of Gas Chromatographic Retention Indices of Alkylbenzene. *Analytica Chimica Acta, 342,* 113–122. doi:10.1016/S0003-2670(96)00578-8

Todeschini, R. (1997). Data Correlation, Number of Significant Principal Components and Shape of Molecules. The K Correlation Index. *Analytica Chimica Acta, 348,* 419–430. doi:10.1016/S0003-2670(97)00290-0

Todeschini, R., Ballabio, D., Consonni, V., Manganaro, A., & Mauri, A.Canonical Measure of Correlation (*CMC*) and Canonical Measure of Distance (*CMD*) between sets of data. Part 1. Theory and simple chemometric applications. *Analytica Chimica Acta, 648,* 45–51. doi:10.1016/j.aca.2009.06.032

Todeschini, R., & Consonni, V. (2000). *Handbook of Molecular Descriptors*. Weinheim, Germany: Wiley-VCH GmbH.

Todeschini, R., & Consonni, V. (2003). Descriptors from Molecular Geometry . In Gasteiger, J. (Ed.), *Handbook of Chemoinformatics* (pp. 1004–1033). Weinheim, Germany: Wiley-VCH. doi:10.1002/9783527618279.ch37

Todeschini, R., & Consonni, V. (2009). *Molecular Descriptors for Chemoinformatics*. Weinheim, Germany: Wiley-VCH GmbH.

Todeschini, R., Consonni, V., Mauri, A., & Pavan, M. (2004). Detecting "bad" regression models: multicriteria fitness functions in regression analysis. *Analytica Chimica Acta, 515,* 199–208. doi:10.1016/j.aca.2003.12.010

Topliss, J. G., & Edwards, R. P. (1979). Chance Factors in Studies of Quantitative Structure-Activity Relationships. *Journal of Medicinal Chemistry, 22,* 1238–1244. doi:10.1021/jm00196a017

Trinajstic, N. (Ed.). (1992). *Chemical Graph Theory*. Boca Raton, FL: CRC Press.

van de Waterbeemd, H., Testa, B., & Folkers, G. (Eds.). (1997). *Computer-Assisted Lead Finding and Optimization*. Weinheim, Germany: Wiley-VCH. doi:10.1002/9783906390406

Varmuza, K., Demuth, W., Karlovits, M., & Scsibrany, H. (2005). Binary Substructure Descriptors for Organic Compounds. *Croatica Chemica Acta, 78,* 141–149.

Wagener, M., Sadowski, J., & Gasteiger, J. (1995). Autocorrelation of Molecular Surface Properties for Modeling *Corticosteroid Binding Globulin* and Cytosolic *Ah* Receptor Activity by Neural Networks. *Journal of the American Chemical Society, 117,* 7769–7775. doi:10.1021/ja00134a023

Wiener, H. (1947). Influence of Interatomic Forces on Paraffin Properties. *The Journal of Chemical Physics, 15,* 766. doi:10.1063/1.1746328

Wildman, S. A., & Crippen, G. M. (1999). Prediction of Physicochemical Parameters by Atomic Contributions. *Journal of Chemical Information and Computer Sciences, 39,* 868–873. doi:10.1021/ci990307l

Wold, S., Jonsson, J., Sjöström, M., Sandberg, M., & Rännar, S. (1993). DNA and peptide sequences and chemical processes multivariately modelled by principal component analysis and partial least-squares projections to latent structures. *Analytica Chimica Acta, 277,* 239–253. doi:10.1016/0003-2670(93)80437-P

Zheng, W., & Tropsha, A. (2000). Novel Variable Selection Quantitative Structure-Property Relationship Approach Based on the *k*-Nearest-Neighbor Principle. *Journal of Chemical Information and Computer Sciences, 40,* 185–194. doi:10.1021/ci980033m

Section 2
Graph–Based Approaches in Chemoinformatics

Chapter 6
Graph Mining in Chemoinformatics

Hiroto Saigo
Kyushu Institute of Technology, Japan

Koji Tsuda
AIST Computational Biology Research Center, Japan

ABSTRACT

In standard QSAR (Quantitative Structure Activity Relationship) approaches, chemical compounds are represented as a set of physicochemical property descriptors, which are then used as numerical features for classification or regression. However, standard descriptors such as structural keys and fingerprints are not comprehensive enough in many cases. Since chemical compounds are naturally represented as attributed graphs, graph mining techniques allow us to create subgraph patterns (i.e., structural motifs) that can be used as additional descriptors. In this chapter, the authors present theoretically motivated QSAR algorithms that can automatically identify informative subgraph patterns. A graph mining subroutine is embedded in the mother algorithm and it is called repeatedly to collect patterns progressively. The authors present three variations that build on support vector machines (SVM), partial least squares regression (PLS) and least angle regression (LARS). In comparison to graph kernels, our methods are more interpretable, thereby allows chemists to identify salient subgraph features to improve the drug-likeliness of lead compounds.

INTRODUCTION

In the first step of drug discovery process, a large number of lead compounds are found by high throughput screening. To identify physicochemical properties of the lead compounds, SAR and QSAR analyses are commonly applied (Gasteiger & Engel, 2003). In machine learning terminology, SAR is understood as a classification task where a chemical compound is given as an input, and the learning machine predicts the value of a binary output variable indicating the activity. In QSAR, the output variable is real-valued and it is a regression task.

For accurate prediction, numerical features that characterize physicochemical properties are

DOI: 10.4018/978-1-61520-911-8.ch006

necessary. A wide range of feature *descriptors* are proposed (Gasteiger & Engel, 2003). A classical example is *structural keys*, where each chemical compound is represented as a binary vector representing the presence or absence of major functional groups such as alcohols and amines. Notice that relevant structural keys differs from problem to problem. For example, structural keys used in pharmaceutics is by far different from the ones in petrochemics. A *fingerprint* is another approach that enumerates all the structures under a given constraint. This approach enumerates all paths up to a certain length in each chemical compound. The paths are used to represent the compounds as a fixed length binary vector. To save the storage, several different patterns have to be assigned to the same bit (*collision*). For this reason, a fingerprint does not always gives us a transparent and interpretable model. Another limitation of the current fingerprinting is the use of path patterns, despite the fact that tree and graph patterns are more informative (Yan, Yu, & Han, 2004).

The use of *frequent subgraphs* as descriptors is studied recently in the data mining community (Wale & Karypis, 2006, Kazius, Nijssen, Kok, Bäck, & Ijzerman, 2006, Helma, Cramer, Kramer, & Raedt, 2004). Frequent subgraph enumeration algorithms such as AGM (Inokuchi, 2005), Gaston (Nijssen & Kok, 2004) and gSpan (Yan & Han, 2002a) can enumerate all the subgraph patterns that appear more than m times in a graph database. The threshold m is called *minimum support*. Frequent subgraph patterns are found by branch-and-bound search in a tree shaped search space (Figure 2). Frequent subgraphs contain good descriptors, but they are often redundant. It is known that, to achieve the best accuracy, the minimum support has to be determined to a small value (e.g., 3-5) (Wale & Karypis, 2006, Kazius et al., 2006, Helma et al., 2004). Such setting creates millions of patterns, which makes subsequent processing difficult. Frequent patterns are not informative, for example, patterns like C-C and C-C-C would be frequent but have almost no information.

Figure 2. Schematic figure of the tree-shaped search space of graph patterns (i.e., the DFS code tree). To find the optimal pattern efficiently, the tree is systematically expanded by rightmost extensions

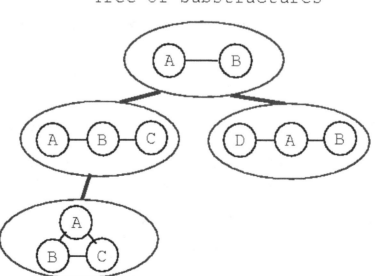

In this chapter, we propose to use *discriminative subgraphs* as descriptors for QSAR problems. We couple frequent graph mining algorithm with learning algorithm to efficiently search for optimal patterns for the data at hand. More precisely, our method is an embedded feature selection method, where mining algorithm is repeatedly called by learning algorithm, and the feature space is expanded progressively (Kohavi & John, 1997). We employ a tree shaped search space similar to that of frequent pattern mining for organizing discriminative subgraph patterns. Instead of frequency counts, however, we store discriminative scores, which we call *gain*, in each node. Based on the gain in each node, we can estimate whether there exists downstream nodes with larger gains than the current node. Our discriminative pattern mining is more efficient than naïve frequent pattern mining, since we use target responses as an additional information source for pruning, that enables us to skip searching for ubiquitous but uninformative patterns such as C-C or C-C-C.

We design classification and regression algorithms based on existing statistical frameworks (i.e., mother algorithms). It allows us to use theoretical results and insights accumulated in the past studies. As the mother algorithm, we employ boosting, partial least squares regression (PLS) and least angle regression (LARS). Our extended algorithms are called graph boosting (gBoost), graph PLS (gPLS) (H. Wold, 1966, 1975, S. Wold, Sjöstöm, & Erikkson, 2001, Rosipal & Krämer, 2006) and graph LARS (gLARS), respectively. Boosting is basically a classification (SAR) a lgorithm, and the other two are for regression (QSAR). The basic idea of extension is common, but they are totally different algorithms. For example, graph boosting chooses features based on \ell_1 norm regularization, whereas PLS extracts orthogonal features via linear projection. LARS is based on the solution tracing of linear programming. To extend such methods to the graph domain, individual treatment is necessary.

In the following, we use gSpan as our graph mining algorithm, however, it is possible to replace gSpan with other graph mining algorithms, or even with different frequent pattern mining algorithms such as tree mining (Zaki, 2005), sequence mining (Pei et al., 2004) or itemset mining (Uno, Kiyomi, & Arimura, 2005).

This chapter is organized as follows. We first show the basic idea of our approach along the graph classification problem. Then we extend it to least squares regression problem, where two different approaches are presented: active set method and partial least squares (PLS). We conclude this chapter with discussion and pointers to other related works. The source codes of all the methods presented in this section are available from the following web pages.

gBoost http://www.kyb.mpg.de/bs/people/nowozin/gboost/

gLARS http://www.kyb.tuebingen.mpg.de/bs/people/tsuda/

gPLS http://www.mpi-inf.mpg.de/%7Ehiroto/

GRAPH CLASSIFICATION

Throughout the chapter, we deal with undirected, labeled and connected graphs. To be more precise, we define the graph and *subgraph isomorphism* as follows:

Definition 1 (Labeled connected graph) *A labeled graph is represented in a 4-tuple $G=(V,E,L,l)$, where V is a set of vertices, $E \subseteq V \times V$ is a set of edges, L is a set of labels, and $l:V \cup E \rightarrow L$ is a mapping that assigns labels to the vertices and edges. A labeled connected graph is a labeled graph such that there is a path between any pair of vertices.*

Definition 2 (Isomorphism) *Let $G'=(V',E',L',l')$ and $G=(V,E,L,l)$ be labeled connected graphs. These two graphs are isomorphic if there exists a bijective function $\phi: V' \rightarrow V$ such that: (1) $\forall v' \in V', l'(v')=l(\phi(v'))$, (2)*

$E = \{\{\phi(v_1'), \phi(v_2')\} \mid \{v_1', v_2'\} \in E'\}$ and (3) $\forall \{v_1', v_2'\} \in E', l'(\{v_1', v_2'\}) = l(\{\phi(v_1'), \phi(v_2')\})$.

Definition 3 (Subgraph)*Given two graphs $G'=(V', E', L', l')$ and $G=(V, E, L, l)$, G' is a subgraph of G if the following conditions are satisfied: (1) $V'\subseteq V$, (2) $E'\subseteq E$, (3) $L'\subseteq L$, (4) $\forall v' \in V', l(v')=l'(v')$ and (5) $\forall e' \in E', l(e')=l'(e')$. If G' is a subgraph of G, then G is a supergraph of G'.*

Definition 4 (Subgraph Isomorphism)*A graph G' is subgraph-isomorphic to a graph G ($G'\subseteq G$) iff it has a subgraph S that is isomorphic to G'.*

We begin with defining the binary classification problem on graph data. In graph classification, the task is to learn a prediction rule from the training examples $\{(G_i, y_i)\}_{i=1}^{n}$, where G_i is a training graph and $y_i \in \{+1, -1\}$ is the associated class label. Let P be the set of all patterns, i.e., the set of all subgraphs included in at least one training graph. Then, each graph G_i is encoded as a $|P|$-dimensional row vector

$$x_{i,p} = \begin{cases} 1 & if \ p \subseteq G_i, \\ -1 & otherwise, \end{cases}$$

and $n \times P$-matrix $\mathbf{X}=[\mathbf{x}_1, \mathbf{x}_2, \ldots, \mathbf{x}_n]^{\mathrm{T}}$. This feature space is illustrated in Figure 1.

Our prediction rule is a convex combination of binary indicators $x_{i,j}$, and has the form

$$f(\mathbf{x}_i) = \sum_{p \in P} \beta_p \mathbf{x}_{i,p}, \tag{1}$$

where β is a $|P|$-dimensional column vector such that $\sum_{p \in P} \beta_p = 1$ and $\beta_p \geq 0$.

This is a linear discriminant function in an intractably large dimensional space. To obtain an interpretable rule, we need to obtain a *sparse* weight vector **b**, where only a few weights are nonzero. In the following, we will present a linear programming approach for efficiently capturing such patterns. Our formulation is based on that of LPBoost (Demiriz, Bennet, & Shawe-Taylor, 2002), and the learning problem is represented as

$$\min_{\mathbf{b}} \| \mathbf{b} \|_1 + \lambda \sum_{i=1}^{n} \left[1 - \mathbf{y}_i f(\mathbf{x}_i)\right]_+, \tag{2}$$

where $\| x \|_1 = \sum_{i=1}^{n} | \mathbf{x}_i |$ denotes the ℓ_1 norm of **x**, λ is a regularization parameter, and the subscript "+" indicates positive part. A soft-margin formulation of the above problem exists (Demiriz et al., 2002), and can be written as

$$\min_{\mathbf{b}, \mathbf{x}, \rho} - \rho + \lambda \sum_{i=1}^{n} \xi_i \tag{3}$$

$s.t. \mathbf{y}^{\mathrm{T}} \mathbf{X} \beta + \xi_i \geq \rho, \ \xi_i \geq 0, \ i=1, \ldots, n$

Figure 1. Feature space based on subgraph patterns. The feature vector consists of binary pattern indicators

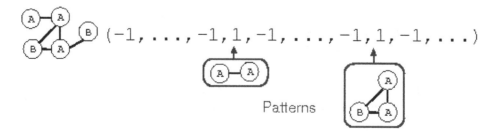

$$\sum_{p \in P} \beta_p = 1, \quad \beta_p \geq 0, \tag{4}$$

where $\mathbf{x}$ are slack variables, ρ is the margin separating negative examples from positives, λ is set to $\dfrac{1}{\nu n}$ where $\nu \in (0,1)$ is a parameter controlling the cost of misclassification which has to be found using model selection techniques, such as cross-validation. It is known that the optimal solution has the following v-property:

Theorem 1 *Assume that the solution of (3) satisfies $\rho \geq 0$. The following statements hold:*

1. ν is an upperbound of the fraction of *margin errors*, i.e., the examples with $\mathrm{y^T X}\,\beta < \rho$
2. ν is a lowerbound of the fraction of the examples such that $\mathrm{y^T X}\,\beta < \rho$

Directly solving this optimization problem is intractable due to the large dimensionality of b. So we solve the following equivalent dual problem instead.

$$\min_{\mathbf{u},v} v \tag{5}$$

$$s.t. \sum_{i=1}^{n} u_i y_i x_{i,p} \leq v, \forall p \in P$$
$$\sum_{i=1}^{n} u_i = 1, \quad 0 \leq u_i \leq \lambda, i = 1,\dots,n. \tag{6}$$

After solving the dual problem, the primal solution β is obtained from the Lagrange multipliers (Demiriz et al., 2002). The dual problem has a limited number of variables, but a huge number of constraints. Such a linear program can be solved by the *column generation* technique (Luenberger, 1969): Starting with an empty pattern set, the pattern whose corresponding constraint is violated the most is identified and added iteratively. Each time a pattern is added, the optimal solution is updated by solving the restricted dual problem.

Denote by $\mathbf{u}(k),\mathbf{v}(k)$ the optimal solution of the restricted problem at iteration $k=0,1,\dots$, and denote by $\widehat{\mathbf{X}}^{(k)} \subseteq P$ the set at iteration k. Initially, $\widehat{\mathbf{X}}^{(0)}$ is empty and $u_i^{(0)} = 1/n$. The restricted problem is defined by replacing the set of constraints (6) with

$$\sum_{i=1}^{n} u_i^{(k)} y_i x_{i,p} \leq v, \forall p \in \hat{\mathbf{X}}^{(k)}.$$

The left hand side of the inequality is called as *gain* in boosting literature. After solving the problem, $\widehat{\mathbf{X}}^{(k)}$ is updated to $\widehat{\mathbf{X}}^{(k+1)}$ by adding a column. Several criteria have been proposed to select the new columns (Merle, Villeneuve, Desrosiers, & Hansen, 1999), but we adopt the most simple rule that is amenable to graph mining: We select the constraint with the largest gain.

$$p^* = argmax_{p \in P} \sum_{i=1}^{n} u_i^{(k)} y_i x_{i,p}. \tag{7}$$

The solution set is updated as $\widehat{\mathbf{X}}^{(k+1)} \leftarrow \widehat{\mathbf{X}}^{(k)} \cup \mathbf{X}_{p^*}$. In the next section, we discuss how to efficiently find the largest gain in detail.

One of the big advantages of our method is that we have a stopping criterion that guarantees that the optimal solution is found: If there is no $p \in P$ such that

$$\sum_{i=1}^{n} u_i^{(k)} y_i x_{i,p} > v^{(k)}, \tag{8}$$

then the current solution is the optimal dual solution. Empirically, the patterns found in the last few iterations have negligibly small weights. The number of iterations can be decreased by relaxing the condition as

$$\sum_{i=1}^{n} u_i^{(k)} y_i x_{i,p} > v^{(k)} + \varepsilon, \tag{9}$$

Let us define the primal objective function as $V = -\rho + \lambda \sum_{i=1}^{n} \xi_i$. Due to the convex duality, we can guarantee that, for the solution obtained from the early termination (9), the objective satisfies $V \leq V^* + \epsilon$, where V^* is the optimal value with the exact termination (8) (Demiriz et al., 2002). In our experiments, $\epsilon = 0.01$ is always used.

Optimal Pattern Search

Our search strategy is a branch-and-bound algorithm that requires a canonical search space in which a whole set of patterns are enumerated without duplication. As the search space, we adopt the DFS (depth first search) code tree (Yan & Han, 2002a). The basic idea of the DFS code tree is to organize patterns as a tree, where a child node has a supergraph of the parent's pattern (Figure 2). A pattern is represented as a text string called the DFS code. The patterns are enumerated by generating the tree from the root to leaves using a recursive algorithm. To avoid duplications, node generation is systematically done by rightmost extensions. See Appendix for details about the DFS code and the rightmost extension.

All embeddings of a pattern in the graphs $\{G_i\}_{i=1}^{n}$ are maintained in each node. If a pattern matches a graph in different ways, all such embeddings are stored. When a new pattern is created by adding an edge, it is not necessary to perform full isomorphism checks with respect to all graphs in the database. A new list of embeddings are made by extending the embeddings of the parent (Yan & Han, 2002a). Technically, it is necessary to devise a data structure such that the embeddings are stored incrementally, because it takes a prohibitive amount of memory to keep all embeddings independently in each node.

As mentioned in (7), our aim is to find the optimal hypothesis that maximizes the gain $g(p)$.

$$g(p) = \sum_{i=1}^{n} u_i^{(k)} y_i x_{i,p}. \tag{10}$$

For efficient search, it is important to minimize the size of the actual search space. To this aim, *tree pruning* is crucially important: Suppose the search tree is generated up to the pattern P and denote by g^* the maximum gain among the ones observed so far. If it is guaranteed that the gain of any supergraph P' is not larger than g^*, we can avoid the generation of downstream nodes without losing the optimal pattern. We employ the following pruning condition.

Theorem 2 *(Morishita, 2001, Kudo, Maeda, & Matsumoto, 2005) Let us define*

$$\mu(p) = 2 \sum_{\{i|y_i=+1, p \subseteq G_i\}} u_i^{(k)} - \sum_{i=1}^{n} y_i u_i^{(k)}.$$

If the following condition is satisfied,

$$g^* > \mu(p) \tag{11}$$

the inequality $g(p') < g^*$ holds for any p' such that $p \subseteq p'$.

Proof 1 *By definition,*

$$\begin{aligned}
g(p') &= \sum_{i=1}^{n} u_i^{(k)} y_i (2I(p' \subseteq G_i) - 1) \\
&= 2 \sum_{\{i|p' \subseteq G_i\}} y_i u_i^{(k)} - \sum_{i=1}^{n} y_i u_i^{(k)} \\
&\leq 2 \sum_{\{i|y_i=+1, p' \subseteq G_i\}} u_i^{(k)} - \sum_{i=1}^{n} y_i u_i^{(k)} \\
&\leq 2 \sum_{\{i|y_i=+1, p \subseteq G_n\}} u_i^{(k)} - \sum_{i=1}^{n} y_i u_i^{(k)}.
\end{aligned}$$

The last line follows from the fact that

$\{i | y_i = +1, p' \subseteq G_i\} \subseteq \{i | y_i = +1, p \subseteq G_i\}$

Therefore, $\mu(p)$ is an upperbound of $g(p')$. If the current maximum gain g^* is more than $\mu(p)$, it is guaranteed that there is no downstream pattern whose gain is larger than g^*.

The gBoost algorithm is summarized in Figures 3 and 4.

Figure 3. gBoost algorithm: main part

Algorithm 1 gBoost algorithm: main part

1: $\hat{\boldsymbol{X}}^{(0)} = \emptyset$, $\boldsymbol{u}_i^{(0)} = 1/n$, $k = 0$
2: **loop**
3:     Find the optimal pattern p^* based on $\boldsymbol{u}^{(k)}$ ▷ Algorithm 2
4:     **if** termination condition (9) holds **then**
5:         break
6:     **end if**
7:     $\hat{\boldsymbol{X}} \leftarrow \hat{\boldsymbol{X}} \cup \boldsymbol{X}_{p^*}$
8:     Solve the restricted dual problem (5) to obtain $\boldsymbol{u}^{(k+1)}$
9:     $k = k + 1$
10: **end loop**

Figure 4. Finding the optimal pattern

Algorithm 2 Finding the Optimal Pattern

1: **procedure** OPTIMAL PATTERN
2:     Global variables: g^*, p^*
3:     $g^* = -\infty$
4:     **for** $p \in$ DFS codes with single nodes **do**
5:         project(p)
6:     **end for**
7:     return p^*
8: **end procedure**
9: **function** PROJECT(p)
10:     **if** p is not a minimum DFS code **then**
11:         return
12:     **end if**
13:     **if** pruning condition (11) holds **then**   ▷ Theorem 2
14:         return
15:     **end if**
16:     **if** $g(p) > g^*$ **then**
17:         $g^* = g(p)$, $p^* = p$
18:     **end if**
19:     **for** $p' \in$ rightmost extensions of p **do**
20:         project(p')
21:     **end for**
22: **end function**

Reusing the Search Space

Our method calls the pattern search algorithm repeatedly with different parameters $\lambda^{(k)}$. In each iteration, the search tree is generated until the pruning condition is satisfied. Creating a new node is time consuming, because the list of embeddings is updated, and the minimality of the DFS code has to be checked (see Appendix). In our previous paper (Saigo, Kadowaki, & Tsuda, 2006), the search tree is erased after the optimal pattern is found, and a new search tree is built from scratch in the next iteration. In this paper, we maintain the whole search tree, including all embeddings, in the main memory for better efficiency. Then, node creation is necessary only if it is not created in previous iterations. Naturally this strategy requires more memory, but we did not experience any overflow problems in our experiments with 8GB memory.

EXTENSION TO LEAST SQUARES REGRESSION

One merit of our mathematical programming-based approach is that a wide range of machine learning problems are solved based on the same pattern search. In this section, we particularly focus on least squares regression. It is possible to apply our approach to, e.g., one-class classification (Rätsch, Mika, Schölkopf, & Müller, 2002), multi-class classification (FreSch97), hierarchical classification (Cai & Hofmann, 2004), 1.5-class classification (Yuan & Casasent, 2003) and knowledge-based support vector machines (quoc06). Mathematical programs are commonly used in machine learning, so certainly there are more applications.

Suppose we are given a training data set $\{(G_n, y_n)\}_{n=1}^{\ell}$, but now y_n may take on any real value. We use the same definition for a hypothesis $h(\mathbf{x}, t, \omega)$ and its corresponding weight $\beta t_{,\omega}$ as those in the classification case. The regression function is defined as

$$f(\mathbf{x}) = \sum_{(\mathbf{t}, \omega) \in T \times \Omega} \beta_{\mathbf{t}, \omega} h(\mathbf{x}; \mathbf{t}, \omega) + b,$$

where b is a newly introduced bias term. The learning problem is written as

$$\min_{\mathbf{b}, b} \quad C \sum_{(\mathbf{t}, \omega) \in T \times \Omega} |\beta_{\mathbf{t}, \omega}| + \frac{1}{2} \sum_{n=1}^{\ell} \left(\sum_{(\mathbf{t}, \omega) \in T \times \Omega} \beta_{\mathbf{t}, \omega} h(\mathbf{x}_n; \mathbf{t}, \omega) + b - y_n \right)^2.$$

Note that we introduced the ℓ_1-norm regularizer to enforce sparsity to parameter vectors. This is exactly the same learning problem as that of LASSO (Tibshrani, 1996). The learning problem above translates to the following quadratic program.

$$\min_{\mathbf{b}, \mathbf{x}, b} C \sum_{(\mathbf{t}, \omega) \in T \times \Omega} (\beta_{t, \omega}^+ + \beta_{t, \omega}^-) + \frac{1}{2} \sum_{n=1}^{\ell} \xi_n^2$$

$$s.t. \sum_{(\mathbf{t}, \omega) \in T \times \Omega} \beta_{\mathbf{t}, \omega} h(\mathbf{x}_n; \mathbf{t}, \omega) + b - y_n \leq \xi_n, \quad n = 1, \dots, \ell$$

$$y_n - \sum_{(\mathbf{t}, \omega) \in T \times \Omega} \beta_{\mathbf{t}, \omega} h(\mathbf{x}_n; \mathbf{t}, \omega) - b \leq \xi_n, \quad n = 1, \dots, \ell$$

$$\beta_{t, \omega}^+, \beta_{t, \omega}^- \geq 0, \quad \forall(t, \omega) \in T \times \Omega,$$

where ξn is a slack variable, $\beta_{t, \omega} = \beta_{t, \omega}^+ - \beta_{t, \omega}^-$. The dual problem is described as

$$\min_{\mathbf{u}} \frac{1}{2} \sum_{n=1}^{\ell} (u_n^+ + u_n^-)^2 - \sum_{n=1}^{\ell} y_n (u_n^+ - u_n^-)$$

$$s.t. -C \leq \sum_{n=1}^{\ell} (u_n^+ - u_n^-) h(\mathbf{x}_n; t, \omega) \leq C, \quad \forall(t, \omega) \in T \times \Omega$$

$$\sum_{n=1}^{\ell} u_n^+ - u_n^- = 0,$$

$$u_n^+, u_n^- \geq 0, n = 1, \cdots, \ell.$$

Unlike the classification case, the dual constraint is two-sided. Therefore, the gain function for regression has a slightly different form:

$$g_{reg}(\mathbf{t}, \omega) = \left| \sum_{n=1}^{\ell} u_n^{(k)} h(\mathbf{x}_n; \mathbf{t}, \omega) \right|.$$

However, we can still use the pruning condition (Theorem 2), because the same proposition holds for g_{reg}.

Computational Experiments

In this subsection, our method is benchmarked with publicly available chemical compound datasets in classification problems.

In this chapter, we use four publicly available chemical datasets, EDKB[4], Mutag[5], CPDB[6], CAS[7] and the AIDS antiviral screen dataset [8]. EDKB is a regression dataset, but the others are classification datasets. The statistics of the datasets are summarized in Table 1. We compared our method (gBoost) with marginalized graph kernel (MGK) (Kashima, Tsuda, & Inokuchi, 2003) and SVM with frequent mining (freqSVM) in 10-fold cross validation experiments. In FreqSVM, the frequent patterns are mined first, and then SVM is applied to the feature space created by the patterns (Helma et al., 2004, Kazius et al., 2006, Wale & Karypis, 2006). These three methods are implemented by ourselves. In addition, we quote the 10 fold cross validation results by Gaston (Kazius et al., 2006), Correlated Pattern Mining (CPM) (Bringmann et al., 2006), and MOLFEA (Helma et al., 2004) from respective papers. The quoted accuracies are all based on 10-fold cross validation. In literature (Kazius et al., 2006, Helma et al., 2004, Bringmann et al., 2006), the accuracies of Gaston, CPM, and MOLFEA are shown for all possible

Table 1. Datasets Summary. The number of positive data (POS) and negative data (NEG) are only provided for classification datasets. Average number of atoms (ATOM) and bonds (BOND) are shown for each dataset

	LABEL	**ALL**	**POS**	**NEG**	**ATOM**	**BOND**
Mutag	binary	188	125	63	45.1	47.1
CPDB	binary	684	341	343	14.1	14.6
CAS	binary	4337	2401	1936	29.9	30.9
AIDS1 (CAvsCM)	binary	1503	422	1081	58.9	61.4
AIDS2 (CACMvsCI)	binary	40939	1324	39615	42.7	44.6
AIDS3 (CAvsCI)	binary	39965	350	39615	42.7	44.5
EDKB-ER	real	131	-	-	19.2	20.7
EDKB-ES	real	59	-	-	18.2	19.7
Regression			ALL	ATOM	BOND	
EDKB-AR			146	19.5	21.1	
EDKB-ER			131	19.2	20.7	
EDKB-ES			59	18.2	19.7	

regularization parameters. For each method, the best test accuracy is taken.

For gBoost, the maximum pattern size (max-pat), which in our case corresponds to the maximum number of nodes in a subgraph, was constrained up to 10, and we did not use the minimum support constraint at all. The regularization parameter v is chosen from {0.01, 0.1, 0.2, 0.3, 0.4, 0.5, 0.6}. We used chemcpp[9] for computing the MGK. For MGK, the termination probability was chosen from {01,01,...,09}, and the λ parameter of SVM was chosen from {01, 1, 10, 100, 1000, 10000}. For freqSVM, frequent patterns are first mined by gSpan with maximum pattern size 10 and minimum support threshold 1% for Mutag and CPDB, 10% for CAS and AIDS. Then SVM was trained with λ parameter from the range {01, 1, 10, 100, 1000, 10000}. To compare with the quoted results as fairly as possible, the best test accuracy is shown for all methods.

Table 2 summarizes the results. Overall, gBoost was competitive among the other state-of-the-art methods. In Mutag, CAS and CPDB, gBoost was the best method, but in AIDS, CPM performed best in accuracy and freq-SVM performed best in AUC. Notice that CPM's good result on AIDS is based on sequence patterns. When subgraph patterns are used, the best result was 0.767. The relatively poor result of gBoost in AIDS in comparison with

freqSVM could be attributed to misselection of features by the ℓ_1 regularizer. It is known that the ℓ_1 regularizer selects too few features occasionally (Zou & Hastie, 2005). One way to weaken sparsity is to introduce an ℓ_2 regularizer in addition to the $\ell1$ regularizer like the elastic net (Zou & Hastie, 2005). The computation time of gBoost is decomposed into *mining time* and *LP time.* The former is used for expanding and traversing the pattern space, and the latter is to solve the series of restricted dual problems. For CAS, the mining time was 1370 seconds and the LP time was 1110 seconds, respectively, on a standard PC with AMD Opteron 2.2GHz and 8GB memory. The computation time can be reduced in several ways: 1) restricting the pattern to simpler ones, e.g., walks or trees, 2) limiting the pattern set a priori, e.g., by the correlation with class labels (Bringmann et al., 2006) or by the minimum support constraints (Kazius et al., 2006). However, in any case, informative patterns might be lost in exchange for better efficiency.

The top 20 discriminative subgraphs for CPDB are displayed in Figure 5. We found that the top 3 substructures with positive weights (0.0672, 0.0656, 0.0577) correspond to known *toxicophores* (Kazius et al., 2006). They correspond to *aromatic amine, aliphatic halide,* and *three-membered heterocycle,* respectively. In ad-

Table 2. Classification performance obtained by 10-fold cross validation in the classification datasets measured by the accuracy (ACC) and the area under the ROC curve (AUC). We obtained the results of MGK, freqSVM and gBoost from our implementations, but the other results are quoted from the literature. The best results are highlighted in bold fonts

Method	Mutag		CAS		CPDB		AIDS1(CAvsCM)	
	ACC	AUC	ACC	AUC	ACC	AUC	ACC	AUC
Gaston [1]	-	-	0.79	-	-	-	-	-
MOLFEA [2]	-	-	-	-	0.785	-	-	-
CPM [3]	-	-	0.801	-	0.760	-	**0.832**	-
MGK	0.808	0.901	0.771	0.763	0.765	0.756	0.762	0.760
freqSVM	0.808	0.906	0.773	0.843	0.778	0.845	0.782	**0.808**
gBoost	**0.852**	**0.926**	**0.825**	**0.889**	**0.788**	**0.854**	0.802	0.774

Figure 5. Top 20 discriminative subgraphs from the CPDB dataset. Each subgraph is shown with the corresponding weight, and ordered by the absolute value from the top left to the bottom right. H atom is omitted, and C atom is represented as a dot for simplicity. Aromatic bonds appeared in an open form are displayed by the combination of dashed and solid lines

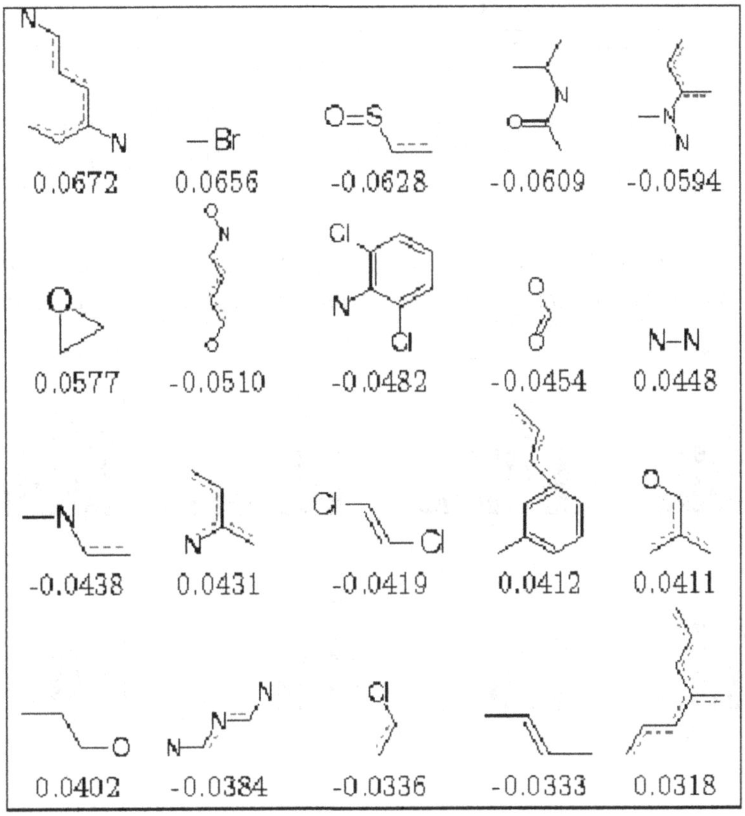

Table 3. Regression performance obtained by leave-one-out cross validation in three assays from the EDKB evaluated by mean absolute error (MAE) and Q^2. Note that for MAE, lower values indicate better prediction, which is vice versa for Q^2. We obtained the results of MGK, freqSVM and gBoost from our implementations, but the other results are quoted from the literature. The best results are highlighted in bold fonts

	Measure	CoMFA(hong03, shi00)	MGK	freqSVM	gBoost
EDKB-AR	MAE	-	0.229	0.193	**0.183**
	Q^2	0.571	0.346	0.465	**0.621**
EDKB-ER	MAE	-	0.320	0.268	0.263
	Q^2	**0.660**	0.267	0.532	0.541
EDKB-ES	MAE	-	0.322	0.248	**0.216**
	Q^2	-	0.522	0.588	**0.753**

Table 4. Influence of the choice of v parameter. P: the number of active patterns, T: the size of the tree-shaped search space, ITR: the number of iterations, ρ: the margin in Equation (3), TrACC: the classification accuracy in the training set

\nu	0.1	0.2	0.3	0.4	0.5	0.6	0.7	0.8
P	317	225	173	106	63	37	19	7
T	21766	21825	19083	16026	11330	6723	2753	1754
ITR	328	235	178	110	65	39	21	9
\rho	0.00859	0.0116	0.0200	0.0340	0.0695	0.121	0.4	0.6
TrACC	0.999	0.965	0.924	0.883	0.838	0.803	0.712	0.709

Figure 6. Total weights of the active patterns of different sizes. The x-axis indicates the pattern size, and the y-axis indicates the sum of weights of the active patterns of the corresponding size

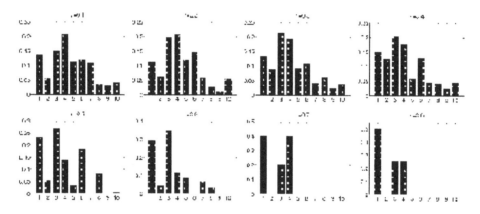

dition, the patterns with weights 0.0431, 0.0412, 0.0411 and 0.0318 seem to be related to *polycyclic aromatic systems*.

To characterize the influence of the regularization parameter v, gBoost is applied to CPDB with $v=\{01,...,08\}$. Table 4 shows the number of "active patterns", the size of the searched space, the number of iterations until convergence, the margin, and the training accuracy. Here, active patterns are defined as those with non-zero weights,

$A=\{p \in P | \beta_p \neq 0\}$

For this experiment, we set the maximum pattern size to 10. When v is low, gBoost creates a complex classification rule with many active patterns so that it can classify the training examples completely. As v is increased, the regularization takes effect and the rule gets simpler with a smaller number of active patterns. At the same time, the active patterns become smaller in size (Figure 6).

Computational Costs

Our method tightly couples the mathematical programming and graph mining. However, the same prediction rule can be obtained by the following "naïve" method:

1. The feature space is completely constructed by frequent substructure mining.
2. The mathematical program is solved by column generation.

Figure 7. Test set accuracy and the number of active patterns against the maxpat parameter. The accuracy is computed by 10-fold cross validation in the CPDB dataset

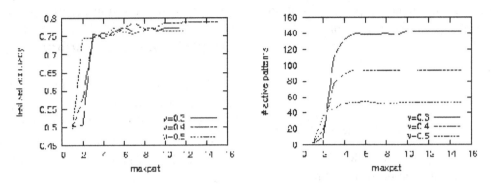

Figure 8. Comparison in computational costs. In the left panel, the tree size refers to the number of nodes in the search space. In the right panel, the mining time is plotted respectively for the progressive and naïve methods. The LP time is identical for the two methods, thus depicted as a single curve

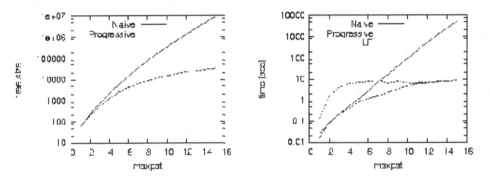

To motivate the use of our method, it is essential to empirically show that our method is more efficient in time. Certainly our method has a smaller search space due to the pruning condition, but the question is how much the pruning condition can contribute to reducing the computation time in real datasets.

We compared the computational time and the test set accuracy of the naïve method and the proposed progressive method. Those two methods produce exactly the same series of reduced linear programs. So we subtracted the time needed to solve the linear programs from the total computational time. The remaining time (i.e., *mining time*) is for constructing and traversing the search space. The results for a range of maxpat constraints in the

CPDB dataset are summarized in Figures 7 and 8. In this experiment, no minimum support constraints are employed. The test accuracy reaches the highest level around maxpat=4, and already at this point, the mining time shows substantial difference. Notice that the time is plotted in the log-scale. Even in the case that small patterns can achieve good accuracy, it makes sense to explore larger patterns for better interpretability. In fact, the toxicophore contains many patterns having more than 4 edges (Kazius et al., 2006). We found that the maxpat is typically set to 10 or more in literature (Wale & Karypis, 2006, Helma et al., 2004, Kazius et al., 2006). In this domain, our mining time is more than 10 times smaller.

GRAPH REGRESSION

In this section we extend graph classification framework to least squares regression:

$$\min_{\mathbf{b}} \| \mathbf{Xb} - \mathbf{y} \|_2^2, \qquad (12)$$

where $\| x \|_2 = \left(\sum_{i=1}^{n} | \mathbf{x}_i |^2 \right)^{1/2}$ denotes the ℓ_2 norm of $\mathbf{x}$, and response variable $\mathbf{y} \in \mathbb{R}^n$ are assumed centered. The solution to this ordinary least squares (OLS) problem is obtained in a closed form as:

$$\beta = (X^T X)^{-1} X^T y \qquad (13)$$

However, this requires the whole design matrix $\mathbf{X}$, which can be intractable in our setting, since it amounts to mining all the frequent subgraphs. Even in the case where we have all the frequent subgraph patterns at hand, the number of subgraphs P easily exceeds the number of graphs n, therefore overfitting problem arises. We therefore desire a regularizer which induces sparsity in β such that it simultaneously avoids overfitting and decreases the size of the actual feature space. Basically, we can use the same ℓ_1 regularizer as presented in the previous section and formulate the mathematical problem, then solve it by column generation (Saigo et al., 2006, Saigo, Nowozin, Kadowaki, Kudo, & Tsuda, 2009). Here, however, we present two more efficient alternatives (PLS and LARS) to solving least squares regression and show their tight coupling with graph mining.

We first present Partial Least Squares (PLS), a popular tool in chemometrics where variables are often highly correlated with each other (i.e., *collinear* setting) (H. Wold, 1966, 1975, S. Wold et al., 2001). Later we solve \ell_1-regularized least squares regression (aka LASSO (Tibshrani, 1996)) by the *active set method* (Osborne, Presnell, & Turlach, 2000), where the entire regularization path is traced. Solving the LASSO problem by the active set method has recently attracted attention of researchers due to its nice statistical properties (Efron, Hastie, Johnstone, & Tibshirani, 2004).

Graph Partial Least Squares (gPLS)

In chemometrics, it is often the case that there exists many measured variables on each of a few observations, and the number of variables P greatly exceeds the number of observations n. PLS was introduced by Wold (H. Wold, 1966) for such a situation, and heavily used in the community as an alternative to OLS. Its statistical property was unclear for a long time, but later recognized that the method is equivalent to conjugate gradient (S. Wold, Ruhe, Wold, & DunnIII, 1984) and shrink coefficients like ridge regression (Frank & Friedman, 1993).

PLS performs dimension reduction similar to PCA, and fits regression function in the low-dimensional space. Let $\mathbf{X}$ be our design matrix and $\mathbf{W}$ be the m-dimensional projection matrix, the problem PLS solves can be written as a constrained least squares problem:

$$\min_{\mathbf{b}} \| \mathbf{y} - \mathbf{Xb} \|_2^2 \qquad (14)$$

$$s.t. \ \beta \in span(\mathbf{W}_m) \qquad (15)$$

where $span(\mathbf{W}_m)$ indicates that the projection matrix $\mathbf{W}$ lies in the m-dimensional *Krylov subspace*

$$K_m(\mathbf{X}^T\mathbf{X}, \mathbf{X}^T\mathbf{y}) = \{\mathbf{X}^T\mathbf{y}, \mathbf{X}^T\mathbf{X}\mathbf{X}^T\mathbf{y}, \ldots, (\mathbf{X}^T\mathbf{X})^{m-1}\mathbf{X}^T\mathbf{y}\}$$

PLS fits regression coefficients β in the low-dimensional projected space $\mathbf{XW}$. The column vectors of the matrix $\mathbf{W} = [\mathbf{w}_1, \mathbf{w}_2, \ldots, \mathbf{w}_m]$ is called as weight vectors in PLS literature. The column vectors of the matrix $\mathbf{XW} = [\mathbf{Xw}_1, \mathbf{Xw}_2, \ldots, \mathbf{Xw}_m]$ is called as latent components, and often denoted as $T = [\mathbf{t}_1, \mathbf{t}_2, \ldots, \mathbf{t}_m]$. Typically, each latent component $\mathbf{t}$ is kept orthogonal to each other. Regression problem in the projected space can be written as

$$\min_{\hat{\mathbf{b}}} \| \mathbf{y} - \mathbf{XW}\hat{\mathbf{b}} \|_2^2, \qquad (16)$$

where $\hat{\mathbf{b}} \in \mathbb{R}^m$ is a m-dimensional regression coefficient vector in the projected space. m controls the number of components in PLS, and can be seen as a regularization parameter. If we set $m=P$, then the OLS solution (13) is recovered. As long as m is chosen such that $m<n$, overfitting problem is naturally avoided. Equation (16) can be interpreted as a covariance maximization procedure (Frank & Friedman, 1993). We give a simple proof that the first weight vector $\mathbf{w}=\mathbf{w}_1$ maximizes the covariance: $cov(\mathbf{Xw,y})=\mathbf{y}^T\mathbf{Xw}$.

Theorem 3 *PLS maximizes squared covariance*

$$\max_{\mathbf{w}} cov^2(\mathbf{Xw, y}) \qquad (17)$$

Proof 2 *By taking the derivative of Equation (16) with respect to* $\hat{\mathbf{b}}$ *and setting it to zero, regression coefficients in the projected space are obtained as*

$$\hat{\mathbf{b}} = \left(\mathbf{W}^T\mathbf{X}^T\mathbf{XW}\right)^{-1}\mathbf{W}^T\mathbf{X}^T\mathbf{y}.$$

Corresponding regression coefficients in the original space are

$$\mathbf{b} = \mathbf{W}\hat{\mathbf{b}} = \mathbf{W}\left(\mathbf{W}^T\mathbf{X}^T\mathbf{XW}\right)^{-1}\mathbf{W}^T\mathbf{X}^T\mathbf{y}. \qquad (18)$$

Replacing this into Equation (16) and solving it with respect to the first weight vector $\mathbf{w}$ obtains

$$\min_{\hat{\mathbf{b}}} \| \mathbf{y} - \mathbf{Xb} \|_2^2 = \min_{\mathbf{w}} \| \mathbf{y} - \mathbf{Xw}\left(\mathbf{w}^T\mathbf{X}^T\mathbf{Xw}\right)^{-1}\mathbf{w}^T\mathbf{X}^T\mathbf{y} \|_2^2$$
$$= \min_{\mathbf{w}} \| \mathbf{y} - \mathbf{Xww}^T\mathbf{X}^T\mathbf{y} \|_2^2,$$

where orthogonal condition $\mathbf{w}^T\mathbf{X}^T\mathbf{Xw}=1$ is used in the last equality. By extracting the last equation,

$$\min_{\mathbf{w}} \| \mathbf{y} - \mathbf{Xww}^T\mathbf{X}^T\mathbf{y} \|_2^2$$
$$= \min_{\mathbf{w}}\left(\mathbf{y} - \mathbf{Xww}^T\mathbf{X}^T\mathbf{y}\right)^T\left(\mathbf{y} - \mathbf{Xww}^T\mathbf{X}^T\mathbf{y}\right)$$
$$= \min_{\mathbf{w}} \mathbf{y}^T\mathbf{y} - 2\mathbf{y}^T\mathbf{Xww}^T\mathbf{X}^T\mathbf{y} + \mathbf{y}^T\mathbf{Xww}^T\mathbf{X}^T\mathbf{Xww}^T\mathbf{X}^T\mathbf{y}.$$

Removing $\mathbf{y}^T\mathbf{y}$ and using the orthogonal condition obtains

$$\min_{\mathbf{w}} - \mathbf{y}^T\mathbf{Xww}^T\mathbf{X}^T\mathbf{y}$$
$$= \max_{\mathbf{w}}(\mathbf{y}^T\mathbf{Xw})(\mathbf{y}^T\mathbf{Xw})^T$$
$$= \max_{\mathbf{w}} cov^2(\mathbf{Xw, y}).$$

A standard NIPALS algorithm (Figure 9) can efficiently compute weight vectors in such a way that they successively maximize covariance. Then

Figure 9. NIPALS

Algorithm 3 NIPALS.

1: Initial: $\tilde{\mathbf{X}}^{(1)} = \mathbf{X}$
2: **for** $k = 1, \ldots, m$ **do**
3:      $\mathbf{w}_k = \tilde{\mathbf{X}}^{(k)\top}\mathbf{y}.$ ▷ Weight vector
4:      $\mathbf{t}_k = \tilde{\mathbf{X}}^{(k)}\mathbf{w}_k$ ▷ Latent components
5:      $\tilde{\mathbf{X}}^{(k+1)} = \tilde{\mathbf{X}}^{(k)} - \mathbf{t}_k\mathbf{t}_k^\top\tilde{\mathbf{X}}^{(k)}$ ▷ Deflation
6: **end for**
7: Convert $\mathbf{W}$ into regression coefficients using equation (18)

---

after m iterations, the regression coefficients β are recovered from weight vectors using Equation (18).

In each iteration of NIPALS, rank-one deflation of design matrix $\mathbf{X}$ is performed to ensure the orthogonality between latent components $\mathbf{t}$. However, this deflation step completely destroys the structure of $\mathbf{X}$, therefore cannot be used in combination with graph mining where we cannot modify the structure of the design matrix. To avoid deflation, we make use of the connection between PLS and the Lanczos method (Lanczos, 1950). The Lanczos method is a *matrix free* method, therefore compatible with graph mining (Saigo & Tsuda, 2008). Moreover, this method is proven to be mathematically equivalent to PLS (Lanczos, 1950, Eldén, 2004). Based on this connection, we give a non-deflation PLS algorithm which does not require deflation at all. Remember that in NIPALS, the first weight vector is obtained as

$$\mathbf{w}_1 = \mathbf{X}^{(1)\mathrm{T}}\mathbf{y} = \mathbf{X}^{\mathrm{T}}\mathbf{y}$$

The second weight vector is obtained as

$$\begin{aligned} \mathbf{w}_2 &= \mathbf{X}^{(2)\mathrm{T}}\mathbf{y} = \left(\mathbf{X}^{(1)} - \mathbf{t}_1\mathbf{t}_1^{\mathrm{T}}\mathbf{X}^{(1)}\right)^{\mathrm{T}}\mathbf{y} = \mathbf{X}^{(1)\mathrm{T}}\left(\mathbf{I} - \mathbf{t}_1\mathbf{t}_1^{\mathrm{T}}\right)\mathbf{y} \\ &= \mathbf{X}^{\mathrm{T}}\left(\mathbf{r}_1 - \left(\mathbf{y}^{\mathrm{T}}\mathbf{t}_1\right)\mathbf{t}_1\right), \end{aligned}$$

where we used *residual* $\mathbf{r}_{k+1} = (\mathbf{r}_k - (\mathbf{y}^{\mathrm{T}}\mathbf{t}_{k-1})\mathbf{t}_{k-1})$ and initialized it as $\mathbf{r}_1 = \mathbf{y}$. The third weight vector is obtained similarly as:

$$\begin{aligned} \mathbf{w}_3 &= \mathbf{X}^{(3)\mathrm{T}}\mathbf{y} = \left(\mathbf{X}^{(2)} - \mathbf{t}_2\mathbf{t}_2^{\mathrm{T}}\mathbf{X}^{(2)}\right)^{\mathrm{T}}\mathbf{y} = \mathbf{X}^{(2)\mathrm{T}}\left(\mathbf{I} - \mathbf{t}_2\mathbf{t}_2^{\mathrm{T}}\right)\mathbf{y} \\ &= \mathbf{X}^{(1)\mathrm{T}}\left(\mathbf{I} - \mathbf{t}_1\mathbf{t}_1^{\mathrm{T}}\right)\left(\mathbf{I} - \mathbf{t}_2\mathbf{t}_2^{\mathrm{T}}\right)\mathbf{y} \\ &= \mathbf{X}^{\mathrm{T}}\left(\mathbf{I} - \mathbf{t}_1\mathbf{t}_1^{\mathrm{T}} - \mathbf{t}_2\mathbf{t}_2^{\mathrm{T}}\right)\mathbf{y} \\ &= \mathbf{X}^{\mathrm{T}}\left(\mathbf{r}_2 - \left(\mathbf{y}^{\mathrm{T}}\mathbf{t}_2\right)\mathbf{t}_2\right). \end{aligned}$$

In the same line, we can represent the k-the weight vector as

$$\mathbf{w}_k = \mathbf{X}^{\mathrm{T}}(\mathbf{r}_{k-1} - (\mathbf{y}^{\mathrm{T}}\mathbf{t}_{k-1})\mathbf{t}_{k-1}$$

Now it is clear that we do not need to update $\mathbf{X}$, but can update $\mathbf{r}$ instead. This non-deflation PLS algorithm is stated in Algorithm 10.

Below, we discuss how to apply the non-deflation PLS algorithm to graph data (Figure 10).

Figure 10. Non-deflation PLS

Algorithm 4 Non-deflation PLS.

1: Initial: $r_1 = y$
2: **for** $k = 1, \ldots, m$ **do**
3: $\quad w_k = X^{\top}r_k$ ▷ Weight vector
4: $\quad$ **if** $k = 1$ **then**
5: $\quad\quad t_1 = Xw_1$
6: $\quad$ **else**
7: $\quad\quad t_k = \left(I - t_{k-1}t_{k-1}^{\top}\right)Xw_k$ ▷ Orthogonalization
8: $\quad$ **end if**
9: $\quad t_k = t_k/\|t_k\|_2$ ▷ Latent component
10: $\quad r_{k+1} = r_k - (y^{\top}t_k)t_k$ ▷ Update residual
11: **end for**
12: Convert W into regression coefficients using equation (18)

Figure 11. gPLS

Algorithm 5 gPLS

1: $r_1 = y$, $\hat{X} = \emptyset$
2: **for** $k = 1, \ldots, m$ **do**
3:     $P_k = \{p \mid \left| X_p^\top r_k \right| \geq \epsilon\}$.   ▷ Pattern search
4:     X_{P_k}: design matrix restricted to P_k
5:     $\hat{X} \leftarrow \hat{X} \cup X_{P_k}$
6:     $w_k = X_{P_k}^\top r_k$ ▷ Weight vector
7:     **if** $k = 1$ **then**
8:         $t_1 = \hat{X} w_1$
9:     **else**
10:         $t_k = \left(I - t_{k-1} t_{k-1}^\top\right) \hat{X} w_k$ ▷ Orthogonalization
11:     **end if**
12:     $t_k = t_k / \|t_k\|_2$ ▷ Latent component
13:     $r_{k+1} = r_k - (y^\top t_k) t_k$ ▷ Update residual
14: **end for**
15: Convert W and $\hat{X}$ into regression coefficients using equation (18).

The feature space has already been illustrated in Figure 1. Since $|P|$ is a large number, we cannot keep the whole design matrix. So we need to set **X** as the empty matrix first, and grow the matrix as the iterations proceed. In each iteration, we obtain the set of patterns P whose weight $w_{ip} = X_p^T r$ is above the threshold, which can be written as

$$P_i = \{p \mid \left| X_p^T r \right| \geq \varepsilon\}, \tag{19}$$

which corresponds to selecting patterns with largest current *correlations*. Then, the design matrix is expanded to include newly introduced patterns. There are two alternative ways to determine the threshold ε: 1) Sort $|w_{ij}|$ in the descending order, take the top-k elements, and set all the other elements to zero. 2) Set ε to a fixed threshold. In the latter case, the number of non-zero elements in w_i may vary. In the experiments, we took the former top-k approach to avoid unbalanced weight vectors and to make efficiency comparisons easier. The pseudocode of gPLS is described in Figure 11. Most numerical computations are carried over from Algorithm 10.

The pattern search problem is exactly the same as the one solved in gBoost (Nowozin, Tsuda, Uno, Kudo, & Bakir, 2007). So we can reuse the same method to enumerate P_i. More specifically, it can be done by gspan function in the gBoost MATLAB toolbox[10]. However, we explain the pattern search algorithm briefly for the completeness of this paper. Our search strategy is the same branch-and-bound algorithm as presented, except that different gain and therefore a different pruning condition is employed. Let us define the gain function as $g(p) = \left| X_p^T r \right|$. Suppose the search tree is generated up to the subgraph pattern P. If it is guaranteed that the gain of any supergraph pattern p' is not larger than ε, we can avoid the generation of downstream nodes without losing the optimal pattern. Our pruning condition is described as follows.

Theorem 4 *Define* $\tilde{y}_i = sgn(r_i)$. *For any pattern* p'P' *such that* p⊆p', g(p')<ε, *if*

$$\max\{g^+(p), g^-(p)\} < \varepsilon \qquad (20)$$

where

$$g^+(p) = 2 \sum_{\{i|\tilde{y}_i=+1, x_{i,j}=1\}} |r_i| - \sum_{i=1}^{n} r_i$$

$$g^-(p) = 2 \sum_{\{i|\tilde{y}_i=-1, x_{i,j}=1\}} |r_i| + \sum_{i=1}^{n} r_i.$$

Other conditions such as the maximum size of pattern (*maxpat*) and the minimum support (*minsup*) can be used in combination with the pruning condition (20).

COMPUTATIONAL EXPERIMENTS

In this subsection, we evaluate the efficiency of gPLS on publicly available chemical datasets. The dataset is already used and the statistics is summarized in Table 1. Among them, the AIDS dataset (Kramer, Raedt, & Helma, 2001, Deshpande, Kuramochi, Wale, & Karypis, 2005) is by far the largest both in the number of examples and the graph size. EDKB is a regression dataset, but the others are classification datasets. In gPLS, we

solved classification problems by regressing the target values to +1 and -1.

Note that for classification problems, gBoost classification, which has been already presented in this chapter, is used. For regression problems, gBoost regression (Saigo, Krämer, & Tsuda, 2008) is used. We omit the detailed explanation of gBoost regression, but it is same as the gBoost classification except for the mathematical programming formulation and gain function. The gain function is the same as that of gPLS.

We set minimum support parameter (*minsup*) to 2 for relatively small datasets (EDKB, CPDB and AIDS1), and to 10% of the number of positives for large datasets (CAS, AIDS2 and AIDS3). Throughout the experiments maximum pattern size (*maxpat*) is set to 10. We used AMD Opteron 2.2GHz system with at most 8GB memory for all experiments.

gPLS vs. gBoost

GPLS is compared with gBoost in five fold cross validation settings. In gPLS, there are two parameters to tune, namely the number of iterations m and the number of obtained patterns per search k. For each dataset, we exhaustively tried all combinations from $m=\{5, 10, 15, 20, 25, 30,$

Figure 12. Regression accuracy (left) and computational time (right) against maximum pattern size (maxpat) in the EDKB dataset

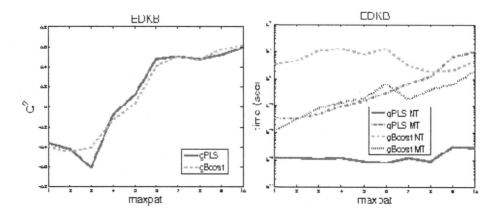

Figure 13. Classification accuracy (left) and computational time (right) against maximum pattern size (maxpat) in the CPDB dataset

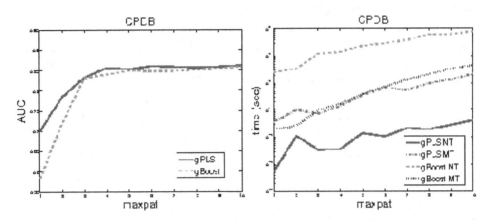

35} and k={5, 10, 15, 20, 25, 30, 35}. In the following, we always report the best test accuracy among all settings. Notice that, for AIDS datasets, the parameter values are changed as m={10, 20, 30, 40, 50}, k={10, 20, 30, 40, 50} to cope with large-scale data. In gBoost, the regularization parameter was varied as v={0.1, 0.2,...,0.9} for classification, and λ={10, 50, 100, 150, 200, 1000} for regression. The number of patterns to add per iteration is set to 50 for CAS and AIDS, and 10 for the other datasets. The accuracy is measured by Q^2 for regression and by the area under the ROC curve (AUC) for classification. The Q^2 score is defined as

$$Q^2 = 1 - \frac{\sum_{i=1}^{n}\left(y_i - f(x_i)\right)^2}{\sum_{i=1}^{n}\left(y_i - \frac{1}{n}\sum_{i=1}^{n}y_i\right)^2}$$

which is close to 1 when the regression function fits good, and takes unbounded negative value for a random input. The interpretation is similar to that of the Pearson correlation coefficient. The results of gPLS and gBoost are compared in Table 5 and Table 6. For EDKB and CPDB datasets, we performed more detailed experiments with different settings of maximum pattern size (Figure 7 and 8). In terms of accuracy, it is difficult to decide which method is better. GPLS was better in EDKB, CPDB and AIDS1, but gBoost was better in CAS. However, in terms of computational time, gPLS is clearly superior. In the table, we distinguish the computational time for pattern search (mining time, MT) and the numerical computations (numerical time, NT). The numerical time of gBoost was significantly larger than that of gPLS in all datasets, showing that gPLS's computational simplicity contributes to reduce the actual computational load. For large datasets (AIDS2 and AIDS3), gBoost did not finish in a reasonable amount of time.

Figure 14 shows the patterns selected by gPLS from the EDKB dataset. It is often observed that similar patterns are extracted together in the same component. This property makes PLS stable, because the regression function is less affected by small changes in graph data. Not only that, this property can help chemists to identify characteristic functional groups in the same latent component.

Table 5. Results of gPLS in various datasets. Values in the parentheses are optimal parameters achieving the best test accuracy. P: the average number of obtained patterns, MT: mining time, NT: numerical time, ITR: the number of iterations required until convergence

	gPLS				
	(m, t)*	P	MT	NT	AUC/$Q^{2\dagger}$
EDKB	(10, 30)	296	16.0	0.0025	$0.647^{\dagger} \pm 0.129$
CPDB	(20, 15)	258	26.8	0.474	0.862 ± 0.0214
CAS	(30, 10)	294	3570	14.1	0.870 ± 0.0098
AIDS1	(10, 10)	99	290	0.0652	0.773 ± 0.0538
AIDS2	(40, 10)	396	50300	167	0.747 ± 0.0266
AIDS3	(50, 20)	946	57100	509	0.883 ± 0.0541

Table 6. Results of gBoost in various datasets. Values in the parentheses are optimal parameters achieving the best test accuracy. P: the average number of obtained patterns, MT: mining time, NT: numerical time, ITR: the number of iterations required until convergence

	gBoost					
	$(\nu / \lambda^{\dagger})$*	P	MT	NT	AUC/$Q^{2\dagger}$	ITR
EDKB	$(100^{\dagger})$	216	15.6	83.3	$0.639^{\dagger} \pm 0.164$	9.2
CPDB	(0.4)	260	22.8	344	0.862 ± 0.0316	18.6
CAS	(0.4)	503	8630	391	0.867 ± 0.000251	13.4
AIDS1	(0.4)	186	783	299	0.752 ± 0.138	19.6
AIDS2	over 24h					
AIDS3	over 24h					

Figure 14. Patterns obtained by gPLS from the EDKB datasets. Each column corresponds to the patterns of each latent component

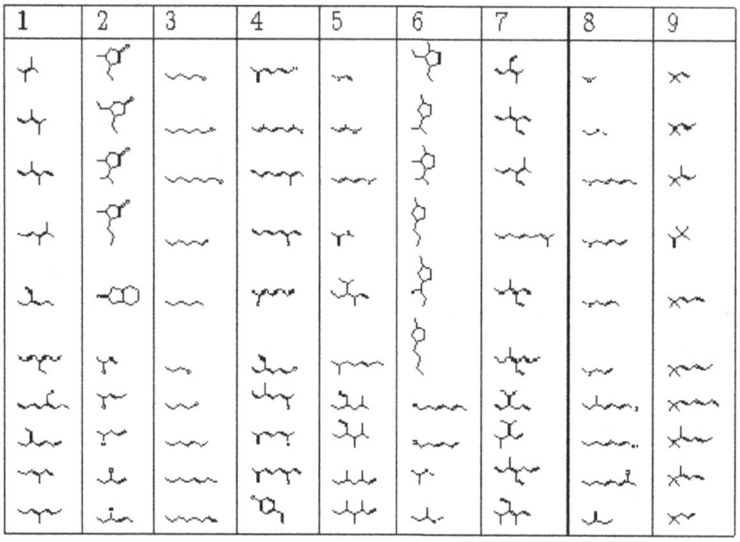

Table 7. Results of gPLS and gBoost in various datasets. Values in the parentheses are optimal parameters achieving the best test accuracy. P: the average number of obtained patterns, MT: mining time, NT: numerical time, ITR: the number of iterations required until convergence

		gPLS				gBoost					
	$(m,r)*$	P	MT	NT	AUC/$Q^{2\dagger}$	$(v/C^\dagger)*$	P	MT	NT	AUC/$Q^{2\dagger}$	ITR
EDKB	(10, 30)	296	16.0	0.0025	$0.647^\dagger$	$(100^\dagger)$	216	15.6	83.3	$0.639^\dagger$	9.2
CPDB	(20, 15)	258	26.8	0.474	0.862	(0.4)	260	22.8	344	0.862	18.6
CAS	(30, 10)	294	3570	14.1	0.870	(0.4)	503	8630	391	0.867	13.4
AIDS1	(10, 10)	99	290	0.0652	0.773	(0.4)	186	783	299	0.752	19.6
AIDS2	(40, 10)	396	50300	167	0.747	over 24h					
AIDS3	(50, 20)	946	57100	509	0.883	over 24h					

Table 8. Frequent mining + PLS vs gPLS in the CPDB dataset

frequent mining + PLS					gPLS			
P	MT	NT	AUC	maxpat	P	MT	NT	AUC
17	0.0927	0.038	0.696	1	15.2	0.308	0.038	0.700
61	0.148	0.0164	0.770	2	45.8	1.20	0.117	0.782
182	0.212	0.0335	0.812	3	73.8	1.09	0.0573	0.833
515	0.282	0.0923	0.842	4	82.6	2.06	0.0488	0.857
1387	0.602	0.221	0.846	5	93.4	1.97	0.0296	0.844
3500	2.55	0.525	0.852	6	85.6	2.67	0.0222	0.833
8215	4.38	1.60	0.848	7	65.4	3.19	0.0146	0.837
18107	7.6	5.32	0.840	8	172	13.1	0.247	0.857
37719	17.9	7.42	0.840	9	209	12.7	0.282	0.859
74857	40.3	51.2	0.842	10	244	26.8	0.474	0.862
143006	70.3	92.8	0.835	11	244	35.4	0.375	0.862
out of memory				12	244	46.3	0.367	0.862
out of memory				13	244	52.4	0.549	0.861
out of memory				∞	244	66.3	0.586	0.861

Efficiency Gain by Iterative Mining

The main idea of iterative mining is to gain efficiency by means of adaptive example weights. We evaluated how large the efficiency gain is by comparing gPLS and a naïve method that enumerates all patterns first and apply PLS afterwards. Table 8 summarizes the results for different maximum pattern sizes (maxpat). In the naïve method, the number of patterns grow exponentially, hence

the computational time for PLS grows rapidly as well. GPLS successfully keeps computational time small in all occasions.

Graph Least Angle Regression (gLARS)

In this subsection, we deal with another regression method which also works good in high dimensional feature space. We consider a least square regression problem with ℓ_1 penalty on the coefficient vectors.

$$\min_{\mathbf{b}} \lambda \parallel \mathbf{b} \parallel_1 + \parallel \mathbf{X}\mathbf{b} - \mathbf{y} \parallel_2^2 . \qquad (21)$$

ℓ_1-regularization has drawn attention of researchers in the last decade. It can shrink most of the coefficients β to zeros, therefore can be used as feature selection. It first appeared in mid 90's (Breiman, 1995, Donoho, Johnsotne, Kerkyacharian, & Picard, 1995, Tibshrani, 1996), and recently recognized that this general idea have been used in signal processing, statistics and machine learning. Tibshirani named this problem (21) as LASSO (Tibshrani, 1996) and showed the solution for a fixed regularization parameter \lambda. In general, regularization parameter depends on a data at hand, and cannot be fixed beforehand. In machine learning, a grid search is often performed to find the best regularization parameter from a wide range of choice between 0 and ∞. Note that setting $\lambda=0$ reduces LASSO to ordinal least squares (OLS), in which β are not regularized at all. Typically setting λ large shrinks β to zeros. Osborne et al. proposed to solve LASSO using an active set method (Osborne et al., 2000). The statistical properties of using the active set method for solving LASSO was recognized by Efron et al. (Efron et al., 2004). They proposed a closely

related method called LARS, and showed that it can trace the entire path of coefficient vectors along the whole regularization parameters. They also showed that the computational complexity of LARS is in the same order as that of OLS. This attractive property is due to the piecewise-linear nature (Rosset & Zhu, 2003) of the LASSO solution path. Namely, the derivative of the LASSO solutions (coefficients) with respect to regularization parameter is linear between special points called knots, so we do not have to make small steps between them (Figure 10). The algorithms starts with the largest possible value of \lambda, then gradually decreases it while tracing the changes in β.

Below, we present a basic procedure of the active set method applied to solving LASSO while iteratively collecting subgraph features by graph mining. First we split $\beta = \beta^+ - \beta^-$ and introduce two nonnegative variables β^+ and β^- The original LASSO problem (21) is rewritten as:

$$\min_{\mathbf{b}^+,\mathbf{b}^-} \lambda \sum \left(\mathbf{b}^+ + \mathbf{b}^- \right)$$
$$+ \frac{1}{2} \left(\mathbf{X}\left(\mathbf{b}^+ - \mathbf{b}^- \right) - \mathbf{y} \right)^{\mathrm{T}} \left(\mathbf{X}\left(\mathbf{b}^+ - \mathbf{b}^- \right) - \mathbf{y} \right)$$
$$\qquad (22)$$

s.t. $\beta^+, \beta^- \geq$ $\qquad (23)$

Figure 15. Schematic figure of the regularization paths in the space of the weight vector β(λ). To follow the path from the starting point λ=λ₀, the direction vector γ and the step size are computed and then one jumps to the next turning point (knot). By repeating this, one can follow the entire regularization path without taking small steps

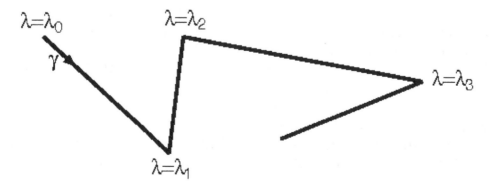

The dual of the above problem is

$$\max_{\mathbf{b}} \frac{1}{2} \mathbf{b}^T \mathbf{X}^T \mathbf{X} \mathbf{b} \qquad (24)$$

$$s.t. -\lambda \le (\mathbf{X}^T \mathbf{r})_p \le \lambda, \quad \forall p \qquad (25)$$

where $\mathbf{r} = \mathbf{X}\beta - \mathbf{y}$ is *residual* in the primal problem. Note that P spans the large column space of X, therefore we have large number of constraints in Equation (25). From the dual problem, it is clear that the optimality of this problem for a fixed parameter λ requires that all the constraints satisfy $|X^T r|_p \le \lambda$. Also, initializing λ as a large value and decreasing it corresponds to controlling the absolute correlation to enter to the solution set. From the KKT condition of the primal problem, following equations hold:

$$X^T X \beta - X^T y + \lambda I - \eta^+ = 0 \qquad (26)$$

$$-X^T X \beta - X^T y + \lambda I - \eta^- = 0 \qquad (27)$$

$$\beta_p^+ \eta_p^+ = 0 \qquad (28)$$

$$\beta_p^- \eta_p^- = 0 \qquad (29)$$

where η^+, $\eta^- \ge$ are Lagrange multipliers corresponding to β^+ and β^- respectively. We call a set of indices of the column of X as active se*t* if they have nonzero coefficients. If initial value $\lambda(=\lambda_0)$ is large enough, then optimization problem (22) forces β to close to zeros. From Equations (28) and (29), this means the occurrence of nonzero η^+ and η^-. Plugging all these relations into Equations (26) and (27) obtains

$$\lambda_0 \ge |X^T y|_p, \forall p$$

Therefore, it suffices to initialize λ_0 as $\|X^T r\|_\infty$ instead of setting it to infinitely large, where $\|x\|_\infty = \max_j |x_j|$ denotes the ℓ_∞ norm of x. Whenever

$\beta \ne 0$, then $\eta = 0$ from Equation (28) and (29), therefore active set satisfies the following conditions

$$\lambda = -(X^T r)_p, \ \{p | \forall p, a_p > 0$$

$$\lambda = (X^T r)_p, \ \{p | \forall p, a_p < 0$$

In other words, all the active set has the same amount of absolute correlation

$$\lambda = \|X^T r\|_\infty$$

From Equation (28), if $\beta^+ > 0$ then $\eta^+ = 0$. Such β^+ is obtained from Equation (26) as:

$$\beta^+ = (X^T X)^{-1}(X^T y - \lambda I) \qquad (30)$$

Similarly,

$$\beta^- = (X^T X)^{-1}(-X^T y + \lambda I) \qquad (31)$$

Assuming that the objective function is locally linear around β^+, we try updating $\beta^+ = \beta^+ + \Delta^+$. Plugging this into Equations (28) and (29) obtains

$$\Delta^+ = (X^T X)^{-1}(X^T(y - X\beta^+) - \lambda I) = (X^T X)^{-1}(X^T r - \lambda I) \qquad (32)$$

Similarly,

$$\Delta^- = -(X^T X)^{-1}(X^T r - \lambda I) \qquad (33)$$

This observation highlights the following fact: β^+ and β^- move along the directions $g^+ = -(X^T X)^{-1}$ and $g^- = (X^T X)^{-1}$, respectively, unless reaching the non-differential point explained below. Since it is desired to take the largest step size until reaching the next event, step size is determined by choosing

$$p^* = \underset{p}{argmax} \left| X^T r \right|_p . \qquad (34)$$

If the full step size $|\|X^T r\|_\infty - \lambda|$ is feasible with respect to all the constraints, the following addition event occurs:

Figure 16. gLARS algorithm

Algorithm 6 gLARS algorithm

1: $p^* = \underset{p}{\mathrm{argmax}} \left| \boldsymbol{X}^\top \boldsymbol{y} \right|_p$ ▷ Initial pattern search

2: $\boldsymbol{\beta} = \left(\hat{\boldsymbol{X}}^\top \hat{\boldsymbol{X}} \right)^{-1} (\boldsymbol{X}^\top \boldsymbol{y})_{p^*}$

3: $\lambda_0 \leftarrow \| \boldsymbol{X}^\top \boldsymbol{y} \|_\infty$

4: $\hat{\boldsymbol{X}} = \emptyset; \ \hat{\boldsymbol{X}} \leftarrow \boldsymbol{X}_{p^*}$

5: $\boldsymbol{r} = \hat{\boldsymbol{X}} \boldsymbol{\beta} - \boldsymbol{y}$

6: $k = 1$

7: **while** $\| \boldsymbol{X}^\top \boldsymbol{r} \|_\infty > 0$ **do**

8: $\quad p^* = \underset{p}{\mathrm{argmax}} \left| \boldsymbol{X}^\top \boldsymbol{r} \right|_p$ ▷ Main pattern search

9: $\quad \Delta = \left(\hat{\boldsymbol{X}}^\top \hat{\boldsymbol{X}} \right)^{-1} \mathrm{sgn}(\boldsymbol{\beta}) | (\boldsymbol{X}^\top \boldsymbol{r})_{p^*} - \lambda_{k-1} |$

10: $\quad$ **if** $(\beta_j + d\Delta)_{p^\dagger} = 0$ for $p^\dagger$ and $0 < d < 1$ in active set **then**

11: $\quad\quad \hat{\boldsymbol{X}} \leftarrow \hat{\boldsymbol{X}} \backslash \hat{\boldsymbol{X}}_{p^\dagger}$ ▷ remove constraint

12: $\quad\quad \boldsymbol{\beta} \leftarrow \boldsymbol{\beta} + d\Delta$

13: $\quad$ **else**

14: $\quad\quad \hat{\boldsymbol{X}} \leftarrow \hat{\boldsymbol{X}} \cup \boldsymbol{X}_{p^*}$ ▷ add constraint

15: $\quad\quad \boldsymbol{\beta} \leftarrow \boldsymbol{\beta} + \Delta$

16: $\quad$ **end if**

17: $\quad \lambda_k \leftarrow \| \hat{\boldsymbol{X}}^\top \boldsymbol{r} \|_\infty$

18: $\quad \boldsymbol{r} = \hat{\boldsymbol{X}} \boldsymbol{\beta} - \boldsymbol{y}$

19: $\quad k = k + 1$

20: **end while**

---

1. make a step $\beta = \beta + \Delta$,
2. add p^* to the active set,

While moving β along this direction, one of the coefficients might reach a non-differential point called knot (See Figure 16). Such a point can be characterized as $\left(\mathbf{b} + d\Delta \right)_{p^\dagger} = 0$, where 0<d<1. In this case the following deletion event occurs:

1. choose the coefficient $p^\dagger$ that reaches a knot first,
2. make a step $\beta = \beta + d\Delta$,
3. delete the $p^\dagger$-th constraint from the active set,

After both addition and deletion events, current correlation is recorded as λ, and residual is updated as $\mathbf{r} = \hat{\mathbf{X}} \mathbf{b} - \mathbf{y}$, where the columns of $\hat{\mathbf{X}}$ stores the current active set.

Below, we describes how to apply active set method to graph data. Our strategy is similar to that of gPLS presented in the previous subsection. First, we initialize $\hat{\mathbf{X}}$ as an empty matrix, and update/downdate the matrix as the iterations proceed. We add/delete only one pattern p in each iteration. Graph mining is embedded in the process of finding the maximum correlation (34), and the same pruning condition as that of gPLS (Theorem (4)) is employed. The pseudocode of this algorithm is described in Figure 16.

Lemma 1. Algorithm 6 traces all the events along a regulatrization path.

Figure 17. Patterns included or excluded in the first 10 events for the EDKB dataset. The 7th event is an exclusion event

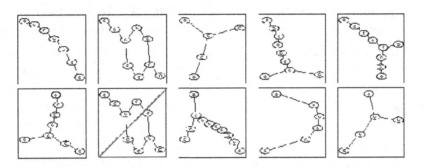

Figure 18. Solution paths for the EDKB dataset. Each curve shows the evolution of the weight parameter of a pattern. A curve can be terminated by an exclusion event. For example, the curve emerging from the 2nd event disappears at the 7th event

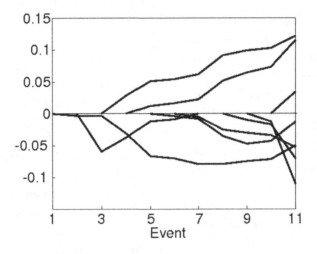

Figure 19. Validation errors of gLARS for the CPDB dataset (maxpat=10). The vertical line indicates the event that achieved the smallest validation error

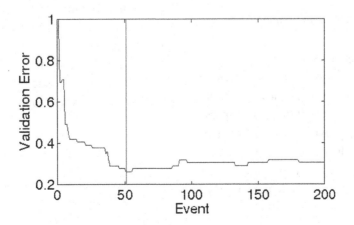

Proof The statement can be proved by verifying that the step size λ_k-λ_{k+1} is the minimum distance between neighboring events. Since we update β in the gradient direction of decreasing r,| $r_i^{(k)}$ |$\geq$| $r_i^{(k+1)}$ | holds for any i. Using this result, we have

$$\|X^T r^{(k)}\|_\infty \geq \|\mathbf{X}^T r^{(k+1)}\|_\infty \qquad (35)$$

So if we choose $\lambda_k = \|X^T r^{(k)}\|_\infty$, then λ_k is always greater than or equal to λ_{k+1}. Moreover, since we start λ as large, choosing the largest $\lambda_{k+1} = \|X^T r^{(k+1)}\|_\infty$ minimizes the distance λ_k-λ_{k+1}.

Computational Experiments

First, we illustrate how our method works using the estrogen receptor dataset from the Endocrine Disruptors Knowledge Base (Table 1). It contains 131 chemical compounds suspected to work as environmental hormones, and our task is to predict its toxicity. The first 10 events of gLARS are shown in Figure 17. As it is just the beginning of the regularization path, most events are inclusions. However, within this short period, an exclusion happens at the 7th event. The evolution of weight parameters is shown in Figure 18. The excluded pattern corresponds to the curve that converges to zero at the 7th event. In real situations, the LASSO path-following algorithm is terminated shortly after the minimum error point to avoid overfitting (Figure 19).

Regression (QSAR Analyses)

We used three datasets (AR, ER and ES) from Endocrine Disruptors Knowledge Base (EDKB) 11. A summary of the datasets in shown in Table 1. We compared gBoost with MGK and Comparative Molecular Field Analysis (CoMFA) (hong03, shi00). We evaluated the performance of MGK, freqSVM and gBoost with our own implementations, and quoted the reported performance

of CoMFA from the literature. For gBoost, the regularization parameter λ is chosen from {0.1, 1, 10, 100, 1000, 10000}. For freqSVM, minimum support threshold is set to 1, and the regularization parameter λ is chosen from {0.1, 1, 10, 100, 1000, 10000}. For MGK, the termination probability is chosen from {0.1...0.9} then obtained kernel is fed into SVR where the tube parameter ε is set to 0.1, the regularization parameter λ is chosen from {0.1, 1, 10, 100, 1000, 10000}. gBoost performed best when setting λ parameter to 100 for all three datasets, and built regressor consisting of 36, 50 and 30 patterns on average for AR, ER and ES, respectively.

Table 3 summarizes the experimental results. The regression models are evaluated by the mean absolute error (MAE) and the Q^2 score:

$$Q^2 = 1 - \frac{\sum_{n=1}^{l}(y_n - f(x_n))^2}{\sum_{n=1}^{l}(y_n - \frac{1}{l}\sum_{n=1}^{l}y_n)^2}.$$

Our method (gBoost) performed constantly better than MGK and freqSVM, and better than CoMFA in AR. On the other hand, in ER, CoMFA was superior to other methods. In CoMFA, all training compounds are aligned to a template compound, which is manually picked up by an expert. Then, each compound is encoded into a high dimensional feature vector describing the steric and electrostatic intersections with the template compound. Thus CoMFA is not a fully automatic method, and it assumes that training compounds are close to each other. The result in ER suggests that the energy features are useful for learning, but at the same time, it seems difficult to incorporate such features into our framework without experts' intervention.

CONCLUSION

We have presented graph mining based machine learning approaches and their application to virtual screening. Our methods automatically create features optimal for a given chemical dataset, so the preparation of physicochemical descriptors is not necessary. On the other hand, we face a problem of selecting subgraph features from a large number of candidates. In order to select small number of subgraph features, we enforced sparsity in all the methods. ℓ_1-regularization is nowadays very popular method (Breiman, 1995, Donoho et al., 1995, Tibshrani, 1996, Kim, Koh, Lustig, Boyd, & Gorinevsky, 2007), and we employed it for solving SVM and least squares. Dimension reduction by PLS has been popular in chemometrics community (Höskuldsson, 1988). In our gPLS approach, highly-correlated subgraph features are successfully treated by building new orthogonal basis. Nevertheless original PLS does not select features (Frank & Friedman, 1993), we introduced sparseness and selected subgraph features in gPLS thanks to non-deflation PLS algorithm.

Other researchers (MorSeS00, Tak06, Bringmann et al., 2006) also proposed similar iterative mining methods for classification based on information gain or correlation. However, to the best of our knowledge, our work is the first one which discussed the mining criterion for regression. Also, the coincidence of the mining criterion in PLS and LARS is not surprising, and indeed is equivalent to that of classical forward stepwise selection (mill90). Our non-deflation PLS algorithm made this connection clearer. In fact, PLS has recently recognized as path-following method (frie2004). Momma et al. discussed PLS as an additive model, and showed its connection to boosting (Momma0601). Although we did not present it here, a cutting edge method of our graph regression setting is Bayesian linear regression (BLR) proposed in (chia09). A Bayesian approach gives us posterior distribution around each predicted response, which can be used for rejecting uncertain predictions to improve performance. Our algorithm consists of two tightly-coupled components: the machine learning part that solves the mathematical program and the graph mining part that finds optimal patterns. This coupling, together with Theorem 2 enables us to prune the search space dramatically without using external constraints such as minimum support. It was shown that our tight coupling of graph mining and machine learning algorithms leads to better efficiency in comparison with the naïve coupling. Our approach is very general, since one can employ different machine learning algorithm for solving various problems. For instance, other supervised learning algorithms are readily described as mathematical programs (Yuan & Casasent, 2003, Cai & Hofmann, 2004, Rätsch et al., 2002), and can be solved by constraint generation in combination with some mining algorithm.

From the mining side, one can try itemset mining (Nowozin, Tsuda, et al., 2007, Saigo, Uno, & Tsuda, 2007), tree mining, sequence mining (Nowozin, Bakir, & Tsuda, 2007) and so forth. From the learning side, we restricted ourselves to solving classification and regression in this chapter. However, unsupervised methods have also been proposed. A binomial mixture model is fitted for graph data by EM algorithm in (Tsuda & Kudo, 2006), whereas variational dirichlet process (DP) is employed for the same problem in (Tsuda & Kurihara, 2008). In (Saigo & Tsuda, 2008), graph mining is coupled with the Lanczos method for collecting subgraph patterns corresponding to the major entries of top principal components.

However, we have to point out that the search technique employed here can further be improved. Combinatorial search is a mature research field and there have been a lot of methods proposed so far. Compared to those sophisticated methods, our pattern search strategy is still fundamental. For example, a standard A^* algorithm can be applied for even better results. In this chapter, we did not use multiple pricing, which was used in our previous paper (Saigo et al., 2006). In multiple pricing, the top *k* patterns are added to the dual problem, not just one. When the search space is not reused, the multiple pricing is effective in reducing the computational time. However, when the search space is progressively expanded, each pattern search becomes much more efficient and the effect of multiple pricing is no longer significant. It also leads to more constraints in the dual problem, which increases the LP time. In (Saigo et al., 2006), we introduced additional constraints to the linear programming to describe chemical compounds with unobserved activities. After publication of (Saigo et al., 2006), we found that the accuracy gain by such additional constraints is not significant empirically. So, we did not include the topic in this paper. Nevertheless, it is an advantage of our mathematical programming-based framework that additional constraints can be incorporated easily.

ACKNOWLEDGMENT

The authors would like to thank Nicole Krämer, Pierre Mahé and Ichigaku Takigawa for fruitful discussion, and Taku Kudo and Sebastian Nowozin for a part of coding.

REFERENCES

Breiman, L. (1995). Better subset regression using the nonnegative garotte. *Technometrics, 37*, 373–384. doi:10.2307/1269730

Bringmann, B., Zimmermann, A., Raedt, L. D., & Nijssen, S. (2006). *Don't be afraid of simpler patterns. In 10th european conference on principles and practice of knowledge discovery in databases (pkdd)* (pp. 55–66).

Cai, L., & Hofmann, T. (2004). Hierarchical document categorization with support vector machines. In *Acm 13th conference on information and knowledge management* (p. 78-87). New York: ACM Press.

Demiriz, A., Bennet, K., & Shawe-Taylor, J. (2002). Linear programming boosting via column generation. *Machine Learning, 46*(1-3), 225–254. doi:10.1023/A:1012470815092

Deshpande, M., Kuramochi, M., Wale, N., & Karypis, G. (2005). Frequent sub-structure-based approaches for classifying chemical compounds. *IEEE Transactions on Knowledge and Data Engineering, 17*(8), 1036–1050. doi:10.1109/TKDE.2005.127

Donoho, D., Johnsotne, I., Kerkyacharian, G., & Picard, D. (1995). Wavelet shrinkage: Asymptopia? (with discussion). *Journal of the Royal Statistical Society. Series B. Methodological, 57*, 301–369.

du Merle, O., Villeneuve, D., Desrosiers, J., & Hansen, P. (1999). Stabilized column generation. *Discrete Mathematics, 194*, 229–237. doi:10.1016/S0012-365X(98)00213-1

Efron, B., Hastie, T., Johnstone, I., & Tibshirani, R. (2004). Least angle regression. *Annals of Statistics, 32*(2), 407–499. doi:10.1214/009053604000000067

Eldén, L. (2004). Partial least squares vs. lanczos bidiagonalization i: Analysis of a projection method for multiple regression. *Computational Statistics & Data Analysis, 46*(1), 11–31. doi:10.1016/S0167-9473(03)00138-5

Frank, E., & Friedman, J. H. (1993). A statistical view of some chemometrics regression tools. *Technometrics, 35*(2), 109–135. doi:10.2307/1269656

Gasteiger, J., & Engel, T. (2003). *Chemoinformatics: a textbook*. New York: Wiley-VCH. doi:10.1002/9783527618279

Helma, C., Cramer, T., Kramer, S., & Raedt, L. (2004). Data mining and machine learning techniques for the identification of mutagenicity inducing substructures and structure activity relationships of noncongeneric compounds. *Journal of Chemical Information and Computer Sciences*, *44*, 1402–1411. doi:10.1021/ci034254q

Höskuldsson, A. (1988). PLS Regression Methods. *Journal of Chemometrics*, *2*, 211–228. doi:10.1002/cem.1180020306

Inokuchi, A. (2005). Mining generalized substructures from a set of labeled graphs. In *Proceedings of the 4th ieee internatinal conference on data mining* (p. 415-418). New York: IEEE Computer Society.

Kashima, H., Tsuda, K., & Inokuchi, A. (2003). Marginalized kernels between labeled graphs. In *Proceedings of the 21st international conference on machine learning* (p. 321-328). Menlo Park, California: AAAI Press.

Kazius, J., Nijssen, S., Kok, J., Bäck, T., & Ijzerman, A. (2006). Substructure mining using elaborate chemical representation. *Journal of Chemical Information and Modeling*, *46*, 597–605. doi:10.1021/ci0503715

Kim, S. J., Koh, K., Lustig, M., Boyd, S., & Gorinevsky, D. (2007). An interior-point method for large-scale \ell_1-regularized least squares. *IEEE Jounal of Selected Topics in Signal Processing*, *1*(4), 606–617. doi:10.1109/JSTSP.2007.910971

Kohavi, R., & John, G. H. (1997). Wrappers for feature subset selection. *Artificial Intelligence*, *1-2*, 273–324. doi:10.1016/S0004-3702(97)00043-X

Kramer, S., Raedt, L., & Helma, C. (2001). Molecular feature mining in HIV data. In *Proceedings of the 7th acm sigkdd international conference on knowledge discovery and data mining*. New York: ACM Press.

Kudo, T., Maeda, E., & Matsumoto, Y. (2005). An application of boosting to graph classification. [Cambridge, MA: MIT Press.]. *Advances in Neural Information Processing Systems*, *17*, 729–736.

Lanczos, C. (1950). An iteration method for the solution of the eigenvalue problem of linear differential and integral operators. *Journal of Research of the National Bureau of Standards*, *45*, 255–282.

Luenberger, D. G. (1969). *Optimization by vector space methods*. New York: Wiley.

Morishita, S. (2001). Computing optimal hypotheses efficiently for boosting. In *Discovery science* (p. 471-481).

Nijssen, S., & Kok, J. (2004). A quickstart in frequent structure mining can make a difference. In *Proceedings of the 10th acm sigkdd international conference on knowledge discovery and data mining* (pp. 647–652). New York: ACM Press.

Nowozin, S., Bakir, G., & Tsuda, K. (2007). Discriminative subsequence mining for action classification. In *Proceedings of the 11th ieee international conference on computer vision (iccv 2007)* (pp. 1919–1923). IEEE Computer Society.

Nowozin, S., Tsuda, K., Uno, T., Kudo, T., & Bakir, G. (2007). Weighted substructure mining for image analysis . In *Ieee computer society conference on computer vision and pattern recognition (cvpr)*. IEEE Computer Society.

Osborne, M., Presnell, B., & Turlach, B. (2000). On the lasso and its dual. *IMA Journal of Numerical Analysis*, *20*, 389–404. doi:10.1093/imanum/20.3.389

Pei, J., Han, J., Mortazavi-asl, B., Wang, J., Pinto, H., & Chen, Q. (2004). Mining sequential patterns by pattern-growth: The prefixspan approach. *IEEE Transactions on Knowledge and Data Engineering*, *16*(11), 1424–1440. doi:10.1109/TKDE.2004.77

Rätsch, G., Mika, S., Schölkopf, B., & Müller, K.-R. (2002). Constructing boosting algorithms from SVMs: an application to one-class classification. *IEEE Transactions on Pattern Analysis and Machine Intelligence, 24*(9), 1184–1199. doi:10.1109/TPAMI.2002.1033211

Rosipal, R., & Krämer, N. (2006). Overview and recent advances in partial least squares . In *Subspace, latent structure and feature selection techniques* (pp. 34–51). New York: Springer. doi:10.1007/11752790_2

Rosset, S., & Zhu, J. (2003). *Piecewise linear regularized solution paths* (Tech. Rep.). Stanford University. (Technical Report HAL:ccsd-00020066)

Saigo, H., Kadowaki, T., & Tsuda, K. (2006). A linear programming approach for molecular QSAR analysis . In Gärtner, T., Garriga, G., & Meinl, T. (Eds.), *International workshop on mining and learning with graphs (mlg)* (pp. 85–96).

Saigo, H., Krämer, N., & Tsuda, K. (2008). Partial least squares regression for graph mining. In *Proceedings of the 14th acm sigkdd international conference on knowledge discovery and data mining (kdd2008)* (pp. 578–586).

Saigo, H., Nowozin, S., Kadowaki, T., Kudo, T., & Tsuda, K. (2009). gBoost: A mathematical programming approach to graph classification and regression. *Machine Learning, 75*(1), 69–89. doi:10.1007/s10994-008-5089-z

Saigo, H., & Tsuda, K. (2008). Iterative subgraph mining for principal component analysis. In *Proceedings of the 8th ieee international conference on data mining (icdm 2008)* (pp. 1007–1012).

Saigo, H., Uno, T., & Tsuda, K. (2007). Mining complex genotypic features for predicting HIV-1 drug resistance. *Bioinformatics (Oxford, England), 23*(18), 2455–2462. doi:10.1093/bioinformatics/btm353

Tibshrani, R. (1996). Regression shrinkage and selection via the LASSO. *Journal of the Royal Statistical Society. Series B. Methodological, 58*(1), 267–288.

Tsuda, K., & Kudo, T. (2006). Clustering graphs by weighted substructure mining. In *Proceedings of the 23rd international conference on machine learning* (p. 953-960). New York: ACM Press.

Tsuda, K., & Kurihara, K. (2008). Graph mining with variational dirichlet process mixture models. In *Siam conference on data mining (sdm)*.

Uno, T., Kiyomi, M., & Arimura, H. (2005). LCM ver.3: collaboration of array, bitmap and prefix tree for frequent itemset mining. In *Osdm '05: Proceedings of the 1st international workshop on open source data mining* (pp. 77–86).

Wale, N., & Karypis, G. (2006). Comparison of descriptor spaces for chemical compound retrieval and classification. In *Proceedings of the 2006 ieee international conference on data mining* (pp. 678–689).

Wold, H. (1966). Estimation of principal components and related models by iterative least squares. In P. R. Krishnaiaah (Ed.), *Multivariate analysis* (pp. 391–420). Maryland Heights, MO: Academic Press.

Wold, H. (1975). Path models with latent variables: The NIPALS approach . In *Quantitative sociology: International perspectives on mathematical and statistical model build ing* (pp. 307–357). Maryland Heights, MO: Academic Press.

Wold, S., Ruhe, A., Wold, H., & Dunn, W. J. III. (1984). The collinearity problem in linear regression. the partial least squares (PLS) approach to generalized inverses. *SIAM J. Sci. Stat. Comput., 5*(3), 735–743. doi:10.1137/0905052

Wold, S., Sjöstöm, M., & Erikkson, L. (2001). PLS-regression: a basic tool of chemometrics. *Chemometrics and Intelligent Laboratory Systems, 58*, 109–130. doi:10.1016/S0169-7439(01)00155-1

Yan, X., & Han, J. (2002a). gSpan: graph-based substructure pattern mining. In *Proceedings of the 2002 ieee international conference on data mining* (p. 721-724). IEEE Computer Society.

Yan, X., & Han, J. (2002b). *gSpan: graph-based substructure pattern mining* (Tech. Rep.). Department of Computer Science, University of Illinois at Urbana-Champaign.

Yan, X., Yu, P. S., & Han, J. (2004). Graph indexing: a frequent structure-based approach. In *Proceedings of the acm sigmod international conference on management of data* (p. 335-346).

Yuan, C., & Casasent, D. (2003). A novel support vector classifier with better rejection performance. In *Proceedings of 2003 ieee computer society conference on pattern recognition and computer vision (cvpr)* (pp. 419–424).

Zaki, M. J. (2005). Efficiently mining frequent trees in a forest: algorithms and applications. In *Ieee transactions on knowledge and data engineering* (pp. 1021–1035).

Zou, H., & Hastie, T. (2005). Regularization and variable selection via the elastic net. *Journal of the Royal Statistical Society. Series B. Methodological, 67*(2), 301–320. doi:10.1111/j.1467-9868.2005.00503.x

ENDNOTES

[1]   (Kazius et al., 2006)

[2]   (Helma et al., 2004)

[3]   (Bringmann, Zimmermann, Raedt, & Nijssen, 2006)

[4]   http://edkb.fda.gov/databasedoor.html

[5]   http://www.predictive-toxicology.org

[6]   Available from the supplementary information of (Helma et al., 2004)

[7]   http://www.cheminformatics.org/datasets/bursi/

[8]   http://dtp.nci.nih.gov/docs/aids/aids_screen.html

[9]   http://chemcpp.sourceforge.net/html/index.html

[10] http://www.kyb.mpg.de/bs/people/nowozin/gboost/

[11] http://edkb.fda.gov/databasedoor.html

APPENDIX

DFS Code Tree

Algorithm 2 finds the optimal pattern which optimizes a gain function. To this end, we need an intelligent way of enumerating all subgraphs of a graph set. This problem is highly nontrivial due to loops: One has to avoid enumerating the same pattern again and again. In this section, we present a canonical search space of graph patterns called *DFS code tree* (Yan & Han, 2002b), that allows to enumerate all subgraphs without duplication. In the following, we assume undirected graphs, but it is straightforward to extend the algorithm for directed graphs.

Figure 20. Depth first search and DFS code of graph. (a) A graph example. (b), (c) Two different depth-first-searches of the same graph. Red numbers represent the DFS indices. Bold edges and dashed edges represent the forward edges and the backward edges respectively

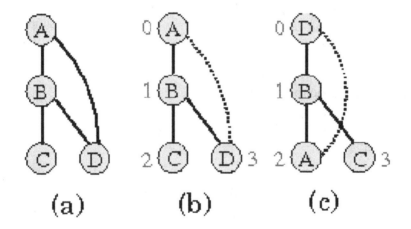

DFS Code

The *DFS code* is a string representation of a graph G based on a depth first search (DFS). According to different starting points and growing edges, there are many ways to perform the search. Therefore, the DFS code of a graph is not unique. To derive a DFS code, each node is indexed from 0 to n-1 according to the discovery time in the DFS. Denote by E^f the *forward edge* set containing all the edges traversed in the DFS, and by E^b the *backward edge* set containing the remaining edges. Figure 8 shows two different indexing of the same graph.

After the indexing, an edge is represented as a pair of indices (i,j) together with vertex and edge labels, $e=(i,j,l_i,l_{ij},l_j) \in V \times V \times L_V \times L_E \times L_V$, where $V=\{0,\ldots,n\text{-}1\}$, L_V and L_E are the set of vertex and edge labels, respectively. The index pair is set as $i<j$, if it is an forward edge, and $i>j$ if backward. It is assumed that there are no self-loop edges. To define the DFS code, a linear order $\prec_T$ is defined among edges. For the two edges $e_1=(i_1,j_1)$ and $e_2=(i_2,j_2)$, $e_1 \prec_T e_2$ if and only if one of the following statements is true:

1. $e_1,e_2 \in E^f$, and $(j_1<j_2$ or $i_1>i_2 \wedge j_1=j_2)$

2. $e_1, e_2 \in E^b$, and ($j_1 < j_2$ or $i_1 = i_2 \wedge j_1 < j_2$).
3. $e_1 \in E^b$, $e_2 \in E^f$, and $i_1 < j_2$.
4. $e_1 \in E^f$, $e_2 \in E^b$, and $j_1 \leq i_2$.

The DFS code is a sequence of edges sorted according to the above order.

Minimum DFS Code

Since there are many possible DFS codes, it is necessary to determine the minimum DFS code as a canonical representation of the graph. Let us define a linear order for two DFS codes $D_1 = (a_0, \ldots, a_m)$ and $D_2 = (b_0, \ldots, b_n)$. By comparing the vertex and edge labels, we can easily build a lexicographical order of individual edges a_i and b_j. Then, the *DFS lexicographic order* for the two codes is defined as follows: $D_1 < D_2$ if and only if either of the following is true,

1. $\exists t, 0 \leq t \leq \min(m, n), a_k = b_k$ for $k < t, a_t < b + t$.
2. $a_k = b_k$ for $0 \leq k \leq m$ and $m \leq n$.

Given a set of DFS codes, the minimum code is defined as the smallest one according to the above order.

Right Most Extension

As in most mining algorithms, we form a tree where each node has a DFS code, and the children of a node have the DFS codes corresponding to the supergraphs. The tree is generated in a depth-first manner and the generation of child nodes of a node is done according to the right most extension (Yan & Han, 2002b). Suppose a node has the DFS code $D_1 = (a_0, a_1, \ldots, a_n)$ where $a_k = (i_k j_k)$. The next edge a_{n+1} is chosen such that the following conditions are satisfied:

1. If a_n is a forward edge and a_{n+1} is a forward edge, then $i_{n+1} \leq j_n$ and $j_{n+1} = j_n + 1$.
2. If a_n is a forward edge and a_{n+1} is a backward edge, then $i_{n+1} = j_n$ and $j_{n+1} < i_n$.
3. If a_n is a backward edge and a_{n+1} is a forward edge, then $i_{n+1} \leq i_n$ and $j_{n+1} = i_n + 1$.
4. If a_n is a backward edge and a_{n+1} is a backward edge, then $i_{n+1} = i_n$ and $j_n < j_{n+1}$.

For every possible a_{n+1}, a child node is generated and the extended DFS code $(a_1, \ldots, a_{n+1})$ is stored. The extension is done such that the extended graph is included in at least one graph in the database.

For each pattern, its embeddings to all graphs in the database are maintained. Whenever a new pattern is created by adding an edge, the list of embeddings is updated. Therefore, it is not necessary to perform isomorphism tests whenever a pattern is extended.

DFS Code Tree

The *DFS code tree*, denoted by $\mathbb{T}$, is a tree-structure whose node represents a DFS code, the relation between a node and its child nodes is given by the right most extension, and the child nodes of the same parent is sorted in the DFS lexicographic order.

It has the following completeness property. Let us remove from $\mathbb{T}$ the subtrees whose root nodes have non-minimum DFS codes, and denote by $\mathbb{T}_{min}$ the reduced tree. It is proven that all subgraphs of graphs in the database are still included in $\mathbb{T}_{min}$ (Yan & Han, 2002b). This property allows us to prune the tree as soon as a non-minimum DFS code is found. In Algorithm 2, the minimality of the DFS code is checked in each node generation, and the tree is pruned if it is not minimum (line 9). This minimality check is basically done by exhaustively enumerating all DFS codes of the corresponding graph. Therefore, the computational time for the check is exponential to the pattern size. Techniques to avoid the total enumeration are described in Section 5.1 of (Yan & Han, 2002b), but still it is the most time consuming part of the algorithm.

Chapter 7
Protein Homology Analysis for Function Prediction with Parallel Sub–Graph Isomorphism

Alper Küçükural
University of Kansas, USA & Sabanci University, Turkey

Andras Szilagyi
University of Kansas, USA

O. Uğur Sezerman
Sabanci University, Turkey

Yang Zhang
University of Kansas, USA

ABSTRACT

To annotate the biological function of a protein molecule, it is essential to have information on its 3D structure. Many successful methods for function prediction are based on determining structurally conserved regions because the functional residues are proved to be more conservative than others in protein evolution. Since the 3D conformation of a protein can be represented by a contact map graph, graph matching, algorithms are often employed to identify the conserved residues in weakly homologous protein pairs. However, the general graph matching algorithm is computationally expensive because graph similarity searching is essentially a NP-hard problem. Parallel implementations of the graph matching are often exploited to speed up the process. In this chapter,the authors review theoretical and computational approaches of graph theory and the recently developed graph matching algorithms for protein function prediction.

INTRODUCTION

Computational assignment of protein function from the *3D protein structure* is one of the important open problems in *structural proteomics*. Currently, many proteins deposited in the Protein Data Bank (PDB) have limited or no biological function annotation. Protein functions are usually derived from evolutionarily related proteins.

DOI: 10.4018/978-1-61520-911-8.ch007

Evolutionary association can be determined from sequence and structural similarities. The methods using sequence information are based on the detection of functional motifs (Huang and Brutlag, 2001; Hulo, et al., 2006; Stark and Russell, 2003), global sequence similarity search (Conesa, et al., 2005; Hawkins, et al., 2006; Martin, et al., 2004), determination of similar loci (Hawkins, et al., 2006), and similarities in phylogeny (Engelhardt, et al., 2005; Storm and Sonnhammer, 2002). However, only around 30% of the protein pairs with less than 50% sequence identity have a similar function. Therefore, sequence similarity itself is not sufficient to develop a robust function prediction (Rost, 2002). In addition, several studies indicate that the inclusion of structural information increases the accuracy of predictions (Devos and Valencia, 2000; Thornton, et al., 2000; Wilson, et al., 2000), because structural features are usually more conserved than sequence.

Similarities between protein structures can be identified by structural alignment methods such as DALI (Holm, et al., 2008), CE (Shindyalov and Bourne, 1998), and TM-align (Zhang and Skolnick, 2005). Several function prediction methods employ structural alignment programs to identify the structurally closest proteins and transfer the functional annotation to the target protein. However, the correlation between function and overall protein fold is weak (Martin, et al., 1998). This can in part be explained by the fact that global structural alignment methods do not always capture locally conserved regions, and the biochemical function of a protein is usually determined by the local structure of a few active residues. Therefore, algorithms that aim to extract local structural information should achieve more robust function prediction (Laskowski, et al., 2005; Weinhold, et al., 2008).

The structures and sequences of remotely homologous protein pairs may have diverged during evolution while local structures involved in protein function may have been preserved. The aim of searching for local structural similarities is to detect these preserved, functionally important structural patterns. To discover local structural motifs, the following methods have been described in the literature. In a method based on 3D templates (Laskowski, et al., 2005), the specific 3D conformations of sets of 2-5 residues were extracted from the structures of functionally significant units. This template set was manually compiled to include four types of templates: the enzyme active site, ligand-binding residues, DNA-binding residues, and reverse templates. Given a target protein, the template set is searched for structures locally matching some part of the target protein, within spheres of a 10 Å radius. The matches are ranked using the SiteSeer scoring function. The degree of overlap between target and template residues is calculated, and the algorithm maximizes the sum of the overlap scores of the matched residues in all possible configurations. The method was tested on various distantly related protein pairs with widely divergent sequences. Significant functional matches were found, e.g. two TIM-barrel proteins with very low sequence identity were found to have a high SiteSeer score, and their functional sites were correctly matched. Moreover, some of the predictions for newly released structures with unknown function have later been experimentally verified. In another study, the combination of sequence and structural features were employed to identify functional similarities, based on the assumption that the preserved amino acids at key sites in similar local structures hint at a functional similarity as well (Friedberg, 2006).

Conserved local regions may contain residues that are not adjacent in sequence. Structurally adjacent residues, however, are preserved in most cases. These structurally conserved patterns have been explored by various tools such as JESS (Barker and Thornton, 2003), PINTS (Stark and Russell, 2003), PDBSiteScan (Ivanisenko, et al., 2004), and PAR-3D (Goyal, et al., 2007). Local structural similarities can be detected with search algorithms based on contact map networks. The algorithms can be described in terms of three major characteristics: representation, scoring, and searching. The contact maps are searched for similar regions with graph matching algorithms. However, the possible mutations, insertions and deletions in the protein structures yield very different contact maps. To discover similarities in the presence of such conformational differences, inexact sub-graph matching algorithms are necessary. Because the problem is NP-hard, parallel computing implementations are often used to speed up the process.

In this chapter, we will review the current status in the field of local structure based function annotation. The topics include contact maps for representation, distance functions with *graph theoretical properties* for scoring, and parallel inexact *sub-graph matching algorithms* for local structural similarity search.

REPRESENTATION SCHEME

There are a number of different ways to represent 3D protein structures as a linear string as a simplification. Pattern search and motif discovery algorithms that use sequential information are then utilized with this representation scheme (Barker and Thornton, 2003; Lo, et al., 2007; Matsuda, et al., 1997). Contact maps constitute a different type of representation method, and can incorporate more information than linear strings. They are extracted from the PDB files using the coordinates of alpha carbon atoms.

Graph Representations of Protein Structure

Protein structures can be converted into a graph where the nodes represent the Cα atoms and the links represent interactions (or contacts) between the corresponding residues (Albert and Barabási, 2002; Strogatz, 2001).

The two most commonly used graph representations of protein structures are contact maps and the *Delaunay tessellated graphs* (Atilgan, et al., 2004; Taylor and Vaisman, 2006). Both types of graphs can be described as an N×N matrix S for a protein with N residues. The definition of a contact differs between the two types. In a contact map, if the distance between Cα atoms of two residues is smaller than a certain cut-off, they are considered to be in contact (Atilgan, et al., 2004).

Delaunay tessellated graphs consist of points connected by edges defined in a special way. A point corresponds to an atom for each residue in the protein. For example, α carbon, β carbon, or the center of mass of the side chain can be used to represent each residue. There is a certain way to connect these points by edges so as to have the Delaunay simplices which form non-overlapping tetrahedra (Taylor and Vaisman, 2006). A Delaunay tessellated graph contains the neighborhood (contact) information of the residues represented by its vertices (Barber, et al., 1996).

Contact maps have been widely used as a representation method of protein structures in the literature (Fariselli and Casadio, 1999; Gupta, et al., 2005; Huan, et al., 2004; Vassura, et al., 2008; Vendruscolo, et al., 2002). This is a convenient way to capture the actual neighborhood information of each residue because the contacting residues are determined on the basis of a distance cut-off. An example of the contact map extraction is illustrated in Figure 1. In the Delaunay tessellation, the neighbors of a residue are defined by the Delaunay tetrahedra. However, these neighbors may be quite far away especially if the residue is on the surface. Using the Delaunay tessellated graphs to represent protein structures does not offer definite advantages compared to contact maps (Huan, et al., 2004; Küçükural, et al., 2008).

In many studies, small molecules (ligands) that bind to the proteins are also represented by graphs. This representation serves as a basis for the construction of a metric for the classification of these small molecules, and is a popular area in chemoinformatics, with many challenges (Schietgat, et al., 2008). This classification can help

Figure 1. a) An illustration of contact map extraction from a protein structure. The distance between residue 18 (methionine) and residue 15 (asparagine) is 5.5 Å. b) If the distance between two residues is below a certain cut-off, the residue pair is represented by a value 1 in the 2D contact map matrix (1s are shown here as grey dots). The elements of the matrix are indexed by the residue numbers of the protein

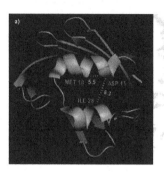

reduce the time to find ligands similar to a given ligand in chemical compound databases, e.g. the Chemical Abstract Service Registry containing 40 million compounds. The graph representations of ligand molecules and proteins differ from each other in how the vertices and edges are defined. In the graph representation of the 3D structure of a protein, vertices represent Cα atoms and edges are defined between neighboring Cα atoms (i.e. those close to each other in space) (Fariselli and Casadio, 1999), whereas in the graph representation of a small molecule, all the atoms are represented by vertices, and chemical bonds between the atoms define the edges (Schietgat, et al., 2008). Here, we only use graphs to represent proteins and their binding sites; we do not use the ligand molecules themselves for our purposes. Our goal is to identify binding sites based on the detection of similar and conserved regions in proteins that are weakly homologous to each other.

SCORING FUNCTION WITH NETWORK PROPERTIES

Defining a suitable scoring function and tuning its parameters are vital to the success of sub-graph matching algorithms. The graph representation of the proteins also provides some features that can be used in matching operations. These properties include intrinsic information about the contribution of each residue to the stability or the function of the protein. Structurally similar parts in two proteins can be found using a target function based on *network properties.*

Graph Theoretical (Network) Properties

A variety of graph theoretical properties are defined in the literature. This chapter presents nine of them. The first property, defined for each vertex of the graph, is the *degree* (k), which is the number of edges incident to the vertex, i.e. for a graph describing a protein, the number of neighbors of each residue (Taylor and Vaisman, 2006). The next measure is the number of *second neighbors*, k_2, which is a measure of the compactness of the graph. Even though k_2 is highly correlated to the degree k, it still provides additional information. For example, if the structure is a single globular unit rather than a number of distinct, small domains, most residues will have a high k_2 value (Küçükural, et al., 2008). The third network property is the *clustering coefficient*, also known as cliquishness, which measures how well the neighbors of a node are connected to each other. The clustering coefficient for each node is calculated as

$$C_n = \frac{2E_n}{k(k-1)} \qquad (1)$$

where E_n is the number of edges connecting the neighbors of the residue n, and k is its degree (Taylor and Vaisman, 2006; Vendruscolo, et al., 2002).

The *characteristic path length L* of the network, i.e. the average of the minimum paths between all node pairs, is also one of the most commonly used network properties (Bagler and Sinha, 2005; Taylor and Vaisman, 2006). Globular proteins yield smaller L values, whereas fibrous proteins yield larger ones. The characteristic path length L_n for residue n is calculated by the average of the shortest paths from the residue to all the other residues given as in (2);

$$L_n = \frac{1}{(N-1)} \sum_{j=1}^{N} \sigma_{nj} \qquad (2)$$

where σ_{nj} is the shortest path length between nodes n and j, and N is the number of residues (Taylor and Vaisman, 2006). The only difference between L and the *weighted characteristic path length w_L* is that weighted edges are used in the calculation of the latter. Contact potentials can be used to assign a weight for each contact (Liang and Dill, 2001; Miyazawa and Jernigan, 1996).

Several more graph theoretical properties can be calculated. The centrality of a node can be described by various measures. Many different centrality measures have been defined in the literature; here we present four of them. Firstly, *betweenness* measures to what extent a node i lies in between other nodes (Freeman, 1977):

$$C_B(i) = \sum_{s \neq i \neq t \in V}^{N} \frac{\sigma_{st}(i)}{\sigma_{st}} \qquad (3)$$

where σ_{st} is the number of shortest paths between nodes s and t, and $\sigma_{st}(i)$ is the number of shortest

paths between nodes s and t that pass through the node i (V denotes the set of all nodes).

The *stress centrality* measures the total number of shortest paths that pass through a node i (Shimbel, 1953):

$$C_S(i) = \sum_{s \neq i \neq t \in V}^{N} \sigma_{st}(i) \qquad (4)$$

The *closeness centrality* is defined as a measure that shows how long the information takes to spread from a given node to other reachable nodes (Sabidussi, 1966):

$$C_C(i) = \frac{1}{\sum_{t \in V} d_G(i,t)} \qquad (5)$$

where $d_G(i, t)$ is the length of the shortest path from node i to node t.

Lastly, *graph centrality* measures the length of the shortest path to reach the node farthest away from the given node (Hage and Harary, 1995):

$$C_G(i) = \frac{1}{\max_{t \in V} d_G(i,t)} \qquad (6)$$

Centrality measures aim to quantify the "importance" of the individual nodes in the network (Brandes, 2001; Newman, 2003). If a node is central to a network representing a protein structure, this node may have an important role in the stability of the protein or in the transduction of signals from one part of the structure to another. A summary of the graph theoretical properties is given in Table 1.

Significance of Network Properties in Protein Structure Characterization

Several recent studies have explored the potential uses of graph theoretical properties calculated from

Table 1.Summary of the graph theoretical properties.

#	Abreviation	Graph Theoretical Property
1	k	Degree
2	k_2	Second Neighbors
3	C	Clustering Coefficient
4	L	Characteristic Path Length
5	w_L	Weighted Characteristic Path Length
6	C_b	Betweenness
7	C_s	Stress Centrality
8	C_c	Clossness Centrality
9	C_g	Graph Centrality

protein structures (Huan, et al., 2004; Küçükural, et al., 2008). Graph theoretical properties, including the average degree, clustering coefficient, and number of second neighbors, have been used to discriminate the native protein structure from artificially created near-native decoys (Taylor and Vaisman, 2006). Besides the average of the properties, moment of their distributions, i.e. standard deviation, skewness, and kurtosis were also used. These features were fed as input vectors to several classification methods implemented in the Pattern Recognition Tools (PRTools) package (Heijden, et al., 2004).

The method was validated on three data sets using five-fold cross validation. The first data set employed was from the PISCES database (Wang and Dunbrack, 2003), and contained 1364 non-homologous proteins with an X-ray structure resolution < 2.2 Å, crystallographic R factor < 0.23, and a maximum pairwise sequence identity < 30%. The second data set consisted of 1364 artificially generated and well-designed decoy set, and the third set contained 101 artificially generated straight helices. Decoy sets were generated by randomly placing Cα atoms at ~3.83Å distance from each other while avoiding their self-intersection, and keeping the globular structure approximately at the size and shape of an average protein (Taylor and Vaisman, 2006).

Further details of the decoy generation stage can be found in Wang et al. (Wang, et al., 2004).

First, the graph representation method was tested on the three data sets. The contact map had a better classification accuracy than the tessellated graph representation, which can be attributed to the fact that it captures better the actual compactness of the protein structure. In some cases, tessellated graphs may incorrectly connect spatially distant residues, leading to a lower classification accuracy.

Second, the discriminatory power of graph theoretical properties was evaluated. When degree, clustering coefficient, number of second neighbors, and contact potential score were used together, the classification accuracy was 99%. Even without the contact potential score, the method had a 98.13% prediction accuracy after the exclusion of outliers.

The degree (k) had the highest discriminatory power; using only the degree distribution, we could distinguish the native and non-native structures with an accuracy of 96.74%. Addition of the number of second neighbors did not improve the accuracy much (96.93%). Cliquishness (C) along with degree (k) yielded a classification accuracy of 98.72% (Küçükural, et al., 2008).

The intrinsic power of graph theoretical properties to distinguish between native and non-native proteins is also exploited in another application: to align protein structures using dynamic programming. The idea behind the approach is that similar structures yield similar contact maps and other network properties derived from contact maps. This similarity is used for the alignment of protein structures (Küçükural and Sezerman, 2009).

Scoring Function

For the problem of inexact sub-graph matching, a suitable scoring function is needed. The scoring function measures the distance or the similarity of nodes, and transforms the problem of inexact sub-graph matching into an optimization problem.

Before a scoring function can be applied, it is advantageous to normalize the input variables, in order to have the data in a defined range (usually 0 to 1). The normalizing constant may be calculated as the difference between the maximum and minimum values of the given property over a representative set of proteins. Then, the normalized property $n(i)$ of node i can be written as

$$n(i) = \frac{v(i) - \min}{\max - \min}, \qquad (7)$$

where $v(i)$ denotes the value before normalization. The distance between corresponding nodes i and j can then be calculated using an appropriate metric applied to the property vectors. For example, the Manhattan distance, normalized to the [0,1] interval, can be used:

$$e(i, j) = \frac{1}{N} \sum_{f=1}^{N} \left| n_f(i) - n_f(j) \right|, \qquad (8)$$

where N is the number of features and $n_f(i)$ is the normalized value of feature f for the node i. A measure of similarity can be introduced as.

$$m(i,j,)=1-e(i,j), \qquad (9)$$

Matching nodes can then be identified based on a cut-off value applied on the distance or the similarity. Figure 2 shows an illustrative example of sub-graphs in two proteins, with a pair of matching nodes. The nodes n_1 and n_2 are very similar to each other by their network properties, which are reflected by their high similarity score.

SEARCH ALGORITHM

Several methods can be designed for local structural similarity search by adapting one of the known computational search algorithms. Graph matching (Kreher and Stinson, 1998) and *parallel graph matching algorithms* (Marek and Wojciech, 1998) are the preferred search methods when the structures are represented by graphs.

Attributed Relational Graphs (ARG) and Graph Matching

An attributed relational graph (ARG) $G = (V, E, A)$ is composed of a set of vertices (nodes) $V = \{v_1, v_2, ..., v_n\}$, a set of edges $E = \{e_1, e_2, ..., e_m\}$, and a set of attributes $A = \{a_1, a_2, ..., a_n\}$ that contain additional information on the nodes and/or the edges. Thus, an ARG contains both syntactic and semantic information: Syntactic information specifies the topological properties of the graph (nodes and edges), while semantic information is contained by attributes assigned to each node in the graph (Cordella, et al., 1998). For protein contact maps, the nodes represent residues and edges represent the neighborhood information on

Figure 2. Contact maps of two proteins and network property vectors for two matching nodes (n_1, n_2)

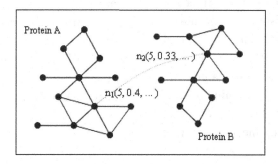

the residues. The attributes can be any additional properties calculated for the residues.

Two varieties of the graph matching problem exist depending on whether errors are allowed: exact and inexact graph matching. Exact matching is equivalent to identifying isomorphous graphs. The exact matching of the two graphs $G_1 = \{V_1, E_1, A_1\}$ and $G_2 = \{V_2, E_2, A_2\}$ yields a mapping between the nodes of G_1 and the nodes of G_2 such as

$$f: V_1 \rightarrow V_2 : \forall (v_i, v_j) \in E_1 \exists (f(v_i, v_j)) \in E_2 \qquad (10)$$

The mapping f yields a set of matched pairs (v_i, v_j) where v_i is from V_1 and v_j from V_2 (Cordella, et al., 1998).

Matching whole graphs is rarely needed in real-world problems. The sizes of the graphs to be matched are often different, and isomorphous subgraphs are sought.

Graph Matching Algorithms

Since graph matching algorithms are computationally expensive, developing more efficient graph matching algorithms is an open and challenging area. The aim is to reduce memory consumption and processing time. Obviously, the brute-force method is very slow and inefficient. In 1974, Ullmann proposed an algorithm based on elimination of successor nodes in tree search (Ullmann, 1976). Today, the most useful and effective algorithms are VF algorithms (named after the inventors of the original algorithm, M. Vento and P. Foggia), in terms of both time requirement and memory consumption. There are various types of exact matching algorithms depending on whether the goal is to find a monomorphism, an isomorphism or a graph-subgraph isomorphism (Cordella, et al., 1999; Cordella, et al., 2001). The VF algorithm was compared with Ullmann's algorithm in a study by Cordella et al. The running time of Ullmann's algorithm is $\Theta(N^3)$ in the best case, where N denotes the size of the graph, while

that of the VF algorithm is $\Theta(N^2)$. In the worst case, the running time of Ullmann's algorithm is $\Theta(N!N^3)$, and that of the VF algorithm is $\Theta(N!N)$. The memory consumption of the two methods also differs: the memory needed by the VF algorithm is uniformly $\Theta(N^2)$, while Ullmann's algorithm needs $\Theta(N^3)$ (Cordella, et al., 1999). The VF algorithm was improved by Cordella et al., resulting in VF2, which has a reduced memory usage can handle large graphs more efficiently (Cordella, et al., 2001).

For large systems, beam search algorithms are preferred, due to their low memory consumption. Beam search is an iterative, heuristic search algorithm that keeps the N best solutions in each iteration, and discards the rest of the matches ranked by a predefined scoring function (Yuehua and Alan, 2007). The algorithm uses two lists, named *parentLists* and *childLists*. The solutions obtained in the previous iteration are kept in *parentLists*, and the possible matches at the current iteration are held in *childLists*. After pruning and constraint checks specific to the matching operation, approved matches in the *childLists* are transferred into *parentLists*. The matching operation starts with a node that is chosen from heavily connected nodes and walks on neighboring nodes that are ranked by their degrees. The pseudo-code of the beam search algorithm is shown in Figure 3.

Numerous graph matching algorithms have been produced in the last three decades. Some of these algorithms are capable of reducing computational complexity and memory by using constraints and restrictions (Yuehua and Alan, 2007). When the task is to match a sample graph against a large set of prototypes, computationally efficient algorithms have an increasing memory consumption (Cordella, et al., 2001). For that reason, scientists have attempted to solve this problem by using parallelizable algorithms such as divide and conquer (Marek and Wojciech, 1998).

Figure 3. Pseudo-code of the beam search algorithm

```
Select the most heavily connected node to start with
while there are more heavily connected nodes in G₁
    if it is a new inital node
        for all the comparable nodes
            find a matching pair
            for each match in the parentList
                if the matching pair is not already included
                    newSolutionSet = new matching pair
                    insertChildList(newSolutionSet)
    else
        for all the solutionSets in the parentList
            if the solutionSet contains any neighbors of currentNode
                Locate the neighbor and its match pair
                for all the neighbors of the match pair in G₂
                    compare neighbor with currentNode
                    if matches
                        solutionSet = solutionSet + new pair
                        insertChildList(solutionSet)
    for all solutionSets in childList
        rank solutionSets
        prune according to scoring function and check constraints
        add the solutions in the childList to parentList
```

Parallel Graph Matching Algorithms

Biological systems can often be described by extremely large graphs. Parallel graph matching algorithms reduce processing time by performing a parallel search on graph trees. Sheng et al. (Sheng, et al., 2003) developed an algorithm for parallel computers, especially those with distributed memory. The implementation uses an asynchronous parallel algorithm, and the processing time was found to be inversely proportional to the number of processors. The main algorithm consists of three steps. In the first step, the master processor broadcasts the two graphs to be matched to all other processors. In the second step, each processor starts to search for matches between subgraphs assigned to it. If any of the processors finds isomorphous subgraphs, it informs the other processors of the match. In the third step, when all the processors have completed their jobs, the search operation for the original graphs can be completed.

Subgraph isomorphism is defined as an exact match between a subgraph from the first graph and a subgraph from the second graph. However, in real-world applications, we are often interested in subgraphs that are similar but not necessarily isomorphous. For example, homologous proteins have the same overall structure but mutations, insertions and deletions result in small differences between the conformations. To find such similarities, graph matching algorithms allowing for errors, i.e. inexact subgraph matching algorithms are needed (Küçükural and Sezerman, 2009).

A Parallel Inexact Graph Matching Algorithm

The beam search algorithm (Yuehua and Alan, 2007) can be to parallelized and adapted to allow for errors. For each process, the initiation nodes are selected from the most heavily connected nodes in the first graph.

Attributed relational graphs describing proteins can be represented as binding residue matrices

(BRM) containing both structural neighborhood information and attributes of residues such as cliquishness, degree, and centrality. These matrices constitute the input to the algorithm.

Master and child processes are designed with different responsibilities. The master process manages the child processes, monitors their states, and sends them the necessary data whenever they need it (Sheng, et al., 2003). The master process itself does not perform any matching operations. Matching is done by the child processes which report their state to the master. Initially, the master process sends the BRM to all child processes, and then each process receives the number of the node with which it should start the matching operation.

Once the initiation nodes have been distributed among child processes, the graph matching operations are run until all the heavily connected nodes have been processed. When the matching operation for a node is finished, the child process sends a signal to the master process to report this event. This process can then accept new initial nodes to start a new matching operation. When all the heavily connected nodes have been processed, the master sends a stop signal to the child processes to close their connections. Each child process employs a beam search algorithm for matching (see Figure 2). To reduce the computation time, problem-specific constraints are used (Küçükural and Sezerman, 2009).

Problem Specific Constraints

Graph matching operations are generally very complex, and their computation time scales at least as $\Theta(N^2)$ (Cordella, et al., 1999). Therefore, some constraints are used to reduce computational complexity.

One of the most important constraints prunes the *childList*. For example, a child list is not allowed to contain more than ten matches. Each residue has at most 15 neighbors for an average protein. When a match between two nodes is found, their neighbors are examined for potential

new matches. However, constraints derived from the parent match are applied. For example, let X_i be a residue number from the first protein and Y_j a residue number from the second protein, and let us suppose that a match has been found between X_i and Y_j. Then new potential match such as between X_{i+1} and Y_{j+1} will have the first match as parent. The following constraint can be introduced for the residue numbers of the parent and child matches:

$$X_i \geq Y_j \Rightarrow X_{i+1} \geq Y_{j+1} \tag{11}$$

For a child match, this constraint has to be met for each parent match leading up to the child, in order to prevent "cross matches".

Another constraint, designed to speed up the algorithm is not to examine matches had been already found by other processes (otherwise, because the algorithm walks the graph following the edges, the same pair of residues could be repeatedly examined for a potential match) (Küçükural and Sezerman, 2009).

Function Prediction Using Sub Graph Matching Algorithms

With appropriate search parameters, the graph matching algorithm finds the most similar parts of two graphs rather than trying to match most of the nodes even though the attributes do not show a significant similarity. Therefore, the graph matching algorithm can detect functional similarity in cases when the global alignment algorithm fails.

When two proteins are compared, all possible structurally similar regions are generated using the inexact sub-graph matching algorithm, the results are ranked by the scoring function given in (9). When the function of a target protein is to be predicted, the target protein is compared with all the proteins in a database. The function of a protein is then predicted on the basis of the functions of the closest protein structures.

A preliminary test of the methodology was performed on a set of 88 proteins (enzymes) with a pairwise sequence identity <30%. The function of each enzyme is described by its EC number. EC (Enzyme Commission) numbers provide a hierarchical classification of enzymes and have four digits (e.g. EC a.b.c.d) (Moss, 2006). The success of a function prediction is judged by counting common EC number digits. If all four EC digits of the predicted function equal the actual function, the prediction is exact. If the first three digits match, the function is still correctly predicted but the enzyme acts on a different ligand than the one hit by the search procedure. The accuracy of the method was evaluated by counting correct predictions based on matching EC numbers. For each protein in the data set, there was a single other enzyme with the same function. During the test, each protein was used as a target, and the remaining structures were scanned for local structural similarities with the target; the function of the target was predicted based on the best hit. When matching first three digits were required for correct prediction, the accuracy was found to be 61.2%, which increased to 76.43% when the best hit in the top 5 hits was considered. When all four EC digits were required to match, the accuracy was 55.28% with the top hit, and 72.87% with the best hit in the top 5 hits considered.

The scores and the ranking depend on the definition of the scoring function and its parameters; therefore, detailed tuning is necessary to obtain optimum results.

CONCLUSION

In this chapter, we have reviewed the use of graph theoretical properties and parallel inexact sub-graph matching algorithms for the prediction of protein function based on finding local structural similarities between the target protein and proteins with known function. Structure representation, scoring, and searching are addressed with contact maps, network properties, and inexact sub-graph matching algorithms, respectively.

These representations and scoring functions are less dependent on the actual coordinates, and therefore more robust than other methods; they also allow for more flexibility. Other methods are highly dependent on the actual coordinates of the protein, and are therefore dependent on the accuracy of the experimental procedure used to determine the structure.

Graph theoretical attribute sets have been used to find better matches between two graphs. Although nine specific network properties are presented in this chapter, other features can be added, weighted according to their contribution to the similarity. The inclusion of these properties into the scoring function has been shown to significantly improve the discovery of similar regions in proteins. The conserved structural patterns that are most informative for function annotation can be extracted using this approach. Moreover, since the matching procedure is based on the preservation of contacts, one can determine the network of contacts that are crucial for the functionality of the given protein. This is the type of information that the experimentalist would be interested in. The conserved network of contacts within a family can be detected via this method, and the functional importance of each contact can then be verified by experiment.

One potential application of this method is to determine the conserved contact networks and the key residues in potential drug targets. This information may be helpful for the design of drug molecules targeting the key residues, which could be a new strategy of drug development. Future development may include adding various amino acid properties, side chain information, exposed surface areas and other features to the scoring function in order to find even better matches between regions. In addition, fuzzy contact maps and more versatile contact definitions may be introduced, allowing for a higher structural flexibility in the represented proteins.

ACKNOWLEDGMENT

The work has been performed under the project HPC-Europa (RII3-CT-2003-506079), with the support of the European Community - Research Infrastructure Action under the FP6 "Structuring the European Research Area" Programme. It is also supported in part by the Alfred P. Sloan Foundation; NSF Career Award (DBI 0746198); and the National Institute of General Medical Sciences (R01GM083107).

REFERENCES

Albert, R., & Barabási, A.-L. (2002). Statistical mechanics of complex networks. *Reviews of Modern Physics*, *74*, 47.

Atilgan, A. R., Akan, P., & Baysal, C. (2004). Small-world communication of residues and significance for protein dynamics. *Biophysical Journal*, *86*, 85–91.

Bagler, G. & Sinha, S. (2005). Network properties of protein structures. *Physica A: Statistical Mechanics and its Applications*, 346, 27-33.

Barber, C. B., David, P. D., & Hannu, H. (1996). The quickhull algorithm for convex hulls . *ACM Transactions on Mathematical Software*, *22*, 469–483.

Barker, J. A., & Thornton, J. M. (2003). An algorithm for constraint-based structural template matching: application to 3D templates with statistical analysis. (pp.1644-1649).

Brandes, U. (2001). A faster algorithm for betweenness centrality. *The Journal of Mathematical Sociology*, *25*, 163–177.

Conesa, A., Gotz, S., Garcia-Gomez, J. M., Terol, J., Talon, M., & Robles, M. (2005). Blast2GO: a universal tool for annotation, visualization and analysis in functional genomics research. *Bioinformatics (Oxford, England)*, *21*, 3674–3676.

Cordella, L. P., Foggia, P., Sansone, C., & Tortorella, F. (1998). Graph Matching: a Fast Algorithm and its Evaluation. In *Proc. 14th Int. Conf. On Pattern Recognition*.

Cordella, L. P., Foggia, P., Sansone, C., & Vento, M. (1999). Performance evaluation of the VF graph matching algorithm. In *Proceedings of the 10th International Conference on Image Analysis and Processing*. IEEE Computer Society.

Cordella, L. P., Foggia, P., Sansone, C., & Vento, M. (2001). An improved algorithm for matching large graphs. In *Proc. of the 3rd IAPR-TC-15 International Workshop on Graph-based Representation*. Italy.

Devos, D., & Valencia, A. (2000). Practical limits of function prediction. *Proteins: Structure, Function, and Bioinformatics*, *41*, 98–107.

Engelhardt, B. E., Jordan, M. I., Muratore, K. E., & Brenner, S. E. (2005). Protein molecular function prediction by Bayesian phylogenomics. *PLoS Computational Biology*, *1*, e45.

Fariselli, P., & Casadio, R. (1999). A neural network based predictor of residue contacts in proteins. *Protein Engineering*, *12*, 15–21.

Freeman, L. (1977). A set of masures of centrality based on betweenness. *Sociometry*, *40*, 35–41.

Friedberg, I. (2006). Automated protein function prediction--the genomic challenge. *Briefings in Bioinformatics*, *7*, 225–242.

Goyal, K., Mohanty, D. & Mande, S.C. (2007). PAR-3D: a server to predict protein active site residues. gkm252.

Gupta, N., Mangal, N., & Biswas, S. (2005). Evolution and similarity evaluation of protein structures in contact map space. *Proteins*, *59*, 196–204.

Hage, P., & Harary, F. (1995). Eccentricity and centrality in networks. *Social Networks*, *17*, 57–63.

Hawkins, T., Luban, S., & Kihara, D. (2006). Enhanced automated function prediction using distantly related sequences and contextual association by PFP. *Protein Science, 15,* 1550–1556.

Heijden, F.v.d., Duin, R.P.W., Ridder, D.d. & Tax, D.M.J. (2004). *Classification. parameter estimation and state estimation - an engineering approach using Matlab.*

Holm, L., Kaariainen, S., Rosenstrom, P. & Schenkel, A. (2008). Searching protein structure databases with DaliLite (vol.3,pp. 2780-2781).

Huan, J., Wang, W., Bandyopadhyay, D., Snoeyink, J., Prins, J., & Tropsha, A. (2004). Mining protein family specific residue packing patterns from protein structure graphs. In *Proceedings of the eighth annual international conference on Resaerch in computational molecular biology.* ACM, San Diego, CA, USA.

Huang, J. Y., & Brutlag, D. L. (2001). The EMOTIF database. *Nucleic Acids Research, 29,* 202–204.

Hulo, N., Bairoch, A., Bulliard, V., Cerutti, L., De Castro, E., & Langendijk-Genevaux, P. S. (2006). The PROSITE database. *Nucleic Acids Research, 34,* D227–D230.

Ivanisenko, V.A., Pintus, S.S., Grigorovich, D.A. & Kolchanov, N.A. (2004). PDBSiteScan: a program for searching for active, binding and posttranslational modification sites in the 3D structures of proteins. W549-554.

Kreher, D. L., & Stinson, D. R. (1998). *Combinatorial Algorithms: Generation, Enumeration and Search.* CRC Press.

Küçükural, A. & Sezerman, O.U. (2009). Protein Strcuture Characterization Using Attributed Sub-Graph Matching Algorithms with Parallel Computing, *(In preperation).*

Küçükural, A., Sezerman, O. U., & Ercil, A. (2008). Discrimination of Native Folds Using Network Properties of Protein Structures. In Brazma, A., Miyano, S., & Akutsu, T. (Eds.), *APBC* (pp. 59–68). Kyoto, Japan: Imperial College Press.

Küçükural, A. & Sezerman, U. (2009). Structural Alignment of Proteins Using Network Properties with Dynamic Programming, *(In preperation).*

Laskowski, R. A., Watson, J. D., & Thornton, J. M. (2005). Protein Function Prediction Using Local 3D Templates. *Journal of Molecular Biology, 351,* 614–626.

Liang, J., & Dill, K. A. (2001). Are Proteins Well-Packed? *Biophysical Journal, 81,* 751–766.

Lo, W.-C., Huang, P.-J., Chang, C.-H., & Lyu, P.-C. (2007). Protein structural similarity search by Ramachandran codes. *BMC Bioinformatics, 8,* 307.

Marek, K., & Wojciech, R. (1998). *Fast parallel algorithms for graph matching problems.* New York: Oxford University Press, Inc.

Martin, A. C., Orengo, C. A., Hutchinson, E. G., Jones, S., Karmirantzou, M., & Laskowski, R. A. (1998). Protein folds and functions. *Structure (London, England), 6,* 875–884.

Martin, D., Berriman, M., & Barton, G. (2004). GOtcha: a new method for prediction of protein function assessed by the annotation of seven genomes. *BMC Bioinformatics, 5,* 178.

Matsuda, H., Taniguchi, F., & Hashimoto, A. (1997). An approach to detection of protein structural motifs using an encoding scheme of backbone conformations. *Proc. of 2nd Pacific Symposium on Biocomputing* (pp280-291).

Miyazawa, S., & Jernigan, R. L. (1996). Residue-residue potentials with a favorable contact pair term and an unfavorable high packing density term. for simulation and threading. *Journal of Molecular Biology,* 256.

Moss, G.P. (2006). Recommendations of the Nomenclature Committee. *International Union of Biochemistry and Molecular Biology on the Nomenclature and Classification of Enzymes by the Reactions they Catalyse.*

Newman, M. E. J. (2003). A measure of betweenness centrality based on random walks *arXiv. org:cond-mat/0309045.*

Rost, B. (2002). Enzyme function less conserved than anticipated. *Journal of Molecular Biology, 318,* 595–608.

Sabidussi, G. (1966). The centrality index of a graph. *Psychometrika, 31,* 581–603.

Schietgat, L., Ramon, J., Bruynooghe, M., & Blockeel, H. (2008). An Efficiently Computable Graph-Based Metric for the Classification of Small Molecules. In *Discovery Science.* 197-209.

Sheng, Y. E., Xicheng, W., Jie, L., & Chunlian, L. (2003). A New Algorithm For Graph Isomorphism And Its Parallel Implementation. *International Conference on Parallel Algorithms and Computing Environments ICPACE.* Hong Kong, China.

Shimbel, A. (1953). Structural parameters of communication networks . *Bulletin of Mathematical Biology, 15,* 501–507.

Shindyalov, I. N., & Bourne, P. E. (1998). Protein structure alignment by incremental combinatorial extension (CE) of the optimal path. *Protein Engineering, 11,* 739–747.

Stark, A., & Russell, R. B. (2003). Annotation in three dimensions. PINTS: Patterns in Non-homologous Tertiary Structures. *Nucleic Acids Research, 31,* 3341–3344.

Storm, C. E. V., & Sonnhammer, E. L. L. (2002). Automated ortholog inference from phylogenetic trees and calculation of orthology reliability. *Bioinformatics (Oxford, England), 18,* 92–99.

Strogatz, S. H. (2001). Exploring complex networks. *Nature, 410,* 268–276.

Taylor, T. J., & Vaisman, I. I. (2006). Graph theoretic properties of networks formed by the Delaunay tessellation of protein structures. *Physical Review E: Statistical, Nonlinear, and Soft Matter Physics, 73,* 041925–041913.

Thornton, J. M., Todd, A. E., Milburn, D., Borkakoti, N., & Orengo, C. A. (2000). From structure to function: Approaches and limitations. *Nature Structural & Molecular Biology, 7,* 991–994.

Ullmann, J. R. (1976). An Algorithm for Subgraph Isomorphism. [JACM]. *Journal of the ACM, 23,* 31–42.

Vassura, M., Margara, L., Di Lena, P., Medri, F., Fariselli, P., & Casadio, R. (2008). FT-COMAR: fault tolerant three-dimensional structure reconstruction from protein contact maps . *Bioinformatics (Oxford, England), 24,* 1313–1315.

Vendruscolo, M., Dokholyan, N. V., Paci, E., & Karplus, M. (2002). Small-world view of the amino acids that play a key role in protein folding. *Physical Review E: Statistical, Nonlinear, and Soft Matter Physics, 65.*

Wang, G., & Dunbrack, R. L. Jr. (2003). PISCES: a protein sequence culling server. *Bioinformatics (Oxford, England), 19,* 1589–1591.

Wang, K., Fain, B., Levitt, M., & Samudrala, R. (2004). Improved protein structure selection using decoy-dependent discriminatory functions. *Bioinformatics (Oxford, England), 4,* 8.

Weinhold, N., Sander, O., Domingues, F. S., Lengauer, T., & Sommer, I. (2008). Local Function Conservation in Sequence and Structure Space. *PLoS Computational Biology, 4,* e1000105.

Wilson, C. A., Kreychman, J., & Gerstein, M. (2000). Assessing annotation transfer for genomics: quantifying the relations between protein sequence, structure and function through traditional and probabilistic scores. *Journal of Molecular Biology, 297,* 233–249.

Yuehua, X., & Alan, F. (2007). On learning linear ranking functions for beam search. *Proceedings of the 24th international conference on Machine learning.* Corvalis, Oregon: ACM. Zhang, Y. & Skolnick, J. (2005). TM-align: a protein structure alignment algorithm based on the TM-score(pp. 2302-2309).

Section 3
Statistical and Bayesian Approaches for Virtual Screening

Chapter 8
Advanced PLS Techniques in Chemometrics and Their Applications to Molecular Design

Kiyoshi Hasegawa
Chugai Pharmaceutical Company, Japan

Kimito Funatsu
University of Tokyo, Japan

ABSTRACT

In quantitative structure-activity/property relationships (QSAR and QSPR), multivariate statistical methods are commonly used for analysis. Partial least squares (PLS) is of particular interest because it can analyze data with strongly collinear, noisy and numerous X variables, and also simultaneously model several response variables Y. Furthermore, PLS can provide us several prediction regions and diagnostic plots as statistical measures. PLS has evolved or changed for copying with sever demands from complex data X and Y structure. In this review article, the authors picked up four advanced PLS techniques and outlined their algorithms with representative examples. Especially, the authors made efforts to describe how to disclose the embedded inner relations in data and how to use their information for molecular design.

INTRODUCTION

Establishing relationships between the chemical structures and their activities or properties are crucial to achieve a goal, doing better and fewer experiments. The quantitative description of the relations is the so-called quantitative structure-activity/property relationships (QSAR and QSPR) [Gedeck et al. 2008, Yap et al. 2007].

QSAR studies express the biological activity of compounds as a function of their various structural descriptors and describe how variation of the biological activity depends on change of the chemical structure. If such a relationship can be derived from the structure-activity data, the model equation allows chemists to say with some

DOI: 10.4018/978-1-61520-911-8.ch008

confidence that which property has an important role in the biological activity. Furthermore, it also allows chemists some level of prediction. By quantifying physico-chemical properties, it should be possible to calculate in advance what the biological activity of a novel compound might be. Even though a compound is discovered which does not fit the model equation, it implies that some other properties are important and provides us new compound for further investigation.

The first and most important work dealing with QSAR was published by Corwin Hansch and his co-workers in the 1960s [Hansch et al. 1964, Hansch 1969]. In this pioneering framework, the Hansch equation was developed for quantitative approach to describe relationship between chemical structure and biological activity as dependent variables y using linear free energy relationships related parameters as independent variables X. The QSAR models are developed using multiple linear regression (MLR) which is a popular classical modeling method.

The MLR model is a powerful technique for optimizing the activity of a chemical compound. With this method a basic assumption is that all the factors involved in variation in biological activity arising from the modification of the molecular structure with a congeneric series can be correlated with concomitant change of physico-chemical parameters. The great advantage of the MLR method is that a causal model is obtained and the physical meaning is obvious. However, the sever conditions must be satisfied to apply MLR. The descriptor variables are orthogonal and the number of compounds is greater than that of descriptors. Otherwise, over fitting results may be obtained and the predictive power of the model is very poor [Hasegawa et al. 2000]. Recently, in QSAR society, new 2D and 3D molecular descriptors have been proposed [Estrada 2008]. The most difficult problem is associated to relatively high uncertainty in molecular descriptors. Thus, search for more informative 2D and/or 3D molecular descriptors has been one of the main

concerns in chemoinformatics [Gasteiger 2003]. If the structural information for the molecules investigated is insufficient, the model is biased. In other words, the model is under-fitting in statistical point of views.

Chemical pattern recognition (CPR) is regarded as another method for modeling QSAR and QSPR [Miyashita et al. 1993, Miyashita et al. 1994]. In scientific research, one certainly hopes to establish a global hard model. In quantum chemistry and molecular mechanics, finding a global hard model of relationships between chemical structure and property is desired. But for some complex molecules, it is difficult to perform quantum mechanical calculations. When a global hard model cannot be obtained, chemists frequently utilize other local soft models, such as analogy and similarity. For instance, an empirical rule "like dissolves like" for solubility and the well-known periodic table for chemical elements are classical analogy methods and are commonly used in chemistry. Chemical phenomena are more complex than physical ones and are affected by many unknown factors. So, the real chemical world is typically multivariate. Facing this multivariate chemical world, we must make many assumptions and/or hypotheses in order to obtain a model and then the model loses its strictness or the generality. The Hansch approach using MLR is regarded as a hard model. In CPR, local soft modeling becomes a powerful approach because soft models can be used to predict the related property and activity. The partial least squares (PLS) method can lead to local soft model solutions to chemical problems [Wold et al. 2001].

BACKGROUND

The PLS method was first developed by Herman Wold in the 1960s and was subsequently adopted by his son Svante Wold in the 1980s for regression problems in chemometrics. PLS is of particular interest because, unlike MLR, it can analyze data with strongly collinear, noisy and numerous X variables, and also simultaneously model several response variables Y. Furthermore, PLS can provide us several prediction regions and diagnostic plots as statistical measures. QSAR scientists can extract the patterns embedded in the data set. PLS is widely employed to solve multivariate calibration and resolution in analytical chemistry [Martens et al. 1989].

PLS has evolved or changed for copying with sever demands from complex X and Y data structure [Wold et al. 2001]. In this review article, we have picked up the following four advanced PLS techniques and outlined their algorithms with representative examples.

- PLS and the hybrid types for variable selection
- Nonlinear PLS with quadratic and kernel functions
- Multi-way PLS and its unfolding type
- Multi-block PLS in hierarchical structure

Although some methods have been invented in process control, spectroscopic modeling, or analytical studies, the representative applications are limited to that in QSAR and QSAR fields. Especially, we made efforts to describe how to disclose the embedded inner relations in data and how to use their information for molecular design.

In life science, a suffix -omics is popular word. Omics represents the discipline across several heterogeneous scientific fields and includes chemogenomics, proteomics, and metabonomics [Ho et al. 2008]. In this newly emerged world, the size of data matrices is huge and each matrix interacts with each other. The advanced PLS techniques would be more frequently employed in the omics field in the near future.

1. PLS AND HYBRID TYPES FOR VARIABLE SELECTION

1.1. Partial Least Squares (PLS)

PLS is of particular interest because, unlike MLR, it can analyze data with strongly collinear, noisy and numerous X variables, and also simultaneously model several response variables Y. Furthermore, PLS can provide us several prediction regions and diagnostic plots as statistical measures.

1.2. Algorithm of PLS

On the basis of the type of activity data to be processed, PLS can be divided into two versions: univariate PLS and multivariate PLS. The PLS algorithm is to sequentially extract the score vectors t, u and the loading vectors p, q from the structural descriptor variables X and multivariate activity variables Y in decreasing order of their corresponding singular values.

$$X = \sum_{h=1}^{A} t_h p_h^{'} + E = TP^{'} + E \qquad (1)$$

$$Y = \sum_{h=1}^{A} u_h q_h^{'} + F = UQ^{'} + F \qquad (2)$$

where A is the number of PLS components determined by cross-validation (CV) [Wiklund et al. 2007]. $T = [t_1|...|t_A]$ and $U = [u_1|...|u_A]$ are score matrices and $P = [p_1|...|p_A]$ and $Q = [q_1|...|q_A]$ are loading matrices for X and Y, respectively. E and F are the model residuals of X and Y. Apostrophe (') on vector or matrix means the transposed form. In the case of univariate PLS, scores u_h and t_h in the h-th component are linked by a linear relationship.

$$u_h = b_h * t_h \qquad (3)$$

where b_h is the inner regression coefficient in h-th component. The weight vectors w and c in W and C matrices are used as the temporary parameters in the iteration of the PLS algorithm. The score vector t is a linear combination of X and the weight vector w from the PLS algorithm.

$$t = Xw \qquad (4)$$

Equations (1) and (2) can be combined to a multiple linear regression (MLR) model via Equation (3).

$$Y = XB_{PLS} \qquad (5)$$

where matrix B_{PLS} contains the PLS regression coefficients and can be calculated as follows:

$$B_{PLS} = X'U(T'XX'U)^{-1}T'Y \qquad (6)$$

Until now, there are three main PLS algorithms. (NIPALS algorithm [Geladi et al. 1986], SIMPLS algorithm [Jong 1993] and Kernel algorithm [Lindgren et al. 1993])

1.3. Applications of PLS

Hasegawa et al. have performed a PLS analysis for sixteen azoxy compounds with antifungal activity [Hasegawa et al. 1994]. Scores are used as chemical descriptors. Scores were calculated from the principal component analysis of 100 aromatic substituents with 9 physico-chemical parameters. The chemical structures were described by three scores ($Z1(R_1)$, $Z2(R_1)$, $Z3(R_1)$) for substituent R_1 and three scores ($Z1(R_2)$, $Z2(R_2)$, $Z3(R_2)$) for substituent R_2. The PLS model with four significant components was obtained by leave-one-out (LOO) and transformed to a MLR equation.

$$y = 0.438\ Z1(R_1) + 1.323\ Z2(R_1) + 0.400$$
$$Z3(R_1) - 0.199\ Z1(R_2) - 0.633\ Z2(R_2) + 0.650$$
$$Z3(R_2) + 3.974 \qquad (7)$$

From the QSAR equation, potent azoxy compounds can be found by increasing the steric bulkiness and hydrophobicity of the R_1 substituent and increasing the electron-withdrawing ability of the R_2 substituent. The score values for a substituent are calculated from principal component loadings and 9 physico-chemical parameters. Even though there is a missing in physico-chemical parameters, NIPALS algorithm can estimate those score values. This is advantage for using scores as chemical descriptors. Authors have proposed some candidate azoxy compounds with higher antifungal activity based on the coefficient values of this QSAR equation and factorial designs.

Miyashita et al. have applied PLS to 83 N-aralkyl thiolcarbamates with fungicidal and herbicidal activities [Miyashita et al. 1992]. The structural descriptors are $E_s^c(R_1)$, $E_s^c(R_2)$, $E_s^c(R_3)$, $\pi(R_1)$ and R_2H. Three component-PLS model was obtained by LOO. The PLS model equations for fungicidal activity Y_F and herbicidal activity Y_H were expressed in terms of the first three latent variables t_1, t_2 and t_3.

$$Y_F = -0.482\ t_1 + 0.278\ t_2 - 0.184\ t_3 + 1.723 \qquad (8)$$

$$Y_H = 0.329\ t_1 + 0.498\ t_2 + 0.467\ t_3 + 2.133 \qquad (9)$$

Y is used to classify a compound into three activity classes. The classification was performed using the calculated Y value. If $Y < 1.5$, $1.5 < Y < 2.5$, and $Y > 2.5$, then the compounds are classified to be weak (Y = 1), moderate (Y = 2) and strong (Y = 3) activities, respectively. Comparing the Equations (8) and (9), the coefficients of t_1 and t_3 for two activities possess opposite signs and for t_2 identical signs. This means that if the first and third latent variables t_1 and t_3 are increased, the herbicidal activity would increase and fungicidal activity would decrease. If the second latent variable t_2 is increased, both activities would increase. The PLS projection of the fungicidal data on the first two scores, t_1 and t_2, is shown in Figure 1. The PLS projection of the herbicidal data is also shown in Figure 2. Big circle means high activity and small circle means low activity. Both of fungicidal and herbicidal activities increase as the t_2 value increases.

PLS has actively been used in three-dimensional (3D) QSAR studies [Xu et al. 2008, Yi et al. 2008, Ravichandran et al. 2008, Tong et al. 2008, Equbal et al. 2008, Akula et al. 2006, Romeiro et al. 2006, Guo et al. 2005, Moro et al. 2005, Peng et al. 2005]. 3D-QSAR aims to construct a relationship between the 3D structures and biological activities. Comparative molecular field analysis (CoMFA), was initially presented by Cramer et al. [Cramer et al. 1988] and this method has become one of the most powerful tools in 3D-QSAR studies. In CoMFA, ligand molecule is described by its steric and electrostatic fields sampled at the intersections of a grid in 3D space. The relation between the field descriptors and biological activity is investigated by PLS. The result of CoMFA is presented as the coefficient contour maps in 3D space.

PLS has also been used in ADME (Absorption, Distribution, Metabolism and Excretion) modeling [Hasegawa et al. 2009 *in press*]. Good ADME profile of compound is needed for drug candidate and their predictions are made by PLS [Obrezanova et al. 2008, Kriegl et al. 2005, Crivori et al. 2004, Eroes et al. 2004, Kratochwil et al. 2002, Zuegge et al. 2002]. Especially, molecular properties such as solubility, logP and pKa are main targets for ADME modeling.

1.4. Genetic Algorithm-Based PLS (GAPLS)

In general, since the explicit mechanism for the biological activity of a compound is not known in advance, many different variable combinations should be investigated in a trial-and-error procedure. This issue is well known as variable selection [Tropsha et al. 2006, Gonzalez et al. 2008]. That is, when the number of descriptors increases, the quality of prediction of the model may decrease. In PLS, as well as MLR, this issue still remains.

Genetic algorithm (GA) is expected to solve this issue. GA has three basic operations (selection, crossover and mutation) and takes the intermediate position between the stepwise and random approaches [Goldberg 1989, Devillers 1996]. GA provides us candidate solutions and QSAR scientists have the opportunity to choose one for further validation. Because of its simple implementation, GA has been used as a promising method for variable selection. Recently, Hasegawa et al. have invented an advanced variable selec-

Figure 1. Plot of score vectors (t_1/t_2) of fungicidal data. Big and small circles mean high and low activities, respectively. Cited from Ref. [Miyashita et al. 1992]

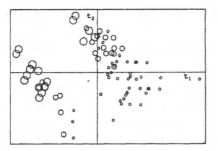

Figure 2. Plot of score vectors (t_1/t_2) of herbicidal data. Big and small circles mean high and low activities, respectively. Cited from Ref. [Miyashita et al. 1992]

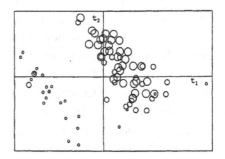

tion method called GA-based PLS (GAPLS) [Hasegawa et al. 1997]. In GAPLS, PLS is used as the statistical method and the best variable combinations are selected by GA based on the LOO R^2 value of the PLS model.

1.5. Algorithm of GAPLS

GA is a simulated method by which a principle of a natural evolution in biology (the struggle for life) is modeled. The struggle for life is that the species having a high fitness under some environmental conditions can prevail in the next generation, and the best species may be reproduced by crossover together with random mutations of genes (chromosome) in the surviving ones.

In GAPLS, the chromosome and its fitness in the species represent a set of variables and predictivity of the derived model, respectively. The GAPLS scheme for variable selection is shown in Figure 3.

GAPLS consists of three basic steps: (1) An initial population of chromosomes is created. Each chromosome is represented by a binary bit string. Bit "1" denotes a selection of the corresponding variable, and bit "0" denotes a non-selection. The values of a binary bit are determined in a random way. (2) A fitness of each chromosome in the population is evaluated by predictivity of the PLS model derived from the binary bit string. The LOO R^2 value is used as predictivity. (3) The population of chromosomes in the next generation is reproduced. First, a better chromosome is selected according to the LOO R^2 value. Second, for any pairs of chromosomes, a random number is drawn to decide whether the crossover operation is undertaken or not. Crossover is an operation that a pair of chromosomes is individually divided, mutually exchanged, and merged. Third, a binary bit string is mutated. For each binary bit, a random number is drawn to decide whether its bit has to be changed or not. The evaluation step (step 2) and reproduction step (step 3) in the

Figure 3. GAPLS scheme for variable selection. Cited from Ref. [Hasegawa et al. 1997]

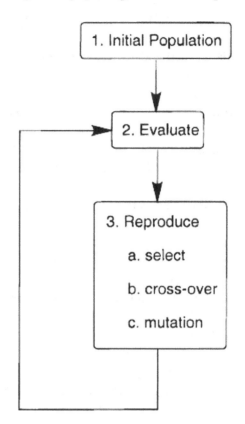

GAPLS scheme are repeated until the number of generations is reached to the given maximum.

1.6. Applications of GAPLS

Hasegawa et al. have reported the first application of GAPLS to QSAR study [Hasegawa et al. 1997]. The data set contains 35 dihydropyridines (DHPs) with calcium channel antagonists. Twelve descriptors for three substituents (R_2, R_3 and R_4) were used to characterize those 35 DHPs. For comparative study with GAPLS, the standard PLS analysis using all parameters was carried out. A four-component PLS model was derived and the R^2 and LOO R^2 values were 0.817 and 0.571, respectively. The GAPLS analysis gave us a four-component PLS model with the selected 6 variables as the best one. The values of R^2 and LOO

R^2 of the model were 0.797 and 0.685, respectively. The value of LOO R^2 was much improved from 0.571 to 0.685 by GAPLS. GAPLS can select the physico-chemical parameters largely contributing to the inhibitory activity. Therefore, the structural requirements for the inhibitory activity can be estimated in an effective manner.

Since the first application of GAPLS, GAPLS has been widely used in QSAR and QSPR studies [Hasegawa et al. 2000, Arakawa et al. 2007, Sabet et al. 2008, Fassihi et al. 2008, Ren et al. 2008, Mohajeri et al. 2008, Sagrado et al. 2008, Edraki et al. 2007, Deeb et al. 2007, Yamashita et al. 2006, Ferreira et al. 2004, Liu et al. 2004, Wanchana et al. 2003, Yamashita et al. 2002, Tropsha et al. 2001, Daren 2001, Hoffman et al. 2000]. Furthermore, other sophisticated optimization methods such as Particle Swarm Optimization [Hu et al. 2007], Ant Colony Optimization [Shen et al. 2005], Genetic Programming [Tang et al. 2002, Arakawa et al. 2006] were implemented in the PLS algorithm and variable selection becomes a hot topic in chemoinformatics [Gasteiger 2003].

1.7. Orthogonal Signal Correction (OSC)

Often the structured noise is present in X, which is defined as systematic variation of X. Examples of the structured noise in QSAR studies are the calculated 2D or 3D molecular descriptors derived from the approximate 3D conformations. It has been shown that this structured noise negatively affects the interpretation of PLS and increases its complexity. Orthogonal signal correction (OSC) has been developed as a preprocessing method in spectroscopic modeling and can suppress this structured noise [Wold et al. 1998].

OSC is a PLS-based filter which uses the properties of the PLS algorithm to find the largest variation in X that is orthogonal to Y. That means that the OSC algorithm succeeds to find a score variable describing the variation in X with minimal correlation to Y. Since only orthogonal variation is removed from the original X data, the filtered X data will constitute an optimal platform for further PLS analysis according to model complexity, predictive ability and interpretation.

1.8. Algorithm of OSC

OSC is a data treatment technique. The goal is to correct the X data matrix by removing the information that is unrelated to the matrix Y. In OSC algorithm, the weight vector (w) is modified to encompass constraints that score t =X*w is orthogonal to the Y matrix. This is realized by minimizing the covariance between t and Y as close as possible. Results of this calculation are the score and loading matrices (T, P) that contain the information not related to the Y matrix. Once the information not correlated with the Y matrix has been modeled, it is removed from the X data. A product of the score (T) and the loading (P) matrices formulated is subtracted from X matrix.

$$X_{OSC} = X - \lambda \sum_{i=1}^{n} T_i P_i^{'} \qquad (10)$$

where X_{OSC} is the OSC corrected X data, n is number of OSC components. λ means a penalty parameter. Two parameter values (n and λ) are determined by LOO CV. A geometric illustration of the difference between the standard PLS and OSC corrected PLS models is shown in Figure 4. Circle and square marks mean active and inactive molecules, respectively. In the left panel, the PLS components cannot separate the between-class variation from the within-class variation, and the resulting PLS component loadings mixes both types of variations. In the right panel, the OSC-PLS components are able to separate these two different variations clearly.

Figure 4. Geometric illustration of difference between PLS and OSC corrected PLS models. Left panel: PLS model, Right panel: OSC corrected PLS model. Circle and square marks mean active and inactive molecules, respectively. Cited from Ref. [Trygg et al. 2007]

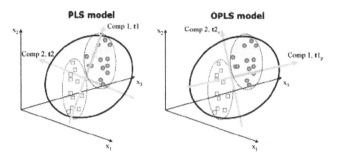

1.9. Applications of OSC

Shen et al. have applied OSC filter to a set of 41 phenylalkyl sulfides as COX2 inhibitors [Shen et al. 2006]. The data set was randomly divided into two groups with 31 compounds used as the training set and the remaining 10 compounds used as the test set. A series of molecular descriptors for phenylalkyl sulfides are structural, spatial, thermodynamic, electronic, topological descriptors, and E-State indices. After variable selection, 19 descriptors were selected as significant ones. The PLS modeling with 19 descriptors provided us 8-component model. The R^2 value for the training set was 0.841 and the LOO R^2 value for the test set was 0.916. By LOO CV, the values of OSC component number (n) and penalty parameter (λ) was optimized to be 5 and 0.6, respectively. The OSC corrected X and PLS resulted in a 7-component model. The R^2 value for the training set was 0.846 and the LOO R^2 value for the test set was 0.943. Performing OSC can improve the PLS regression model by reducing both model complexity (the number of components) and prediction errors (the LOO R^2 value). Authors have pointed out that OSC filter is necessary for interpreting the model equation correctly.

Although OSC has been originated from spectroscopic modeling, it was used as a preprocessing procedure in CoMFA [Bohac et al. 2002]. Due to dramatic reduction of the noisy field descrip-

tors, the corresponding contour maps in CoMFA become clearer for chemical interpretation. Furthermore, OSC has been used in the collaboration to the PLS algorithm directly. This method is called orthogonal PLS (OPLS) [Trygg et al. 2002]. OPLS and its multivariate version O2PLS [Trygg et al. 2003] are used in metabonomics and proteomics studies in recent years [Trygg et al. 2007, Bylesjoe et al. 2007].

2. NONLINEAR PLS WITH QUADRATIC AND KERNEL FUNCTIONS

The PLS method is useful, however, its major restriction is that only linear relation can be extracted from data. Since many structure-activity data are inherently nonlinear in nature, it is desirable to have a flexible method which can model any nonlinear relations. There has been a considerable interest in support vector machine (SVM) for nonlinear modeling [Vapnik 1995, Doucet et al. 2007, Han et al. 2007]. The SVM method can approximate any functions, but search for the best combinations of SVM parameters is time-consuming step. Another shortcoming of the SVM method is that interpretation of the model is rather complicated.

Several nonlinear versions of PLS have been proposed so far. Quadratic PLS (QPLS) [Wold et al. 1989], Kernel PLS (KPLS) [Rosipal et al. 2001], Spline version of PLS [Wold 1992, Eriksson et al. 2000, Berglund et al. 2001] and Implicit nonlinear latent variable regression [Berglund et al. 1997] are available.

2.1. Quadratic PLS (QPLS)

Wold et al. have proposed quadratic PLS (QPLS) as a nonlinear version of PLS [Wold et al. 1989]. In this approach, a quadratic inner relation is used instead of a linear relation in PLS. Unfortunately, the mathematical descriptions underlying the algorithm are not so clear, and few applications of QPLS method have been reported in the literatures. Hasegawa et al. have pointed out minor modifications of the QPLS algorithm and applied the corrected version to QSAR studies [Hasegawa et al. 1996]. After Hasegawa's work, Baffi et al. have proposed more revised QPLS algorithm. [Baffi et al. 1999]

2.2. Algorithm of QPLS

QPLS is an extension of PLS to deal with nonlinear data. X and Y variables in QPLS can be decomposed in the same way as PLS. Two scores t_h and u_h in the h-th component are linked by a quadratic function instead a linear relationship.

$$u_h = c_2 * t_h^2 + c_1 * t_h + c_0 \qquad (11)$$

where c_0, c_1 and c_2 represent the coefficients in the quadratic form. If both the quadratic coefficient (c_2) and constant (c_0) are equal to zero, the inner relation becomes a linear one in the PLS model.

2.3. Applications of QPLS

Hasegawa et al. have evaluated the performance of QPLS using 12 phenyl alkylamines with monoamine oxidase (MAO) inhibitory activities [Hasegawa et al. 1996]. The data consists

of one *in vitro* MAO activity and three *in vivo* MAO activities. The phenyl alkylamines were represented by 29 chemical descriptors. The *in vitro* and *in vivo* MAO inhibitory activities were analyzed separately by QPLS. For *in vitro* MAO inhibitory activity, one-component QPLS model was obtained by LOO. The coefficients of the QPLS model (c_0, c_1 and c_2) were 0.026, 1.025, and -0.028, respectively. Because the quadratic coefficients (c_2) almost equals zero, the relation between the *in vitro* MAO inhibitory activity and structural descriptors is considered to be linear. In analogy, the QPLS model was developed for three *in vivo* MAO inhibitory activities. A one-component QPLS model was obtained by LOO. The coefficients of the QPLS model (c_0, c_1 and c_2) were -0.153, 2.262, and 0.309, respectively. There is a weak nonlinearity between the *in vivo* MAO inhibitory activities and structural descriptors. The structural requirements for the MAO inhibitory activities can be estimated from the PLS loadings. The QPLS method can give us compact solution with one component, and the chemical interpretation is easily made.

Hasegawa et al. have analyzed twelve *E. coli* dihydrofolare reductase (DHFR) inhibitors [Hasegawa et al. 1997]. At first, the CoMFA field data of DHFR inhibitors was analyzed by the standard PLS. The inhibitory activity was explained by the electrostatic field only. The plot of score (t) against score (u) in the first component of the PLS model suggests a nonlinear relationship between the electrostatic field variables and inhibitory activity. Further analysis of DHFR inhibitors was carried out by QPLS. The QPLS model had higher predictivity with only one component and showed superior performance as compared to the PLS model (LOO $R^2 = 0.940$ vs. 0.722). The coefficients (c_0, c_1, and c_2) of the QPLS model were 0.068, 0.269 and -0.024, respectively. There is a weak nonlinearity between the inhibitory activity and electrostatic field variables. The score plot in the first component of the QPLS model is shown in Figure 5. In nonlinear CoMFA, the loading value

(p) is used instead of the coefficient value of the standard CoMFA. Authors have recommended the score plot of PLS is carefully examined before any conclusions are drawn from the PLS analysis.

Kimura et al. have applied QPLS to 89 synthetic elastase substrates [Kimura et al. 1996]. They have transformed the QPLS models into the MLR equations with the squared and cross terms of descriptors. Owing the transparent MLR equations, they have successfully estimated the structural requirements for the binding affinities without seeing any score plots.

In QPLS, determination of the coefficient values is performed based on a linearlization of the inner quadratic function, so that optimization is not always carried out properly. Therefore, the result of QPLS depends on the initial guess of parameter values. Yoshida et al. have proposed a new hybrid method that integrates GA and QPLS, in which GA is introduced to optimize the inner relation function instead of linearlization [Yoshida et al. 1997]. They have shown that the hybrid method can lead a significant improvement over the conventional QPLS method.

2.4. Kernel PLS (KPLS)

Kernel PLS (KPLS) is an alternative approach to deal with nonlinear relationship between X and Y. KPLS projects the original input data to a feature space in arbitrary dimension where a linear PLS model is created. KPLS has been shown to perform better than linear PLS in nonlinear systems. The KPLS method is resemble to SVM on the kernel philosophy. KPLS has been first introduced by Lindgren et al. in the context of working with

Figure 5. Plot of score vectors t_1 and u_1 scores of QPLS model. Cited from Ref. [Hasegawa et al. 1997]

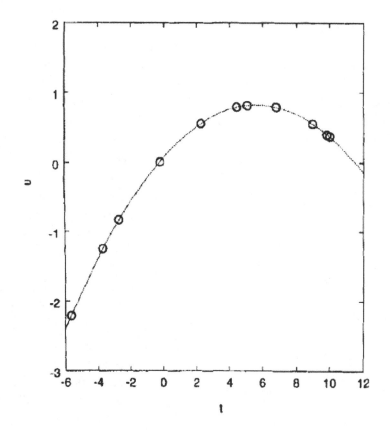

linear kernel on data set [Lindgren et al. 1994]. They have devised kernel functions in order to make the PLS modeling more efficient.

2.5. Algorithm of KPLS

Define Φ as a nonlinear map which projects the input variables from the original space to the feature space, $\Phi = [\Phi(x_1)|...|\Phi(x_N)]$. Where N is the number of samples. Define a matrix K as the kernel matrix to represent $\Phi\Phi'$, where $K_{ij} = K(x_i, x_j) = <\Phi(x_i), \Phi(x_j)>$, i, j = 1, 2,..., N. Just like the standard PLS method, the KPLS algorithm is to sequentially extract the latent vectors t, u from the Φ and Y matrices.

$$t = \Phi\Phi' = Ku \qquad (12)$$

Note that the kernel matrix K is treated instead of the original nonlinear map Φ. This reflects fast and smart description in the KPLS algorithm owing to so called kernel trick. Several kinds of kernel functions are available, polynomial kernel, sigmoid kernel, Gaussian kernel etc. Among different types of kernels, Gaussian kernel is common used.

$$K(x_i, x_j) = \exp(-\frac{\left\| x_i - x_j \right\|^2}{\sigma}) \qquad (13)$$

where σ is a parameter controlling the shape of Gaussian function.

2.6. Applications of KPLS

Jalai et al. have combined GA and KPLS (GAKPLS). They have applied GAKPLS to a data set consisting of 114 substituted aromatic sulfonamides as carbonic anhydrase II inhibitors [Jalai et al. 2007]. A total of 275 descriptors were calculated for each compound. GAKPLS gave us the model with 8 descriptors as the best one. (Not be described as to the PLS components and the Gaussian parameters in the original article.)

Authors have compared KPLS with PLS and concluded the former model is superior over the latter model. (LOO $R^2 = 0.800$ vs. 0.713) The selected 8 descriptors are simple and also are able to encode different aspects of the inhibition mechanism of sulfonamides.

KPLS is not so popular compared to the standard nonlinear modeling such as SVM. However, recently, KPLS has been applied to the metabolism and scoring function problems and the promising results have been reported [Wang et al. 2007, Deng et al. 2004]. KPLS does not have any stochastic characters and the same results can be reproduced any time. This is a big contrast to the SVM method [Vapnik 1995].

3. MULTI-WAY PLS AND ITS UNFOLDING TYPE

3.1. Multi-Way PLS

The multi-way PLS method is a generalization of the standard PLS for analyzing multi-way data. It maintains the concept of obtaining multi score variables such that on the one hand one can fit the three-way array $\underline{X}$ with least squares and on the other maximize the covariance with the matrix Y. Multi-way, especially three-way PLS has been born in spectroscopic modeling [Bro 2006]. In spectroscopic field, sample*wavelength*time is a typical three-way data. There are four main algorithms dealing with three-way data (Bro method [Bro 1996], PARAFAC PARAFAC PARAFAC [Bro 1997], Tucker1 [Stanimirova et al. 2006] and Tucker3 [Andersson et al. 1998]) Among those, Bro method is introduced in the next section.

3.2. Algorithm of Multi-Way PLS

For simplicity, we focus the case of one activity descriptor y. The goal of the method is to make a decomposition of the three-way array $\underline{X}$ into a set of triads. A triad consists of one score vector t and two weight vectors, the vector one in the second mode called w^j and in the third mode called w^k. The model of $\underline{X}$ is given by the equation.

$$\underline{X} = \sum_{h=1}^{A} t_h \otimes w_h^j \otimes w_h^k + E \qquad (14)$$

$$y = \sum_{h=1}^{A} b_h t_h + f \qquad (15)$$

where symbol $\otimes$ is the Kronecker product. E and f are the model residuals of X and y, respectively. By a simple algebraic manipulation, the problem of finding the vector w^j and w^k leads to a singular value decomposition (SVD) of Z. The matrix Z is a J by K matrix with the jk'th element being the inner product of y and the column obtained by fixing the second and third modes of $\underline{X}$ at j and k, respectively.

The regression coefficient vector b_{PLS} can be calculated by modified weight matrix W*, score matrix T and y.

$$b_{PLS} = W^*(T'y) \qquad (16)$$

The modified weight matrix W* is defined as follows:

$$W^* = [w_1 \left| (I - w_1 w_1') w_2 \right| \cdots \left| (I - w_1 w_1')(I - w_2 w_2') \cdots (I - w_{A-1} w_{A-1}') w_A \right|] \qquad (17)$$

$$w = w^k \otimes w^j \qquad (18)$$

The algorithm of four-way PLS is similar to that of three-way PLS. The four-way array can be decomposed into one score vector and three weight vectors. The regression coefficient vector b_{pls} can be calculated from modified matrix W*, score matrix T and y. The difference is a weight vector w in W*, which is derived from Kronecker products of three weight vectors.

3.3. Applications of Multi-Way PLS

A choice of an active conformation and the corresponding alignment rule is an important problem for determining the success of 3D-QSAR studies. This problem becomes the most difficult one especially when no crystal structures are available [Hasegawa et al. 2005]. Hasegawa et al. have investigated the ability of the three-way PLS method for selecting bioactive conformation in 3D-QSAR studies [Hasegawa et al. 2000]. The series of 20 styrene derivatives were used as data set. The biological activity is an inhibition of protein tyrosine kinase (PTK). The 3D structure of each compound was built up and it was then subjected to conformational analysis. After conformation analysis, cluster analysis was used to select unique conformations. The resulting unique conformations were superimposed onto the lowest energy-conformation of the most-active compound No. 1 and were characterized by 3D field variables in CoMFA. One sample-variable sheet was formed by selecting the conformations that are closest to each unique conformation of compound No. 1 in the field variables. Since compound No. 1 has 12 unique conformations, the total number of sample-variable sheets becomes 12. The three-way arrays was constructed by collecting all sample-variable sheets. Figure 6 shows the scheme for constructing three-way array $\underline{X}$. The three-way PLS analysis with LOO gave us a five-component model. In order to rank sample-variable sheets and select the best one, the regression coefficient vector was examined. The regression coefficient vector contains information how each sheet contributes to inhibitory activity. Sheet 11 gave us the highest regression coefficient value and hence, the conformations in sheet 11 were presumed to be the bioactive conformations of PTK inhibitors. Authors have stressed that the three-way PLS model can select the bioactive conformation among huge candidates in rational way, which is free from the subjective determination.

Figure 6. Construction of three-way array from conformations and field descriptors

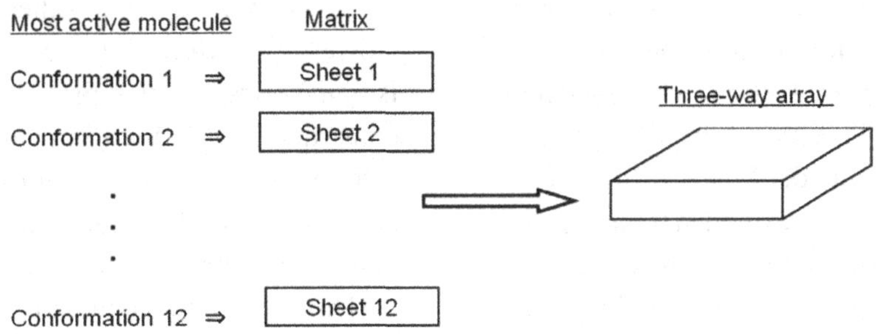

Hasegawa et al. have applied this technique to more general cases where both the conformation and alignment variables are simultaneously varied. In this case, the sample-variable sheets under each alignment rule are defined and by collecting all alignment rules, one can obtain the complete four-way array from sample, variable, conformation and alignment vectors. The relationship between the four-way array and activity vector can be investigated by four-way PLS. This type of analysis attracts many attentions because the errors originated from modern theoretical calculations are further minimized by taking into account many nodes. Authors have applied four-way PLS to benzodiazepine derivatives as antagonists of cholecystokinin-B (CCK-B) [Hasegawa et al. 2003]. From weight vectors they have made an elucidation of bioactive conformations and alignment rule.

Bhonsle et al. have proposed the quasi-multi way PLS approach to both develop highly QSR models and pick up bioactive conformations [Bhonsle et al. 2008, Bhonsle et al. 2005]. Their approach is based on several sequential two-way PLS analyses. By monitoring the prediction values, they have reduced the number of conformations for each compound step by step and in the final stage, the best combination of conformations could be obtained. Their approach contains some biased estimations because the result heavily depends on the first cycle in the sequential procedure.

Another interesting application using three-way PLS is the study of customizing empirical protein-ligand scoring function. Catana et al. have constructed three-way array consisting of ligand-protein complex, interaction descriptors, and binding modes [Catana et al. 2007]. By analyzing relationship between the three-way array and binding affinities, they have obtained high predictive scoring function over the previous ones. They have concluded that this technique can be applied in numerous practical situations. For example, one could develop a precise customized scoring function on a congeneric series bound to a certain protein, or to different proteins, or on a diverse set of ligands bound to a certain protein.

Hasegawa et al. have proposed a new surface-based 3D-QSAR method using Kohonen neural network (KNN) and three-way PLS [Hasegawa et al. 2002]. They have applied this approach to 25 dopamine2 (D2) receptor antagonists. After superimposition of the molecular structures, the molecular electrostatic potential (MEP) value on van der Waals surface was calculated. The Cartesian coordinates and the associated MEP values of all points sampled on van der Waals surface were used for KNN training. As a result of KNN training, 2D KNN map with the MEP values was generated. The three-way array was constructed by collecting all KNN maps. The correlation between the D2 receptor antagonist activities and three-way array was investigated by three-way PLS. The four-component three-way PLS model was obtained by LOO. The established three-way PLS model was converted into the MLR model and the regression coefficients were back-projected on molecular surface. Authors have concluded that the three-way PLS model can describe more real interaction on van der Waals with enzyme.

Hasegawa et al. have extended their approach to more general case where both the electrostatic and lipophilic potentials on molecular surface simultaneously change [Hasegawa et al. 2003]. In this case, the electrostatic and lipophilic KNN maps are generated for each compound and four-way array is constructed by collecting two KNN maps of all samples. The correlation between the four-way array and biological activity is examined via four-way PLS. For validation, the structure-activity data of estrogen receptor antagonists was investigated. The regression coefficients regarding to electrostatic potentials and lipophilic potentials were nicely projected on van der Waals surface. The visual inspection of projection map indicated that the 3D distribution of the standard PLS model is more messy and complex than that of the four-way PLS model. In that sense, the four-way PLS model is superior to the standard PLS model for presenting the chemical features on molecular surface.

Polanski et al. have employed similar strategy and proposed so called comparative molecular surface analysis [Polanski 2006]. They have defined chemical descriptors by unfolding all the nodes with the MEP values in KNN map. The relationship between the activities and chemical descriptors was analyzed by the standard PLS. Authors have constructed the 3D-QSAR model but the model lost the important neighboring relationship between the nodes in KNN map.

3.4. Unfolding PLS (UPLS)

The three-way data $\underline{Y}$ is often encountered in analytical chemistry, a prime example being the use of various hyphenated techniques, like LC-MS, GC-MS, and 2D-NMR on a series of analytical samples. In the QSAR field, three-way data $\underline{Y}$ is not as abundant as in other areas of research. However, pharmacokinetic study on compounds is time-series, and the three-way $\underline{Y}$ matrix can be constructed in QSAR field. Three-way data offers different ways of unfolding to produce two-way array, which will help out to elucidate the dominant patterns in data set from different angles.

Unfolding PLS (UPLS) has been proposed by Eriksson et al. behind this background [Eriksson et al. 2004]. In UPLS, interpretation is sometimes difficult as the primary and secondary variable modes are not modeled explicitly, but mixed up during the unfolding step. However, several interesting diagnostic and interpretative options (score and loading plot) are available as in the standard PLS. This provides us an opportunity to enhance our ability to reliably detect clusters, trends, outliers and other abnormalities.

3.5. Algorithm of UPLS

There are two relevant ways of unfolding the three-way $\underline{Y}$ matrix. The first preserves the x-direction (M) and the second the z-direction (L). The former way captures x-direction features in the data. The latter approach indicates whether there are features in the data critically influenced by z-direction. The resulting two-way matrices have either N*M rows and L columns or, N*L rows and M columns. (Figure 7) The scores of the resulting PLS models on this level (the lower level) may later be used on the next level (the upper level).

The rearrangement of the PLS scores on the lower level is next made. The arranged new matrix is Y_T matrix where in each row, all t1 values of one compound are followed by all t2 values of the same compound, which are followed by all t3 values of the same compound, and so on. Subsequently, the PLS model is established where the X matrix is related to the Y_T matrix. Depending on the choice of unfolding principle on the lower lever, the upper level PLS modeling interrogates the chemical descriptors from different

Figure 7. Two ways of unfolding of three-way array Y

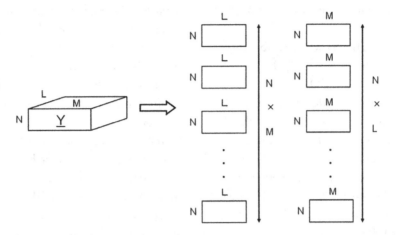

view points, i.e., to x-direction or z-direction relationships. The basic idea of the UPLS method was originated from batch-process in chemical industry [Chiang et al. 2006].

3.6. Application of UPLS

Eriksson et al. have performed the UPLS analysis to the data set comprising 35 4-pyranones with toxic data [Eriksson et al. 2004]. To describe the physico-chemical properties a set of 26 descriptors was compiled. The biological data were acquired as a function of changing time and pH. The $\underline{Y}$ matrix is built up from N = 35 compounds, L = 5 time points (2, 4, 6, 8 and 10 hours) and M = 6 pH-values (5.6, 6.0, 6.6, 7.0, 7.6 and 8.0). The PLS model was calculated for the two-way matrix consisting of N*L rows (35*5 = 175) and M = 6 columns (toxicity measurements at six pH-values). After removing some compounds, the Y_T matrix consisting of 14 rows (compounds) and 5 columns (t_1 score vector with 5 time points) was constructed. The resulting PLS model had only one significant component with the following statistics: $R^2 = 0.75$, and LOO $R^2 = 0.66$. From the PLS weights, an increased toxicity is coupled to an increased hydrophobicity, molecular weight and molar volume. To investigate the relation between toxicity data and pH, the Y matrix was unfolded.

The resulting two-way matrix of toxicity data has N*M rows (35*6 = 210) and L = 5 columns (toxicity measurements at five time points). After removing some compounds, the resulting Y_T matrix constituted 10 rows (compounds) and 6 columns (t_1 score vector with 6 pH values). The resulting PLS model contained only one component with the following performance statistics: $R^2 = 0.72$ and LOO $R^2 = 0.53$. According to the PLS weights, HB-Donors, Sum_donor, Sum_total, LUMO energy, and molecular weight are most appropriate to capture the pH-influenced change in toxicity data. Authors have concluded that UPLS seems to be promising tool because compounds that have time- or pH-sensitive toxicological profiles are rapidly identified.

4. MULTI-BLOCK PLS

4.1. Hierarchical PLS (HPLS)

When dealing with very many variables, plots and lists of loadings, weights and coefficients, tend to become messy and the results are often difficult to overview. Instead of reducing the number of variables and thus reducing the validity of the model, a better alternative is to divide the variables into conceptually meaningful blocks. Blocking the descriptors does not necessarily give better predictions, but certainly, it generally simplifies chemical interpretation.

Hierarchical PLS (HPLS) was developed for this purpose [Westerhuis et al. 1998]. The HPLS method operates on two levels. On the lower level the details of each block are modeled. This analysis provides us the block score vectors, which are used to construct the data matrices on the upper level. On the upper level, a relationship between relatively few super variables is developed. (Figure 8) Mathematical relationships between HPLS and the standard PLS were discussed in the literature [Westerhuis et al. 1998].

4.2. Algorithm of HPLS

HPLS is an extension to traditional PLS regression. With HPLS, variables are first divided into multi X blocks. Using the standard PLS, each block of variables is summarized by a few X scores

(super variables) that are concatenated into a new X block. The new X block is used for final PLS modeling at the top level. The advantage of using PLS at the base level is that the projection of each X block is oriented in a way relevant to the description of the Y block. This has the effect of stabilizing the PLS model at the top level, resulting in models with increased predictive ability. HPLS is an alternative to potentially variable selection as well as to the smarter block-scaling, both of which can be carried out in a multitude of ways. Another advantage of HPLS is enhanced model interpretability. Since parameters are distributed over the levels of the model hierarchy, there are fewer parameters to interpret at each level. Statistics and diagnostics that apply to standard PLS also work in HPLS.

Figure 8. Schematic overview of HPLS. Cited from Ref. [Eriksson et al. 2006]

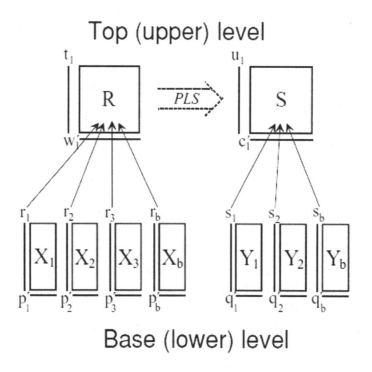

4.3. Applications of HPLS

Larsen et al. have applied HPLS to affinities data for diverse human H+/peptide Symporter ligand chemotypes [Larsen et al. 2008]. The data set contained 114 compounds. Three sets of molecular descriptors were considered: 110 VolSurf descriptors, 500 GRid-Independent Descriptors (GRIND), and 204 MOE descriptors. The data set was partitioned into 76 training and 38 test sets. For the base-level models, CV returned three significant components for all three blocks. Transferring the X scores from the base-level models to the top level resulted in one-component model with R^2 = 0.77 and LOO R^2 = 0.75. The chemical interpretation can be inspected from the first weight vectors along VolSurf, GRIND and MOE blocks on the base level. Authors have described that the model is derived from alignment-free descriptors and operates on a computationally efficient and fast protocol, requiring neither time-consuming molecular superimposition nor conformational sampling.

Eriksson et al. have analyzed a set of haloalkanes with toxicity data using HPLS [Eriksson et al. 2006]. Due to the complexity of the toxicity data, Y matrix was sub-divided in four blocks. From the first component in the top-level modeling, they have concluded that toxicity correlates with hydrophobicity and size of molecules. From the second component, toxicity is related to the reactivity of hydroxyl radical of compounds.

Lindstrom et al. have presented for predicting ligand binding to proteins using HPLS [Lindstrom et al. 2006]. The protein-ligand complexes were characterized by four different descriptor blocks: 3D structural descriptors of the proteins, protein-ligand interactions based on scoring function, binding site surface area, and ligand 2D/3D descriptors. The results of HPLS showed that all four descriptor blocks contributes to the HPLS model. An inspection of the weight vectors of the four different blocks disclosed that ligand descriptors are more important for binding to proteins.

CONCLUSION

PLS and its advanced techniques were reviewed with their algorithms and representative examples. PLS can analyze data with strongly collinear, noisy and numerous X variables, and also simultaneously model several response variables Y. Furthermore, PLS can provide us several prediction regions and diagnostic plots as statistical measures. Nonlinear PLS has been invented to deal with nonlinear relationships. Especially, QPLS can afford us compact solution with less components, and the chemical interpretation is easily made. Multi-way PLS is a generalization of PLS for analyzing multi-way data. The algorithm of multi-way PLS has the advantage compared to the unfolding approach with PLS in views of fewer parameters and simplicity of model. Multi-way PLS was used for selecting bioactive conformations and its alignment rule in rational way. Multi-block PLS in hierarchical type has great potential for dealing with the omics data.

The PLS algorithm is very flexible and many types of methods can be merged into its framework. For example, logistic PLS has been developed for modeling logistic function in data [Liu et al. 2008, Fort et al. 2005, Ohgaru et al. 2008]. PLS/SIMCA hybrid type has also been developed in classification problems [Bylesjoe et al. 2007]. Therefore, more unique hybrid types of algorithm would be proposed and applied in chemometrics.

REFERENCES

Akula, N., Lecanu, L., Greeson, J., & Papadopoulos, V. (2006). 3D QSAR studies of AChE inhibitors based on molecular docking scores and CoMFA. *Bioorganic & Medicinal Chemistry Letters, 16*(24), 6277–6280.

Andersson, C.A., & Bro, R. (1998). Improving the speed of multi-way algorithms: Part I. Tucker3. *Chemometrics and Intelligent Laboratory Systems, 42*, 93–103.

Arakawa, M., Hasegawa, K., & Funatsu, K. (2006). QSAR study of anti-HIV HEPT analogues based on multi-objective genetic programming and counter-propagation neural network. *Chemometrics and Intelligent Laboratory Systems*, *83*(2), 91–98.

Arakawa, M., Hasegawa, K., & Funatsu, K. (2007). The recent trend in QSAR modeling -variable selection and 3D-QSAR methods. *Current Computer-aided Drug Design*, *3*(4), 254–262.

Baffi, G., Martin, E. B., & Morris, A. J. (1999). Non-linear projection to latent structures revisited: the quadratic PLS algorithm. *Computers & Chemical Engineering*, *23*, 395–411.

Berglund, A., Kettaneh, N., Uppgard, L., Wold, S., Bendwell, N., & Cameron, D. R. (2001). The GIFI approach to non-linear PLS modeling. *Journal of Chemometrics*, *15*(4), 321–336.

Berglund, A., & Wold, S. (1997). INLR, implicit non-linear latent variable regression. *Journal of Chemometrics*, *11*(2), 141–156.

Bhonsle, J. B., & Huddler, D. (2008). Novel method for mining QSPR-relevant conformations. *Chemical Engineering Communications*, *195*(11), 1396–1423.

Bhonsle, J. B., Wang, Z., Tamamura, H., Fujii, N., Peiper, S. C., & Trent, J. O. (2005). A simple, automated quasi-4D-QSAR, quasi-multi way PLS approach to develop highly predictive QSAR models for highly flexible CXCR4 inhibitor cyclic pentapeptide ligands using scripted common molecular modeling tools. *QSAR & Combinatorial Science*, *24*(5), 620–630.

Bohac, M., Loeprecht, B., Damborsky, J., & Schuurmann, G. (2002). Impact of orthogonal signal correction (OSC) on the predictive ability of CoMFA models for the ciliate toxicity of nitrobenzenes. *Quantitative Structure-Activity Relationships*, *21*(1), 3–11.

Bro, R. (1996). Multiway calibration. Multilinear PLS. *Journal of Chemometrics*, *10*(1), 47–61.

Bro, R. (1997). Tutorial PARAFAC. Tutorial and applications. *Chemometrics and Intelligent Laboratory Systems*, *38*, 149–171.

Bro, R. (2006). Review on multiway analysis in chemistry-2000-2005. *Critical Reviews in Analytical Chemistry*, *36*(3-4), 279–293.

Bylesjoe, M., Eriksson, D., Kusano, M., Moritz, T., & Trygg, J. (2007). Data integration in plant biology: the O2PLS method for combined modeling of transcript and metabolite data. *The Plant Journal*, *52*(6), 1181–1191.

Bylesjoe, M., Rantalainen, M., Cloarec, O., Nicholson, J. K., Holmes, E., & Trygg, J. (2007). OPLS discriminant analysis: combining the strengths of PLS-DA and SIMCA classification. *Journal of Chemometrics*, *20*(8-10), 341–351.

Catana, C., & Stouten, P. F. W. (2007). Novel, Customizable Scoring Functions, Parameterized Using N-PLS, for Structure-Based Drug Discovery. *Journal of Chemical Information and Modeling*, *47*(1), 85–91.

Chiang, L. H., Leardi, R., Pell, R. J., & Seasholtz, M. B. (2006). Industrial experiences with multivariate statistical analysis of batch process data. *Chemometrics and Intelligent Laboratory Systems*, *81*(2), 109–119.

Cramer, R. D., Patterson, D. E., & Bunce, J. D. (1988). Comparative molecular field analysis (CoMFA). 1. Effect of shape on binding of steroids to carrier proteins. *Journal of the American Chemical Society*, *110*(18), 5959–5967.

Crivori, P., Zamora, I., Speed, B., Orrenius, C., & Poggesi, I. (2004). Model based on GRID-derived descriptors for estimating CYP3A4 enzyme stability of potential drug candidates. *Journal of Computer-Aided Molecular Design*, *18*(3), 155–166.

Daren, Z. (2001). QSPR studies of PCBs by the combination of genetic algorithms and PLS analysis. *Computers & Chemistry, 25*(2), 197–204.

Deeb, O., Hemmateenejad, B., Jaber, A., Garduno-Juarez, R., & Miri, R. (2007). Effect of the electronic and physicochemical parameters on the carcinogenesis activity of some sulfa drugs using QSAR analysis based on genetic-MLR and genetic-PLS. *Chemosphere, 67*(11), 2122–2130.

Deng, W., Breneman, C., & Embrechts, M. J. (2004). Predicting protein-ligand binding affinities using novel geometrical descriptors and machine-learning methods. *Journal of Chemical Information and Computer Sciences, 44*(2), 699–703.

Devillers, J. (Ed.). (1996). *Genetic Algorithms in Molecular Modeling.* New York: Academic Press.

Doucet, J., Barbault, F., Xia, H., Panaye, A., & Fan, B. (2007). Nonlinear SVM approaches to QSPR/QSAR studies and drug design. *Current Computer-aided Drug Design, 3*(4), 263–289.

Edraki, N., Hemmateenejad, B., Miri, R., & Khoshneviszade, M. (2007). QSAR study of phenoxypyrimidine Derivatives as Potent Inhibitors of p38 Kinase using different chemometric tools. *Chemical Biology & Drug Design, 70*(6), 530–539.

Equbal, T., Silakari, O., & Ravikumar, M. (2008). Exploring three-dimensional quantitative structural activity relationship (3D-QSAR) analysis of SCH 66336 (Sarasar) analogues of farnesyltransferase inhibitors. *European Journal of Medicinal Chemistry, 43*(1), 204–209.

Eriksson, L., Andersson, P. L., Johansson, E., & Tysklind, M. (2006). Megavariate analysis of environmental QSAR data. Part II - Investigating very complex problem formulations using hierarchical, non-linear and batch-wise extensions of PCA and PLS. *Molecular Diversity, 10*(2), 187–205.

Eriksson, L., Gottfries, J., Johansson, E., & Wold, S. (2004). Time-resolved QSAR: an approach to PLS modeling of three-way biological data. *Chemometrics and Intelligent Laboratory Systems, 73*(1), 73–84.

Eriksson, L., Johansson, E., Lindgren, F., & Wold, S. (2000). GIFI-PLS: modeling of non-linearities and discontinuities in QSAR. *Quantitative Structure-Activity Relationships, 19*(4), 345–355.

Eroes, D., Keri, G., Koevesdi, I., Szantai-Kis, C., Meszaros, G., & Oerfi, L. (2004). Comparison of predictive ability of water solubility QSPR models generated by MLR, PLS and ANN methods. *Mini Reviews in Medicinal Chemistry, 4*(2), 167–177.

Estrada, E. (2008). How the parts organize in the whole? A top-down view of molecular descriptors and properties for QSAR and drug design. *Mini Reviews in Medicinal Chemistry, 8*(3), 213–221.

Fassihi, A., & Sabet, R. (2008). QSAR study of p56lck protein tyrosine kinase inhibitory activity of flavonoid derivatives using MLR and GA-PLS. *International Journal of Molecular Sciences, 9*(9), 1876–1892.

Ferreira da Cunha, E. F., Martins, R. C. A., Albuquerque, M. G., & Bicca de Alencastro, R. (2004). LIV-3D-QSAR model for estrogen receptor ligands. *Journal of Molecular Modeling, 10*(4), 297–304.

Fort, G., & Lambert-Lacroix, S. (2005). Classification using partial least squares with penalized logistic regression. *Bioinformatics (Oxford, England), 21*(7), 1104–1111.

Gasteiger, J. (Ed.). (2003). *Handbook of Chemoinformatics.* Amsterdam: Wiely-VCH.

Gedeck, P., & Lewis, R. A. (2008). Exploiting QSAR models in lead optimization. *Current Opinion in Drug Discovery & Development, 11*(4), 569–575.

Geladi, P., & Kowalski, B. R. (1986). Partial least-squares regression: a tutorial. *Analytica Chimica Acta, 185*, 1–17.

Goldberg, D. E. (Ed.). (1989). *Genetic Algorithm in Searcg, Optimization and MachineLearning.* New York: Addison Wesly.

Gonzalez, M. P., Teran, C., Saiz-Urra, L., & Teijeira, M. (2008). Variable selection methods in QSAR: an overview. *Current Topics in Medicinal Chemistry, 8*(18), 1606–1627.

Guo, Y., Xiao, J., Guo, Z., Chu, F., Cheng, Y., & Wu, S. (2005). Exploration of a binding mode of indole amide analogues as potent histone deacetylase inhibitors and 3D-QSAR analyses. *Bioorganic & Medicinal Chemistry, 13*(18), 5424–5434.

Han, L. Y., Zheng, C. J., Xie, B., Jia, J., Ma, X. H., & Zhu, F. (2007). Support vector machines approach for predicting druggable proteins: recent progress in its exploration and investigation of its usefulness. *Drug Discovery Today, 12*(7 & 8), 304–313.

Hansch, C. (1969). Quantitative approach to biochemical structure-activity relationships. *Accounts of Chemical Research, 2*(8), 232–239.

Hansch, C., & Fujita, T. (1964). ρ-σ-π Analysis; method for the correlation of biological activity and chemical structure. *Journal of the American Chemical Society, 86*(8), 1616–1626.

Hasegawa, K., Arakawa, M., & Funatsu, K. (2000). Rational choice of bioactive conformations through use of conformation analysis and 3-way partial least squares modeling. *Chemometrics and Intelligent Laboratory Systems, 50*(2), 253–261.

Hasegawa, K., Arakawa, M., & Funatsu, K. (2003). Simultaneous determination of bioactive conformations and alignment rules by multi-way PLS modeling. *Computational Biology and Chemistry, 27*(3), 211–216.

Hasegawa, K., Arakawa, M., & Funatsu, K. (2005). Novel computational approaches in QSAR and molecular design based on GA, multi-way PLS and NN. *Current Computer-aided Drug Design, 1*(2), 129–145.

Hasegawa, K., Deushi, T., Yoshida, H., Miyashita, Y., & Sasaki, S. (1994). Chemometric QSAR studies of antifungal azoxy compounds. *Journal of Computer-Aided Molecular Design, 8*(4), 449–456.

Hasegawa, K., & Funatsu, K. (2000). Partial least squares modeling and genetic algorithm optimization in quantitative structure-activity relationships. *SAR and QSAR in Environmental Research, 11*(3-4), 189–209.

Hasegawa, K., & Funatsu, K. (2009 in press). Data Modeling and Chemical Interpretation of ADME Properties Using Regression and Rule Mining Techniques . In Gary, W. C. (Ed.), *Frontier in Drug Design & Discovery 4.* Bentham Science Publisher.

Hasegawa, K., Kimura, T., & Funatsu, K. (1997). Nonlinear CoMFA using QPLS as a novel 3D-QSAR approach. *Quantitative Structure-Activity Relationships, 16*(3), 219–223.

Hasegawa, K., Kimura, T., Miyashita, Y., & Funatsu, K. (1996). Nonlinear Partial Least Squares Modeling of Phenyl Alkylamines with the Monoamine Oxidase Inhibitory Activities. *Journal of Chemical Information and Computer Sciences, 36*(5), 1025–1029.

Hasegawa, K., Matsuoka, S., Arakawa, M., & Funatsu, K. (2002). New molecular surface-based 3D-QSAR method using kohonen neural network and 3-way PLS. *Computers & Chemistry, 26*(6), 583–589.

Hasegawa, K., Matsuoka, S., Arakawa, M., & Funatsu, K. (2003). Multi-way PLS modeling of structure-activity data by incorporating electrostatic and lipophilic potentials on molecular surface. *Computational Biology and Chemistry, 27*(3), 381–386.

Hasegawa, K., Miyashita, Y., & Funatsu, K. (1997). GA Strategy for Variable Selection in QSAR Studies: GA Based PLS Analysis of Calcium Channel Antagonists. *Journal of Chemical Information and Computer Sciences, 37*(2), 306–310.

Ho, R. L., & Lieu, C. A. (2008). Systems biology: an evolving approach in drug discovery and development. *Drugs in R&D., 9*(4), 203–216.

Hoffman, B. T., Kopajtic, T., Katz, J. L., & Newman, A. H. (2000). 2D QSAR Modeling and Preliminary Database Searching for Dopamine Transporter Inhibitors Using Genetic Algorithm Variable Selection of Molconn Z Descriptors. *Journal of Medicinal Chemistry, 43*(22), 4151–4159.

Hu, L., Wu, H., Lin, W., Jiang, J., & Yu, R. (2007). Quantitative structure-activity relationship studies for the binding affinities of imidazobenzodiazepines for the α6 benzodiazepine receptor isoform utilizing optimized blockwise variable combination by particle swarm optimization for partial least squares modeling. *QSAR & Combinatorial Science, 26*(1), 92–101.

Jalali-Heravi, M., & Kyani, A. (2007). Application of genetic algorithm-kernel partial least square as a novel nonlinear feature selection method: Activity of carbonic anhydrase II inhibitors. *European Journal of Medicinal Chemistry, 42*(5), 649–659.

Jong, S. (1993). SIMPLS: an alternative approach to partial least squares regression. *Chemometrics and Intelligent Laboratory Systems, 18*, 251–263.

Kimura, T., Miyashita, Y., Funatsu, K., & Sasaki, S. (1996). Quantitative Structure-Activity Relationships of the Synthetic Substrates for Elastase Enzyme Using Nonlinear Partial Least Squares Regression. *Journal of Chemical Information and Computer Sciences, 36*(2), 185–189.

Kratochwil, N. A., Huber, W., Muller, F., Kansy, M., & Gerber, P. R. (2002). Predicting plasma protein binding of drugs: a new approach. *Biochemical Pharmacology, 64*(9), 1355–1374.

Kriegl, J. M., Eriksson, L., Arnhold, T., Beck, B., Johansson, E., & Fox, T. (2005). Multivariate modeling of cytochrome P450 3A4 inhibition. *European Journal of Pharmaceutical Sciences, 24*(5), 451–463.

Larsen, S. B., Jorgensen, F. S., & Olsen, L. (2008). QSAR Models for the Human H+/Peptide Symporter, hPEPT1: Affinity Prediction Using Alignment-Independent Descriptors. *Journal of Chemical Information and Modeling, 48*(1), 233–241.

Lindgren, F., Geladi, P., & Wold, S. (1993). The kernel algorithm for PLS. *Journal of Chemometrics, 7*(1), 45–59.

Lindgren, F., Geladi, P., & Wold, S. (1994). Kernel-based pls regression; cross-validation and applications to spectral data. *Journal of Chemometrics, 8*(6), 377–389.

Lindstrom, A., Pettersson, F., Almqvist, F., Berglund, A., Kihlberg, J., & Linusson, A. (2006). Hierarchical PLS Modeling for Predicting the Binding of a Comprehensive Set of Structurally Diverse Protein-Ligand Complexes. *Journal of Chemical Information and Modeling, 46*(3), 1154–1167.

Liu, H. X., Zhang, R. S., Yao, X. J., Liu, M. C., Hu, Z. D., & Fan, B. T. (2004). Prediction of the Isoelectric Point of an Amino Acid Based on GA-PLS and SVMs. *Journal of Chemical Information and Computer Sciences, 44*(1), 161–167.

Liu, J., Kern, P. S., Gerberick, G. F., Santos-Filho, O. A., Esposito, E. X., Hopfinger, A. J., & Tseng, Y. J. (2008). Categorical QSAR models for skin sensitization based on local lymph node assay measures and both ground and excited state 4D-fingerprint descriptors. *Journal of Computer-Aided Molecular Design, 22*(6-7), 345–366.

Martens, H., & Naes, T. (Eds.). (1989). *Multivnriate Calibration*. New York: Wiley.

Miyashita, Y., Li, Z., & Sasaki, S. (1993). Chemical pattern recognition and multivariate analysis for QSAR studies. *Trends in Analytical Chemistry, 12*(2), 50–60.

Miyashita, Y., Ohsako, H., Takayama, C., & Sasaki, S. (1992). Multivariate structure-activity relationships analysis of fungicidal and herbicidal thiolcarbamates using partial least squares method. *Quantitative Structure-Activity Relationships, 11*(1), 17–22.

Miyashita, Y., & Sasaki, S. (Eds.). (1994). *Chemical Pattern Recognition and Multivariate Aanalysis*. Tokyo: Kyoritsu Publisher.

Mohajeri, A., Hemmateenejad, B., Mehdipour, A., & Miri, R. (2008). Modeling calcium channel antagonistic activity of dihydropyridine derivatives using QTMS indices analyzed by GA-PLS and PC-GA-PLS. *Journal of Molecular Graphics & Modelling, 26*(7), 1057–1065.

Moro, S., Bacilieri, M., Cacciari, B., & Spalluto, G. (2005). Autocorrelation of Molecular Electrostatic Potential Surface Properties Combined with Partial Least Squares Analysis as New Strategy for the Prediction of the Activity of Human A3 Adenosine Receptor Antagonists. *Journal of Medicinal Chemistry, 48*(18), 5698–5704.

Obrezanova, O., Gola, J. M. R., Champness, E. J., & Segall, M. D. (2008). Automatic QSAR modeling of ADME properties: blood-brain barrier penetration and aqueous solubility. *Journal of Computer-Aided Molecular Design, 22*(6-7), 431–440.

Ohgaru, T., Shimizu, R., Okamoto, K., Kawashita, N., Kawase, M., & Shirakuni, Y. (2008). Enhancement of ordinal CoMFA by ridge logistic partial least squares. *Journal of Chemical Information and Modeling, 48*(4), 910–917.

Peng, Y., Keenan, S. M., Zhang, Q., Kholodovych, V., & Welsh, W. J. (2005). 3D- QSAR Comparative Molecular Field Analysis on Opioid Receptor Antagonists: Pooling Data from Different Studies. *Journal of Medicinal Chemistry, 48*(5), 1620–1629.

Polanski, J. (2006). Drug design using comparative molecular surface analysis. *Expert Opinion on Drug Discovery, 1*(7), 693–707.

Ravichandran, V., Kumar, B. R. P., Sankar, S., & Agrawal, R. K. (2008). Comparative molecular similarity indices analysis for predicting anti-HIV activity of phenyl ethyl thiourea (PET) derivatives. *Medicinal Chemistry Research, 17*(1), 1–11.

Ren, Y., Chen, G., Hu, Z., Chen, X., & Yan, B. (2008). Applying novel Three-Dimensional Holographic Vector of Atomic Interaction Field to QSAR studies of artemisinin derivatives. *QSAR & Combinatorial Science, 27*(2), 198–207.

Romeiro, N. C., Albuquerque, M. G., Bicca de Alencastro, R., Ravi, M., & Hopfinger, A. J. (2006). Free-energy force-field three-dimensional quantitative structure-activity relationship analysis of a set of p38-mitogen activated protein kinase inhibitors. *Journal of Molecular Modeling, 12*(6), 855–868.

Rosipal, R., & Trejo, L. J. (2001). Kernel Partial Least Squares Regression in Reproducing Kernel Hilbert Space. *Journal of Machine Learning Research, 2*, 97–123.

Sabet, R., & Fassihi, A. (2008). QSAR study of antimicrobial 3-hydroxypyridine-4-one and 3-hydroxypyran-4-one derivatives using different chemometric tools. *International Journal of Molecular Sciences, 9*(12), 2407–2423.

Sagrado, S., & Cronin, M. T. D. (2008). Application of the modelling power approach to variable subset selection for GA-PLS QSAR models. *Analytica Chimica Acta, 609*(2), 169–174.

Shen, Q., Jiang, J., Shen, G., & Yu, R. (2006). Ridge estimated orthogonal signal correction for data preprocessing prior to PLS modeling: QSAR studies of cyclooxygenase-2 inhibitors. *Chemometrics and Intelligent Laboratory Systems, 82*(1-2), 44–49.

Shen, Q., Jiang, J., Tao, J., Shen, G., & Yu, R. (2005). Modified Ant Colony Optimization Algorithm for Variable Selection in QSAR Modeling: QSAR Studies of Cyclooxygenase Inhibitors. *Journal of Chemical Information and Modeling, 45*(4), 1024–1029.

Stanimirova, I., Zehl, K., Massart, D. L., Heyden, Y., & Einax, J. W. (2006). Chemometric analysis of soil pollution data using the Tucker N-way method. *Analytical and Bioanalytical Chemistry, 385*(4), 771–779.

Tang, K., & Li, T. (2002). Combining PLS with GA-GP for QSAR. *Chemometrics and Intelligent Laboratory Systems, 64*(1), 55–64.

Tong, J., & Liu, S. (2008). Three-Dimensional Holographic Vector of atomic interaction field applied in QSAR of anti-HIV HEPT analogues. *QSAR & Combinatorial Science, 27*(3), 330–337.

Tropsha, A. (2006). Variable selection QSAR modeling, model validation, and virtual screening. *Annual Reports in Computational Chemistry, 2*, 113–126.

Tropsha, A., & Zheng, W. (2001). Identification of the descriptor pharmacophores using variable selection QSAR: applications to database mining. *Current Pharmaceutical Design, 7*(7), 599–612.

Trygg, J., Holmes, E., & Lundstedt, T. (2007). Chemometrics in Metabonomics. *Journal of Proteome Research, 6*(2), 469–479.

Trygg, J., & Wold, S. (2002). Orthogonal projections to latent structures (O-PLS). *Journal of Chemometrics, 16*(3), 119–128.

Trygg, J., & Wold, S. (2003). O2-PLS, a two-block (X-Y) latent variable regression (LVR) method with an integral OSC filter. *Journal of Chemometrics, 17*(1), 53–64.

Vapnik, V. N. (Ed.). (1995). *The Nature of Statistical Learning Theory*. Berlin: Springer.

Wanchana, S., Yamashita, F., & Hashida, M. (2003). QSAR Analysis of the Inhibition of Recombinant CYP 3A4 Activity by Structurally Diverse Compounds Using a Genetic Algorithm-Combined Partial Least Squares Method. *Pharmaceutical Research, 20*(9), 1401–1408.

Wang, Y., Li, Y., & Wang, B. (2007). An in silico method for screening nicotine derivatives as cytochrome P450 2A6 selective inhibitors based on kernel partial least squares. *International Journal of Molecular Sciences, 8*(2), 166–179.

Westerhuis, J. A., Kourti, T., & Macgregor, J. F. (1998). Analysis of multiblock and hierarchical PCA and PLS models. *Journal of Chemometrics, 12*(5), 301–321.

Wiklund, S., Nilsson, D., Eriksson, L., Sjoestrom, M., Wold, S., & Faber, K. (2007). A randomization test for PLS component selection. *Journal of Chemometrics, 21*(10-11), 427–439.

Wold, S. (1992). Nonlinear partial least squares modeling II. Spline inner relation. *Chemometrics and Intelligent Laboratory Systems, 14,* 71–84.

Wold, S., Antti, H., Lindgren, F. & Ohman, J. (1998). Orthogonal signal correction of near-infrared spectra. *Chemometrics and Intelligent Laboratory Systems, 44*(1,2), 175-185.

Wold, S., Kettaneh-Wold, N., & Skagerberg, B. (1989). Nonlinear PLS modeling. *Chemometrics and Intelligent Laboratory Systems, 7*(1-2), 53–65.

Wold, S., Sjostrom, M., & Eriksson, L. (2001). PLS-regression: a basic tool of chemometrics. *Chemometrics and Intelligent Laboratory Systems, 58*(2), 109–130.

Wold, S., Trygg, J., Berglund, A., & Antti, H. (2001). Some recent developments in PLS modeling. *Chemometrics and Intelligent Laboratory Systems, 58*(2), 131–150.

Xu, L., Liang, G., Li, Z., Wang, J., & Zhou, P. (2008). Three-dimensional holographic vector of atomic interaction field for quantitative structure-activity relationship of Aza-bioisosteres of anthrapyrazoles (Aza-APs*). Journal of Molecular Graphics & Modelling, 26*(8), 1252–1258.

Yamashita, F., Fujiwara, S., Wanchana, S., & Hashida, M. (2006). Quantitative structure/activity relationship modelling of pharmacokinetic properties using genetic algorithm-combined partial least squares method. *Journal of Drug Targeting, 14*(7), 496–504.

Yamashita, F., Wanchana, S., & Hashida, M. (2002). Quantitative structure/property relationship analysis of Caco-2 permeability using a genetic algorithm-based partial least squares method. *Journal of Pharmaceutical Sciences, 91*(10), 2230–2239.

Yap, C. W., Li, H., Ji, Z. L., & Chen, Y. Z. (2007). Regression methods for developing QSAR and QSPR models to predict compounds of specific pharmacodynamic, pharmacokinetic and toxicological properties. *Mini Reviews in Medicinal Chemistry, 7*(11), 1097–1107.

Yi, P., Fang, X., & Qiu, M. (2008). 3D-QSAR studies of Checkpoint Kinase Weel inhibitors based on molecular docking, CoMFA and CoMSIA. *European Journal of Medicinal Chemistry, 43*(5), 925–938.

Yoshida, H., & Funatsu, K. (1997). Optimization of the inner relation function of QPLS using genetic algorithm. *Journal of Chemical Information and Computer Sciences, 37*(6), 1115–1121.

Zuegge, J., Fechner, U., Roche, O., Parrott, N. J., Engkvist, O., & Schneider, G. (2002). A fast virtual screening filter for cytochrome P450 3A4 inhibition liability of compound libraries. *Quantitative Structure-Activity Relationships, 21*(3), 249–256.

Chapter 9
Nonlinear Partial Least Squares:
An Overview

Roman Rosipal
Medical University of Vienna, Austria & Pacific Development and Technology, LLC, USA

ABSTRACT

In many areas of research and industrial situations, including many data analytic problems in chemistry, a strong nonlinear relation between different sets of data may exist. While linear models may be a good simple approximation to these problems, when nonlinearity is severe they often perform unacceptably. The nonlinear partial least squares (PLS) method was developed in the area of chemical data analysis. A specific feature of PLS is that relations between sets of observed variables are modeled by means of latent variables usually not directly observed and measured. Since its introduction, two methodologically different concepts of fitting existing nonlinear relationships initiated development of a series of different nonlinear PLS models. General principles of the two concepts and representative models are reviewed in this chapter. The aim of the chapter is two-fold i) to clearly summarize achieved results and thus ii) to motivate development of new computationally efficient nonlinear PLS models with better performance and good interpretability.

INTRODUCTION

Two-block linear partial least squares (PLS) has been proven to be a valuable method for modeling relationships between two data sets (data blocks). This method was developed in chemometrics and has received a great deal of attention in the fields of analytical chemistry, organic and bio-organic chemistry, medicinal chemistry and chemical engineering. PLS has also been successfully applied in other scientific areas including bioinformatics (Boulesteix & Strimmer, 2007), food research (Martens & Martens, 1986), medicine (Worsley, 1997), pharmacology (Leach & Gillet, 2003, Nilsson, Jong, & Smilde, 1997), social sciences (Hulland, 1999), physiology and neurophysiology (Lobaugh, West, & McIntosh, 2001, Trejo, Rosipal, & Matthews, 2006), to name a few.

DOI: 10.4018/978-1-61520-911-8.ch009

PLS models relationships between sets of observed variables by means of latent variables. It can serve for regression and classification tasks as well as dimension reduction techniques and modeling. The underlying assumption of all PLS methods is that the observed data is generated by a system or process which is driven by a small number of latent (not directly observed or measured) variables. This projection of the observed data onto a subspace of usually orthogonal latent variables has been shown to be a powerful technique when observed variables are highly correlated, noisy and the ratio between the number of observations (data samples) and observed variables is low. The basic assumption of linear PLS is that the studied relation between observed data sets is linear and the same assumption of linearity then holds for modeling the relation in the projected subspace; that is, between latent variables.

However, in many areas of research and industrial situations a strong nonlinear relation between sets of data may exist. Although linear PLS can be used to approximate this nonlinearity, in many situations such approximation may not be adequate and the use of a nonlinear model is needed.

This chapter introduces the main concepts of nonlinear PLS and provides an overview of its application to different data analysis problems. The aim is to present a concise introduction that is a valuable guide for anyone who is concerned with nonlinear data analysis.

BACKGROUND

The concept of nonlinear PLS modeling was introduced by S. Wold, Kettaneh-Wold, and Skagerberg (1989). Already in this seminal work, the authors distinguished and described two basic principles for modeling curved relationships between sets of observed data. The first principle, here denoted as Type I, is well-known and used in mathematical statistics and other research fields. The principle applies first a nonlinear transformation to

observed variables. In the new representation a linear model is constructed. This principle can be easily applied to PLS, and indeed several different nonlinear PLS models were proposed and applied to real data sets. The first nonlinear PLS models in this category were constructed by using simple polynomial transformations of the observed data (Berglund & Wold, 1997, 1999). However, the proposed polynomial transformation approach possesses several computational and generalization limitations. To overcome these limitations, a computationally elegant kernel PLS method was proposed by Rosipal and Trejo (2001). The powerful concept of a kernel mapping function allows to construct highly flexible but still computationally simple nonlinear PLS models. However, in spite of the ability of kernel PLS to fit highly complex nonlinear data relationships, the model represents a 'black-box' with limited possibility to interpret the results with respect to the original data.

It is the second, here denoted as Type II, general principle for constructing nonlinear PLS models which overcomes the problem of loss of interpretability, but this is achieved at the expense of computational cost and optimization complexity. In contrast to the Type I principle, a nonlinear relation between latent variables is assumed and modeled, while the extracted latent vectors themselves are kept to be a linear combination of the original, not transformed, data. A quadratic function was used to fit relationship between latent variables in the first Type II nonlinear PLS approaches (Höskuldsson, 1992, S. Wold et al., 1989). Later, smoothing splines (Frank, 1990, S. Wold, 1992), artificial neural networks (Baffi, Martin, & Morris, 2000, Qin & McAvoy, 1992), radial basis neural networks (Wilson, Irwin, & Lightbody, 1997) or genetic programming methods (Hiden, McKay, Willis, & Montague, 1998) were used to fit more complex nonlinear relationships. Computational and optimization difficulties of the approach arise at the point when initially estimated weights for projecting observed data into latent vectors need to be corrected. The initial weights are estimated in the first step of the approach when a linear relationship between latent vectors is assumed. However, this linear assumption is violated by considering a nonlinear relation between the extracted latent vectors in the second step. Several methods were proposed to iteratively update both, the initial weights estimate and the latent variables relation fitting function, but a simple optimization and computational methodology is still missing (Baffi, Martin, & Morris, 1999, Searson, Willis, & Montague, 2007, S. Wold et al., 1989, Yoshida & Funatsu, 1997).

Before directly jumping into the area of nonlinear PLS, a section devoted to fundamentals of linear PLS is provided. The description uses nomenclature and originates from a recently published detailed survey of the variants of linear PLS and advances in the domain (Rosipal & Krämer, 2006). For readers interested in detailed understanding of linear PLS the survey can be a good starting point before reading this chapter. Next, an overview of the nonlinear PLS concepts is presented. Special emphasis is placed on kernel learning and the kernel PLS description. Detailed procedural and algorithmic implementation of each method mentioned goes beyond the scope of the chapter, but the limitation is compensated by an extensive literature survey. The technical overview of the nonlinear PLS approaches is followed by the section discussing selected positive and negative aspects of the presented methods. Ideas for further comparison, evaluation and extension of the current nonlinear PLS development status are also briefly sketched. A few concluding remarks close the chapter.

LINEAR PARTIAL LEAST SQUARES

Consider the general setting of a linear PLS algorithm to model the relation between two data sets (blocks of observed variables). Denote by $x \in \mathcal{X} \subset \mathbb{R}^N$ an N-dimensional vector of variables in the first set of data and similarly by $y \in \mathcal{Y} \subset \mathbb{R}^M$

a vector of variables from the second set. PLS models the relationship between the two data sets by means of latent vectors (score vectors, components). Denote by n the number of data samples and let X be the $(n \times N)$ matrix of centered (zero-mean) variables sampled from the Ξ-space. Similarly, let the Ψ-space data are represented by the $(n \times M)$ zero-mean matrix Y. PLS decomposes the **X** and **Y** matrices into the form

$$X = TPT + E$$

$$\mathbf{Y} = \mathbf{UQ}^T + \mathbf{F} \tag{1}$$

where **T** and **U** are the $(n \times p)$ matrices of the p extracted score (latent) vectors, the $(N \times p)$ matrix **P** and the $(M \times p)$ matrix **Q** represent matrices of loadings and the $(n \times N)$ matrix **E** and the $(n \times M)$ matrix **F** are the matrices of residuals. The PLS method, which in its classical form is based on the nonlinear iterative partial least squares (NIPALS) algorithm (H. Wold, 1975), finds weight vectors **w,c** such that

$$[cov(\mathbf{t},\mathbf{u})]^2 = [cov(\mathbf{Xw},\mathbf{Yc})]^2$$
$$= max_{|\mathbf{r}|=|\mathbf{s}|=1}[cov(\mathbf{Xr},\mathbf{Ys})]^2 \tag{2}$$

where $cov(\mathbf{t}, \mathbf{u}) = \mathbf{t}^T\mathbf{u} / n$ denotes the sample covariance between the score vectors **t** and **u**. The NIPALS algorithm starts with random initialization of the Ψ-space score vector **u** and repeats a sequence of the following steps until convergence

1) $w = X^T u / (u^T u)$ 4) $c = Y^T t / (t^T t)$
2) $\| w \| \to 1$ 5) $\| c \| \to 1$
3) $t = Xw$ 6) $u = Yc$

$$\tag{3}$$

where $\|\cdot\| \to 1$ denotes transformation of a vector to unit norm. Once the score vectors **t** and **u** are extracted, the vectors of loadings **p** and **q** from (1) can be computed by regressing **X** on **t** and **Y** on **u**, respectively

$p=X^T t/(t^T t)$ and $q=Y^T u/(u^T u)$

Note that different numerical techniques can be used to extract weight vectors and corresponding score vectors, some of which can be more efficient than NIPALS (De Jong, 1993, Höskuldsson, 1988).

PLS is an iterative process. After the extraction of the score vectors t and u, the matrices X and Y are deflated by subtracting their rank-one matrix approximations. Different forms of deflation exist, and each form defines a certain variant of PLS (for example, see Rosipal & Krämer, 2006). The most frequently used variant of linear PLS is based on two assumptions i) the score vectors $\{t_i\}_{i=1}^p$ are good predictors of Y, and ii) a linear *inner relation* between the scores vectors t and u exists; that is,

$$U = TD + H \tag{4}$$

where D is the $(p \times p)$ diagonal matrix and H denotes the matrix of residuals. The asymmetric assumption of the input–output (predictor–predicted) variables relation is transformed into a deflation scheme where the input space score vectors $\{t_i\}_{i=1}^p$ are good predictors of Y. The score vectors are then used to deflate Y; that is, a component of the regression of Y on t is removed from Y at each iteration of PLS

$$X \leftarrow X-tt^T X/(t^T t) \text{ and } Y \leftarrow Y-tt^T Y/(t^T t)=Y-tc^T \tag{5}$$

where c is the weight vector defined in step 4 of the NIPALS algorithm (3).

PLS Regression and Classification

By considering the assumption (4) of a linear relation between the scores vectors t and u, the decomposition of the Y matrix in (1) can be rewritten as

$$Y = TDQ^T + (HQ^T + F)$$

and this defines the linear PLS regression model

$$Y = TC^T + F^* \tag{6}$$

where $C^T = DQ^T$ denotes the $(p \times M)$ matrix of regression coefficients and $F^* = HQ^T + F$ is the residual matrix. Equation (6) is simply the decomposition of Y using ordinary least squares regression with orthogonal predictors T; that is, the estimate of C is given as $C=(T^T T)^{-1} T^T Y$. The PLS regression model (6) can also be expressed using the originally observed data X and written as (Höskuldsson, 1988, Manne, 1987, Rännar, Lindgren, Geladi, & Wold, 1994, Rosipal & Krämer, 2006)

$$Y = XB + F^* \tag{7}$$

where the estimate of B has the following form

$$B=X^T U(T^T XX^T U)^{-1} T^T Y$$

The PLS classification model closely follows the regression model (6). However, in the classification scenario the Y matrix is a class membership matrix uniquely coding each class of data. The close connection between Fischer discriminant analysis, canonical correlation analysis (CCA) and PLS has been discussed in Barker and Rayens, (2003), De Bie, Cristianini, and Rosipal (2005), Rosipal, Trejo, and Matthews (2003), and Rosipal and Krämer, (2006).

The major focus of the chapter is the nonlinear extraction of the PLS score vectors and the followed up regression or classification step using the extracted vectors is somehow irrelevant to us. Moreover, variants of PLS similar to CCA for modeling symmetric relationships between sets of data exist (see Rosipal and Krämer (2006) or Wegelin (2000) for overview). There is no theoretical limitation to apply the principles of nonlinear PLS to these variants as well.

NONLINEAR PARTIAL LEAST SQUARES

Several different nonlinear PLS methods have been proposed. In principle these methods can be categorized into two groups. The Type I group of approaches consists of models where the observed **X** matrix of independent variables is projected onto a nonlinear surface. The inner relation (4) between score vectors **t** and **u** is kept linear. In contrast, it is the Type II group of approaches where the nonlinear relation between **X** and **Y** data sets is modeled by replacing the linear relation (4) with a nonlinear form. In what follows, both types of nonlinear PLS are described in detail.

Type I: Nonlinear Projection of X

Methods of this group of nonlinear PLS are based on a principle of mapping the original data by means of a nonlinear function to a new representation (data space) where linear PLS is applied. A simple example would be the extension of the **X** matrix by considering component-wise square terms x_i^2 and cross-terms $x_i x_j$ of the input vector

x. It has been shown by Gnanadesikan (1977) and pointed out by S. Wold et al. (1989) that such an extension corresponds to the projection of the original data space onto a quadratic surface. In fact, this result was used by S. Wold et al. (1984) for PLS response surface modeling. To better understand this idea consider a simple binary classification problem depicted in the left part of Figure 1. An ellipsoidal boundary between two classes depicted by circles and crosses would be impossible to properly model with a linear line.

Now consider a simple transformation of the original 2D data into a 3D data space denoted by $\mathcal{F}$ and defined by the following mapping

$$\Phi: \quad \mathcal{X} = \mathcal{R}^2 \to \mathcal{F} = \mathcal{R}^3$$
$$\mathrm{x} = (x_1, x_2) \to \Phi(\mathrm{x}) = (x_1^2, \sqrt{2}x_1 x_2, x_2^2)$$

$$(8)$$

It can be easily observed (right part of Figure 1) that the original nonlinear classification problem in 2D was transformed into a linear problem in 3D.

However, in practice it might be very difficult to find such a simple nonlinear transformation or

Figure 1. An example of transformation of a nonlinear classification problem with an ellipsoidal decision boundary (left) to a problem where a linear hyperplane can be used to separate two classes (right). Two classes are defined by circles and crosses, respectively. The original 2D data (left) were transformed into a 3D space (right) using the nonlinear mapping defined in Equation (8)

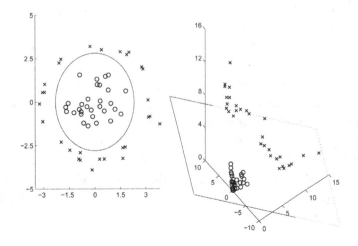

extension of the original data space into a new space where the original nonlinear problem becomes linear. This is mainly due to the reason that we do not know the exact shape of the nonlinear boundary in the classification scenario or we do not know the exact form of the nonlinear relationship between predictor and predicted space in regression. Nevertheless, following the idea that in a higher dimensional space, where original data are mapped, the nonlinear problem becomes easier to solve or a less complex boundary can be found, many very useful techniques and methods have been developed in different research communities. Polynomial regression (for example see Seber and Lee (2003)), generalized additive models (Hastie & Tibshirani, 1990), projection pursuit regression (Friedman & Stuetzle, 1981), higher order splines or multivariate adaptive regression splines (Friedman, 1991, Wahba, 1990) are a few of the popular models developed in the statistical community. Regularization networks (Girosi, Jones, & Poggio, 1995), artificial neural networks (Haykin, 1999) or recently developed theory and algorithms of kernel learning and support vector machines are examples of nonlinear model developments in the machine learning community (Cristianini & Shawe-Taylor, 2000, Schölkopf & Smola, 2002, Shawe-Taylor & Cristianini, 2004).

The *nonlinear PLS* model with quadratic projection of **X** was firstly discussed by S. Wold et al. (1989), but the principle of extending **X** with nonlinear terms was elaborated later by Berglund and Wold (1997). In their approach, called implicit nonlinear latent variables regression (INLR), Berglund and Wold proposed to extend the **X** matrix with squared x_i^2, cubic x_i^3 or higher order polynomial terms while ignoring cross-terms. To support this idea they showed that by "expanding the **X**-block with square terms in some ways corresponds to using a model that includes not only the squares of the latent variables, t_1^2, t_2^2, t_3^2, etc., but also the cross-products of these, $t_1 t_2, t_1 t_3, t_2 t_3$, etc." (Berglund & Wold, 1997).

On the other hand, their expansion of the **X** matrix with squared or higher order terms only, was strongly motivated by scaling limits of the full expansion. In the case of a high-dimensional predictor data and a small number of observations, that is, $n \ll N$, expanding **X** with both squared and cross-terms would result into extreme increase of the number of new variables in comparison to the existing number of observations. For an N-dimensional input space, there exist $\dfrac{(d + N - 1)!}{d!(N - 1)!}$ different the d-th order products (monomials) of the elements x_i.

For example, consider a standard problem from near-infrared reflectance spectroscopy where the predictor space consists of 250 absorbances at different wavelengths. Using the second-order polynomial expansion would result in a new expanded space with 31375 variables. Cubic expansion ($d = 3$) would increase this number to more than 2.6 million monomial terms and the problem quickly becomes computationally intractable.

The other problem associated with such an expansion is related to the one known as 'curse of dimensionality'. The curse of dimensionality is the term associated with the problem of an exponential increase of parameters which need to be estimated when mapping original data into a high-dimensional space, while the number of observed samples remains unchanged. For example, a regression model fitted by ordinary least squares method would result in an unbiased estimate of regression coefficients, but the variance of the estimate can be unacceptably high. Thus, although such a model will fit training data well, in the sense of sum of square errors, its generalization ability to fit previously unobserved (testing) data can be very poor. To avoid the problem, different forms of regularization are usually applied. One approach that is often used is based on diminishing the influence of regression coefficients with high absolute values. This can be done by penalizing a properly selected norm of the vector of estimated regression coefficients.

However, the new concept of statistical learning theory developed in its origin by Vapnik and Chervonenkis (Vapnik, 1998, 1995) provided new insights and perspectives into the problems of curse of dimensionality and bad generalization of learning machines. Based on the theory new support vector machines and kernel learning algorithms and concepts were developed. Among other models the nonlinear kernel PLS method was proposed. The basic principles of kernel learning are described in the following subsection.

KERNEL LEARNING AND KERNEL PLS

Statistical learning theory was introduced by Vapnik and Chervonenkis in the early 1970's (Vapnik, 1998, Vapnik & Chervonenkis, 1974). However, it was not until the middle of the 1990's when this theory inspired development of new types of learning algorithms. The core principle of the new algorithms is the mapping of originally observed data into a high-dimensional feature space where simple linear models are constructed. The generalization abilities of the constructed models are controlled by considering theoretical results of the structural risk minimization (SRM) principle. SRM is an inductive principle where a trade-off between good-model fitting and the complexity of the model is appropriately balanced. In other words, having a finite set of data, a model selection step defined by the principle consists of two aspects which needs to be considered in parallel i) the quality of fitting training data (empirical error) and ii) the complexity of the hypothesis space where the final model is constructed. Over-complex models would fit training data perfectly well, but they may show high prediction or classification errors when dealing with new, previously unobserved, testing data. Vapnik and Chervonenkis (1971) introduced the concept of the VC dimension (for

Vapnik-Chervonenkis dimension) which can be used as a measure of the complexity of a hypothesis space in which learning machines are constructed. Motivated by this concept of the VC dimension very powerful support vector machines (SVM) for classification were constructed (Cristianini & Shawe-Taylor, 2000, Schölkopf & Smola, 2002, Vapnik, 1998). For real-valued function approximations, including regression tasks, the SRM principle motivated the development of new, and renewal of previous theoretical and practical results of regularization theory (Girosi et al., 1995). Regularization theory was proposed in the 1960's by Tichonov and Ivanov (Ivanov, 1976, Tikhonov, 1963). Later, Kimeldorf and Wahba (1971) proved the important *representer theorem*, which also applies to kernel learning algorithms, including SVM. Using a less rigorous mathematical language, the representer theorem defines the form of an approximation or regression function $f(\mathbf{x})$ constructed in a functional space H and minimizing the following functional form

$$\min_{f \in \mathcal{H}} R_{reg} = \sum_{i=1}^{n} L(y_i, f(\mathbf{x}_i)) + \xi \parallel f \parallel_{\mathcal{H}}^2 \qquad (9)$$

where $L(y_i, f(\mathbf{x}_i))$ is a loss function measuring point-wise difference between observed values $\{y_i\}_{i=1}^n$ and their approximations $\{f(\mathbf{x}_i)\}_{i=1}^n$. A typical example of L would be the squared loss $(y_i - f(\mathbf{x}_i))^2$. The regularization term ξ is a positive number controlling the trade-off between approximating properties and the smoothness of f. The squared norm $\parallel f \parallel_{\mathcal{H}}^2$ is sometimes called the 'stabilizer'. The representer theorem states that the solution of (9) can be always written in the form

$$f(\mathbf{x}) = \sum_{i=1}^{n} d_i k(\mathbf{x}, \mathbf{x}_i) \qquad (10)$$

175

where $\{d_i\}_{i=1}^n$ are real-value coefficients and $k(\mathbf{x}, \mathbf{x}_j)$ is a symmetric positive definite function of two variables called the *kernel function*. Next, the properties and specific relations between the space H and the kernel function k will be discussed.

First, return to the mapping defined in (8) and consider the mappings $\Phi(\mathbf{x})$, $\Phi(\mathbf{y}) \in F$ of the two points $\mathbf{x}, \mathbf{y} \in \mathcal{X}$

$$x = (x_1, x_2) \quad \rightarrow \quad \Phi(x) = (x_1^2, \sqrt{2}x_1 x_2, x_2^2)$$
$$y = (y_1, y_2) \quad \rightarrow \quad \Phi(y) = (y_1^2, \sqrt{2}y_1 y_2, y_2^2)$$

Now compute a dot product between these two mappings

$$\begin{aligned}\langle \Phi(x), \Phi(y) \rangle &= (x_1^2, \sqrt{2}x_1 x_2, x_2^2)^T (y_1^2, \sqrt{2}y_1 y_2, y_2^2) \\ &= ((x_1, x_2)^T (y_1, y_2))^2 \\ &= \langle x, y \rangle^2 \end{aligned}$$

It becomes clear that $\langle \mathbf{x}, \mathbf{y} \rangle^2$ corresponds to a canonical (Euclidean) dot product in the space F where the input data were transformed by Φ. This is an important concept where computation of a dot product between two vectors in the feature space F can be replaced by evaluating the function $\langle \mathbf{x}, \mathbf{y} \rangle^2$ instead; that is, by computing a square value of a dot product between the two original points $\mathbf{x}$, $\mathbf{y}$ in $\mathcal{X}$. In our previous example with the 250-dimensional input space of absorbances this would mean to evaluate square values of the dot product between two 250-dimensional vectors instead of computing the dot product of two 31375-dimensional vectors in the mapped space. More interestingly the function $\langle \mathbf{x}, \mathbf{y} \rangle^2$ satisfies the Mercer theorem conditions (Cristianini & Shawe-Taylor, 2000, Mercer, 1909) and it is a valid kernel function known as the second order polynomial kernel $k(\mathbf{x}, \mathbf{y}) = \langle \mathbf{x}, \mathbf{y} \rangle^2$. The Moore-Aronszajn theorem establishes the fact that for any kernel function, there exists a unique functional space, called a reproducing kernel Hilbert space (RKHS) (Aronszajn, 1950). Now, return to the representer theorem and define H to be a RKHS corresponding to a kernel function k. The link between the RKHS space H defined by k and a space of mapped features vectors F follows from Mercer's theorem. Any kernel function can be written in the form

$$k(\mathbf{x}, \mathbf{y}) = \sum_{i=1}^{\mathcal{D}_{\mathcal{H}}} \alpha_i \psi_i(\mathbf{x}) \psi_i(\mathbf{y}) = \Phi(\mathbf{x})^T \Phi(\mathbf{y}) = \langle \Phi(\mathbf{x}), \Phi(\mathbf{y}) \rangle$$

where $\{\psi_i\}_{i=1}^{\mathcal{D}_{\mathcal{H}}}$ is a sequence of linearly independent functions, $\{\alpha_i\}_{i=1}^{\mathcal{D}_{\mathcal{H}}}$ are positive numbers and $\mathcal{D}_{\mathcal{H}} \leq \infty$ is the dimension of the space $\mathcal{H}$. Following this relation the feature map Φ can be written as

$$\begin{aligned}\Phi : \quad \mathcal{X} \quad &\rightarrow \quad \mathcal{F} \\ \mathbf{x} \quad &\rightarrow \quad \Phi(\mathbf{x}) = (\sqrt{\alpha_1}\psi_1(\mathbf{x}), \sqrt{\alpha_2}\psi_2(\mathbf{x}), \ldots, \sqrt{\alpha_{\mathcal{D}_{\mathcal{H}}}}\psi_{\mathcal{D}_{\mathcal{H}}}(\mathbf{x}))\end{aligned}$$

Thus, if we are only interested in the computation of dot products in F, it does not matter how F was constructed and simply all dot products can be replaced by a unique kernel function associated with F. This important to note because different feature spaces associated with the same kernel function can be constructed (Schölkopf & Smola, 2002). In literature, this replacement of a dot product with the kernel function value is known as the *kernel trick* method.

The polynomial kernel function $k(\mathbf{x}, \mathbf{y}) = \langle \mathbf{x}, \mathbf{y} \rangle^d$ is a simple extension of the above mentioned second order polynomial mapping by considering a feature space of all monomials of d-th order. Another kernel function widely used in practice is the Gaussian kernel function $k(\mathbf{x}, \mathbf{y}) = e^{\left(-\frac{\|\mathbf{x}-\mathbf{y}\|^2}{\delta}\right)}$, where $\delta > 0$ determines the width of the function. Different kernel functions have been used and constructed (Cristianini & Shawe-Taylor, 2000,

Saitoh, 1997, Schölkopf & Smola, 2002, Shawe-Taylor & Cristianini, 2004). Interestingly, a linear kernel $k(\mathbf{x},\mathbf{y})=\langle\mathbf{x},\mathbf{y}\rangle=\mathbf{x}^T\mathbf{y}$ is also an admissible kernel function. Thus, the linear kernel principal component analysis (PCA) and linear kernel PLS methods (Rännar et al., 1994, Wu, Massarat, & De Jong, 1997), previously developed to reduce computational costs in the case where the input data dimension exceeds the number of observed samples ($n<N$), can be considered belonging to the framework of kernel learning.

The recently developed theory of kernel learning has also been applied to PLS. The nonlinear kernel PLS methodology for modeling relations between sets of observed variables, regression and classification problems was proposed by Rosipal and Trejo (2001) and Rosipal et al. (2003). The idea of kernel PLS is based on a nonlinear mapping of the original data from $\mathcal{X}$ into a high-dimensional feature space F where a linear PLS model is constructed.

Define the Gram matrix $\mathbf{K}$ of the cross dot products between all mapped input data points, that is, $\mathbf{K}=\Phi\Phi^T$, where Φ denotes the matrix of the mapped $\mathcal{X}$-space data $\{\Phi(\mathbf{x}_i)\in\mathcal{F}\}_{i=1}^n$. The kernel trick implies that the elements i,j of $\mathbf{K}$ are equal to the values of the kernel function $k(\mathbf{x}_i,\mathbf{x}_j)$. Further consider that the mapped data were centered; that is, Φ is a zero-mean matrix. This can be easily achieved by directly centering $\mathbf{K}$ and explicit manipulation of Φ is not needed (Schölkopf, Smola, & Müller, 1998, Rosipal & Trejo, 2001). Denote by $\mathbf{I}_n$ an n-dimensional identity matrix and by $\mathbf{1}_n$ a vector of ones with the length of n. The centered Gram matrix can be then computed as

$$\mathrm{K} \leftarrow (\mathrm{I}_n - \frac{1}{n}\mathbf{1}_n\mathbf{1}_n^T)\mathrm{K}(\mathrm{I}_n - \frac{1}{n}\mathbf{1}_n\mathbf{1}_n^T).$$

Now, consider a modified version of the NIPALS algorithm where steps 1 and 3 are merged and the score vectors $\mathbf{t}$ and $\mathbf{u}$ are scaled to unit norm instead of scaling the weight vectors $\mathbf{w}$

and $\mathbf{c}$. The obtained kernel form of the NIPALS algorithm is as follows (Rosipal & Trejo, 2001)[1]

1) $\mathrm{t} = \phi\phi^T\mathrm{u} = \mathrm{Ku}$ 4) $\mathrm{u} = \mathrm{Yc}$

2) $\|\mathrm{t}\| \to 1$ 5) $\|\mathrm{u}\| \to 1$

3) $\mathrm{c} = \mathrm{Y}^T\mathrm{t}$

Although step 2 guarantees orthonormality of the score vectors, the score vectors can be rescaled to follow the standard linear NIPALS algorithm with the unit norm weight vectors $\mathbf{w}$ (Rännar et al., 1994). In the following the unit norm orthonormal score vectors will be considered; that is, $(\mathbf{T}^T\mathbf{T})^{-1}=\mathbf{I}$.

Note that steps 3 and 4 can be further merged which may become useful in applications where an analogous kernel mapping Ψ of the $\mathcal{Y}$-space data is considered; that is, the Gram matrix $\mathrm{Ky}=\Psi\Psi\mathrm{T}$ of the cross dot products between all mapped output data is constructed. Then, the kernel NIPALS algorithm consists of the following four steps

1) $\mathrm{t} = \mathrm{Ku}$ 3) $\mathrm{u} = \mathrm{K}_y\mathrm{t}$

2) $\|\mathrm{t}\| \to 1$ 4) $\|\mathrm{u}\| \to 1$

This form of kernel PLS can be useful when one is interested in modeling symmetric relationships between two sets of data. This is known as the PLS Mode A method (Rosipal & Krämer, 2006, Wegelin, 2000, H. Wold, 1985). The above mentioned kernel form then represents the kernel PLS Mode A form which has a close connection to the kernel CCA method (De Bie et al., 2005).

The important part of the iterative PLS algorithm is the deflation step. In the kernel PLS approach the elements of a feature space F where data are mapped are usually not accessible and the deflation scheme (5) needs to be replaced by its kernel variant (Rosipal & Trejo, 2001)

$$\mathbf{K}\leftarrow\mathbf{K}\text{-}\mathbf{tt}^T\mathbf{K}\text{-}\mathbf{Ktt}^T+\mathbf{tt}^T\mathbf{Ktt}^T=(\mathbf{I}\text{-}\mathbf{tt}^T)\mathbf{K}(\mathbf{I}\text{-}\mathbf{tt}^T)$$

Now continue with the kernel analogy of the linear PLS model defined in the previous section. While the kernel from of the model (6) remains the same,[2] the kernel variant of the model (7) has the following form

$$\mathbf{Y} = \mathbf{\Phi}\mathbf{B} + \mathbf{F}^*$$

where the estimate of $\mathbf{B}$ is now

$$\mathbf{B} = \mathbf{\Phi}^T \mathbf{U}(\mathbf{T}^T \mathbf{K} \mathbf{U})^{-1} \mathbf{T}^T \mathbf{Y}$$

Denote by $\mathbf{d}^m = \mathbf{U}(\mathbf{T}^T \mathbf{K} \mathbf{U})^{-1} \mathbf{T}^T \mathbf{y}^m$, $m = 1, \dots, M$, where the $(n \times 1)$ vector $\mathbf{y}^m$ represents the m-th output variable. Then the kernel PLS regression estimate of the m-th output for a given input sample $\mathbf{x}$ can be written in the form

$$\hat{y}^m = \Phi(\mathbf{x})^T \mathbf{d}^m = \sum_{i=1}^{n} d_i^m k(\mathbf{x}, \mathbf{x}_i)$$

which resembles the solution (10) of the representer theorem. However, the form of regularization in PLS differs from the penalized regression models where a direct penalization of regression coefficients is the part of an optimized formula. Therefore, in the case of kernel PLS regression, the functional form (9) of the representer theorem cannot be straightforwardly formulated. Penalization proprieties of PLS have been discussed and compared with other penalized regression models elsewhere (for example see Lingjærde & Christophersen, 2000, Rosipal & Krämer, 2006).

By considering the kernel variant of the model (6), the kernel PLS regression estimate $\hat{y}^m$ can be also written as

$$\hat{y}^m = c_1^m t_1(\mathbf{x}) + c_2^m t_2(\mathbf{x}) + \dots + c_p^m t_p(\mathbf{x}) = \sum_{i=1}^{p} c_i^m t_i(\mathbf{x})$$

where $\mathbf{c}^m = \mathbf{T}^T \mathbf{y}^m$ is the estimate of a vector of regression coefficients for the m-th regression model. The notation $\{t_i(\mathbf{x})\}_{i=1}^{p}$ stresses the fact that the score vectors can now be understood as nonlinear functions sampled at the data points $\mathbf{x}$. The following example demonstrates this point.

In Figure 2 an example of kernel PLS regression is depicted. One hundred uniformly spaced points in the range [0, 3.25] were taken and the corresponding values of the function $z(\mathbf{x}) = 4.26(e^{-x} - 4e^{-2x} + 3e^{-3x})$ were computed. The function was used by Wahba (1990) to demonstrate smoothing spline properties. An additional sample of one hundred points representing noise was generated following the Gaussian distribution with zero-mean and variance equal to 0.04. These points were added to the computed values of z and subsequently the values were centered by the mean. The Gaussian kernel with the width parameter δ equal to 1.8 was used. The extracted score vectors plotted as a function of the input points x are depicted in Figure 3.

Note the following very important aspect about the number of kernel PLS score vectors. In contrast to linear PLS or Type II nonlinear PLS described next, the number of possible kernel PLS score vectors is given by the rank of $\mathbf{K}$, not by the rank of $\mathbf{X}$. The rank of $\mathbf{K}$ is either given by n, the number of different sample points, or by the dimensionality $\mathcal{D}_{\mathcal{H}}$ of $\mathcal{F}$ if $n < \mathcal{D}_{\mathcal{H}}$. However, in practice this is usually not the case because data are mapped in a way that $n \ll \mathcal{D}_{\mathcal{H}}$. Centering of K will remove one degree of freedom, therefore the rank of K is in general equal to $\min(\mathcal{D}_{\mathcal{H}}, n^{-1})$. In the particular example depicted in Figure 2, the original input space dimension is equal to one and linear or Type II nonlinear PLS can work with one PLS score vector only. However, the kernel PLS method can in theory extract up to 99 nonlinear PLS score vectors. Indeed, to illustrate this point the first five and the ninth score vectors are depicted in Figure 3. This important aspect of finding a 'finer' data decomposition and representation is similar to the properties of kernel PCA as already discussed in (Schölkopf et al., 1998).

Figure 2. An example of kernel PLS regression. The generated function z(x)=4.26(e⁻ˣ-4e⁻²ˣ+3e⁻³ˣ) is shown as a thick solid line. Dot markers represent noisy form of z used as training output points in kernel PLS regression. Kernel PLS regression using the first one, four and nine score vectors is shown as a dashed, thin solid, and dash-dotted line, respectively. The individual score vectors are plotted in Figure 3

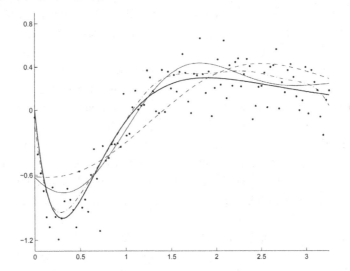

Returning to Figure 3, it can be observed that while the first three to four score vectors follow the shape of the approximated function itself, the higher number score vectors start to model the 'wiggling' noisy part of the investigated function.

This becomes evident in the last (bottom right) subplot where the ninth score vector is depicted. Note that the score vectors are extracted such that they increasingly describe overall variance in the input data space and more interestingly they also

Figure 3. The first five and the ninth (from top left to bottom right) F-space score vectors $t_i(x), i=1,...,5,9$ computed from noisy signal described in Figure 2. The generated function z without noise is shown dash-dotted

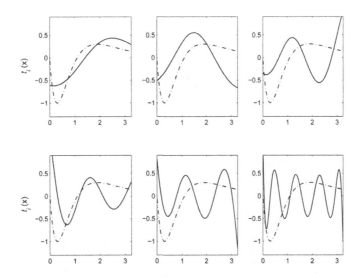

describe the overall variance of the observed output data samples. Smola, Schölkopf, and Müller (1998) has theoretically shown that by choosing the fla*ttest fu*nction in a feature space F, conditioned by 'good' smoothing properties of the selected kernel function, a smooth function in the input space can be obtained. The term flat function needs to be understood as a linear regression model where the penalization of, in absolute value, large regression coefficients is applied. Shrinkage (or regularized least-square) regression methods like ridge regression, principal component regression or PLS belong to this category, although each of these methods applies a specific principle of penalization (Butler & Denham, 2000, Frank & Friedman, 1993, Goutis, 1996, Krämer, 2007, Lingjærde & Christophersen, 2000, Rosipal & Krämer, 2006).

Finally, I discuss the smoothing proprieties of a kernel mapping. By using different kernel functions feature spaces with different approximation properties are induced. Using the cubic polynomial kernel will induce a feature space with higher ap-

proximation abilities to fit polynomial functions of the third or higher orders in comparison to the second order polynomial kernel. Similarly, by selecting the Gaussian kernel with different values of the width term δ corresponds to choosing feature spaces with different smoothing properties (Girosi, 1998). In general, wider Gaussian kernels will more strongly penalize higher frequency components resulting in smoother estimates. This is demonstrated in Figure 4. Thus, the kernel PLS approach is associated not only with a proper selection of the final number of score vectors but also with a proper selection of the kernel function. This needs to be balanced through the model selection method used. Although cross-validation is usually applied, recently, estimating the effective degrees of freedom in kernel PLS, Krämer and Braun (2007), have discussed and compared several other model selection criteria.

Figure 4. Examples of kernel PLS regression using different values δ of the Gaussian kernel function but the same number of nine score vectors. A dash-dotted line represents the model depicted in Figure 2 with the value of δ=1.8. A thin solid line represents the model with the extremely small value δ=0.01. Tendency of this model to follow noise elements represented by dot markers is clearly observed. The generated function without noise is shown as a thick solid line

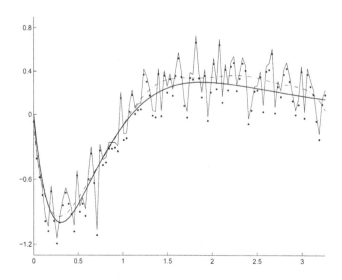

Type II: Nonlinear Inner Relation

S. Wold et al. (1989) were the first to extend the linear PLS model to its nonlinear form. They have done this by replacing the linear inner relation (4) between the score vectors **t** and **u** by a nonlinear model

$$\mathbf{u} = g(\mathbf{t}) + \mathbf{h} = g(\mathbf{X}, \mathbf{w}) + \mathbf{h} \tag{11}$$

where g represents a continuous nonlinear function. Again, **h** denotes a vector of residuals. The relation between each pair of latent variables is modeled separately; that is, in general a different form of the nonlinear function g can be used for each pair **t** and **u**. Polynomial functions, smoothing splines, artificial neural networks or radial basis function networks have been used to model g. Importantly, in contrast to the Type I nonlinear PLS models, the assumption that the score vectors **t** and **u** are linear projections of the original variables is kept. This has a significant consequence towards the modification of the original NIPALS algorithm described in the previous linear PLS section. It can be observed that the vector of weights **w**, computed in the first step of the NIPALS algorithm (3), represents the sample covariance between the output space score vector **u** and the input space data matrix **X**. However, the use of a nonlinear model to relate the score vectors in the inner relation affects the computation of **w**. Although **w** represents the association among variables of **X** and **u** also in nonlinear PLS, this association will be closely related to the covariance values only if the nonlinear mapping between latent variables is monotonic and slightly nonlinear (curved). If this is not the case, an update of **w** needs to be considered. Thus, two different approaches have been proposed and used.

In the first approach no update of **w** is applied. S. Wold et al. (1989) named this approach the *quick and dirty* nonlinear PLS algorithm and it consists of the standard NIPALS steps until the convergence and the subsequent nonlinear fitting

of the inner relation between the extracted pair of the **t** and **u** vectors. Different nonlinear models were used to fit this relation: higher order polynomial regression (S. Wold et al., 1989), smoothing splines or different types of smoothing estimators (Frank, 1990, 1995), artificial neural networks (Qin & McAvoy, 1992) and radial basis function networks (Wilson et al., 1997).

In the second group of approaches a nonlinear function modeling the inner relation is used to update an initial linear PLS estimate of the weight vector **w**. S. Wold et al. (1989) proposed to update **w** by means of a Newton-Raphson-like linearization of g. The procedure thus consists of a first-order Taylor series expansion of g, followed by the calculation of the correction term $\Delta\mathbf{w}$ which is used to update **w**. So, consider the nonlinear inner relation (11) where $g(\mathbf{t}) = g(\mathbf{X}, \mathbf{w})$ is continuous and differentiable with respect to **w**. The second-order Taylor expansion of (11) has the form

$$\hat{\mathbf{u}} = \mathbf{u}_{00} + \left.\frac{\partial g}{\partial \mathbf{w}}\right|_{00} \Delta\mathbf{w} \tag{12}$$

where $u_{00} = g(\mathbf{t})$ is the value of g at the known value of **t**. Similarly, $\left.\frac{\partial g}{\partial \mathbf{w}}\right|_{00}$ stands for the partial derivatives of g numerically evaluated at the same known value of **t**. The second term of (12) can be written element-wise as

$$\left.\frac{\partial g}{\partial \mathbf{w}}\right|_{00} \Delta\mathbf{w} = \sum_{i=1}^{N} \left.\frac{\partial g}{\partial w_i}\right|_{00} \Delta w_i$$

At this point several different methods to compute the correction $\Delta\mathbf{w}$ were proposed. To simplify further notation consider the matrix form of the linear approximation $\hat{\mathbf{u}}$

$$\hat{\mathbf{u}} = Z\mathbf{v}$$

where $Z = [u_{00} \; \frac{\partial g}{\partial w}|_{00}]$ and $v=[1 \; \Delta w]^T$. The following variants to compute Δw were suggested:

S. Wold et al. (1989) proposed to compute Δw by reflecting the first steps of the ordinary NIPALS algorithm

1. $v = (Z^T Z)^{-1} Z^T \hat{u}$ (*S. Wold et al. (1989) original formula:* $v = Z^T \hat{u} / (\hat{u}^T \hat{u})$)
2. $\|v\| \to 1$
3. $s = Zv$
4. $b = u^T s/(s^T s)$
5. *for Δw take the corresponding elements of bv; that is, elements at the $2,...,(N+1)$ positions*

Correspondence of the first three steps of the original S. Wold et al. (1989) algorithm (for step 1 see formula in brackets) with the first steps of NIPALS can be easily observed by replacing v with w and s with t. However, this original idea of the regression of v on Z, that is, assuming the relation $Z = \hat{u}v$ instead of $\hat{u} = Zv$, was later questioned by Hasegawa, Kimura, Miyashita, and Funatsu (1996). In a personal communication the authors verified this error with S. Wold and they have applied the correct formula $v = (Z^T Z)^{-1} Z^T \hat{u}$. This point and the correction was later also repeated by Baffi et al. (1999).

Baffi et al. (1999), initiated by the S. Wold et al. (1989) idea of nonlinear PLS, proposed a different way for updating w. The approach, denoted *error-based*, considers the difference $\hat{u} - u_{00} = \frac{\partial g}{\partial w}\Big|_{00} \Delta w$ at the first step. Next, the authors replace the Newton-Raphson linearization estimate $\hat{u}$ by the actual value of u computed in the last step of the NIPALS loop as $u = Yc$. This leads to the following definition of a mismatch e

$$e = u - u_{00} = \frac{\partial g}{\partial w}\Big|_{00} \Delta w = Z_w \Delta w \qquad (13)$$

where the matrix $Z_w = [\frac{\partial g}{\partial w}|_{00}]$ consists of the partial derivatives $\{\frac{\partial g}{\partial w_i}|_{00}\}_{i=1}^N$. From this mismatch relation Δw can be directly computed by regressing the mismatch e on Z_w, that is,

$$\Delta w = (Z_w^T Z_w)^{-1} Z_w^T e$$

It needs to be mentioned that both S. Wold et al. (1989) and Baffi et al. (1999) considered the case where g is also dependent on variables different than w. More precisely, they considered the quadratic relation $u = b_0 + b_1 t + b_2 t^2 + h$. In this case partial derivatives of g with respect to the elements of the vector $b = [b_0, b_1, b_2]$ can be added to Z or Z_w matrix, respectively, and Δb to the vector v.

Finally, the whole nonlinear PLS method consisting of the NIPALS steps and the steps for updating w is summarized below. The method starts with a random initialization of u. The following steps are repeated until convergence

1) $w = X^T u / (u^T u)$ 7) $\|c\| \to 1$
2) $\|w\| \to 1$ 8) $u = Yc$
3) $t = Xw$ 9) *compute* Δw
4) *fit* $g(.)$ *using* u, t 10) $w = w + \Delta w$
5) $u_{00} = g(t)$ 11) *go to step* 2)
6) $c = Y^T u_{00} / (u_{00}^T u_{00})$

Detailed description of these steps can be found in S. Wold et al. (1989), Baffi et al. (1999) or S. Wold (1992).

DISCUSSION AND FUTURE RESEARCH DIRECTIONS

Both Type I and Type II represent two general principles for constructing nonlinear PLS models. Many variants of nonlinear PLS can be constructed by applying the wide variety of existing nonlinear methods developed outside of the chemometrics research field. An example is the GIFI approach to nonlinear PLS (Berglund et al., 2001). The method developed in mathematical statistics consists of quantizing the original **X** variables into bins. A subsequent representation of each bin with a 1/0 dummy variable then reflects whether the observed continuous value falls into a bin or not (Michailidis & De Leeuw, 1998). The method fits into Type I nonlinear PLS where the quantization represent a mapping of the observed variables into a dummy representation. Similarly, different nonlinear models can be used in Type II nonlinear PLS to represent the inner relation g function. When considering quick and dirty nonlinear PLS, that is, nonlinear PLS without updating of **w**, almost any nonlinear model can be used. This is different from the situation when **w** is updated by any of the rules described in the previous subsection. In this case the Newton-Raphson procedure requires that the nonlinear function g is continuous and differentiable with respect to **w**. Moreover, a simultaneous mathematical optimization of the inner relation function and computation of new **w** from a linearization of the function is not always efficient and convergence to an optimal solution is sometimes not achieved. To relieve these limits Yoshida and Funatsu (1997) and Li, Mei, and Cong (1999) proposed to use a genetic algorithm based optimization technique to simultaneously update g and **w**. Interestingly, in the work of Yoshida and Funatsu (1997) this was done without a linearization of g. Therefore, a wider set of nonlinear functions to represent g can be used because the criterion of continuity and differentiability can be dropped. In theory, recently extensively used and now-popular con-

cept of kernel learning cannot only by used in the quick and dirty nonlinear PLS scenario, but also in the case where an appropriate update of **w** would be advantageous. An example can be a support vector regression model for g. Finally, Hiden et al. (1998) and Searson et al. (2007) extended the concept of genetic programming to represent and estimate the inner relation model itself.

Comparing the two types of nonlinear PLS it is difficult to define the favorable methodology. While Type I is easily implementable, often computationally less demanding and capable to model complex nonlinear relations, it usually leads to a loss of the interpretability of the results with respect to the original data.[3] On the other hand it is not difficult to construct data situations where the Type II approach of keeping latent variables to be linear projections of the original data may not be adequate. Consider the situation depicted in the left part of Figure 1. Taking any one-dimensional linear projection would create a score vector where both classes would be mixed up with a low level of separability. Thus, a better choice can be an appropriate initial nonlinear mapping of the data. However, consider the same task but add to the problem several dimensions with random variables possessing no separability proprieties with respect to the two classes. In this case an appropriate projection to a lower, optimally two-dimensional original problem space, followed by a nonlinear mapping of the obtained score vectors can be a better choice. In practice a researcher needs to decide about the adequacy of using a particular method based on the problem in hands and requirements like simplicity of the solution, implementation difficulties or interpretation of the results.

Since the work of S. Wold et al. (1989), there have been twenty years of development of the nonlinear PLS concepts, which have resulted in a wide variety of different models. Extensive experimental work has been carried out to compare the studied nonlinear PLS models with the existing concepts of nonlinear modeling developed outside of chemometrics. However, much less research has been carried out to compare the nonlinear PLS models themselves. Within the area of Type II models, Baffi et al. (1999) numerically compared the approach of S. Wold et al. (1989) with their error-based approach. On a synthetic and a real data set they demonstrated better prediction abilities of the error-based approach; in addition, this was achieved with a lower number of latent variables. Similarly, in the area of the Type I models, the kernel PLS method was compared with approaches within the area of kernel learning. The method was proved to be competitive with the classification and regression approaches like SVM, kernel ridge regression or kernel Fischer discriminant analysis (Rosipal et al., 2003, Rosipal & Trejo, 2001). However, a rigorous and thorough experimental and theoretical comparison of different Type I and Type II nonlinear PLS models is lacking.

Finally, there is no restriction on combining the Type I and Type II nonlinear PLS concepts. This can be done by mapping observed data first (Type I) and applying a Type II nonlinear PLS model on transformed data. Would this concept of applying two nonlinear maps be beneficial? In a specific task of smoothing electroencephalographic (EEG) signals it has been observed that better results can be obtained by applying locally weighted PLS regression in a feature space, that is, the nonlinear kernel PLS method was modified by applying a nonlinear PLS model in the mapped data space (Rosipal & Trejo, 2004). Recall that locally weighted PLS regression approximates nonlinearity by a series of local linear models (Cleveland, 1979, Næs & Isaksson, 1992).

CONCLUSION

There is no doubt that nonlinear PLS modeling represent an important concept for analysis of data sets with nonlinear relationships. However, until recently and in contrast to linear PLS, nonlinear

PLS has been mainly used in the chemical data analysis domain. It was the new concept of nonlinear kernel PLS, representing an elegant way of dealing with nonlinear aspects of measured data, which has considerably extended the applicability of nonlinear PLS into a wider area of research fields (Hardoon, Ajanki, Puolamaki, Shawe-Taylor, & Kaski, 2007, Lee, Wu, Huntbatch, & Yang, 2007, Mu, Nandi, & Rangayyan, 2007, Saunders, Hardoon, Shawe-Taylor, & Widmer, 2008, Trejo et al., 2006). The main reason is the fact that the kernel PLS method keeps computational and implementation simplicity of linear PLS while providing a powerful modeling, regression, discrimination or classification tool. Moreover, kernel PLS has been proven to be competitive with the state-of-the-art SVM and other kernel regression and classification methods. However, it would be a big mistake to prefer the kernel method over the other nonlinear PLS approaches, especially Type II nonlinear PLS. The PLS method projects original data onto a more compact space of latent variables. Among many advantages of such an approach is the ability to analyze and interpret the importance of individual observed variables. The feature which is somehow lost in kernel learning, where usually not easily interpretable nonlinear kernel mapping is applied. The interpretabillity is an important factor not only in chemical data analysis but also in other research and application domains. For example, in an experimental design where many insignificant terms are measured, PLS results can guide the practitioner into more compact experimental settings with a significant cost reduction and without a high risk associated with the 'blind' variables deletion.

The chapter reviewed two main concepts of nonlinear PLS. The aim of the review was to provide fundamental and unifying insights into the understanding of individual methods. Hopefully, this review will draw more attention to developing new nonlinear PLS methods that may overcome drawbacks of the existing approaches.

At the time of writing, I was not aware of a comprehensive software covering all of the described nonlinear PLS methods. A set of Matlab® routines for kernel PLS is available upon request.

ACKNOWLEDGMENT

The author's research was supported by the funds from the Austrian Science Fund FWF project P19857 carried out at the Medical University of Vienna, Vienna, Austria and by the EC project BAMOD, No LSHC-CT-2005-019031, Workpackage 6: Statistical Algorithms, carried out at the Institute of Measurement Science, Slovak Academy of Sciences, Bratislava, Slovakia. Partial support for the production of this review was provided by Pacific Development and Technology, LLC, Palo Alto, CA, USA. I also wish to acknowledge constructive comments by Dr. Leonard J. Trejo.

REFERENCES

Aronszajn, N. (1950). Theory of reproducing kernels. *Transactions of the American Mathematical Society, 68,* 337–404.

Baffi, G., Martin, E., & Morris, A. (1999). Nonlinear projection to latent structures revisited: the quadratic PLS algorithm. *Computers & Chemical Engineering, 23,* 395–411.

Baffi, G., Martin, E., & Morris, A. (2000). Nonlinear dynamic projection to latent structures modelling. *Chemometrics and Intelligent Laboratory Systems, 52,* 5–22.

Barker, M., & Rayens, W. (2003). Partial least squares for discrimination. *Journal of Chemometrics, 17,* 166–173.

Berglund, A., Kettaneh, N., Uppgåd, L., Wold, S., Bendwell, N., & Cameron, D. (2001). The GIFI approach to non-linear PLS modeling. *Journal of Chemometrics, 15,* 321–336.

Berglund, A., & Wold, S. (1997). INLR, Implicit Non-linear Latent Variable Regression. *Journal of Chemometrics, 11*(2), 141–156.

Berglund, A., & Wold, S. (1999). A Serial Extension of Multiblock PLS. *Journal of Chemometrics, 13,* 461–471.

Boulesteix, A. L., & Strimmer, K. (2007). Partial Least Squares: A Versatile Tool for the Analysis of High-Dimensional Genomic Data. *Briefings in Bioinformatics, 8*(1), 32–44.

Butler, N., & Denham, M. (2000). The peculiar shrinkage properties of partial least squares regression. *Journal of the Royal Statistical Society. Series B. Methodological, 62,* 585–593.

Cleveland, W. (1979). Robust Locally Weighted Regression and Smoothing Scatterplots. *Journal of the American Statistical Association, 74*(368), 829–836.

Cristianini, N., & Shawe-Taylor, J. (2000). *An Introduction to Support Vector Machines and Other Kernel-based Learning Methodsq.* New York: Cambridge University Press.

De Bie, T., Cristianini, N., & Rosipal, R. (2005). Eigenproblems in Pattern Recognition. In Bayro-Corrochano, E. (Ed.), *Handbook of Geometric Computing: Applications in Pattern Recognition, Computer Vision, Neuralcomputing, and Robotics* (pp. 129–170). New York: Springer.

De Jong, S. (1993). SIMPLS: an alternative approach to partial least squares regression. *Chemometrics and Intelligent Laboratory Systems, 18,* 251–263.

Frank, I. (1990). A nonlinear PLS model. *Chemolab, 8,* 109–119.

Frank, I. (1995). Modern nonlinear regression methods. *Chemometrics and Intelligent Laboratory Systems*, *27*, 1–9.

Frank, I., & Friedman, J. (1993). A Statistical View of Some Chemometrics Regression Tools. *Technometrics*, *35*, 109–147.

Friedman, J. (1991). Multivariate Adaptive Regression Splines (with discussion). *Annals of Statistics*, *19*, 1–141.

Friedman, J., & Stuetzle, W. (1981). Projection pursuit regression. *Journal of the American Statistical Association*, *76*(376), 817–823.

Girosi, F. (1998). An Equivalence Between Sparse Approximation and Support Vector Machines. *Neural Computation*, *10*(6), 1455–1480.

Girosi, F., Jones, M., & Poggio, T. (1995). Regularization theory and neural networks architectures. *Neural Computation*, *7*(2), 219–269.

Gnanadesikan, R. (1977). *Methods for Statistical Data Analysis of Multivariate Observations*. New York: Wiley.

Goutis, C. (1996). Partial least squares yields shrinkage estimators. *Annals of Statistics*, *24*, 816–824.

Hardoon, D., Ajanki, A., Puolamaki, K., Shawe-Taylor, J., & Kaski, S. (2007). Information Retrieval by Inferring Implicit Queries from Eye Movements. In *Proceedings of the 11th International Conference on Artificial Intelligence and Statistics (AISTATS)*.

Hasegawa, K., Kimura, T., Miyashita, Y., & Funatsu, K. (1996). Nonlinear Partial Least Squares Modeling of Phenyl Alkylamines with the Monoamine Oxidase Inhibitory Activities. *Journal of Chemical Information and Computer Sciences*, *36*(5), 1025–1029.

Hastie, T., & Tibshirani, R. (1990). *Generalized Additive Models*. New York: Chapman & Hall/CRC.

Haykin, S. (1999). *Neural Networks: A Comprehensive Foundation* (2nd ed.). Upper Saddle River, NJ: Prentice-Hall.

Hiden, H., McKay, B., Willis, M., & Montague, G. (1998). Non-linear partial least squares using genetic programming. In J. Koza (Ed.), *Genetic Programming: Proceedings of the Third Annual Conference*. San Francisco: Morgan Kaufmann.

Höskuldsson, A. (1988). PLS Regression Methods. *Journal of Chemometrics*, *2*, 211–228.

Höskuldsson, A. (1992). Quadratic PLS regression. *Journal of Chemometrics*, *6*(6), 307–334.

Hulland, J. (1999). Use of partial least squares (PLS) in strategic management research: A review of four recent studies. *Strategic Management Journal*, *20*, 195–204.

Ivanov, V. (1976). *The Theory of Approximate Methods and Their Application to the Numerical Solution of Singular Integral Equations*. Nordhoff International.

Kimeldorf, G., & Wahba, G. (1971). Some results on Tchebycheffian spline functions. *Journal of Mathematical Analysis and Applications*, *33*, 82–95.

Krämer, N. (2007). An overview on the shrinkage properties of partial least squares regression. *Computational Statistics*, *22*(2), 249–273.

Krämer, N., & Braun, M. (2007). Kernelizing PLS, Degrees of Freedom, and Efficient Model Selection. In Z. Ghahramani (Ed.), *Proceedings of the 24th International Conference on Machine Learning* (pp. 441–448).

Leach, A., & Gillet, V. J. (2003). *An Introduction to Chemoinformatics*. New York: Springer.

Lee, S., Wu, Q., Huntbatch, A., & Yang, G. (2007). Predictive K-PLSR Myocardial Contractility Modeling with Phase Contrast MR Velocity Mapping . In Ayache, N., Ourselin, S., & Maeder, A. (Eds.), *Medical Image Computing and Computer-Assisted Intervention – MICCAI 2007* (pp. 866–873). New York: Springer.

Li, T., Mei, H., & Cong, P. (1999). Combining nonlinear PLS with the numeric genetic algorithm for QSAR. *Chemometrics and Intelligent Laboratory Systems, 45,* 177–184.

Lingjærde, O., & Christophersen, N. (2000). Shrinkage Structure of Partial Least Squares. *Scandinavian Journal of Statistics, 27,* 459–473.

Lobaugh, N., West, R., & McIntosh, A. (2001). Spatiotemporal analysis of experimental differences in event-related potential data with partial least squares. *Psychophysiology, 38,* 517–530.

Manne, R. (1987). Analysis of Two Partial-Least-Squares Algorithms for Multivariate Calibration. *Chemometrics and Intelligent Laboratory Systems, 2,* 187–197.

Martens, M., & Martens, H. (1986). Partial Least Squares Regression . In Piggott, J. (Ed.), *Statistical Procedures in Food Research* (pp. 293–359). London: Elsevier Applied Science.

Mercer, J. (1909). Functions of positive and negative type and their connection with the theory of integral equations. *Philosophical Transactions of the Royal Society of London, Series A, 209,* 415–446.

Michailidis, G., & De Leeuw, J. (1998). The Gifi System of Descriptive Multivariate Analysis. *Statistical Science, 13*(4), 307–336.

Momma, M. (2005). Efficient Computations via Scalable Sparse Kernel Partial Least Squares and Boosted Latent Features. In *Proceedings of SIGKDD International Conference on Knowledge and Data Mining* (pp. 654–659). Chicago, IL.

Mu, T., Nandi, A., & Rangayyan, R. (2007). Classification of breast masses via nonlinear transformation of features based on a kernel matrix. *Medical & Biological Engineering & Computing, 45*(8), 769–780.

Næs, T., & Isaksson, T. (1992). Locally Weighted Regression in Diffuse Near-Infrared Transmittance Spectroscopy. *Applied Spectroscopy, 46*(1), 34–43.

Nilsson, J., de Jong, S., & Smilde, A. (1997). Multiway Calibration in 3D QSAR. *Journal of Chemometrics, 11,* 511–524.

Qin, S., & McAvoy, T. (1992). Non-linear PLS modelling using neural networks. *Computers & Chemical Engineering, 16*(4), 379–391.

Rännar, S., Lindgren, F., Geladi, P., & Wold, S. (1994). A PLS kernel algorithm for data sets with many variables and fewer objects. Part 1: Theory and algorithm. *Chemometrics and Intelligent Laboratory Systems, 8,* 111–125.

Rosipal, R., & Krämer, N. (2006). Overview and Recent Advances in Partial Least Squares . In Saunders, C., Grobelnik, M., Gunn, S., & Shawe-Taylor, J. (Eds.), *Subspace, Latent Structure and Feature Selection Techniques.* New York: Springer.

Rosipal, R., & Trejo, L. (2001). Kernel Partial Least Squares Regression in Reproducing Kernel Hilbert Space. *Journal of Machine Learning Research, 2,* 97–123.

Rosipal, R., & Trejo, L. (2004). Kernel PLS Estimation of Single-trial Event-related Potentials. *Psychophysiology, 41,* S94. (Abstracts of The 44th Society for Psychophysiological Research Annual Meeting, Santa Fe, NM)

Rosipal, R., Trejo, L., & Matthews, B. (2003). Kernel PLS-SVC for Linear and Nonlinear Classification. In *Proceedings of the Twentieth International Conference on Machine Learning* (pp. 640–647). Washington, DC.

Saitoh, S. (1997). *Integral Transforms, Reproducing Kernels and Their Applications*. New York: Addison Wesley Longman.

Saunders, C., Hardoon, D., Shawe-Taylor, J., & Widmer, G. (2008). Using String Kernels to Identify Famous Performers from their Playing Style. *Intelligent Data Analysis, 12*(4), 425–440.

Schölkopf, B., Smola, A., & Müller, K. (1998). Nonlinear Component Analysis as a Kernel Eigenvalue Problem. *Neural Computation, 10,* 1299–1319.

Schölkopf, B., & Smola, A. J. (2002). *Learning with Kernels – Support Vector Machines, Regularization, Optimization and Beyond*. Cambridge, MA: The MIT Press.

Searson, D., Willis, M., & Montague, G. (2007). Co-evolution of non-linear PLS model components. *Journal of Chemometrics, 21*(12), 592–603.

Seber, G. A. F., & Lee, A. J. (2003). *Linear Regression Analysis* (2nd ed.). New York: Wiley-Interscience.

Shawe-Taylor, J., & Cristianini, N. (2004). *Kernel Methods for Pattern Analysis*. Cambridge, UK: Cambridge University Press.

Smola, A., Schölkopf, B., & Müller, K. (1998). The connection between regularization operators and support vector kernels. *Neural Networks, 11,* 637–649.

Tikhonov, A. (1963). On solving ill-posed problem and method of regularization. *Doklady Akademii Nauk USSR, 153,* 501–504.

Trejo, L., Rosipal, R., & Matthews, B. (2006). Brain-Computer Interfaces for 1-D and 2-D Cursor Control: Designs using Volitional Control of the EEG Spectrum or Steady-State Visual Evoked Potentials. *IEEE Transactions on Neural Systems and Rehabilitation Engineering, 14*(2), 225–229.

Vapnik, V. (1995). *The Nature of Statistical Learning Theory*. New York: Springer-Verlag.

Vapnik, V. (1998). *Statistical Learning Theory*. New York: Wiley.

Vapnik, V., & Chervonenkis, A. (1971). On the uniform convergence of relative frequencies of events to their probabilities. *Theory of Probability and Its Applications, 16*(2), 264–280.

Vapnik, V., & Chervonenkis, A. (1974). *Theory of Pattern Recognition [in Russian]*. Nauka, Moscow. ((German Translation: W.N. Vapnik and A.J. Cherwonenkis (1979). Theorie der Zeichenerkennung. Akademia-Verlag, Berlin))

Wahba, G. (1990). *Splines Models of Observational Data* (*Vol. 59*). Philadelphia: SIAM.

Wegelin, J. (2000). *A survey of Partial Least Squares (PLS) methods, with emphasis on the two-block case (Tech. Rep.)*. Seattle: Department of Statistics, University of Washington.

Wilson, D., Irwin, G., & Lightbody, G. (1997). Nonlinear PLS modeling using radial basis functions. In *American Control Conference*. Albuquerque, New Mexico.

Wold, H. (1975). Path models with latent variables: The NIPALS approach. In H. B. et al. (Ed.), *Quantitative Sociology: International perspectives on mathematical and statistical model building* (pp. 307–357). Maryland Hieghts, MO: Academic Press.

Wold, H. (1985). Partial least squares . In Kotz, S., & Johnson, N. (Eds.), *Encyclopedia of the Statistical Sciences* (*Vol. 6*, pp. 581–591). New York: John Wiley & Sons.

Wold, S. (1992). Nonlinear partial least squares modeling. II. Spline inner relation. *Chemolab, 14,* 71–84.

Wold, S., Albano, C., Wold, H., Dunn, W. III, Edlund, U., & Esbensen, K. (1984). Multivariate data analysis in chemistry . In Kowalski, B. (Ed.), *Chemometrics. Mathematics and Statistics in Chemistry. The Netherlands: Reidel.* Dordrecht.

Wold, S., Kettaneh-Wold, N., & Skagerberg, B. (1989). Nonlinear PLS Modeling. *Chemometrics and Intelligent Laboratory Systems, 7,* 53–65.

Worsley, K. (1997). An overview and some new developments in the statistical analysis of PET and fMRI data. *Human Brain Mapping, 5,* 254–258.

Wu, W., Massarat, D., & De Jong, S. (1997). The kernel PCA algorithms for wide data. Part I: theory and algorithms. *Chemometrics and Intelligent Laboratory Systems, 36,* 165–172.

Yoshida, H., & Funatsu, K. (1997). Optimization of the Inner Relation Function of QPLS Using Genetic Algorithm. *Journal of Chemical Information and Modeling, 37*(6), 1115–1121.

ENDNOTES

[1] In the case of the one-dimensional $\mathcal{Y}$-space computationally more efficient kernel PLS algorithms have been proposed (Momma, 2005, Rosipal et al., 2003).

[2] Now considering the matrices of score vectors T and regression coefficients C to be computed in $\mathcal{F}$.

[3] One needs to be very careful about proper balancing interpretability and prediction ability of the used model. If the model fits and predicts data poorly interpretation of the observed relations can often be misleading.

Chapter 10
Virtual Screening Methods Based on Bayesian Statistics

Martin Vogt
Rheinische Friedrich-Wilhelms-Universität, Germany

Jürgen Bajorath
Rheinische Friedrich-Wilhelms-Universität, Germany

ABSTRACT

Computational screening of in silico-formatted compound libraries, often termed virtual screening (VS), has become a standard approach in early-phase drug discovery. In analogy to experimental high-throughput screening (HTS), VS is mostly applied for hit identification, although other applications such as database filtering are also pursued. Contemporary VS approaches utilize target structure and/or ligand information as a starting point. A characteristic feature of current ligand-based VS approaches is that many of these methods differ substantially in the complexity of the underlying algorithms and also of the molecular representations that are utilized. In recent years, probabilistic VS methods have become increasingly popular in the field and are currently among the most widely used ligand-based approaches. In this contribution, the authors will introduce and discuss selected methodologies that are based on Bayesian principles.

INTRODUCTION

During the 1990ies, *virtual screening* (VS) approaches became increasingly popular in pharmaceutical research (Walters, Stahl, & Murcko, 1998). The early 1990s were also the time when rational drug design (Blundell, Jhoti, & Abell, 2002, Kuntz, 1992), or better structure-based drug design (because ligand-based design methods

are no less "rational"), became a new paradigm in drug discovery, with all the promises coming along with it. The ensuing substantial increase in the number of X-ray structures of target proteins that became available from then on also opened the door for extensive applications of computational structure-based screening methods that are conventionally termed protein-ligand "docking" (Brooijmans & Kuntz, 2003, Kuntz, 1992) and that were originally developed during the 1980s (Kuntz, Blaney, Oatley, Langridge, & Ferrin,

DOI: 10.4018/978-1-61520-911-8.ch010

1982). Thus, a first wave of VS that drew significant attention outside the expert community was largely structure-oriented, although ligand-based approaches to chemical database mining had also been extensively applied since the 1980s. These early ligand-based approaches (that play an important role to this date) included compound clustering algorithms (R. D. Brown & Martin, 1996, Willett, 1987), quantitative structure-activity relationship (QSAR) models (Hopfinger, 1980, Hopfinger et al., 1997), 2D molecular fingerprint representations (Willett, 2005) for similarity searching, and pharmacophore models for 3D database searching (Martin, 1992, Willett, 2005).

Ligand-based virtual screening approaches were further boosted during the 1990s by two important developments. Computational methods focusing on the assessment of whole-molecule similarity as an indicator of similar biological activity were increasingly introduced and widely recognized (Johnson & Maggiora, 1990). Furthermore, during the mid 1990s, the structure-based drug design paradigm was succeeded by another one focusing on high-throughput technologies and the pharmaceutical "number's game", i.e. more and more compounds were to be made and tested in a highly efficient manner in order to identify sufficient numbers of new candidates. Combinatorial chemistry (Czarnik, 1997) and HTS (Fox, Farr-Jones, & Yund, 1999) took center stage, compound libraries began to grow at unprecedented rates, and novel computational approaches were required for synthesis planning, library design and analysis, hence introducing the new field of molecular diversity analysis (Martin, 2001). Together with molecular similarity methods (Johnson & Maggiora, 1990), methodologies for diversity analysis including cell-based partitioning techniques (Pearlman & Smith, 1998) have substantially influenced the field of ligand-based VS by complementing more traditional QSAR and pharmacophore approaches. The term "chemoinformatics" first appeared in the literature in 1998 (F. K. Brown, 1998), and the majority of

ligand-based VS methods are at present covered under the chemoinformatics umbrella (Bajorath, 2001). By the end of the 1990s, both structure- and ligand-based VS approaches were widely applied (Bajorath, 2001, 2002) and the popularity of ligand-based methods has grown ever since (Stahura & Bajorath, 2005).

Today, the leading publication venues for chemoinformatics and VS approaches including the *Journal of Chemical Information and Modeling* and the *Journal of Medicinal Chemistry* report novel methods and practical applications in this area on a regular basis and the field is steadily branching out and becoming increasingly heterogeneous from a methodological point of view. During the past few years the clear trend could be observed that machine learning methods originally developed in computer or information science were, and continue to be, adopted for ligand-based VS (Stahura & Bajorath, 2005). In addition to neural networks (Livingstone, Manallack, & Tetko, 1997) and self-organizing maps (Kohonen, 1989), currently especially popular approaches include kernel-based methodologies (Schölkopf & Smola, 2002) such as support vector machines (Jorissen & Gilson, 2005, Warmuth et al., 2003) and binary kernel discrimination (Harper, Bradshaw, Gittins, Green, & Leach, 2001, D. J. Wilton et al., 2006) and in addition probabilistic methods, in particular, Bayesian modeling (Duda, Hart, & Stork, 2000).

In this contribution, we will discuss different methodologies of varying complexity that are based on Bayesian statistics and that are relevant for ligand-based VS. These approaches include relevance weighting for substructural analysis, binary kernel discrimination, binary QSAR, and Bayesian classification. The theory underlying these methodologies and exemplary application examples will be presented. Special emphasis will be put on Bayesian VS methods that incorporate information-theoretic components in order to prioritize descriptors for the analysis of compounds with specific biological activity, distinguish between active and database compounds, or predict the potential recall of similarity search calculations.

BACKGROUND

The general task in *ligand-based VS* is to distinguish between small numbers of active molecules and exceedingly large numbers of inactive database compounds. Importantly, ligand-based VS is generally only applicable if known active compounds are available as screening templates (Bajorath, 2001). Then, either global (i.e. whole molecule) or local (i.e. pharmacophore) similarity between templates and test compounds is evaluated as an indicator of potentially similar bioactivity. It is currently not unusual to search through databases containing several million compounds in order to identify perhaps 10 or 20 potential hits that are hidden in a haystack of decoys. Accordingly, VS methods suitable for this task must not only be computationally efficient but must also have significant potential to distinguish between active or inactive compounds or assign high priority to potentially active molecules.

In principle, two categories of methods exist; similarity search and compound classification techniques (Bajorath, 2002). *Similarity searching* utilizes 2D fingerprints (Willett, 2006), 3D pharmacophore fingerprints (Mason et al., 1999), or suitable molecular graph representations (Gillet, Willett, & Bradshaw, 2003) as queries, compares them in a pair-wise manner with database compounds using a similarity metric, and produces a compound ranking in the order of decreasing molecular similarity. Irrespective of their specific design, *fingerprints* are bit string representations of structural elements or molecular properties. In contrast to similarity searching, *compound classification* methods do not provide a compound ranking but predict compound class labels (e.g. active vs. inactive) (Bajorath, 2001). All machine

learning methods utilized in this context are classification methods.

Clustering and partitioning algorithms and machine learning methods applied for VS generally rely on the use of numerical molecular descriptors, i.e. more or less complex mathematical functions describing molecular structure and properties (Livingstone, 2000, Xue & Bajorath, 2000). Such descriptors constitute chemical feature spaces of varying dimensionality into which test compounds are projected for classification analysis. All compound classification approaches are whole-molecule similarity methods and distance between descriptor vectors representing compounds in however generated chemical reference spaces serves as a measure of molecular similarity.

Machine learning methods generally require the availability of learning sets of known active and inactive (or assumed to be inactive) compounds for model building. It follows that distinguishing between descriptor value distributions of active and inactive compounds is a major task in VS, irrespective of the methods used. Probabilistic modeling techniques such as Bayesian statistics are particularly well-suited to compare feature distributions of different classes of compounds (e.g. active vs. inactive), derive classification models, and assign class label probabilities to test compounds. Together with their computational efficiency, this explains at least in part the increasing popularity of *Bayesian models* in the VS arena. Their intrinsic drawback is that some fundamental assumptions about the relationships between features and the nature of their value distributions must be made. In particular, it must be assumed that descriptors are independent and value distributions are normal (Duda et al., 2000). For many compound datasets, chemical reference spaces, and feature value distributions, these assumptions only represent rough approximations, which might limit prediction accuracy. Another drawback of many (but not all) machine learning methods used in compound classification and VS is that their classification results can not be interpreted in a chemically intuitive way (Bajorath, 2002), in part because they operate in complex and high-dimensional feature spaces. Despite these potential limitations, Bayesian models have often been found to produce high prediction accuracy in VS applications. These models are typically derived as binary classifiers for specific biological activities (Jenkins, Bender, & Davies, 2006, Nigsch, Bender, Jenkins, & Mitchell, 2008).

However, classifiers derived for many different compound activity classes are often also combined in order to predict potential activities of test compounds using classifier arrays and derive activity profiles (Nidhi, Glick, Davies, & Jenkins, 2006, Paolini, Shapland, Hoorn, & Mason, 2006). In addition to these classifiers, *Bayesian models* can also be combined with Kullback-Leibler divergence analysis from information theory (Kullback, 1997) in order to generate methods to screen database or predict similarity search performance, as described in detail in the following.

STATISTICAL APPROACHES TO VIRTUAL SCREENING

The statistical methods discussed herein are based on a typical VS scenario meeting the following criteria:

1. A *compound database* or library to be searched for active compounds is given. Such a library will be denoted by the symbol B.
2. A set of *training compounds* known to be biologically active for a specific target is given. This set as well as the target will be denoted by the symbol A. A common designation for compounds active against a target is the term *activity class*. Typically, the size of a set of training compounds ranges from tens to a few hundred molecules.
3. The compound library B may contain some molecules that are active against target A,

but most of the compounds are expected to be inactive.

In an HTS experiment, all compounds of a library are tested in order to identify all compounds of the compound library B that are active against target A. Using computational methods, this objective can be reformulated in probabilistic terms: identify a *small subset* of the compound database that shows increased probability for compounds to be active over a random selection.

Based on this objective, the success of VS trials can be quantified. A VS trial can be considered successful if a certain percentage p of all active compounds of the database can be recovered in a relatively small percentage q of selected compounds. Popular measures for the success found in the literature are p, the *recovery rate*, p/q, the *enrichment factor* and the *hit rate*, which is defined as the ratio of the number of active compounds in the selection set to the overall size of the selection set. In practical applications, the hit rate is often the only measure that can be determined for VS calculations because usually only the selected set of compounds are tested for their activity. On the other hand, enrichment factors and recovery rates are often more meaningful to account for the success of a method because the sensitivity of a method determines its ability to identify active compounds within a subset of selected molecules. For example, it might be more desirable to have a hit rate of 0.1 for a selection set size of 100 compounds than a hit rate of 0.3 for a selection set size of 10. In the former case, 10 active compounds are detected compared to only three in the latter, which increases the chances of identifying a successful compound for further study.

In light of the objective defined above, *Bayesian modeling* uses statistical methods to assess the likelihood that a compound is active. That is, given a representation of a molecule either by a fingerprint or a vector of continuous-valued descriptors, ideally one would like to determine the probability $P(A|\mathbf{x})$ that a compound is active, given its descriptor representation **x**. This is in general not possible but when a set of active compounds are available for training, the conditional probability distribution $P(\mathbf{x}|A)$ of active molecules to possess representation **x** may be estimated from the training data. Likewise, the probability distribution for inactive compounds $P(\mathbf{x}|B)$ can be estimated from either a set of known inactive compounds or from the compound database B under the assumption that only very few of the compounds in B show the desired activity for target A. *Bayesian modeling* is based on the estimation of these probabilities and has its name from Bayes theorem, which relates the unknown probability $P(A|\mathbf{x})$ to $P(\mathbf{x}|A)$:

$$P(A \mid \mathbf{x}) = \frac{P(\mathbf{x} \mid A)P(A)}{P(\mathbf{x})} \qquad (1)$$

The prior probability $P(A)$ of a compound being active is in general unknown but is usually very small, i.e. the vast majority of compounds of a compound library will be inactive against target A. As can be seen from (1), the probability $P(A|\mathbf{x})$ is directly proportional to $P(\mathbf{x}|A)$. The likelihood that a compound is active can thus be defined as

$$L(A|\mathbf{x})=\alpha P(\mathbf{x}|A), \ \alpha>0 \qquad (2)$$

$L(A|\mathbf{x})$ is a function proportional to $P(\mathbf{x}|A)$, a probability distribution that can be usually estimated from the training data. Note that the likelihood will not yield an absolute probability measure, instead a relative measure for the likelihood of activity is obtained according to which compounds may be ranked or classified. Even when prior probabilities can be estimated Bayesian modeling is often found to produce poor probability estimates but performs well in classification and ranking (Domingos & Pazzani, 1996, Zhang & Su, 2004). Relating the likelihood $L(A|\mathbf{x})$ of a compound being active given the descriptor **x** to the likelihood $L(B|\mathbf{x})$ that the compound is inactive by considering the ratio of both values

$$R(\mathbf{x}) = \frac{L(A \mid \mathbf{x})}{L(B \mid \mathbf{x})} = \frac{P(\mathbf{x} \mid A)}{P(\mathbf{x} \mid B)} \qquad (3)$$

yields a score for each compound given by its representation **x** according to which the compounds of the database can be ranked in decreasing order of likelihood. Note that it is very common to apply the logarithm to the ratio so the score becomes $S(\mathbf{x})=\log R(\mathbf{x})$. This is usually computationally more efficient and numerically more stable. The methods introduced in the following either use Equations (2) or (3) for classification according to target-specific activities or prioritization of chemical compounds for a single biological target. The methods discussed in the following differ in two main points from each other. First, different multi-dimensional representations might be used. Substructural analysis, naïve Bayesian classification, and binary kernel discrimination make use of binary fingerprint representations, whereas binary QSAR and the Bayesian modeling approach presented later make use of continuous valued descriptors. Second, different assumptions about the descriptor value distributions may be incorporated into the methods. *Naïve Bayesian classification* assumes independence of individual features while binary kernel discrimination uses a Parzen-window technique to estimate the joint distributions thus making no such assumption. On the other hand, continuous-valued descriptor distributions might be estimated using histograms of discretized values as in binary QSAR or might be described by the sample mean and variation assuming a normal distribution for descriptor values.

SUBSTRUCTURAL ANALYSIS AND NAÏVE BAYESIAN CLASSIFIERS FOR FEATURE DATA

Among the most established and widespread uses of Bayesian statistics are weighting schemes for *substructural analysis* and the *naïve Bayesian classifiers* as applied to fingerprint representations. In contrast to similarity measures for fingerprints (Willett, 1998), the probabilistic approach described below yields a specific weighting scheme for fingerprint bits according to their relevance for activity. Combining multiple Bayesian classifiers for different activity classes allows the construction of so-called multiple-category models (Nidhi et al., 2006, Nigsch et al., 2008).

When given a fingerprint $\mathbf{f}=(f_i)_{i1\ldots n}$ the probabilities $P(A|f_i=1)$ for a compound to be active if it possesses a feature f_i cannot be estimated reliably for a database given just a reference set of known active compounds. Instead, from a set of active compounds it is only possible to estimate the probabilities $P(f_i=1|A)$ by counting the relative frequencies of the occurance of f_i for compounds of the training set. Obviously, the estimation of these probabilities depends on the composition of the set of reference compounds. For practical VS applications, reference sets should be carefully selected so that they are not, for instance, dominated by a single analog series of compounds stemming from a single lead optimization process. Such series could seriously bias feature distributions of "typical" actives for a given target. On the other hand, using such reference sets for benchmarking purposes of VS methods can greatly influence the results and give overly optimistic recall performance.

Substructural analysis was pioneered by Cramer, Redl, and Berkoff (1974) as an additive weighting scheme for substructural features to assess the likelihood of activity of chemical compounds. In substructural analysis weights are associated to single fingerprint features that reflect their importance for a target-specific activity. Ormerod, Willett, and Bawden (1989) introduced the so-called *relevance weights* for VS originally developed by Robertson and Jones (1976) in the context of document retrieval. The weights are derived from a training set A of active and a training set I of inactive compounds. Let N be the total number of training compounds, N_A be the

number of active, and N_I be the number of inactive training compounds. Furthermore $N^{f=1}$ denotes the number of compounds in the training set possessing a feature f and likewise $N_A^{f=1}$, and $N_I^{f=1}$ denote the number of actives and inactives with feature f, respectively. The probabilities

$$P(f=1) = \frac{N^{f=1}}{N}, P(f=1\mid A) = \frac{N_A^{f=1}}{N_A}, and P(f=1\mid I) = \frac{N_I^{f=1}}{N_I}$$

comment: in the equation, "and" should not be in italics can then be estimated from the training data.

Four different weighting schemes are derived by considering alternative strategies. The weighting scheme can be based on *normalized likelihoods*,

$$L(A\mid f) = \frac{P(f\mid A)}{P(f)}$$

or on the *likelihood ratios*

$$\frac{L(A\mid f)}{L(I\mid f)} = \frac{P(f\mid A)}{P(f\mid I)}.$$

If the training set consists of much more inactive than active compounds, which is quite common in VS, the differences between the two weights will only be marginal. Furthermore, only confirmed active compounds might be available for training and it is only possible to estimate the probability distributions $P(f)$, but not $P(f|I)$ from the compound library to be screened. Another aspect considers the relevance for activity if a compound does not possess a certain feature, i.e. the question whether the absence of a feature f might be an indication that a compound is more likely to show activity A or if it is simply irrelevant to activity. Answering this question from a theoretical point of view is difficult and certainly also depends on the type of features used for the representation. For example, for substructure-type fingerprints, the absence of features in an activity class that occur frequently in the screening library might indeed be an indication of activity. So, weights are assigned either by only considering the probabilities for a feature to be present or also for a feature to be absent.

Weighting schemes R_1 and R_2 are based on the normalized likelihood and the likelihood ratio as given above and only consider the presence of features as indicators for activity. Applying the naivety assumption of *feature independence* yields the following scoring schemes:

$$R_1(\mathbf{f}) = \log \frac{\prod_{f_i=1} P(f_i = 1 \mid A)}{\prod_{f_i=1} P(f_i = 1)} = \sum_{i=1}^{n} f_i \log \frac{N_A^{f_i=1} N}{N_A N^{f_i=1}}$$

(4)

In a similar way, when considering the likelihood ratios we obtain:

$$R_2(\mathbf{f}) = \log \frac{P(\mathbf{f} \mid A)}{P(\mathbf{f} \mid I)} = \log \frac{\prod_{f_i=1} P(f_i = 1 \mid A)}{\prod_{f_i=1} P(f_i = 1 \mid I)} = \sum_{i=1}^{n} f_i \log \frac{N_A^{f_i=1} N_I}{N_A N_I^{f_i=1}}$$

(5)

Similarly the weights R_3 and R_4 are based on the normalized likelihood and the likelihood ratio incorporating weights for features absent in a fingerprint $\mathbf{f}$:

$$R_3(\mathbf{f}) = \log L(A \mid \mathbf{f}) + C = \log \frac{P(\mathbf{f} \mid A)}{P(\mathbf{f})} + C$$

and

$$R_4(\mathbf{f}) = \log \frac{L(A \mid \mathbf{f})}{L(I \mid \mathbf{f})} + C = \log \frac{P(\mathbf{f} \mid A)}{P(\mathbf{f} \mid I)} + C$$

where C is a constant. For R_3 we obtain:

$$R_3(\mathbf{f}) = \log \frac{\prod_{i=1}^{n} P(f_i \mid A)}{\prod_{i=1}^{n} P(f_i)}$$

$$R_3(\mathbf{f}) = \sum_{i=1}^{n} f_i \log \frac{N_A^{f_i=1} N}{N_A N^{f_i=1}} + \sum_{i=1}^{n} (1 - f_i) \log \frac{N_A^{f_i=0} N}{N_A N^{f_i=0}} + C$$

(6)

Note that in the above equation, weights are assigned to features that are not present. This is often computationally not efficient and might not even be feasible for sparsely populated fingerprints like pharmacophore fingerprints or connectivity fingerprints. Using a constant offset of

$$C = -\sum_{i=1}^{n} \log \frac{N_A^{f_i=0} N}{N_A N^{f_i=0}}$$

avoids this problem and only features f_i with $f_i=1$ are assigned a non-zero weight:

$$R_3(\mathbf{f}) = \sum_{i=1}^{n} f_i \left(\log \frac{N_A^{f_i=1} N}{N_A N^{f_i=1}} - \log \frac{N_A^{f_i=0} N}{N_A N^{f_i=0}} \right)$$

$$R_3(\mathbf{f}) = \sum_{i=1}^{n} f_i \log \frac{N_A^{f_i=1} N^{f_i=0}}{N_A^{f_i=0} N^{f_i=1}}$$

(7)

Similarily for $R_4(\mathbf{f})$ we get:

$$R_4(\mathbf{f}) = \sum_{i=1}^{n} f_i \log \frac{N_A^{f_i=1} N_I^{f_i=0}}{N_A^{f_i=0} N_I^{f_i=1}}$$

(8)

Laplacian Correction

An important aspect for the estimation of probabilities of individual features is that, due to small sample sizes and the low marginal probabilities of at least some if not most of the features, the probabilities $P(f_i=1|A)$ as estimated by the relative frequencies $N_A^{f_i} / N_A$ comment: "f_i" must be

"$f_i = 1$" might be overconfident. To overcome this problem a number K_i of virtual samples are added where each has the baseline probability $P(f_i = 1) = N^{f_i=1} / N$ of possessing feature f_i:

$$P_K(f_i = 1 \mid A) = \frac{N_A^{f_i=1} + (N^{f_i=1} / N)K_i}{N_A + K_i}$$

$$(9)$$

When K_i is set to be $1 / P(f_i = 1) = N / N^{f_i=1}$ we obtain a *Laplacian correction* of the estimate:

$$P_L(f_i = 1 \mid A) = \frac{N_A^{f_i=1} + 1}{N_A + N / N^{f_i=1}}$$

$$(10)$$

The likelihood $L(A|f_i)$ can then be obtained by normalizing with $P(f_i)$ accordingly:

$$L(A \mid f_i = 1) = \frac{P_L(f_i = 1 \mid A)}{P(f_i = 1)} = \frac{(N_A^{f_i=1} + 1)N}{(N_A + N / N^{f_i=1})N^{f_i=1}}$$

$$L(A \mid f_i = 1) = \frac{N_A^{f_i=1} + 1}{N^{f_i=1}(N_A / N) + 1}$$

$$(11)$$

Naïve classification models are then constructed from a fingerprint $\mathbf{f}=(f_i)_{i=1\dots n}$ under the independence assumption:

$$S(A \mid \mathbf{f}) = \log L(A \mid \mathbf{f}) = \sum_{i=1}^{n} \log L(A \mid f_i) = \sum_{i=1}^{n} \frac{N_A^{f_i} + 1}{N^{f_i}(N_A / N) + 1}$$

comment: "f_i" in the numerator and denominator must be

"$f_i = 1$" $\qquad\qquad(12)$

Especially Equations (5) and (12) have found frequent use in compound prioritization (Bender, Mussa, Glen, & Reiling, 2004, Glick, Jenkins, Nettles, Hitchings, & Davies, 2006) and classification of compounds (Nidhi et al., 2006, Nigsch et al., 2008).

BINARY KERNEL DISCRIMINATION

The figures show the non-parametric estimation of probability densities with Parzen windows using different kernel-functions. (a) shows the influence of the smoothing parameter σ of a Gaussian kernel function on the estimation of a one-dimensional probability distribution. (b) schematically illustrates the binary kernel function. Points correspond to compounds embedded in a 2D-plane. The influence of different parameter settings on the smoothness of the resulting probability distribution is shown at varying gray-scaled intensity levels.

The relevance weights for substructural analysis and the *naïve Bayesian classifiers* discussed in the previous sections make the explicit assumption of feature independence of both the prior and conditional distributions. It is well known that these assumptions in general are approximations (Lewis, 1998, Zhang, 2004) but often produce good classification performance, although the estimation of actual probabilities might be poor. An advantage of the feature independence assumption, which should not be underestimated, is the possibility to very efficiently train a model using only very few training samples. In the case of fingerprints, for each bit (or feature), a single probability parameter needs to be estimated and, as features are assumed to be independent, no correlations between features need to be taken into account. Hence, the joint distribution can be easily estimated. In some sense, the independence assumption is a very pragmatic one, allowing the construction of computationally efficient and mathematically rigorous models under the given assumption.

An alternative to this approach avoiding such assumptions is to use a non-parametric model for the estimation of the probability distributions. Non-parametric methods make no prior assumptions about the feature distributions, especially feature independence. A common technique from machine learning is to apply *kernel functions* for estimating probability densities, an approach also known as the Parzen-window method (Duda et al., 2000).

Harper et al. (2001) introduced Parzen-windows applied to binary fingerprints for VS purposes under the name *binary kernel discrimination*.

In analogy to Bayesian classification the likelihood ratio

$$R(x) = \frac{L(A \mid v)}{L(I \mid v)} \propto \frac{p(v \mid A)}{p(v \mid I)} \qquad (13)$$

is considered and compounds can be prioritized accordingly.

Suppose training sets A and I of N_A and N_I compounds, respectively, are given where A contains compounds with the desired property like target-specific activity and I contains the others. The probability densities $p(v|A)$ and $p(v|I)$ are estimated using *kernel functions:*

$$p(v \mid A) = \frac{1}{N_A} \sum_{i=1}^{N_A} K_\lambda(v - v_i^A), \qquad (14)$$

$$p(v \mid B) = \frac{1}{N_I} \sum_{i=1}^{N_I} K_\lambda(v - v_i^I). \qquad (15)$$

Here v_i^A and v_i^I are the descriptor values (or fingerprints) of the compounds from A and I, respectively. K_λ is a symmetric multidimensional density function and λ is a smoothing parameter. The estimate $p(v|A)$ is a linear combination of probability density functions centered at each point of the training set. The parameter λ controls the "smoothness" of the estimate, i.e. the range of influence of the kernel around each data point. Under continuity assumptions the estimate can be shown to converge to the true density function (Duda et al., 2000). Figure 1 (a) shows the probability density estimate using Gaussian kernel functions with varying standard deviations σ as smoothing parameter for a sample of seven data points. The quality of the estimates will mainly depend on two factors: a) the number of compounds in the training set and b) the unbiased nature of the training data. That is, the training data ideally should be a representative subset of compounds with respect to the descriptor space used for representing them. As mentioned above, this might be a problem for methods like Bayesian modeling and binary kernel discrimination, because frequently analog series of active compounds might dominate a training set, thereby skewing the distribution towards a few single chemotypes.

For fingerprints, i.e. binary vectors v, w of length n, the following kernel function has been suggested:

$$K_{\lambda(}v, w) = \lambda n^{-\|v-w\|}(1-\lambda)^{\|v-w\|} \qquad (16)$$

where $\|\cdot\|$ is the Hamming distance and the smoothing parameter λ ranges from 1 to 0.5. This kernel function can be regarded as the equivalent of the symmetric Gaussian kernel function for independent binary features where the smoothing parameter σ is replaced by the parameter λ. Figure 1 (b)

Figure 1. Density estimation using Parzen-windows. (a) Gaussian kernel function. (b) Binary kernel function

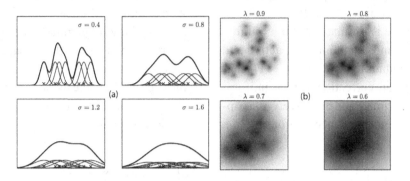

illustrates the effect of the smoothing parameter of the binary kernel function on an embedding of data points in a two-dimensional Euclidean plane. This approach requires a training process in order to find an appropriate value for λ. Harper et al. (2001) suggest a stepwise increase for λ from 0.50 to 0.99 and using a form of leave one out validation to determine the optimal parameter. Typically one can expect the parameter λ to be larger when more training data is available, as each data point needs to cover less range.

Binary kernel discrimination has been shown to be an effective approach to VS by ranking compounds according to their likelihood of being active (D. Wilton, Willett, Lawson, & Mullier, 2003). An interesting variation has been introduced by Chen et al. (2006) by replacing the Hamming distance measures with other similarity measures. Their experiments showed an overall favorable performance of the Tanimoto coefficient and the Dice coefficient over the Hamming distance.

BINARY QSAR

Descriptor values are binned using a Gaussian kernel function for smoothing. The figure illustrates how a value of 0.3 is split among multiple bins of unit width by adding a fraction to the bin count of each bin corresponding to the area under the curve

intersecting the respective bin. In this example, a smoothing parameter of $\sigma=0.5$ was chosen.

The following methods are based on the numerical representation of molecules via real-valued property descriptors. The main ideas of Bayesian modeling easily carry over to the continuous case, but additional assumptions about descriptor distributions need to be made.

Binary QSAR (Labute, 1999) has its origins in quantitative structure activity relationship analysis typically performed by constructing a linear regression model relating a number of real-valued descriptors to compound potency (Hansch, Maloney, Fujita, & Muir, 1962, Hansch & Fujita, 1964). Binary QSAR can be seen as a variation of these applications where quantitative information of activity is not available or not reliable. This is typically the case for VS where the primary goal is the detection of hits that are promising starting candidates for optimization with the aim to develop highly potent compounds with favorable properties.

Binary QSAR uses Bayes theorem to estimate the activity of a given molecule in the form

$$P(A \mid \mathbf{x}) = \frac{p(\mathbf{x} \mid A)P(A)}{p(\mathbf{x} \mid A)P(A) + p(\mathbf{x} \mid I)P(I)}$$

(17)

where I denotes inactivity so $P(I)=1-P(A)$. Here a lower case p is used to indicate probability density functions. The prior distribution $P(A)$ of a compound being active is estimated from a training set S of N compounds containing N_A active and N_I inactive compounds to be $P(A)=(N_A+1)/(N+2)$ using the Laplacian correction with a uniform prior. Using this estimate and the naivety assumption of descriptor independence a reformulation of (17) yields

$$P(A \mid \mathbf{x}) \approx \left(1 + \frac{N_I+1}{N_A+1} \prod_{i=1}^{n} \frac{p(x_i \mid I)}{p(x_i \mid A)}\right)^{-1} \quad (18)$$

Note that this estimator for the probability is monotonic with the likelihood ratio $\dfrac{P(x_i \mid A)}{P(x_i \mid I)}$ discussed in the next section which is used as a scoring function in order to rank the database compounds. An important aspect of *Bayesian modeling* using continuous descriptors is how the probability density function may be estimated.

Similar to the binary kernel discrimination method, *kernel functions* are used in conjunction with binning, in this case one-dimensional Gaussian distributions parameterized by the standard deviation σ as smoothing parameter. Given a descriptor-specific discretization scheme for K bins defined by the boundary values $(b_0,b_1],\ldots,(b_{K-1},b_K]$, the contribution of a descriptor value z to bin k is given by

$$\int_{b_{k-1}}^{b_k} \frac{1}{\sigma\sqrt{2\pi}} \exp \frac{-(x-z)^2}{2\sigma^2} \, dx.$$

Figure 2 illustrates this approach. A histogram is obtained by summing over the contributions of all samples of a training set:

$$B_k = \sum_{j=1}^{N} \int_{b_{k-1}}^{b_k} \frac{1}{\sigma\sqrt{2\pi}} \exp \frac{-(x-z_j)^2}{2\sigma^2} \, dx, \quad k=1,\ldots,K \quad (19)$$

Figure 2. Binning using Gaussian kernel functions

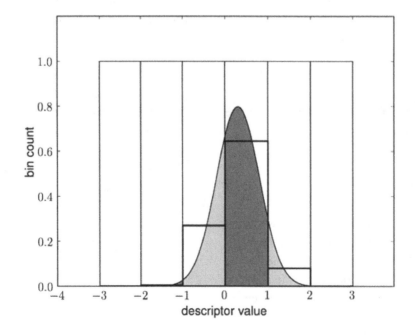

So, each sample contributes to each bin by a fraction determined by the area under a Gaussian distribution centered at the sample value.

Again a *Laplacian correction* is applied to the probability estimation of the discrete probability $P(z \in (b_{k-1}, b_k])$:

$$P(z \in (b_{k-1}, b_k]) \approx \frac{B_k + 1/c}{c + B/c}, c = \sum_{j=1}^{K} B_j$$

(20)

comment: "B" in the denominator must be "K" the continuous density function $p(z)$ can then be estimated as:

$$p(z) \approx \sum_{k=1}^{K} \frac{B_k + 1/c}{c + B/c} \int_{b_{k-1}}^{b_k} \frac{1}{\sigma\sqrt{2\pi}} \exp\frac{-(x-z)^2}{2\sigma^2} dx$$

(21)

comment: "B" in the denominator must be "K"

The density estimates according to (21) given a set of active compounds and a set of inactive compounds can be substituted in (18) to yield the binary QSAR model.

In accordance with the independence assumption Labute (1999) suggested to use principal components based on the training data of a set of descriptors to yield a set of uncorrelated variables x_i for the binary QSAR model.

The *binary QSAR* method is a classification method for the prediction of compound activity. It estimates a probability for activity and compounds with a probability greater than 0.5 are classified as being active. The quality of the probability estimate is influenced by three main factors:

1.   *The accuracy of the estimated probabilities.* This is mainly a question of the quantity and the quality of the training data. The use of the Gaussian-smoothed histograms and the Laplacian corrections alleviate the problem of generating accurate histograms from few training samples. On the other hand,

as mentioned earlier, overrepresentation of a single chemotype in the training set can seriously bias the probability estimates.

2.   *The independence of the descriptor variables.* For the probability estimate to be as accurate as possible the descriptor variables need to be independent. From this point of view, the use of principal component analysis to yield uncorrelated variables is vital to binary QSAR. However, uncorrelated variables are often not independent.

3.   *The estimation of the prior probabilities of a compound to be active.* Binary QSAR basically assumes a training set with known activities and splits the set into actives according to a potency threshold for activity. An assumption inherent in the method is that the probability of activity according to the ratio of actives in the training set carries over to the data set to be screened.

Binary QSAR has proven to be an accurate prediction tool for target specific activity that is robust for noisy data (Gao, 2001, Labute, 1999). In the context of HTS, estimating prior probabilities based upon available data is in general not possible, making an accurate estimation of the probability of activity impossible. On the other hand, the probability estimation, as has been noted, is monotonic with the likelihood ratio and thus binary QSAR might be used successfully in prioritizing compounds of a database and can also be used in multiple category classification by building multiple models and assigning a category based upon the highest predicted probability.

BDACCS - BAYESIAN MODELING

The Euclidean distance is a simple and straightforward measure for chemical similarity in high-dimensional descriptor spaces of continuous-valued descriptors (Godden & Bajorath, 2006). The use of the Euclidean metric can be rationalized from a probabilistic point of view by noting that this metric corresponds to a log-likelihood measure based upon the assumption of normally and independently distributed descriptor values (Vogt, Godden, & Bajorath, 2007), as detailed below. When using a parametric model for the estimation of descriptor values, the absence of further knowledge, i.e. when only mean and standard deviation are known, makes the assumption of Gaussian distributions the most unbiased one (Shannon & Weaver, 1963) and yields good results in practice (Vogt & Bajorath, 2008b, Vogt et al., 2007).

The likelihood ratio

$$R(\mathbf{x}) = \frac{L(A \mid \mathbf{x})}{L(B \mid \mathbf{x})} = \frac{P(\mathbf{x} \mid A)}{P(\mathbf{x} \mid B)} \qquad (22)$$

will give a relative measure of a compound *c* having the desired property when compared to other molecules. If compounds are represented by continuous descriptors $\mathbf{x}=(x_i)_{i=1\ldots n}$ in an *n*-dimensional chemical space and the assumptions of descriptor independence and Gaussian distributions are made, then from

$$L(A \mid \mathbf{x}) \propto p(\mathbf{x} \mid A) = \prod_{i=1}^{n} p(x_i \mid A) \propto \prod_{i=1}^{n} \frac{1}{\sqrt{2\pi\sigma_i^2}} \exp \frac{-(x_i - \mu_i)^2}{2\sigma_i^2}, \qquad (23)$$

where μ_i and σ_i are the mean and standard deviation of descriptor *i*, it follows by considering the negative log-likelihood that:

$$-\log L(A \mid \mathbf{x}) \propto \sum_{i=1}^{n} \frac{(x_i - \mu_i)^2}{\sigma_i^2}. \qquad (24)$$

Hence, considering normalized Euclidean distance (Godden & Bajorath, 2006) in chemical space corresponds to assumptions about Gaussian distributions of the descriptor values. The ability to describe similarity metrics in the context of

basic assumptions concerning the independence of descriptors and their distributions is a valuable tool in assessing the quality of these measures.

The logarithm of the likelihood ratio (3) yields a *log-oddsscoring function:*

$$\log R(x) = \sum_{i=1}^{n} \Big(\log p(x_i \mid A) - \log p(x_i \mid B)\Big). \qquad (25)$$

This approach can be seen as the continuous analog of the relevance weighting for substructural analysis and naïve Bayesian classifiers discussed at the beginning of this section. The scoring function can be formulated as:

$$\log R(x) = \sum_{i=1}^{n} \left(\frac{(x - \nu_i)^2}{\tau_i^2} - \frac{(x - \mu_i)^2}{\sigma_i^2} \right) \qquad (26)$$

Here μ_i and σ_i are the sample mean and standard deviation for descriptor *i* and a set of training compounds *A* with a desired property like bioactivity against a specific target and ν_i and τ_i are the sample mean and standard deviation of descriptor *i* of training compounds *B* not possessing that property.

In a typical VS scenario, given a relatively small set of active reference structures, the overwhelming majority of library compounds will be inactive and only very few compounds will show activity. In this case, the training set for estimating the probability distributions of active compounds consists of the reference structures. Furthermore, for all practical purposes, the distributions of the inactive compounds can be safely estimated by considering the total compound library, including potential actives, because they only marginally influence the estimates.

The Bayesian approach as described above is not limited to a single type of representation, but can also successfully be applied to the combination of different representations like continuous descriptors and binary fingerprints. The MetaBDACCS approach (Vogt & Bajorath, 2008a) unifies descriptors and fingerprints in a single model and shows a significant increase in VS performance for a number of biological activity classes.

Predicting Virtual Screening Performance

Given the statistical nature of the approach, its success relies on the difference in distribution of some descriptors between sets of compounds A and B. In short, the more the distributions of a descriptor differ the larger the discriminatory power of the descriptor becomes. A quantitative measure for the divergence of distributions is the *Kullback-Leibler divergence*

$$D\big[p(\mathbf{x}\mid A)\mid\mid p(\mathbf{x}\mid B)\big] = \int p(\mathbf{x}\mid A)\log\frac{p(\mathbf{x}\mid A)}{p(\mathbf{x}\mid B)}\,d\mathbf{x}. \tag{27}$$

The Kullback-Leibler divergence corresponds to the expected score of the log-likelihood-ratio $\log R(\mathbf{x})$ for compounds of class A. Given estimates for the conditional distributions $p(\mathbf{x}|A)$ and $p(\mathbf{x}|B)$ the Kullback-Leibler divergence can be calculated analytically. For normally distributed descriptors it is

$$D\big[p(\mathbf{x}\mid A)\mid\mid p(\mathbf{x}\mid B)\big] = \sum_{i=1}^{n}\left(\log\frac{\sigma_i}{\tau_i} + \frac{\tau_i^2 - \sigma_i^2 + (\mu_i - \nu_i)^2}{2\sigma_i^2}\right). \tag{28}$$

However it may equally well be applied to binary features:

$$D\big[p(\mathbf{f}\mid A)\mid\mid p(\mathbf{f}\mid B)\big] = \sum_{i=1}^{n}\left(P(f_i=1\mid A)\log\frac{P(f_i=1\mid A)}{P(f_i=1\mid B)} + P(f_i=0\mid A)\log\frac{P(f_i=0\mid A)}{P(f_i=0\mid B)}\right). \tag{29}$$

In practice, Equation (27) can be used to analyze the fitness of given chemical descriptor spaces

for the purpose of virtual screening for specific biological targets. Low-dimensional chemical subspaces that are well-suited for distinguishing between active and inactive compounds might be identified, which reflect the significance of specific descriptor sets to discriminate specific activities (Vogt & Bajorath, 2008b).

EXEMPLARY APPLICATION

Forty activity classes (dots) were used in benchmarking trials in order to establish a quantitative relationship between the Kullback-Leibler divergence and the expected recall of a VS trial using four different representations. The quality of the linear regression model was assessed using seven test classes (indicated by differing symbols). The graphs on the left display the results for the MACCS fingerprint and its combination with the property descriptors while the graphs on the right show the corresponding results for the TGD fingerprint.

In the previous sections, we have discussed Bayesian modeling techniques that represent a widely applicable approach to VS. Here, we demonstrate how the probabilistic approach can be used to assess the performance of VS trials in a prospective manner. It has been explained that the *Kullback-Leibler divergence* is a quantitative measure of the divergence between probability distributions. Analyzing this formalism, it is possible to estimate how well active and inactive molecules can be separated in a chosen chemical descriptor space. It is well worth noting that the Kullback-Leibler divergence exactly corresponds to the expected score of the *log-odds scoring function* (26) (Vogt & Bajorath, 2007), thus emphasizing the quantitative relationship between the divergence measure and VS performance.

We present an exemplary application to illustrate how this relationship can be exploited to make predictions about the activity class-specific performance of varying Bayesian models based upon different representations, i.e. different fingerprints or combinations of fingerprints with property descriptors. It is a well-known phenomenon that the performance of VS methods displays a strong activity class dependence (Martin, Kofron, & Traphagen, 2002, Sheridan & Kearsley, 2002). Therefore, the ability to predict if a VS might be successful for a given activity class and a chosen type of fingerprint or descriptor representation is in general of high value. Essentially, this type of analysis makes it possible to estimate whether a VS trial could in principle separate active from database compounds, provided unknown active molecules have property distributions that are similar known reference molecules. Of course, whether or not a screening database contains molecules with the desired activity is unclear and could not be predicted.

For this application we have chosen as a screening library ZINC 8 (Irwin & Shoichet, 2005), a public-domain database of commercially available compounds containing 5.8 million molecules (after applying an in-house drug-likeness filter). All compounds in the database are assumed to be inactive. Although the database might actually contain moleclules that are active against one or the other target, their number would be exceedingly small compared to the remainder of the screening database and would thus not influence the results of Kullback-Leibler divergence calculations. As property descriptors, 71 descriptors with limited correlation were selected from a pool of 142 1D and 2D descriptors implemented in the Molecular Operating Environment (MOE) (MOE, 2007) were used. Two different fingerprints were used in our calculations; MACCS consisting of 166 structural keys that represent substructures consisting of one to ten non-hydrogen atoms and TGD, a two-point 2D topological fingerprint (Sheridan, Miller, Underwood, & Kearsley, 1996). TGD consists of 420 bits that capture pairs of atoms where each atom is assigned to one of seven atom types and inter-atomic distances are divided into 15 different bond distances.

In order to establish a quantitative relationship between the *Kullback-Leibler divergence* and the recall of VS, 40 activity classes assembled from various sources were used (Vogt & Bajorath, 2007). Linear regression models relating the Kullback-Leibler divergence to the recovery rate of each of four variations of Bayesian screening using the two fingerprints either alone or in combination with the 71 property descriptors are obtained by the following procedure. For each activity class, a number of randomized trials are performed by 'hiding' a part of the active compounds in the screening database and using the remaining ones as reference for the Bayesian model. The Kullback-Leibler divergence obtained from the model and the corresponding recall of active compounds for the top ranked 1,000 database molecules are then used as a data point for building the regression model. Here, we carried out 100 trials for each class, always using ten compounds for training. From the resulting 4,000 data points, a linear regression model was calculated relating the logarithm of the Kullback-Leibler divergence to the recall. Figure 3 shows the average Kullback-Leibler divergence and the average recovery rate achieved over 100 trials for each class as a data point.

In order to test the quality of the model seven additional activity classes were benchmarked and their average recovery rate plotted against their Kullback-Leibler-divergence in Figure 3. The curves show a good correlation between predicted and achieved recovery rates, hence allowing a prospective assessment of the method for different activity classes and molecular representations.

FUTURE RESEARCH DIRECTIONS

There is little doubt that *Bayesian modeling* techniques will play a significant future role in the chemoinformatics area. *Naïve Bayesian classifiers* have for considerable time been a part

Figure 3. Prediction of recall performance for four different types of molecular representations

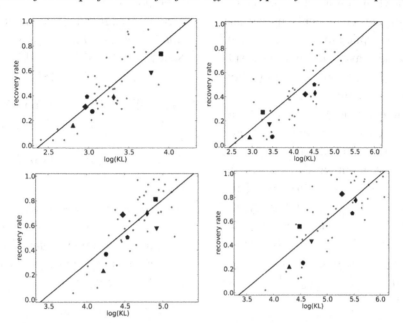

of the spectrum of machine learning methods applied in chemical database mining. Recently, there has been a clear trend to utilize multiple Bayesian classifiers for systematic ligand-target prediction and compound profiling. There are reasons to believe that the use of multi-target classifiers and Bayesian networks will further expand in the future, in particular, in the context of computer-aided chemogenomics research that is tightly linked to chemoinformatics. Multi-target *Bayesian models* are also readily applicable to conventional VS. Another high-potential growth area of Bayesian statistics is descriptor profiling and prioritization. Here a combination of Bayesian statistics and information-theoretic methods such as Kullback-Leibler divergence analysis is thought to be particularly promising. As discussed herein, this combination makes it possible to identify descriptors that are particularly predictive for given compound classes. Moreover, future research will likely increasingly focus on combining very different types of descriptors through this methodological framework, for example, numerical property descriptors and binary finger-

prints that are typically applied independently of each other. Such descriptor combinations might significantly improve the prediction accuracy of virtual screening methods. We have also shown that the BDACCS approach in combination with Kullback-Leibler divergence analysis makes it possible to predict the performance of fingerprint similarity searching for a given compound class and screening database, for which other methods have thus far not been available. These calculations have significant future application potential in order to prioritize alternative fingerprints for practical virtual screening applications.

CONCLUSION

Bayesian methods are currently among the most popular approaches in chemoinformatics and utilized for a number of applications ranging from compound class label prediction to systematic ligand-target profiling. Together with support vector machines and decision tree methods, Bayesian classifiers presently dominate machine learning techniques used in chemoinformatics. In this contribution, we have discussed in detail four Bayesian approaches that are applied to virtual compound screening including Bayesian classifiers, substructural analysis, binary kernel discrimination, and binary QSAR. In addition, we have focused on methods that combine Bayesian statistics and information theory. These approaches also have significant potential for database screening and, moreover, for combining and prioritizing different types of chemical descriptors. Thus, these combined methods go beyond the typical application domain of Bayesian classifiers and can also be used to estimate the probability of success of practical virtual screening campaigns. Taken together, these methodological examples illustrate the potential of Bayesian approaches for compound classification, feature selection, and database mining.

REFERENCES

Bajorath, J. (2001). Selected concepts and investigations in compound classification, molecular descriptor analysis, and virtual screening. *Journal of Chemical Information and Computer Sciences*, *41*, 233–245. doi:10.1021/ci0001482

Bajorath, J. (2002). Integration of virtual and high-throughput screening. *Nature Reviews. Drug Discovery*, *1*, 882–894. doi:10.1038/nrd941

Bender, A., Mussa, H. Y., Glen, R. C., & Reiling, S. (2004). Molecular similarity searching using atom environments, information-based feature selection, and a naïve Bayesian classifier. *Journal of Chemical Information and Computer Sciences*, *44*, 170–178. doi:10.1021/ci034207y

Blundell, T. L., Jhoti, H., & Abell, C. (2002). High-throughput crystallography for lead discovery in drug design. *Nature Reviews. Drug Discovery*, *1*, 45–54. doi:10.1038/nrd706

Brooijmans, N., & Kuntz, I. D. (2003). Molecular recognition and docking algorithms. *Annual Review of Biophysics and Biomolecular Structure*, *32*, 335–373. doi:10.1146/annurev. biophys.32.110601.142532

Brown, F. K. (1998). Chemoinformatics: What is it and how does it impact drug discovery. *Annual Reports in Medicinal Chemistry*, *33*, 375–384. doi:10.1016/S0065-7743(08)61100-8

Brown, R. D., & Martin, Y. C. (1996). Use of structure-activity data to compare structure-based clustering methods and descriptors for use in compound selection. *Journal of Chemical Information and Computer Sciences*, *36*, 572–584. doi:10.1021/ci9501047

Chen, B., Harrison, R. F., Pasupa, K., Willett, P., Wilton, D. J., & Wood, D. J. (2006). Virtual screening using binary kernel discrimination: Effect of noisy training data and the optimization of performance. *Journal of Chemical Information and Modeling*, *46*, 478–486. doi:10.1021/ci0505426

Cramer, R., Redl, G., & Berkoff, C. (1974). Substructural analysis. A novel approach to the problem of drug design. *Journal of Medicinal Chemistry*, *17*, 533–535. doi:10.1021/jm00251a014

Czarnik, A. W. (1997). Encoding methods for combinatorial chemistry. *Current Opinion in Chemical Biology*, *1*, 60–66. doi:10.1016/S1367-5931(97)80109-3

Domingos, P., & Pazzani, M. (1996). Beyond independence: Conditions for the optimality of the simple Bayesian classifier . In *Machine learning* (pp. 105–112). San Francisco: Morgan Kaufmann.

Duda, R. O., Hart, P. E., & Stork, D. G. (2000). *Pattern classification* (2nd ed.). New York: Wiley-Interscience.

Fox, S., Farr-Jones, S., & Yund, M. A. (1999). High-throughput screening for drug discovery: Continually transitioning into new technologies. *Journal of Biomolecular Screening, 4*, 183–186. doi:10.1177/108705719900400405

Gao, H. (2001). Application of BCUT metrics and genetic algorithm in binary QSAR analysis. *Journal of Chemical Information and Computer Sciences, 41*, 402–407. doi:10.1021/ci000306p

Gillet, V., Willett, P., & Bradshaw, J. (2003). Similarity searching using reduced graphs. *Journal of Chemical Information and Computer Sciences, 43*, 338–345. doi:10.1021/ci025592e

Glick, M., Jenkins, J. L., Nettles, J. H., Hitchings, H., & Davies, J. W. (2006). Enrichment of high-throughput screening data with increasing levels of noise using support vector machines, recursive partitioning, and Laplacian-modified naive Bayesian classifiers. *Journal of Chemical Information and Modeling, 46*, 193–200. doi:10.1021/ci050374h

Godden, J., & Bajorath, J. (2006). A distance function for retrieval of active molecules from complex chemical space representations. *Journal of Chemical Information and Modeling, 46*, 1094–1097. doi:10.1021/ci050510i

Hansch, C., & Fujita, T. (1964). p-σ-πanalysis. A method for the correlation of biological activity and chemical structure. *Journal of the American Chemical Society, 86*, 1616–1626. doi:10.1021/ja01062a035

Hansch, C., Maloney, P. P., Fujita, T., & Muir, R. M. (1962). Correlation of biological activity of phenoxyacetic acids with Hammett substituent constants and partition coefficients. *Nature, 194*, 178–180. doi:10.1038/194178b0

Harper, G., Bradshaw, J., Gittins, J. C., Green, D. V. S., & Leach, A. R. (2001). Prediction of biological activity for high-throughput screening using binary kernel discrimination. *Journal of Chemical Information and Computer Sciences, 41*, 1295–1300. doi:10.1021/ci000397q

Hopfinger, A. J. (1980). A QSAR investigation of dihydrofolate reductase inhibition by Baker triazines based upon molecular shape analysis. *Journal of the American Chemical Society, 102*, 7196–7206. doi:10.1021/ja00544a005

Hopfinger, A. J., Wang, S., Tokarski, J. S., Jin, B., Albuquerque, M., & Madhav, P. J. (1997). Construction of 3D-QSAR models using the 4D-QSAR analysis formalism. *Journal of the American Chemical Society, 119*, 10509–10524. doi:10.1021/ja9718937

Irwin, J., & Shoichet, B. (2005). ZINC - a free database of commercially available compounds for virtual screening. *Journal of Chemical Information and Modeling, 45*, 177–182. doi:10.1021/ci049714+

Jenkins, J. L., Bender, A., & Davies, J. W. (2006). In silico target fishing: Predicting biological targets from chemical structure. *Drug Discovery Today. Technologies, 3*, 413–421. doi:10.1016/j.ddtec.2006.12.008

Johnson, M., & Maggiora, G. (Eds.). (1990). *Concepts and applications of molecular similarity.* New York: John Wiley & Sons.

Jorissen, R. N., & Gilson, M. K. (2005). Virtual screening of molecular databases using a support vector machine. *Journal of Chemical Information and Modeling, 45*, 549–561. doi:10.1021/ci049641u

Kohonen, T. (1989). *Self-organization and associative memory.* Berlin: Springer.

Kullback, S. (1997). *Information theory and statistics.* Mineola, MN: Dover Publications.

Kuntz, I. D. (1992). Structure-based strategies for drug design and discovery. *Science, 257,* 1078–1082. doi:10.1126/science.257.5073.1078

Kuntz, I. D., Blaney, J. M., Oatley, S. J., Langridge, R., & Ferrin, T. E. (1982). A geometric approach to macromolecule-ligand interactions. *Journal of Molecular Biology, 161,* 269–288. doi:10.1016/0022-2836(82)90153-X

Labute, P. (1999). A new method for the determination of quantitative structure activity relationships . In *Pacific symposium on biocomputing (Vol. 4,* pp. 444–455). Binary QSAR.

Lewis, D. D. (1998). Naive (Bayes) at forty: The independence assumption in information retrieval. In *Lecture notes in computer science* [Berlin: Springer.]. *Machine Learning, ECML-98,* 4–15.

Livingstone, D. J. (2000). The characterization of chemical structures using molecular properties. a survey. *Journal of Chemical Information and Computer Sciences, 40,* 195–209. doi:10.1021/ci990162i

Livingstone, D. J., Manallack, D. T., & Tetko, I. V. (1997). Data modelling with neural networks: advantages and limitations. *Journal of Computer-Aided Molecular Design, 11,* 135–142. doi:10.1023/A:1008074223811

Martin, Y. C. (1992). 3D database searching in drug design. *Journal of Medicinal Chemistry, 35,* 2145–2154. doi:10.1021/jm00090a001

Martin, Y. C. (2001). Diverse viewpoints on computational aspects of molecular diversity. *Journal of Combinatorial Chemistry, 3,* 231–250. doi:10.1021/cc000073e

Martin, Y. C., Kofron, J. L., & Traphagen, L. M. (2002). Do structurally similar molecules have similar biological activity? *Journal of Medicinal Chemistry, 45,* 4350–4358. doi:10.1021/jm020155c

Mason, J., Morize, I., Menard, P., Cheney, D., Hulme, C., & Labaudiniere, R. (1999). New 4-point pharmacophore method for molecular similarity and diversity applications: Overview of the method and applications, including a novel approach to the design of combinatorial libraries containing privileged substructures. *Journal of Medicinal Chemistry, 42,* 3251–3264. doi:10.1021/jm9806998

Molecular operating environment (MOE 2007.09). (2007). Montreal, Canada: Chemical Computing Group.

Nidhi, Glick, M., Davies, J. W., & Jenkins, J. L. (2006). Prediction of biological targets for compounds using multiple-category Bayesian models trained on chemogenomics databases. *Journal of Chemical Information and Modeling, 46,* 1124–1133. doi:10.1021/ci060003g

Nigsch, F., Bender, A., Jenkins, J. L., & Mitchell, J. B. O. (2008). Ligand-target prediction using winnow and naive Bayesian algorithms and the implications of overall performance statistics. *Journal of Chemical Information and Modeling, 48,* 2313–2325. doi:10.1021/ci800079x

Ormerod, A., Willett, P., & Bawden, D. (1989). Comparison of fragment weighting schemes for substructural analysis. *Quantitative Structure-Activity Relationships, 8,* 115–129. doi:10.1002/qsar.19890080207

Paolini, G. V., Shapland, R. H. B., van Hoorn, W. P., & Mason, J. S. (2006). Global mapping of pharmocological space. *Nature Biotechnology, 24,* 805–815. doi:10.1038/nbt1228

Pearlman, R. S., & Smith, K. (1998). Novel software tools for chemical diversity. *Perspectives in Drug Discovery and Design, 9,* 339–353. doi:10.1023/A:1027232610247

Robertson, S. E., & Jones, K. S. (1976). Relevance weighting of search terms. *Journal of the American Society for Information Science American Society for Information Science*, *27*, 129–146. doi:10.1002/asi.4630270302

Schölkopf, B., & Smola, A. (2002). *Learning with kernels*. Cambridge, MA: MIT Press.

Shannon, C. E., & Weaver, W. (1963). *The mathematical theory of communication*. Urbana and Chicago, IL: University of Illinois Press.

Sheridan, R. P., & Kearsley, S. K. (2002). Why do we need so many chemical similarity search methods? *Drug Discovery Today*, *7*, 903–911. doi:10.1016/S1359-6446(02)02411-X

Sheridan, R. P., Miller, M. D., Underwood, D. J., & Kearsley, S. K. (1996). Chemical similarity using geometric atom pair descriptors. *Journal of Chemical Information and Computer Sciences*, *36*, 128–136. doi:10.1021/ci950275b

Stahura, F. L., & Bajorath, J. (2005). New methodologies for ligand-based virtual screening. *Current Pharmaceutical Design*, *11*, 1189–1202. doi:10.2174/1381612053507549

Vogt, M., & Bajorath, J. (2007). Introduction of an information-theoretic method to predict recovery rates of active compounds for Bayesian in silico screening: Theory and screening trials. *Journal of Chemical Information and Modeling*, *47*, 337–341. doi:10.1021/ci600418u

Vogt, M., & Bajorath, J. (2008a). Bayesian screening for active compounds in high-dimensional chemical spaces combining property descriptors and molecular fingerprints. *Chemical Biology & Drug Design*, *71*, 8–14.

Vogt, M., & Bajorath, J. (2008b). Bayesian similarity searching in high-dimensional descriptor spaces combined with Kullback-Leibler descriptor divergence analysis. *Journal of Chemical Information and Modeling*, *48*, 247–255. doi:10.1021/ci700333t

Vogt, M., Godden, J., & Bajorath, J. (2007). Bayesian interpretation of a distance function for navigating high-dimensional descriptor spaces. *Journal of Chemical Information and Modeling*, *47*, 39–46. doi:10.1021/ci600280b

Walters, W. P., Stahl, M. T., & Murcko, M. A. (1998). Virtual screening - an overview. *Drug Discovery Today*, *3*, 160–178. doi:10.1016/S1359-6446(97)01163-X

Warmuth, M. K., Liao, J., Rätsch, G., Mathieson, M., Putta, S., & Lemmen, C. (2003). Active learning with support vector machines in the drug discovery process. *Journal of Chemical Information and Computer Sciences*, *43*, 667–673. doi:10.1021/ci025620t

Willett, P. (1987). *Similarity and clustering in chemical information systems*. Letchworth: Research Studies Press.

Willett, P. (1998). Chemical similarity searching. *Journal of Chemical Information and Computer Sciences*, *38*, 983–996. doi:10.1021/ci9800211

Willett, P. (2005). Searching techniques for databases of two- and three-dimensional chemical structures. *Journal of Medicinal Chemistry*, *48*, 4183–4199. doi:10.1021/jm0582165

Willett, P. (2006). Similarity-based virtual screening using 2D fingerprints. *Drug Discovery Today*, *11*, 1046–1053. doi:10.1016/j.drudis.2006.10.005

Wilton, D., Willett, P., Lawson, K., & Mullier, G. (2003). Comparison of ranking methods for virtual screening in lead-discovery programs. *Journal of Chemical Information and Computer Sciences*, *43*, 469–474. doi:10.1021/ci025586i

Wilton, D. J., Harrison, R. F., Willett, P., Delaney, J., Lawson, K., & Mullier, G. (2006). Virtual screening using binary kernel discrimination: Analysis of pesticide data. *Journal of Chemical Information and Modeling, 46*, 471–477. doi:10.1021/ci050397w

Xue, L., & Bajorath, J. (2000). Molecular descriptors for effective classification of biologically active compounds based on principal component analysis identified by a genetic algorithm. *Journal of Chemical Information and Computer Sciences, 40*, 801–809. doi:10.1021/ci000322m

Zhang, H. (2004). The optimality of naive Bayes. In *Proceedings of the seventeenth Florida artificial intelligence research society conference* (pp. 562–567). Menlo Park, CA: The AAAI Press.

Zhang, H., & Su, J. (2004). Naive Bayesian classifiers for ranking. In *Lecture notes in computer science* [Berlin: Springer.]. *Machine Learning, ECML-04*, 501–512.

ADDITIONAL READING

Aitchison, J., & Aitken, C. G. G. (1976). Multivariate binary discrimination by the kernel method. *Biometrika, 63*, 413–420. doi:10.1093/biomet/63.3.413

Bajorath, J. (2006). Chemoinformatics methods for systematic comparison of molecules from natural and synthetic sources and design of hybrid libraries. *Journal of Computer-Aided Molecular Design, 16*, 431–439. doi:10.1023/A:1020868022748

Bender, A., Mussa, H. Y., & Glen, R. C. (2005). Screening for dihydrofolate reductase inhibitors using MOLPRINT 2D, a fast fragment-based method employing the naïve Bayesian classifier: limitations of the descriptor and the importance of balanced chemistry in training and test sets. *Journal of Biomolecular Screening, 10*, 658–666. doi:10.1177/1087057105281048

Cover, T. M., & Thomas, J. A. (2006). *Elements of information theory* (2nd ed.). New York: Wiley-Interscience.

Gao, H., Williams, C., Labute, P., & Bajorath, J. (1999). Binary quantitative structure-activity relationship (QSAR) analysis of estrogen receptor ligands. *Journal of Chemical Information and Computer Sciences, 39*, 164–168. doi:10.1021/ci980140g

Glick, M., Klon, A. E., Acklin, P., & Davies, J. W. (2004). Enrichment of extremely noisy high-throughput screening data using a naïve Bayes classifier. *Journal of Biomolecular Screening, 9*, 32–36. doi:10.1177/1087057103260590

Harper, G., & Pickett, S. D. (2006). Methods for mining HTS data. *Drug Discovery Today, 11*, 694–699. doi:10.1016/j.drudis.2006.06.006

Liu, Y. (2004). A comparative study on feature selection methods for drug discovery. *Journal of Chemical Information and Computer Sciences, 44*, 1823–1828. doi:10.1021/ci049875d

Ripley, B. D. (1996). *Pattern recognition and neural networks*. Cambridge, UK: Cambridge University Press.

Stahura, F. L., & Bajorath, J. (2004). Virtual screening methods that complement HTS. *Combinatorial Chemistry & High Throughput Screening, 7*, 259–269.

Sun, H. (2006). An accurate and interpretable Bayesian classification model for prediction of hERG liability. *ChemMedChem, 1*, 315–322. doi:10.1002/cmdc.200500047

Vogt, M., & Bajorath, J. (2007). Introduction of a generally applicable method to estimate retrieval of active molecules for similarity searching using fingerprints. *ChemMedChem, 2*, 1311–1320. doi:10.1002/cmdc.200700090

Vogt, M., & Bajorath, J. (in press). Data mining approaches for compound selection and iterative screening . In Balakin, K. V. (Ed.), *Pharmaceutical data mining: Approaches and applications for drug discovery* (pp. 115–143). Hoboken, NJ: John Wiley & Sons.

Xia, X., Maliski, E., Gallant, P., & Rogers, D. (2004). Classification of kinase inhibitors using a Bayesian model. *Journal of Medicinal Chemistry, 47*, 4463–4470. doi:10.1021/jm0303195

Xue, L., & J., B. (2000). Molecular descriptors in chemoinformatics, computational combinatorial chemistry, and virtual screening. *Combinatorial Chemistry & High Throughput Screening, 3*, 363–372.

Chapter 11
Learning Binding Affinity from Augmented High Throughput Screening Data

Nicos Angelopoulos
Edinburgh University, UK

Andreas Hadjiprocopis
Higher Technical Institute, Cyprus

Malcolm D. Walkinshaw
Edinburgh University, UK

ABSTRACT

In high throughput screening a large number of molecules are tested against a single target protein to determine binding affinity of each molecule to the target. The objective of such tests within the pharmaceutical industry is to identify potential drug-like lead molecules. Current technology allows for thousands of molecules to be tested inexpensively. The analysis of linking such biological data with molecular properties is thus becoming a major goal in both academic and pharmaceutical research. This chapter details how screening data can be augmented with high-dimensional descriptor data and how machine learning techniques can be utilised to build predictive models. The pyruvate kinase protein is used as a model target throughout the chapter. Binding affinity data from a public repository provide binding information on a large set of screened molecules. The authors consider three machine learning paradigms: Bayesian model averaging, Neural Networks, and Support Vector Machines. The authors apply algorithms from the three paradigms to three subsets of the data and comment on the relative merits of each. They also used the learnt models to classify the molecules in a large in-house molecular database that holds commercially available chemical structures from a large number of suppliers. They discuss the degree of agreement in compounds selected and ranked for three algorithms. Details of the technical challenges in such large scale classification and the ability of each paradigm to cope with these are put forward. The application of machine learning techniques to binding data augmented by high-dimensional can provide a powerful tool in compound testing. The emphasis of this work is on making very few assumptions or technical choices with regard to the machine learning techniques. This is to facilitate application of such techniques by non-experts.

DOI: 10.4018/978-1-61520-911-8.ch011

INTRODUCTION

High throughput screening (HTS) is now a standard approach used in the pharmaceutical industry to identify potential drug-like lead molecules. Typically thousands, sometimes running up to a million, small molecule compounds are tested in a *binding* assay and affinity data is generated for each compound. More recently in a major initiative led by NCBI (the National Centre for Biotechnology Information, http://www.ncbi.nlm. nih.gov/) information on the results of more than 800 bioassays has been collated and disseminated. The analysis linking biological data with molecular properties is a major goal in both academic and pharmaceutical research.

One such assay is the protein pyruvate kinase (PYK), (Inglese et al., 2006). It is a potential anti-cancer and antiparasitic drug target. PYK acts as a tetramer and catalyses the last step in the breakdown of sugar to form pyruvate. This pathway is required for survival of the trypanasomatid parasites that cause diseases including sleeping sickness and leishmaniasis (Nowicki et al., 2008). Human PYK is also implicated in cancer pathogenesis and is crucial for the altered metabolism observed in tumour cells (Christofk et al., 2008). Inhibitors specific to the various pyruvate kinases are therefore of significant medical interest (Chan, Tan, & Sim, 2007).

We describe an approach for analysing this activity data in terms of *molecular descriptor* properties. This approach is based on *Bayesian model averaging* techniques. The results of the Bayesian approach are compared to that of two well-established machine learning techniques, namely Neural Networks and Support Vector Machines. A number of specific algorithms from the broad area of each technique are evaluated on one partition of the data. The best performing algorithm from each paradigm is selected and further, comprehensive comparison is carried out.

In order to realise our main objective of ligand discovery we retrained the best performing algorithms on all available known binder data and screened the 3.7 million unique compounds contained in the *Eduliss molecular database* for potentially active compounds.

BACKGROUND

Recent estimates suggest that there are about 3,000 possible druggable proteins (Zheng et al., 2006). These have structural features that allow the binding of small molecules. When bound, these adjust the biological function of their target. The object of ligand discovery is to identify molecules that will have a desired effect to the function of the protein.

Descriptor based approaches to ligand discovery have a long history in the field of QSAR (Hansch, Hoekman, & Gao, 1996). In early approaches simple models built by experts were proposed as predictors of *chemical activity*. The well known Lipinski *Rule of Fives* (Lipinski, Lombardo, Dominy, & Feeney, 2001) describes four properties (descriptors) common to the most successful *drug molecules*. The model depends on just four descriptors: solubility, molecular weight and the number of hydrogen bond donor and acceptor atoms in the molecule. There are however hundreds of characterising descriptors for every molecule (Todeschini, Consonni, Mannhold, Kubinyi, & Timmerman, 2000) covering calculated physical properties like polarizability or shape properties describing for example the presence of ring structures or particular chemical groups in the molecule. More modern approaches use statistical methods to build models based on large numbers of descriptors. In most cases preprocessing is used to do feature selection, that is to reduce the original number of descriptors to a subset that can be used to provide predictions which are approximately as good as those achieved by using the whole set of descriptors.

Descriptors such as the weighted holistic invariant molecular (WHIM) descriptors (Todeschini, Lasagni, & Marengo, 1994) have been used in compound modelling for the *prediction* of molecular properties (Tonga, Liu, Zhou, Wu, & Li, 2008, Tabarakia, Khayamiana, & Ensafia, 2006, Todeschini, Gramatica, & Navas, 1998). In this work we use a very extensive set of 1600 descriptors calculated for each molecule with the program DRAGON 5.4 (http://www.talete.mi.it/). The algorithms we employ build models which correlate the extensive set of descriptors to biological activity. By identifying the key properties common to those molecules showing activity in the bioassay we have a potentially useful method for predicting and identifying new active compounds.

From a machine learning perspective the problem is formulated as a classical decision problem albeit in a space of extremely high dimensions. In this chapter we evaluate two standard techniques namely Support Vector Machines (SVM) and Neural Networks (NN) and one less common technique; Bayesian learning of classification trees. NN is a massively connected network of simple processing units interacting locally by adjusting the strength of its internal connections in order to learn the mapping from input to output. SVM is a technique based on statistical learning theory aiming at projecting the input data into a high-dimensional feature space where it is quite likely that it will become linearly separable. The hyperplane offering maximal separation between the data points of the two classes is selected as a decision boundary.

The Bayesian model averaging techniques we examine here build *classification trees* (Breiman, Friedman, Olshen, & Stone, 1984, Ripley, 1996) using an *MCMC* approach. These trees are also known as decision trees and utilise the descriptors by creating split rules on their values providing predictive models of activity. Large chains of trees are constructed which approximate the posterior probability of models. Averaging allows for consensus decisions to be taken across a whole chain.

DATASETS

Biochemical Data

The data used in this analysis was taken from results published by the NIH Chemical Genomics Center (http://ncgc.nih.gov/db/). In this assay purified pyruvate kinase from the bacterium *Bacillus stearothermophilus* was used to generate ATP from ADP using its substrate phospoenolpyruvate. The generation of ATP was measured by the emission of light when ATP reacted with the enzyme luciferase. Over 50,000 small molecule compounds were tested for their inhibitory (and stimulatory) effects and an IC_{50} value for active compounds was measured. The IC_{50} value is the concentration of compound required to reduce the measured PYK activity by 50%. The curated database contains information classifying the compounds as active, inconclusive or inactive. There are 812 compounds labelled as inconclusive which were removed from the set for the purposes of this study.

The Eduliss database is a web accessible collection of 3.7 million unique molecules, mainly commercially available, small molecule compounds (http://eduliss.bch.ed.ac.uk/, September 2008). For each compound the 3D structure is stored as an SDF file along with the calculated descriptor values.

We used the Dragon software to produce chemical descriptors for the molecules both in the assay experiments and the database. There are 1665 descriptors that can be calculated for each molecule. Each descriptor belongs to one of the 20 groups shown in Table 1. However, the descriptor producing software is sometimes unable to calculate all the descriptors for an input molecule. For the assay molecules 1571 of those descriptors were reliably produced. In addition, from the 3,776,760 unique compounds in Eduliss, only 1,537,089 have a complete set of the 1571 descriptors used here. It is thus important that classification can cope with missing values.

Table 1. Descriptor groups produced by the Dragon software

ID	Descriptor name	Population
1	constitutional descriptors	48
2	topological descriptors	119
3	walk and path counts	47
4	connectivity indices	33
5	information indices	47
6	2D autocorrelations	96
7	edge adjacency indices	107
8	BCUT descriptors	64
9	topological charge indices	21
10	eigenvalue based indices	44
11	Randic molecular profiles	41
12	geometrical descriptors	74
13	RDF descriptors	150
14	3D MoRSE descriptors	160
15	WHIM descriptors	99
16	GETAWAY descriptors	197
17	functional group counts	154
18	atom centred fragments	120
19	charge descriptors	14
20	molecular properties	30

Figure 1 charts the missing values in the unique compounds dataset. The area of very high missing percentages around the $x = 1500$ region correspond to the 13 descriptors in the "charge descriptors" group. The area of medium missing values ($y = 623791$) around $x = 900-1000$ corresponds to the whole set of descriptors in the 14th group ("3D MoRSE descriptors").

ALGORITHMS

Bayesian Learning

Bayesian model averaging (BMA) is a *statistical machine learning* technique where the full posterior distribution over possible models is approximated. That is, the result of learning is a large number of models each with an associated probability. In this respect BMA is unlike SVM and NN which seek or built a single model. BMA exploits Markov Chain Monte Carlo (MCMC) algorithms to stochastically approximate the full Bayesian posterior. Another important aspect of BMA is that it facilitates the incorporation of *prior knowledge*.

Figure 1. Number of unique molecules in Eduliss missing a specific descriptor (y) against descriptors (x)

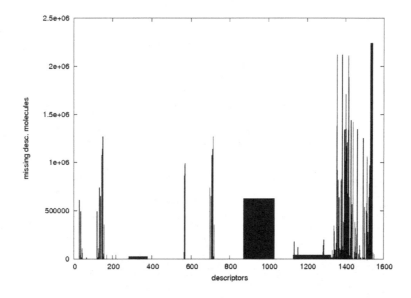

BMA for learning classification and regression trees was introduced by (Chipman, George, & McCulloch, 1998) and applied to the analysis of breast cancer data. A more detailed exposition can be found in (Denison, Holmes, Mallick, & Smith, 2002). These employ standard MCMC methods which depend on the definition of both a prior and a set of functions for stochastically moving between models. In this paper we follow (Angelopoulos & Cussens, 2005b) as their approach only depends on the definition of a prior. Proposal steps are performed by the software as the model space is perceived as a partially constructed search structure. We develop their programs to cope with the high-dimensionality of the data and employ multiple chains to improve performance.

Classification and regression trees use a number of decisions on features, here descriptors, to classify each datum to one of a number of possible classes. An example is shown in Figure 2. In this paper we limit discussion to classification trees as we are interested in classifying into two discrete alternatives: active and inactive. Regression trees allow classification over ranges of values.

Within each tree, nodes are numbered sequentially. Internal nodes are labelled by a descriptor and an associated split value. Molecules whose value for the label descriptor is less or equal to the split value follow the left branch, where the rest follow the right branch. Leaf nodes classify a molecule as active with probability proportional to the number of training set actives in that leaf (see Figure 2 for an example). Tree T and its parameters Θ when applied to molecules M define a distribution over classes C, $p(C|T,\Theta,M)$. Given that we are interested in two classes: active (A) and inactive (I), then for the tree T with b number of leaves, Θ is the matrix of parameters with b rows each of the form $[p_{i,A}, p_{i,I}]$ where $1 \leq i \leq b$. So Θ defines a distribution over the possible classes at every leaf whereas, each tree in T assigns each molecule to a leaf according to its descriptor values.

Figure 2. Tree at iteration 1000 of a Metropolis-Hastings (MH) run. Oval nodes are decisions on a single descriptor and the edges downwards from the node indicate a decision. The left edge is followed by molecules which have a value of less or equal to the split value for this descriptor, whereas all other molecules follow the right edge. Leaf nodes present a distribution over classes

As the number of possible trees grow exponentially on the number of descriptors and their possible values, a variety of approximate algorithms have been proposed to learn such trees from data (for example see (Quinlan, 1986, Buntine, 1992)). *Metropolis-Hastings* (MH) is a stochastic algorithm that has been used extensively in the literature to sample from the posterior probability distributions that cannot be fully determined. MH is an MCMC algorithm that uses stochastic moves to construct a Markov chain. Under weak conditions the chain approximates the desired posterior. Here we are interested in the probability of a tree in explaining our data: $P(T|M)$, where M is a set of descriptors for a set of molecules. MH constructs a chain of visited trees by stochastically proposing at each iteration a new tree that is either accepted or rejected. The move at iteration i is with accordance to proposal distribution $q(T_i, T_{i'})$ and the proposed tree $T_{i'}$ is accepted with probability $\alpha(T_i, T_{i'}) = min(1, R)$ where

$$R = \frac{p(T_{i'})P(C \mid T_{i'}, M)q(T_i, T_{i'})}{p(T_i)P(C \mid T_i, M)q(T_{i'}, T_i)} \qquad (1)$$

The quotient in (1) has three main components. The first is based on the prior probability of the trees (i.e. $p(T)$). The second is based on the goodness of fit for a tree in explaining the data (i.e. <<:P(C|T,M):>>) and is known as the marginal likelihood (Chipman et al., 1998). The third term is the quotient of transition probabilities.

Under the usual assumptions of a likelihood independent of M and independence of θ_i within Θ with each θ_i Dirichlet $(\alpha_{i1}, \alpha_{i2})$ and letting n_{ic} be the number of training molecules of class c in a leaf i (with $n_i = \sum_i n_{ik}$), the standard formulation for the marginal is

$$P(C \mid T, M) = \left(\frac{\Gamma(\sum_k \alpha_k)}{\prod_k \Gamma(\alpha_k)} \right) \prod_{i=1}^{b} \frac{\prod_k \Gamma(n_{ik} + \alpha_k)}{\Gamma(n_i + \sum_k \alpha_k)} \qquad (2)$$

The α_{ic} represent prior information on relative class frequencies. In this paper all α_{ic} are set to 1 except where explicitly stated.

Similar to (Chipman et al., 1998) we use a program which implicitly generates trees according to a prior that depends on the following splitting rule $\Psi_{d\eta} = \zeta(1 + d_\eta)^{-\xi}$ where d_η is the depth of node η and ζ and ξ are user defined parameters controlling the size of the trees. Thereafter we assume $\zeta = .95$ and $\xi = 1$. This prior penalises deep trees, which tend to over fit. Our approach follows the theory of (Angelopoulos & Cussens, 2005b). The prior and the proposal are tightly coupled and the proposal ratio becomes easy to compute. This is achieved by encoding the prior as a *Distributional Logic Program* (Angelopoulos & Cussens, 2005a) and letting the main component of the proposal be a backward jump to one of the probabilistic choices of the branch that led to T_i. From that choice onwards $T_{i'}$ is re-sampled according to the prior.

In effect, a node in T_i is selected uniformly and the node along with any branches from that node are regrown. We briefly explain how this is achieved within our framework. Following the logic programming convention of variables starting with a capital letter and term functors and predicate names with a lower case, the main part of the probabilistic program used is as follows:

```
<<:(C_0):>>              <<:cart(M,
Cart):-:>>

                              <<:\
psi_0 \mbox{ is } \zeta:>>,
                    <<:\psi_{0}:>>
:       <<:split(0,M,Cart):>>.
<<:(C_1):>>             <<:\
psi_D:>>:       <<:split(D,
M_B, c(F,Val,L,R)):-:>>
```

```
                        <<:\
psi_{D+1} \mbox{ is } \zeta *
(1+D)^{-\xi}:>>,
                        <<:D_1
\mbox{ is } D + 1:>>,
                        <<:r\_
select(F,Val,M_B,L_B,R_B):>>,
                <<:\psi_
{D+1}:>>:         <<:split(D_1,
L_B, L):>>,
                <<:\psi_
{D+1}:>>:         <<:split(D_1,
R_B, R):>>.
<<:(C_2):>>          <<:1- \
psi_D:>>:         <<:split(D,
M_B, 1(M_B)):>>
```

Clause C_0 is non-stochastic and serves as a convenient entry point. It is called once at each MCMC run and it declares that *Cart* is a valid representation of a C&RT derived for set of training molecules M and sampled according to the prior specified above. At depth D each node is considered in turn. Clauses C_1 and C_2 correspond to the two possibilities. The node will either become an internal node for T_i by use of C_1 or a leaf node by application of C_2.

Clause C_1 declares that a subtree is initiated at node B by randomly selecting ($r\_select/5$)) a feature F and an associated splitting value *Val* and using those to partition M_B to two distinct parts L_B and R_B. It then increases the depth by one and recursively calls itself on the left and right partitions of M_B. Clause C_2 constructs a leaf node at level D and populates with all input the training data. Each time a split is consider for a node C_1 is selected with probability Ψ_D and C_2 with the complementary probability $1-\Psi_D$. For instance at $D=0$ $\Psi_D=\zeta=0.95$.

Sampling in this fashion produces a tree according to the described prior. Each *split* / 3 call leaves a probabilistic choice point. Our approach exploits the one-to-one correspondence of probabilistic choices to nodes in the tree to effect a uniform backtracking strategy. The probabilistic part of this program is similar to that presented in (Angelopoulos & Cussens, 2005b). The one point in which our prior differs is in selecting a feature and associated value. Due to the high-dimensionality of our data the approach taken here is to construct the set of possible split values on-the-fly and after a split descriptor is selected rather than using cached values. The Monte Carlo part of MH allows approximation of the posterior from the constructed chain. The posterior probability of tree T is equal to the number of times it appears in the chain ($\#T$) over the total number of iterations I, $\left(P(T \mid M,C) = \dfrac{\#T}{I}\right)$. More importantly the framework assigns probability to features of the trees. By model averaging principles the class of a datum can be computed over the whole chain and over a number of independent chains.

Support Vector Machines

A Support Vector Machine (SVM) (Cortes & Vapnik, 1995, Fradkin & Muchnik, 2006) is a supervised learning method for separating data into two classes using hyperplanes. Firstly, the input data is mapped into a high-dimensional feature space where it hopefully becomes linearly separable. This mapping is achieved via the use of a kernel function such as one of following four families of functions: linear, polynomial, radial basis or sigmoid. Note however, that a plethora of other kernels exist, such as graph and string kernels.

Once linearly separable, the two classes of the input data can be separated in this feature space by a hyperplane. In fact, there is an infinite number of such hyperplanes and SVM chooses the hyperplane which maximises the margin between the two classes on the assumption that the larger the margin the lower the error of the classifier when dealing with unknown data. If the mapping to feature space fails to transform the data to linearly separable, SVM selects the hyperplane which minimises the degree of misclassification and maximises the margin between the properly classified points.

The main advantage of SVM is that because it is formulated as a quadratic programming problem, there are no local minima (a serious disadvantage for Neural Networks); the algorithm is guaranteed to reach the globally optimum solution. On the other hand, SVM, again as a problem of quadratic programming, is associated with a training complexity (convergence time and memory) which is highly dependent on the number of training data vectors. This is not the case with feed-forward neural networks whose performance depends mainly on the dimensionality of the data. Additionally, SVM's performance depends on model selection; that is choice of kernel function and its parameters and the margin parameter.

A tutorial about SVM is provided by (Burges, 1998). (Meyer, Leisch, & Hornik, 2003) compare the performance of SVM with a large number of other classification methods. The results of our tests were obtained using an implementation of SVM from (Joachims, 1998) with default settings.

Feed-Forward Neural Networks

Feed-forward neural networks (FFNN) are computational techniques inspired by the physiology of the brain, epitomising the connectionist view of distributing knowledge and decision-making among a network of massively connected, simple processing units interacting at a local level. The mathematical foundation of FFNN lies on Kolmogorov's general representation theorem which states that any continuous multivariate function can be represented by superpositions and compositions (a network) of univariate functions (of neurons). Furthermore, FFNN are universal function approximators; a FFNN with a single hidden layer employing as many non-linear units as required, can approximate any continuous function arbitrarily well (Cybenko, 1989).

FFNN training (Bishop, 1995) is exercised through a supervised process by which the network is presented with a sequence of inputs and respective desired outputs. "Learning" occurs by adjusting the free parameters of the network (the weights) in a way that the discrepancy between the actual and desired responses (the error) is minimised. In general, the adjustment of the weights follows the steepest-descent direction of the error-weights surface. Other, non gradient-descent training methods do exist – for example based on genetic algorithms or borrowed from thermodynamics, e.g. simulated annealing.

The ability of a trained FFNN to classify or map correctly a previously unseen input vector is called Generalisation. Largely, generalisation depends on how accurately the training data captures the underlying data-generation model and the architecture of the FFNN. The latter being a major problem since there is no systematic method for selecting the most appropriate network model for the data at hand other than rules of thumb. Additionally, a FFNN may demand excessive resources when dealing with high-dimensional data under the strain of adjusting a huge number of parameters (weights) due to combinatorial explosion.

FFNN entities (FFNNE) (Hadjiprocopis & Smith, 1997) have been proposed as a means of dealing with high-dimensional data within a neural network framework but without the side effects usually associated with monolithic FFNN. A FFNNE consists of a large number of FFNN forming a highly inter-connected network arranged in layers. Each input-layer FFNN deals with a relatively small subset of the parameters of the input data, sending signals to FFNN of the layers ahead.

EXPERIMENTS

Our overall objective is to extract descriptor based models from the datasets described in the previous section. We approach this in three stages. Firstly we select a subset of the inactives and add them to the active compounds. We run learning experiments on a 90% portion of the derived dataset holding 10% as a test set. This stage allows a general overview of how well a variety of algorithms from the three paradigms perform on data of known class. Then, the second stage validates three of the algorithms on a ten-fold cross validation for the derived dataset and on two additional datasets which are constructed from alternative selections for the inactives. In the third stage we use 100% of the first dataset to retrain the main algorithm and use the output to classify 3.7 million molecules of unknown class.

The 50,000 molecules of the experimentally determined affinities have a vast bias towards non-binding compounds. Precisely 582 molecules for which we can calculate a reasonably full set of descriptors have been classified to be active (positive). To avoid dominance of the inactive (negative) examples and to reduce the computational power needed we randomly selected 582 inactives for the original dataset and added them to the active molecules. We will refer to this collection of data as dataset *A*. We then choose 524 molecules from each active and inactive sets to form the training set with the remaining 58 molecules from each class as the test set.

Single Fold Validation

We first ran 1 million long iterations of the MH algorithm. We will refer to these as MH runs. The algorithm did well to explore branches from depth 2 or more. However, it rejected the vast majority of proposed trees that radically changed the current tree, i.e. when backtracking to depth 0 or 1. The trees in Figure 2 and Figure 3 are from the same run and display the chain tree at iteration 1,000 and 10,000 respectively. It can been seen that the 3 leftmost nodes near the root of the tree are the same in both of the trees. A major contributing factor to this behaviour is the multi-dimensionality of the data. The longer the chain stays at the same root the harder it becomes for nodes within the tree to be improved upon by a single move.

Table 2 shows results for the experiments we have ran on the first fold selection of dataset *A*. Each row provides the mean and standard deviation over 10 runs. Each run starts from different initial conditions by selecting a different seed to the random number generator. Two sets of values are shown: one for the correctly identified active

Figure 3. Tree at iteration 10,000 of same MH run as that of the tree in Figure 2. The two trees share a common branch which starts at the root and continues along the leftmost branches down to depth 3

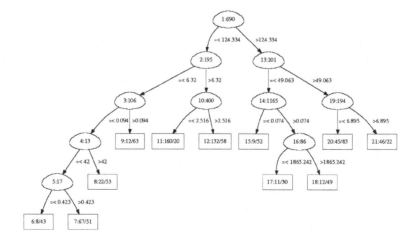

molecules and one for the inactive molecules. Note that our test set contains 58 active and 58 inactive examples. The first row in Table 2, MH, gives the values for the 1 million iterations of the MH algorithm which we have already described. Note that the chain seems to explore local maxima reasonably well. On average out of the 58 active test molecules it correctly classifies 44.7, while it does slightly worst in classifying inactives.

In a second approach (MH-Aw2) we increased the relative importance of actives to be twice that of inactives. The results are similar to the standard runs and are shown in the MH-Aw2 row of Table 2. The lack of significant increase in true actives classification possibly points to the marginal effect the likelihood has in improving the search beyond that threshold. Furthermore, the slight increase noted, comes at the price of decreased accuracy in predicting inactives.

We also ran tempering chains (MH-T3) on dataset A. In tempering a number of hot chains are run in addition to the main chain. The main motivation is that the hot chains move in smoother likelihood spaces and thus might be less likely to get stuck in local maxima. Following standard practise we use temperatures in the region of $0<\tau<1$ that are used as exponents to the likelihood function. Thus the cold chain corresponds to tem-

perature $\tau=1$. In the tempering experiments we use $T=(0833,0.714,0625)$. The MH-T3 row in Table 2 shows that tempering helps to slightly improve classification of inactive test data. Inspection of the chains, although showing improvement of mixing at the root, indicates that chains still get stuck to a single root after a few thousand iterations.

To deal with the high-dimensionality of the feature (here descriptors) space in the Bayesian approach we propose restarts of the MH algorithm. We ran 100 chains at 100,000 iterations each and combine their outputs to a single chain. We refer to these as runs of the Mixed algorithm. Bayesian model averaging provides a robust theoretical framework for this approach. In effect the original MH algorithm is interleaved with an independent sampler.

The results from running the Mixed algorithm are shown in Table 2 on the row labelled Mixed. There is a clear improvement in the quality of the results achieved and the robustness of the system. Similarly to the standard MH we also ran experiments with varying: actives weight equal to 2 and three hot chains, Mixed-Aw2 and Mixed-T3 respectively. By using a weight of 2 for actives we see marginal improvements in the results. In the case of tempering we also see some but no substantial improvement, which we attribute to the fact that the independent sampler complements the MH well, so there is no extra benefit in using tempering for the Mixed algorithm.

We ran two sets of NN-based experiments: FFNN and FFNNE (entities). For the former a network with 1513 inputs was employed. The network consists of 4 hidden layers with 1001, 699, 351 and 51 units each respectively and it is fully connected. Training was stopped after 1000 iterations. The FFNNE consists of 499 FFNN, each configured to have a number of inputs between 350 and 850 and three hidden layers, yielding a total of about 100 million neurons.

Finally we ran SVM experiments with the *SVMlight* library using the linear kernel function. We have also experimented with alternative

Table 2. Classification results for experiments ran on a train-and-test partition of dataset A

	true +ve		true -ve	
	mean	std err	mean	std err
MH	44.7	1.65	42.33	1.18
MH-Aw2	45.7	0.72	39.33	1.90
MH-T3	45.3	1.44	43.67	1.52
FFNN	46.8	0.98	42.6	1.09
FFNNE	48	1	42.66	1.53
SVM	49	-	43	-
Mixed	49.3	0.54	45	0.47
Mixed-Aw2	49	0.471	45.33	0.272
Mixed-T3	49.33	0.728	46.33	0.272

kernels, using default parameters, but the linear version as reported here produced the best results consistently. This is in agreement with existing views in the literature (Ben-Hur, Ong, Sonnenburg, Scholkopf, & Ratsch, 2008) that the linear kernel is the more appropriate one for high dimensional spaces.

In Table 3 we present the sensitivity and specificity (Altman & Bland, 1994) results for all the experiments described so far. Letting T^+, T^-, F^+ and F^- be the numbers of true positive, true negative, false positive and false negative, then:

$$sensitivity = \frac{T^+}{T^+ + F^-} \qquad specificity = \frac{T^-}{T^- + F^+}$$

$$(3)$$

Table 3 also incorporates average runtimes for each algorithm. Note that there is little variation in all cases for restarting each algorithm with a different random seed. The SVM is by far the fastest algorithm as it only required 13 seconds of CPU time. The Bayesian algorithms exhibit linear increases between the standard and the T3 versions. For instance Mixed-T3 is four times that of Mixed as it uses 3 extra hot chains. In cross-paradigm comparison FFNN is comparable to Mixed and FFNNE to Mixed-T3. However, it is

worth noting that Mixed chains can easily be run in a distributed fashion. In our experiments a cluster of 8 modest performance units have been utilised to linearly reduce running times. The numbers reported here refer to total times.

Cross Validation on Three Datasets

To validate the results on dataset A we further select from it nine disjoint training and testing subsets and perform ten-fold cross validation (original selection being the first fold). Each training set contains 90% of the original data for training purposes and the remaining 10% is used for testing. In what follows we present results for the SVM, FFNN and Mixed algorithms. In Table 4 we present the results for all fold segments of dataset A. We show means of true positives and true negatives (out of $\{58\}$) along with their standard error. Finally the last six rows show the sensitivity and specificity results derived from the mean true positive and negative values.

In terms of sensitivity, results for all predictors are comparable, with SVM doing marginally better. On average, FFNN: 0.745, SVM: 0.781, Mixed: 0.751. In terms of specificity, Mixed performs consistently better (on average, FFNN: 0.774, SVM: 0.771, Mixed: 0.824).

The positive likelihood ratio (LR-positive) (Swets, 1988) incorporates sensitivity and specificity into one single metric providing an estimate of how well positives are predicted. In the application considered here the ability to predict active compounds (positives) is of paramount importance and LR-positive can be regarded as an overall indicator. It is given by $\frac{Sensitivity}{1 - Specificity}$ and for the three predictors, FFNN, SVM and Mixed, it has the values of 3.30, 3.41 and 4.27 respectively.

Another important measure is stability of prediction. FFNN in most cases have much larger spread in the number of prediction within one

Table 3. Sensitivity, specificity, LR-positive and runtime figures for fold 1 dataset A

	Sensitivity	Specificity	LR-positive	Time (sec)
MH	0.7707	0.7298	2.85233	18888
MH-Aw2	0.7879	0.6781	2.44765	24636
MH-T3	0.7810	0.7529	3.16066	61921
FFNN	0.8069	0.7345	3.03917	96106
FFNNE	0.8276	0.7355	3.12892	1121671
SVM	0.8448	0.7414	3.26682	13
Mixed	0.8500	0.7759	3.79295	166175
Mixed-Aw2	0.8448	0.7816	3.86813	214474
Mixed-T3	0.8505	0.7988	4.22714	637423

Table 4. Ten-fold cross validation of dataset A for FFNN, SVM and Mixed

	FFNN				SVM		Mixed			
	T_μ^+	T_ϵ^+	T_μ^-	T_ϵ^-	T^+	T^-	T_μ^+	T_ϵ^+	T_μ^-	T_ϵ^-
1	46.8	0.98	42.6	1.09	49	43	49.3	0.54	45	0.47
2	42.5	1.19	48.9	0.27	44	48	41.7	0.27	50	0
3	43.8	1.19	44.9	0.54	44	46	46.3	0.98	49.7	0.72
4	38.8	1.66	39.8	1.25	39	41	40	0.94	43.7	0.27
5	40.4	0.94	41.8	0.94	41	39	37.3	0.54	42	0
6	44.5	0.98	46.3	0.27	47	47	42.7	0.54	51	0
7	46.2	0.72	45.6	1.91	49	45	48.3	0.27	47	0.47
8	42.7	2.16	44.8	0.27	47	49	45.3	0.54	49.3	0.54
9	41.4	1.44	48.2	0.98	42	48	39	0.47	52	0
10	44.8	0.47	46.0	1.09	51	41	45.7	0.27	48.3	0.27
Avg	43.2		44.9		45.3	44.7	43.6		47.8	

	Sensitivity			Specificity			LRpos		
	FFNN	SVM	Mixed	FFNN	SVM	Mixed	FFNN	SVM	Mixed
1	0.81	0.84	0.85	0.73	0.74	0.78	3	3.23	3.86
2	0.73	0.76	0.72	0.84	0.83	0.86	4.56	4.47	5.14
3	0.76	0.76	0.79	0.77	0.79	0.85	3.3	3.62	5.27
4	0.67	0.67	0.69	0.69	0.71	0.75	2.16	2.31	2.76
5	0.70	0.71	0.64	0.72	0.67	0.72	2.5	2.15	2.29
6	0.77	0.81	0.73	0.80	0.81	0.88	3.85	4.26	6.08
7	0.80	0.84	0.83	0.79	0.78	0.81	3.81	3.82	4.37
8	0.74	0.81	0.78	0.77	0.84	0.85	3.22	5.06	5.2
9	0.71	0.72	0.67	0.83	0.83	0.90	4.18	4.24	6.7
10	0.77	0.88	0.78	0.79	0.71	0.83	3.67	3.03	4.59
Avg	0.745	0.781	0.751	0.774	0.771	0.824	3.42	3.62	4.63

segment. That is, they have larger standard error entries in Table 4. This reinforces the current view in the literature that FFNN performance is prone to starting conditions and local minima. Variation of prediction is not an issue for SVM because it is purely deterministic in the sense that it does not depend on any random choices.

In terms of computational effort, SVM is by far the most efficient algorithm - its training over all datasets/folds takes just a few seconds (see Table 3). As far as the other algorithms are concerned, FFNN takes 27 hours, Mixed takes 77% more and FFNNE takes 117% more (on a 2.13 GHz cpu with 4 GB RAM).

Finally as the inactive molecules in dataset *A* were randomly selected from a much larger pool, we repeated the ten-fold cross validation on two further selections of negatives (datasets *B* and *C*). Sensitivity and specificity for these are shown in Figures 4 and 5. As it can be seen, in absolute terms both algorithms show results in the same region as that of dataset *A*. When comparing the algorithms against each other, our conclusions from *A* are also valid for *B* and *C*. SVM, FFNN and Mixed do roughly equally well in terms of sensitivity but Mixed performs better in terms of specificity. In predicting active compounds as measured by LR-positive, Mixed also does better. For dataset *B* the LR-positive values for FFNN, SVM and Mixed are: 3.07 and 3.77, while for dataset *C* the corresponding values are: 2,96, 3.01 and 3.40.

Classifying Eduliss Molecules

To identify possible leads for pyruvate kinase we trained the Mixed algorithm on all 1164 examples of dataset *A* and used the produced chains to classify molecules in a large database. As each classified molecule is also given a probability of activity, the whole database can be ranked for potential ligands.

The produced chains had 477,930 models of which 373,469 were distinct across chains. Of those, only trees visited over a threshold number of times are retained and merged into a single chain. In Figure 6 the distribution of models (x-axis) against the logarithm of times of appearance (y-axis) in the chain is shown. The trees are sorted in descending order of the logarithm. As it can be seen, the chains appear to move well in

Figure 4. Sensitivity for datasets B (left) and C. Averages: FFNN$_B$=0.742, FFNN$_C$=0.729, SVM$_B$=0.776, SVM$_C$=0.753, Mixed$_B$=0.754, Mixed$_C$=0.731

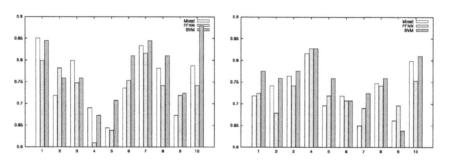

Figure 5. Specificity for datasets B (left) and C. Averages: FFNN$_B$=0.758, FFNN$_C$=0.750, SVM$_B$=0.747, SVM$_C$=0.750, Mixed$_B$=0.800, Mixed$_C$=0.785

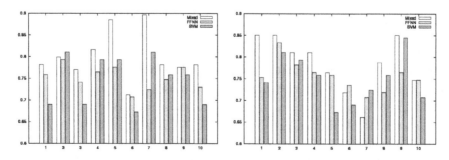

Figure 6. Log-occurrence values for each model in chain of Mixed algorithm which was trained with all 1164 examples of dataset A. The horizontal lines mark log(20) = 2.9957 and log(100) = 4.6052

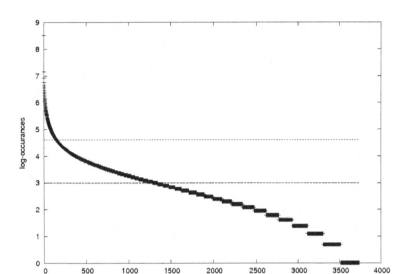

the space in that there are no models with large number of visits.

We ran classifications of the database for two values of a cut-off threshold U: 100 and 20. The former reduces the population of the voting chain to 15032 whereas the latter reduces the chain to 134431 trees. In what follows we present results for $U= 20$ which in Figure 6 corresponds to removing all trees below the $y=log(20)=2.9957$ line. Even if included, these trees would have small individual contributions. It is worth noting that the most visited tree appeared in the chain 5040 times and its log-likelihood is $log(P(C|T,M))=-6161$.

The 3.7 million unique compounds of Eduliss, (Hsin, 2009), are classified according to the averaged prediction of all the models in the chain. Let E be the set of molecules in the database, T the set of classification trees with associated parameters Θ, C the set of classes, with τ and θ are members of the respective sets, then:

$$P(C \mid E, T, \Theta) = \sum_{\tau \in T, \theta \in \Theta} P(\tau \mid M, C) \; P(C \mid E, \tau, \theta)$$

The probability of activity for a molecule in the database is thus dependant on the summation of contributions from all trees. Each consists of two parts: the importance of a tree (measured by its posterior, which is simply the times it appears in the chain over the length of the chain) and the classification of the molecule by that tree $(P(C|E,\tau,\theta))$.

When there are missing descriptors we use an extra layer of averaging. In general, each tree that contains decisions based on missing descriptors can be used to produce K decisions. This is by means of choosing both branches of the tree when such a decision is encountered. The overall probability of activity is derived by summing all K decisions on activity and dividing by K.

For FFNN and SVM, missing descriptors are substituted by the average value of the specific descriptor over all data.

The probability distribution over classes for each molecule can be regarded as an activity ranking. In Figure 7 the first six active molecules from our dataset are presented side-by-side with the top six hits from the Eduliss database. There is clear molecular similarity between the top hits

Figure 7. **A**: *Six active molecules from the training set.* **B**: *The six molecules scoring highest probability of activity. From top to bottom their probabilities are 0.904738, 0.904384, 0.902034, 0.901443, 0.901044 and 0.900508*

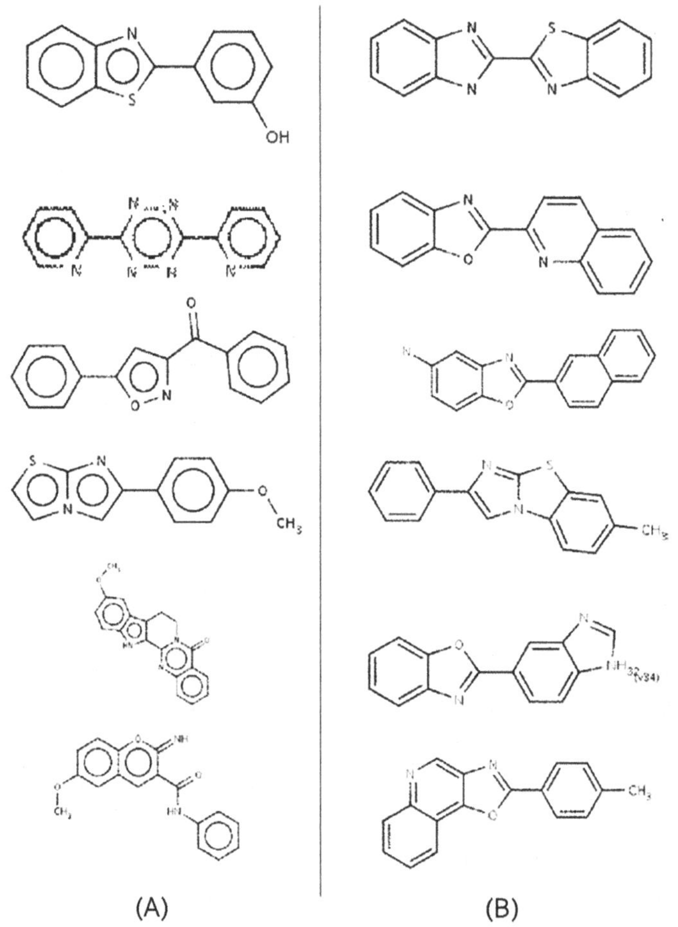

(A)　　　　　　　　　　(B)

and known active molecules, which without being conclusive evidence, is a promising sign. Both families consist of between two and four planar ring systems. The prediction of such related and similar looking molecules out of the database of 3.7 million molecules suggests that the algorithm provides a self consistent selection procedure with hit-molecules occupying the same region of chemical space as the known active compounds. It is worth noting that in virtual screening the objective is to enrich the chances of finding bioactive molecules by considering number of successes in large sets (enrichment factors).

Figure 8 shows the frequencies for each descriptor in our set. Each occurrence to a tree with $U>20$ has been counted, thus multiple appearances on a single tree are accounted for. For instance the counts of some descriptors exceed the number of classifying trees used. In Figure 8 we can see a good mix for a large number of descriptors which is further evidence to the stochastic activity of the algorithm.

Figure 9 depicts a plot of number of molecules (y) against probability of activity (x). The latter has been sorted into discrete bins of width 0.01. The probability curve has a very regular well

Figure 8. The classification of all unique molecules of Eduliss. Frequencies for all the descriptors

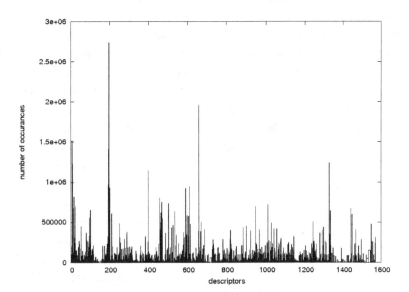

Figure 9. The classification of all unique molecules of Eduliss by the Mixed algorithm. Probability bins (width = 0.01) against number of molecules

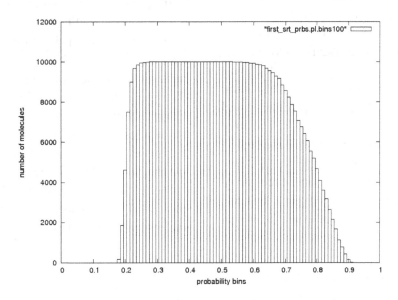

spread contour. The final high probability bin contains seven compounds with probabilities between 0.90 and 0.91. The top six compounds are shown in Figure 7.

CROSS CLASSIFIER AGREEMENT

The three best performing algorithms on the full dataset are fed with the 3.7 million unique compounds contained in Eduliss and assess potentially active compounds. At the end, each unique compound will be associated with a set of probabilities of being positive corresponding to each of these three algorithms.

In order to get an indication about the extent of the agreement between the predictions of the different classifiers on the unknown test set comprising all the unique molecules in the Eduliss database the root mean square (RMS) difference between the outcomes of each pair of classifiers, $A,B \in \{Mixed, FFNN, SVM\}$, was calculated using:

$$RMS(A, B) = \sqrt{\frac{1}{3.7 \times 10^6} \sum_{i=1}^{3.7 \times 10^6} (A_i - B_i)^2}$$

The smaller the RMS value, the closer the two classifiers in their predictions. In this respect, the two algorithms Mixed and FFNN agree the most with an RMS difference of 0.180, followed by the the pair Mixed and SVM (0.203). The algorithms FFNN and SVM agree the least (0.231).

Additionally, Table 5 gives the number of positive (+ve:A_i>0.5) and negative (-ve:$A_i \leq 0.5$) predictions for each classifier as well as the count of agreements and disagreements between them. So, Mixed has 1,113,699 positive predictions and 2,663,061 negative, SVM has 1,251,715 positive and 2,525,045 negative and FFNN has 500,057 positive and 3,276,703 negative.

This table also provides a count of the agreements and disagreements of predictions between each pair of classifiers. So, for example, Mixed

and SVM predicted the same 809,153 molecules as positive and the same 2,220,499 molecules as negative. On the other hand, the same 442,562 molecules predicted as negative by Mixed, were predicted as positive by SVM and, conversely, the same 304,546 molecules predicted as positive by Mixed but as negative by SVM. In summary, Mixed and SVM agree 3,029,652 times and disagree 747,108 which makes them the most consistent pair. Followed by SVM and FFNN (agree 2,777,354 times and disagree 999,406). The least consistent pair is Mixed and FFNN (agree 2,750,064 times and disagree 1,026,696 times).

The number of molecules predicted positive by all three classifiers was 270,779 and those predicted negative was 2,119,376.

Figure 10 shows a plot of the probability for each molecule in the database and for each classifier, sorted in descending order. In this plot we can see that FFNN has the least number of positives and SVM the most. For FFNN and Mixed, the number of molecules classified with either very high or very low probability decreases very rapidly, with most molecules being in the uncertainty zone of 0.15 (0.3 for Mixed) to 0.6. The picture for SVM is that probabilities are distributed among molecules more or less evenly as the curve resembles a line.

For each pair of algorithms, the predicted probability of activity for each unknown molecule is plotted in a 2D scatter plot. For example, in plot 11(a), each point $p_i(x_i, y_i)$ represents the ith molecule in the database which Mixed assigned

Table 5. Summary of agreement for the three main classifiers

		Mixed		SVM		FFNN	
		+ve	-ve	+ve	-ve	+ve	-ve
Mixed	+ve	1,113,699	-	809,153	304,546	293,530	820,169
	-ve	-	2,663,061	442,562	2,220,499	206,527	2,456,534
SVM	+ve			1,251,715	-	376,183	875,532
	-ve			-	2,525,045	123,874	2,401,171
FFNN	+ve					500,057	-
	-ve					-	3,276,703

Figure 10. The probability for each compound of each classifier

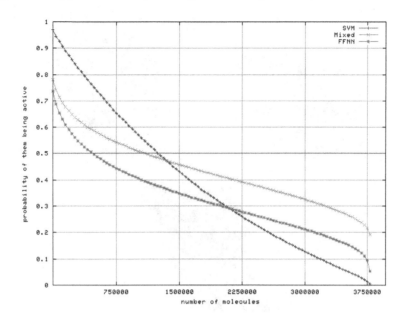

it a probability of x_i and FFNN a probability of y_i. If the two classifiers behaved identically, then all the points will be along the $f(x) = x$ line. Disagreements are manifested by perturbations from this line. For this particular pair, we can see that they agree more on the negative predictions and disagree quite strongly on the positive ones.

Figure 11(b) shows the scatter plot for SVM versus FFNN and 11(c) that for SVM versus Mixed. From a visual inspection, it can be confirmed that the pair SVM and Mixed agree the most.

FUTURE RESEARCH DIRECTIONS

Unlike many traditional machine learning tasks, ligand discovery is inherently difficult to verify. It is unlikely that for the top N predictions to be all positives, instead success is measured on the improvement against a random selection. The ratio of a predicting technique to that of a random selection is known in ligand discovery as the enrichment factor. An in-silico approach to

validation would be via docking algorithms such as simplified non-quantum mechanical methods (Woo & Roux, 2005), FLEXX, (Kramer, Rarey, & Lengauer, 1999), and Lideaus, (Taylor et al., 2008). Although such an analysis could provide additional insights, it must be treated very cautiously as the artifacts of two algorithms, those of the trainer and the validator, must be taken into consideration.

In comparative studies across classifiers that provide confidence measures on their predictions, one can either select one predictor or create a hybrid system where the individual predictions are somehow incorporated into a final outcome with the possibility of a gating algorithm. A promising future research direction would be to apply existing research within machine learning about multi-classifier systems in the field of ligand discovery.

In our study, the conservative approach to a hybrid classifier would be to restrict attention to the 270,779 consensus molecules. This can be generalised by varying the 0.5hreshold to an arbitrary value (usually a larger value). For instance, when this threshold is set to 0.65 the

Figure 11. Pairwise scatter plots of classifier agreement. (a) FFNN versus Mixed, (b) SVM versus FFNN, and (c) SVM versus Mixed

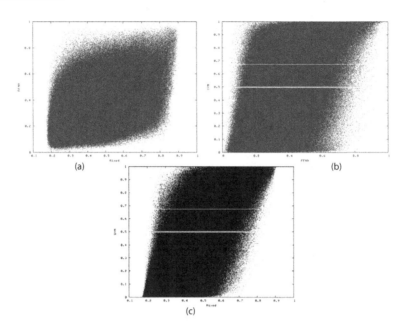

consensus set only contains 45,985 molecules. Further refinements can be made by ranking on a weight function and selecting a cutoff point based on practical considerations regarding the number of molecules that can be tested. That would be particularly suitable for in-house libraries held by large corporations and research institutions.

Principal Components Analysis (PCA) is a method to correlate features (descriptors) and transform them into new, uncorrelated variables ranked according to their variability, (Jolliffe, 2002). In this way, related features and redundancy among features can be revealed. PCA can be used as a dimensionality reduction technique by selecting the first N of these new variables explaining best the data at hand. Other methods such as Random Forests (RF) can also be used for feature selection and dimensionality reduction based on random sampling from the given data and performing sensitivity analysis, (Breiman, 2001). Public domain software for RF is available for R, (Liaw & Wiener, 2002). It would be of great interest to combine the algorithms proposed here in tandem with such variable selection approaches.

The techniques presented here can be readily applied to other protein targets. In addition, different molecular databases can be searched for ligands. For instance, the publicly available ZINC database (http://zinc.docking.org/), (Irwin & Shoichet, 2005), which contains over 4.6 million commercially available compounds and the PubChem Compound database (http://pubchem. ncbi.nlm.nih.gov/) which contains more that 19 million molecules. As more high-throughput results are becoming available there will be new challenges and opportunities in the area for combining the biases of the different libraries to model the chemical space more fully.

CONCLUSION

This chapter provides a new methodology that can be applied quite generally to analyse the results of large scale biological and pharmacological screens and is therefore of broad interest and utility. By using machine learning techniques, we have shown how existing software, datasets and descriptors can be combined and utilised for lead discovery.

In addition to examining machine learning techniques that have been used before in ligand discovery, such as FFNN and SVM we also present results of Bayesian model averaging for classification trees which is a less standard approach. The Bayesian methodology allows further analysis of the results particularly in identifying distinct biochemical groups among top hits and explaining which are the major contributors to best scoring molecules. The framework can be used to this end, as features of the models can be averaged similarly to the way chains of trees have been averaged in this paper.

The results of the Bayesian approach have shown promise and were at least as good if not better to SVM and Neural Networks. LR-positive averages over datasets *A*, *B* and *C* were FFNN:3.09, SVM:3.16 and Mixed:3.81. It has to be noted that SVM require far less computational time than any of the other methods. However, computational power is not the most common bottleneck in ligand discovery. The costs of testing the compounds far outweigh the modest expense for the computational needs described here and the time needed for the FFNN and Mixed algorithms to train.

Bayesian model averaging over classification trees is an appealing framework for ligand discovery as in-vitro screening results can be used to provide a simple model of bioactivity which in turn provides a convenient method for predicting and selecting sets of new compounds with a higher chance of having the required properties. There are four notable features of this approach. Firstly, mixing chain generation methodologies can address the multi-dimensionality of the data. Secondly, molecules with missing descriptor values can still be classified and included in compound testing. Thirdly, probabilities can readily be interpreted as a rank of ligand-binding activity. Finally, the MCMC experiments described here are inherently easy to run in a distributed environment and thus can be efficiently run on clusters of computers.

ACKNOWLEDGMENT

Thanks to Paul Taylor for helpful discussions, Kun-Yi Hsin for help with the data and to James Cussens for discussions on priors. This work was in part supported by UK's BBSRC project D00604 X/1.

REFERENCES

Altman, D. G., & Bland, J. M. (1994). Diagnostic tests 1: sensitivity and specificity. *British Medical Journal, 308*, 1552.

Angelopoulos, N., & Cussens, J. (2005a, July-August). Exploiting informative priors for Bayesian classification and regression trees. In *Artificial Intelligence; Proceedings of the 19th International Joint Conference* (p. 641-646). Edinburgh, UK: IJCAI.

Angelopoulos, N., & Cussens, J. (2005b, August). Tempering for Bayesian C&RT. In *Machine Learning; Proceeding of the 22nd International Conference* (p. 17-24). Bonn, Germany: ACM.

Ben-Hur, A., Ong, C. S., Sonnenburg, S., Scholkopf, B., & Ratsch, G. (2008). Support vector machines and kernels for computational biology. *PLoS Computational Biology, 4*(10). doi:10.1371/journal.pcbi.1000173

Bishop, C. (1995). *Neural networks for pattern recognition*. Oxford: Clarendon Press.

Breiman, L. (2001). Random forests. *Machine Learning, 45*(1), 5–32. doi:10.1023/A:1010933404324

Breiman, L., Friedman, J. H., Olshen, R. A., & Stone, C. J. (1984). *Classification and regression trees*. Belmond, CA, USA: Wadsworth International. 22

Buntine, W. L. (1992, June). Learning classification trees. *Statistics and Computing, 2*(2), 63–73. doi:10.1007/BF01889584

Burges, C. J. C. (1998). A tutorial on support vector machines for pattern recognition. *Data Mining and Knowledge Discovery, 2,* 121–167. doi:10.1023/A:1009715923555

Chan, M., Tan, D. S., & Sim, T. S. (2007). Plasmodium falciparum pyruvate kinase as a novel target for antimalarial drug-screening. *Travel Medicine and Infectious Disease, 5*(2), 125–131. doi:10.1016/j.tmaid.2006.01.015

Chipman, H., George, E., & McCulloch, R. (1998). Bayesian CART model search (with discussion). *Journal of the American Statistical Association, 93,* 935–960. doi:10.2307/2669832

Christofk, H. R., Vander Heiden, M. G., Harris, M. H., Ramanathan, A., Gerszten, R. E., & Wei, R. (2008). The M2 splice isoform of pyruvate kinase is important for cancer metabolism and tumour growth. *Nature, 452,* 230–233. doi:10.1038/nature06734

Cortes, C., & Vapnik, V. (1995, September). Support-vector networks. *Machine Learning, 20*(3), 273–297. doi:10.1007/BF00994018

Cybenko, G. (1989). Approximation by superpositions of a sigmoidal function. *Mathematics of Control, Signals, and Systems, 2*(4), 303–314. doi:10.1007/BF02551274

Denison, D. G. T., Holmes, C. C., Mallick, B. K., & Smith, A. F. M. (2002). *Bayesian methods for nonlinear classification and regression.* Chichester, UK: Wiley.

Fradkin, D., & Muchnik, I. (2006). Support vector machines for classification. *Discrete Methods in Epidemiology, 70,* 13–20.

Hadjiprocopis, A., & Smith, P. (1997). Feed forward neural network entities. In J. Mira, R. Moreno-Diaz, & J. Cabestany (Eds.), *Biological and artificial computation: From neuroscience to technology* (pp. 349{359). Spain: Springer-Verlag.

Hansch, C., Hoekman, D., & Gao, H. (1996). Comparative QSAR: Toward a deeper understanding of chemicobiological interactions. *Chemical Reviews, 96,* 1045–1076. doi:10.1021/cr9400976

Hsin, K.-Y. (2009). *Eduliss 2, Edinburgh University Ligand Selection System.* Retrieved from http://eduliss.bch.ed.ac.uk/

Inglese, J., Auld, D. S., Jadhav, A., Johnson, R. L., Simeonov, A., & Yasgar, A. (2006). Quantitative high-throughput screening: a titration-based approach that e±ciently identifies biological activities in large chemical libraries. *Proceedings of the National Academy of Sciences of the United States of America, 103,* 11473–11478. doi:10.1073/pnas.0604348103

Irwin, J. J., & Shoichet, B. K. (2005). ZINC{a free database of commercially available compounds for virtual screening. *Journal of Chemical Information and Modeling, 45*(1), 177–182. doi:10.1021/ci049714+

Joachims, T. (1998). Making large-scale support vector machine learning practical. In B. Schaolkopf, C. J. Burges, & A. J. Smola (Eds.), *Advances in kernel methods: Support vector machines* (p. 169-184). Cambridge, MA: MIT Press. Retrieved from http://citeseer.ist.psu.edu/joachims98making.html

Jolli®e, I. T. (2002). *Principal component analysis* (2nd ed.). New York: Springer.

Kramer, B., Rarey, M., & Lengauer, T. (1999). Evaluation of the FLEXX incremental construction algorithm for proteinligand docking. *Proteins, 37*(2), 228–241. doi:10.1002/(SICI)1097-0134(19991101)37:2<228::AID-PROT8>3.0.CO;2-8

Liaw, A., & Wiener, M. (2002). Classification and regression by randomForest. *R News*, *2*(3), 18.

Lipinski, C. A., Lombardo, F., Dominy, B. W., & Feeney, P. J. (2001). Experimental and computational approaches to estimate solubility and permeability in drug discovery and development settings. *Advanced Drug Delivery Reviews*, *46*(1-3), 3–26. doi:10.1016/S0169-409X(00)00129-0

Meyer, D., Leisch, F., & Hornik, K. (2003, September). The support vector machine under test. *Neurocomputing*, *55*, 169–186. Retrieved from http://dx.doi.org/10.1016/S0925-2312(03)00431-4. doi:10.1016/S0925-2312(03)00431-4

Nowicki, M. W., Tulloch, L. B., Worralll, L., McNae, I. W., Hannaert, V., & Michels, P. A. (2008). Design, synthesis and trypanocidal activity of lead compounds based on inhibitors of parasite glycolysis. *Bioorganic & Medicinal Chemistry*, *16*(9), 5050–5061. doi:10.1016/j.bmc.2008.03.045

Quinlan, J. R. (1986). Induction of decision trees. *Machine Learning*, *1*, 81–106. doi:10.1007/BF00116251

Ripley, B. D. (1996). *Pattern recognition and neural networks*. Cambridge, UK: Cambridge University Press.

Swets, J. A. (1988, June). Measuring the accuracy of diagnostic systems. *Science*, *240*(4857), 1285–1293. doi:10.1126/science.3287615

Tabarakia, R., Khayamiana, T., & Ensafia, A. A. (2006, September). Wavelet neural network modeling in QSPR for prediction of solubility of 25 anthraquinone dyes at di®erent temperatures and pressures in supercritical carbon dioxide. *Journal of Molecular Graphics & Modelling*, *25*(1), 46–54. doi:10.1016/j.jmgm.2005.10.012

Taylor, P., Blackburn, E., Sheng, Y. G., Harding, S., Hsin, K.-Y., & Kan, D. (2008). Ligand discovery and virtual screening using the program LIDAEUS. *British Journal of Pharmacology*, *153*, S55S67. doi:10.1038/sj.bjp.0707532

Todeschini, R., Consonni, V., Mannhold, R., Kubinyi, H., & Timmerman, H. (2000). *Handbook of molecular descriptors*. Weinheim, New York: Wiley-VCH.

Todeschini, R., Gramatica, P., & Navas, N. (1998). 3D-modeling and predication by WHIM descriptors. Part 9. Chromatographic relative retentation time and physicochemical properties of polychlorinated biphenyls (PCBs). *Chemometrics and Intelligent Laboratory Systems*, *40*(1), 53–63. doi:10.1016/S0169-7439(97)00079-8

Todeschini, R., Lasagni, M., & Marengo, E. (1994). New molecular descriptors for 2D and 3D structures. Theory. *Journal of Chemometrics*, *8*(4), 263–272. doi:10.1002/cem.1180080405

Tonga, J., Liu, S., Zhou, P., Wu, B., & Li, Z. (2008, July). A novel descriptor of amino acids and its application in peptide QSAR. *Journal of Theoretical Biology*, *253*(1), 90–97. doi:10.1016/j.jtbi.2008.02.030

Woo, H. J., & Roux, B. (2005). Calculation of absolute proteinligand binding free energy from computer simulations. *Proceedings of the National Academy of Sciences of the United States of America*, *102*(19), 6825–6830. doi:10.1073/pnas.0409005102

Zheng, C. J., Han, L. Y., Yap, C. W., Ji, Z. L., Cao, Z. W., & Chen, Y. Z. (2006). Therapeutic targets: progress of their exploration and investigation of their characteristics. *Pharmacological Reviews*, *58*, 259–279. doi:10.1124/pr.58.2.4

Section 4
Machine Learning Approaches for Drug Discovery, Toxicology, and Biological Systems

Chapter 12
Application of Machine Learning in Drug Discovery and Development

Shuxing Zhang
The University of Texas at M.D. Anderson Cancer Center, USA

ABSTRACT

Machine learning techniques have been widely used in drug discovery and development, particularly in the areas of cheminformatics, bioinformatics and other types of pharmaceutical research. It has been demonstrated they are suitable for large high dimensional data, and the models built with these methods can be used for robust external predictions. However, various problems and challenges still exist, and new approaches are in great need. In this Chapter, the authors will review the current development of machine learning techniques, and especially focus on several machine learning techniques they developed as well as their application to model building, lead discovery via virtual screening, integration with molecular docking, and prediction of off-target properties. The authors will suggest some potential different avenues to unify different disciplines, such as cheminformatics, bioinformatics and systems biology, for the purpose of developing integrated in silico drug discovery and development approaches.

INTRODUCTION

Drug discovery and development is regarded as one of the most complex research areas encompassing many disciplines (Cohen et al., 2004; Gomeni et al., 2001; Martin, 1991; Tropsha & Zheng, 2001). This might be part of the reasons why it is extremely expensive and time-consuming, and why the current pharmaceutical business is hitting

a wall with its severely stalled discovery engine (Tralau-Stewart et al., 2009). In seeking efficient and costly effective approaches, computer-aided methods are gaining more and more attentions, and recent years have witnessed dramatic progress in computer-aided drug design (Cohen et al., 2004; Gomeni et al., 2001; Martin, 1991; Tropsha & Zheng, 2001). This is partially due to the significant advances in the analysis of explosively growing biological and chemical data using modern machine learning techniques (Mitchell et

DOI: 10.4018/978-1-61520-911-8.ch012

al., 1989; Mjolsness & DeCoste, 2001; Schneider & Downs, 2003). *Machine learning* has been referred to the development of algorithms that improve their performance in pattern recognition, classification, regression and prediction based on the models derived from existing data. Therefore, it is closely related to fields such as data mining, pattern recognition, theoretical computer science, and many other areas (Mitchell et al., 1989). For instance, algorithms for classification have been used frequently to identify active and inactive compounds, while regression approaches are applied to the training and prediction of continuous data. Despite being widely used in other biomedical research such as bioinformatics, we will focus herein on its application to small molecule drug discovery and development.

Currently there is a variety of existing implementations of machine learning. However, it is often difficult to assess the usefulness and limitations of a particular method for the problems at hand (Mitchell et al., 1989; Schneider & Downs, 2003). In drug discovery and development, machine learning has been widely used in *quantitative structure-activity relationship (QSAR)*, ligand-based *virtual screening, in silico ADMET* (absorption, distribution, metabolism, excretion, and toxicity) studies, and many other areas (King et al., 1992; King et al., 1996; Mitchell et al., 1989; Mjolsness & DeCoste, 2001; Schneider & Downs, 2003). Different QSAR approaches have been developed during the past few decades (Hansch et al., 1963; Klein et al., 1986; Kubinyi, 1986). In a generalized classification or regression problem, modern QSAR are characterized by the use of multiple descriptors of chemical structures combined with the application of both linear and non-linear optimization techniques, and a strong emphasis on rigorous model validation to afford robust and predictive QSAR models (Tropsha, 2005; Tropsha, 2006). *Molecular descriptors* are used for representing structural and physicochemical properties of compounds. More than 3000 thousand descriptors have been

developed to date, ranging from constitutional descriptors, such as molecular weight, to more complex 2D and 3D descriptors representing different topologic, geometric, connectivity, and physicochemical properties (Li et al., 2007). Frequently used descriptors in QSAR modeling include constitutional descriptors (e.g., counts of atoms, bonds, etc.), property-based descriptors (e.g., logP), BCUT descriptors, topological descriptors, geometrical descriptors, electrostatic, quantum chemical descriptors, thermodynamic descriptors, and many others. These descriptors can be calculated by several popular programs such as DRAGON (Tetko et al., 2005), Molconn-Z (Kellogg, 2002), MOE (Chemical Computing Group, Quebec, Canada), CODESSA (http://www.codessa-pro.com/index.htm), ADMET Predictor (Simulation Plus, Lancast, CA), JOELib (http://www.ra.cs.uni-tuebingen.de/software/joelib/), and PowerMV (http://www.niss.org/PowerMV/).

Once descriptors are obtained for the molecules, statistical modeling techniques are required to establish correlation between the descriptors and activities. For instance, a comprehensive review of the application of neural network in a variety of QSAR problems has been presented, which discussed how NNs can be applied to the prediction of physicochemical and pharmacokinetic properties (Baskin et al., 2008). In addition, SVM was found to yield improved performance compared to multiple linear regressions (MLR) and radial basis functions (RBF) (Yao et al., 2004). Various version of SVM programs have been developed and they were used in calculating the activity of enzyme inhibitors as well as in many other studies of similar types (Duch et al., 2007). In addition to the activity prediction of molecules, QSAR models are also frequently used in virtual screening for hit discovery (Hansch & Fujita, 1995; Kubinyi, 1990; Tropsha & Golbraikh, 2007). Virtual screening is usually applied to the identification of those that are potentially active in the biological tests of interest. The ultimate goal is to reduce the molecules to be tested from a large dataset to a small, manageable size, and thus make the test cost-effective. There are two types of virtual screening approaches: ligand-based and structure-based. The latter requires the knowledge of 3D structures of the biological targets. QSAR models are usually applied in the ligand-based virtual screening. A machine learning technique first takes the training set compounds, which have been tested previously, to develop rules (QSAR models). The rules (models) are then used to make prediction of the molecules in the database, for either *classification* (e.g., active/inactive) or *regression* (continuous activity values). Such protocols, which are being applied to the analysis of various chemical datasets, have been rapidly developed, including decision tree, SVM, kNN, recursive partition, *lazy learning,* binary kernel discrimination, *inter alia* (Chen et al., 2007).

Similarly, machine learning is also being frequently used to conduct ADMET predictions (Clark & Grootenhuis, 2002; Davis & Riley, 2004; Li et al., 2007). *In silico* prediction of pharmacokinetic/pharmacodynamic (PK/PD) and ADMET properties are critical in facilitating drug discovery and evaluation (Clark & Grootenhuis, 2002; Davis & Riley, 2004; Li et al., 2007). This early computational study will help in the selection of better lead compounds and potentially reduce the failure rate in the late stages of drug development. In addition, many chemical compounds are promiscuous inhibitors of multiple targets (Feng & Shoichet, 2006). As part of the effort of drug discovery and safety evaluation, machine learning can also be used to predict their adverse drug reactions or off-target properties. For instance, Gaussian kernel SVM was used to successfully classify a set of drugs in terms of their potential to cause an adverse drug reaction TdP (Yap et al., 2004). Although TdP is involved in multiple mechanisms, the SVM prediction accuracy on an independent set of molecules was 90% more than that with ANN and decision tree methods.

This Chapter will first review the current progress of machine learning techniques by focusing on several machine learning methods that are commonly used in drug discovery, particularly those developed by us. We will also visit the application of machine learning to drug discovery as well as the underlying difficulties in QSAR model developing, ligand-based virtual screening for lead identification, and ADMET *in silico* predictions. Importance of *model validation* and applicability along with their new application in molecular docking/scoring and ligand off-target property predictions will also be addressed. We will finally provide our insight on some emerging trends and the future direction of the field so as to foster its further expanding by the research community.

COMMONLY USED MACHINE LEARNING TECHNIQUES

Several machine learning methods have been widely used for the classification of pharmaceutical relevance. These methods include linear discriminant analysis (LDA), binary kernel discrimination (BKD), decision tree (DT), neural network (NN), and support vector machine (SVM), k nearest neighbor (kNN), lazy learning, among many others. For comparison, we here also discuss several other linear methods widely used in drug discovery such as partial least square (PLS), principle component analysis (PCA) and multiple linear regression (MLR).

Multiple Linear Regression (MLR)

Linear regression is a form of regression analysis in which the relationship between one or more independent variables and the dependent variable is modeled by a least squares function (Mandel, 1985). This function is a linear combination of one or more model parameters, called regression coefficients. A disturbance term ε, which is a random variable, is added to this assumed relationship to capture the influence of everything else on Y other than X. Hence, a multiple linear regression model can be formulated as: $Y = \beta X + \varepsilon$, where Y is a column vector that includes the observed values of $Y_1, Y_2, \ldots Y_n$; β is the coefficients needing to be determined ($\beta_1, \beta_2, \ldots, \beta_p$); ε includes the unobserved stochastic components $\varepsilon_1, \varepsilon_2, \ldots \varepsilon_n$; and the matrix X is (Mandel, 1985):

$$X = \begin{pmatrix} 1 & x_{11} & \cdots & x_{1p} \\ 1 & x_{21} & \cdots & x_{2p} \\ \vdots & \vdots & \cdots & \vdots \\ 1 & x_{n1} & \cdots & x_{np} \end{pmatrix} \qquad (1)$$

Partial Least Square (PLS)

PLS was first introduced by the Swedish statistician Herman Wold (1984), and has been widely applied in the field of chemometrics, sensory evaluation, and chemical engineering process data. PLS tries to find a linear model describing the target variables (e.g., activity) in terms of other observable variables (e.g., descriptors) (Kubinyi, 1996; Wold et al., 1984). The general model of PLS is $X = TA^T + E$ and $Y = TB^T + F$, where X is a matrix of predictors with $n \times m$ dimensions; T is an $n \times l$ factor matrix; Y is an $n \times p$ matrix of responses; and A and B are $n \times m$ and $n \times m$ loading matrices, respectively. E and F are the error terms. In drug discovery PLS has long been used in popular 3D QSAR approaches such as CoMFA (Cramer, III et al., 1988) and CoMSIA (Ducrot et al., 2001).

Principle Component Analysis (PCA)

PCA involves a mathematical procedure that transforms a number of possibly correlated variables into a smaller number of uncorrelated variables called principal components, accounting for as much of the variability in the data as possible (Joliffe, 2002). PCA involves the calculation of the eigenvalue decomposition of a data covari-

ance matrix or singular value decomposition of a data matrix, and it is the simplest of the true eigenvector-based multivariate analyses. PCA is mathematically defined as an orthogonal linear transformation of the data to a new coordinate system such that the greatest variance by any projection of the data comes to lie on the new components. For a data matrix X^T, where each row represents a different repetition of the experiment and each column gives the results from a particular probe, the PCA transformation is given by: $Y^T = X^T W = V \Sigma W^T$, where $V \Sigma W^T$ is the singular value decomposition of X^T (Joliffe, 2002).

Linear Discriminant Analysis (LDA)

LDA (McLachlan, 2004) separates two classes of vectors by constructing a hyperplane defined by the function $F = \sum_i^n w_i x_i$, where F is the classification score and w_i is the weight associated with the corresponding descriptor x_i. A positive F score indicates that a vector x belongs to one class, otherwise belongs to the other class.

Binary Kernel Discrimination (BKD)

BKD was first used in the prediction of biological activity (Harper et al., 2001) and later in virtual screening (Chen et al., 2007). The kernel function is $F_\lambda(i,j) = [\lambda^{n-dij}(1-\lambda)^{dij}]^{\beta/n}$, where λ is a smoothing parameter, n is the length of the binary descriptor vectors, d_{ij} is the Hamming distance between the descriptor vectors for compounds i and j, and β is a user-defined constant. Training set compounds are ranked with the optimum value of λ found from the analysis of the training set, and this optimum value is then used for scoring the compounds in the test set.

Neural Network (NN)

NNs can generate arbitrary nonlinear decision boundaries by addition of many simple functions (Baskin et al., 2008). This is typically achieved by a multi-stage transformation using a network (directed graph) of interconnected layers of "computing" nodes that integrate input signals from previous layers. The input features, X_i, are represented by individual nodes in the input layer and are subsequently transformed into a new set of features using several hyperplanes, $\mathbf{W}_k$, corresponding to the hidden layer nodes. The hidden layer nodes transform these signals further using $h_k(\mathbf{X})=\sigma(\mathbf{W}_k\mathbf{X}+W_{k0})$, where the scalar functions $\sigma(x)$ are usually chosen to be logistic functions. As a result, the outputs are in general non-linear functions of inputs (Duch et al., 2007).

Decision Tree (DT)

A decision tree (Hong et al., 2005; Li et al., 2005) is a predictive model which maps observations to target values. More descriptive names are classification trees or regression trees. In these tree structures, leaves represent classifications and branches represent conjunctions of features that lead to those classifications. A tree can be learned by splitting the data into subsets based on descriptors and repeating the process recursively. A random forest classifier uses a number of decision trees, in order to improve the classification rate. In decision tree, data comes in the form of $(x, y) = (x_1, x_2,..., x_n, y)$. The dependent variable, y, is the variable that we are trying to classify or generalize. The other variables, x_1, x_2, etc., are the variables that will help with that task.

Support Vector Machine (SVM)

SVMs are a set of learning methods used for classification and regression (Vapnik, 1995). They construct a hyperplane separating two different classes of feature vectors with a maximum margin. Intuitively, a good separation is achieved by the hyperplane that has the largest distance to the neighboring data points of both classes. This hyperplane is constructed by finding a vector w and a parameter b that minimizes $\|w\|^2$ which

satisfies the following conditions: $w \cdot x_i + b \geq -1$ for $y_i = -1$ for and $w \cdot x_i + b \leq -1$ for $y_i = -1$. Here x_i is a feature vector, y_i is the class group index, w is a vector normal to the hyperplane, $\|w\|^2$ is the Euclidean norm of w. The support vector methods can also be applied to regression, termed Support vector regression (SVR), by maintaining all the main features that characterize the maximal margin algorithm. As in the classification, the regression learning minimizes a cost function. This approach is numerically very efficient and can be applied to large-scale problems. SVR also offers flexibility of models if in conjunction with kernel approaches. Particularly, the ε-insensitive SVR model assumes that the measured error $M(r)$ is set to zero if $r<\varepsilon$, where $r = |\, Y_k^{obs} - Y(X_k)\,|$. This allows one to define the error function in a flexible way, reflecting the expected level of errors by varying error bars ε.

K Nearest Neighbor (KNN)

kNN is among the simplest machine learning methods for classification and regression based on the closest training examples in the feature space (Zheng & Tropsha, 2000). Predictions are made based on the k nearest neighbors of the object: $y_i=w_1y_1+ w_2y_2+ ... + w_ky_k$. k is a positive integer, typically small. The neighbors are usually found based on their distance, such as Euclidean distance. The distance can be defined by

$$d(x_i, x_j) = \sqrt{(x_{i1} - x_{j1})^2 + (x_{i2} - x_{j2})^2 + \cdots + (x_{in} - x_{jn})^2}$$

where n is the number of the dimensions.

Here we focus on variable selection k nearest neighbor (kNN) QSAR approach (Zheng & Tropsha, 2000) as we have been intensively involved in the development of QSAR framework based on this method (Oloff et al., 2005; Oloff et al., 2006; Zhang et al., 2006b; Zhang et al., 2007). In our procedure the model optimization is driven by simulated annealing. It is aimed at the development of models with the highest leave-one-out (LOO) cross-validated correlation coefficient q^2 for the training set (Zheng & Tropsha, 2000).

$$q^2 = 1 - \frac{\sum_{i=1}^{N}(y_i - \hat{y}_i)^2}{\sum_{i=1}^{N}(y_i - \overline{y})^2}, \qquad (2)$$

where N and $\overline{y}$ are the number of compounds and the average observed activity of the training set, and y_i and $\hat{y}_i$ are the observed and predicted activities of the i-th compound. The procedure starts with the random selection of a predefined number of descriptors from all descriptors. Activity of a compound y_i excluded in the LOO cross-validation procedure is predicted as the weighted average of activities of its nearest neighbors according to the following formula (Zheng & Tropsha, 2000):

$$y_i = \frac{\sum_{j=1}^{k} y_j \exp(-d_{ij} / \sum_{l=1}^{k} d_{il})}{\sum_{j=1}^{k} \exp(-d_{ij} / \sum_{l=1}^{k} d_{il})}, \qquad (3)$$

where d_{ij} are distances between the i-th compound and its k nearest neighbors ($j=1,...,k$). The optimal number of nearest neighbors that yields the highest q^2 value is defined as part of the LOO cross-validation process as well. After each run of the LOO procedure, a predefined number of descriptors are randomly changed, and the new value of q^2 is defined. If q^2 (new) > q^2(old), the new set of descriptors is accepted. If q^2 (new) $\leq$ q^2(old), the new set of descriptors is accepted with probability p = exp(q^2(new) - q^2(old))/T, and rejected with probability (1-p), where T is a simulated annealing "temperature" parameter. During the process, T is decreasing until the predefined value, and when this value is achieved the optimization process is terminated.

Lazy Learning

In most current applications of machine learning approaches to QSAR problems, a single global linear or non-linear model is typically developed to fit all of the training set data. However, for large and/or diverse datasets it is practically impossible to establish a single linear or even non-linear relationship between descriptor variables and the target property in a high dimensional descriptor space that would be able to approximate the response variable satisfactorily. There are two ways to solve this problem (Atkeson et al., 1997a): using a more complex global model or fitting a simple model to local patterns instead of the whole region of interest. Herein, we discuss the application of locally weighted linear regression (Atkeson, 1992; Atkeson et al., 1997b) to building local linear models based on compounds in the vicinity of a query compound. This vicinity depends upon the local density of compounds in the training set. Rather than building a single global model for the entire dataset this approach produces multiple local linear models that collectively form a global model, either linear or non-linear. Our implementation was termed Automated Lazy Learning QSAR (ALL-QSAR) method. Lazy learning methods (Aha, 1997; Armengol & Plaza, 2003; Atkeson et al., 1997a; Wettschereck et al., 1997), also known as instance/memory-based learning defer processing of the training data until a query needs to be answered. There are several available programs, such as LOWESS and LOESS, that have become standard mathematical statistical tools included in the S statistical package (Cleveland, 1981). Despite their apparent popularity as general purpose statistical applications, lazy learning approaches have been rarely used in the QSAR analysis so far. The problem can be formulated as follows:

$$\hat{y}(\mathrm{x}) = ^2{}^{\mathrm{T}} f(\mathrm{x}) \qquad (4)$$

where $\beta^T = (\beta_1, \beta_2, ..., \beta_M)$ and $f^T(x) = f_1(x), f_2(x), ..., f_M(x)$. After applying the weighting the equation for the residual will be established (Atkeson et al., 1997a):

$$\varepsilon = \sum_{k=1}^{N} w_k^2 (y_k - {}^{2\,T} f_k)^2 \qquad (5)$$

The coefficient (β) for each polynomial term (descriptor) is obtained. Using Cholesky decomposition (Atkeson et al., 1997a; Press et al., 1992), β can be obtained with (Rencher, 2002):

$$\beta = (X^T X)^{-1} X^T y \qquad (6)$$

where $X^T X$ is an M×M matrix and $X^T y$ is an M-column vector. They can be calculated by the following formulas (Atkeson et al., 1997a; Rencher, 2002):

$$X^T X = \sum_{k=1}^{N} w_k^2 f_k f_k^T \qquad (7)$$

$$X^T y = \sum_{k=1}^{N} w_k^2 f_k y_k \qquad (8)$$

For every new query, the nearest neighbor data points and their weights change, so a different local linear model is built. Once the coefficient (β) for each descriptor is determined, the descriptor values for the query compound are entered in the regression model to obtain its activity value. Although each local model is linear, the global model can be either linear or non-linear. ALL-QSAR is demonstrated in Figure 1.

Figure 1. Locally weighted regression. The straight strong black line is the global linear regression and the long dotted gray line is the weighted linear regression (showing for one region). The thickness of gray lines indicates the strength of the weight. The strong black curve line is the final function obtained after combining local linear regressions for all the points

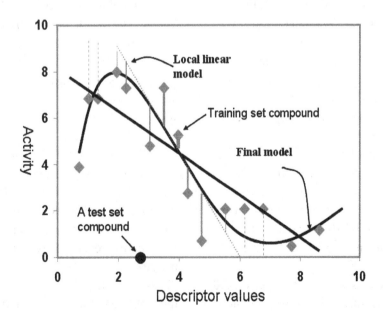

APPLICATION OF MACHINE LEARNING IN DRUG DISCOVERY AND DEVELOPMENT

Machine Learning in QSAR

As discussed above, various machine learning techniques have been implemented for QSAR studies, including *k*NN (Oloff et al., 2005; Oloff et al., 2006; Shen et al., 2004; Zhang et al., 2006b; Zheng & Tropsha, 2000), SVM (Kovatcheva et al., 2004; Xue et al., 2004; Yao et al., 2004), ALL-QSAR (Aha, 1997; Armengol & Plaza, 2003; Helma, 2006; Zhang et al., 2006a), etc. For instance, SVMs have become popular in recent years. In an effort of modeling and prediction of the induction of apoptosis by 4-aryl-4-H-chromenes (Fatemi & Gharaghani, 2007), SVM was utilized to construct the quantitative structure-activity relationship models. The results revealed that the SVM models were much better than those obtained with other methods. The root-mean-square errors of the training set and the test set for the best SVM model are 0.181 and 0.241, and the correlation coefficients were 0.950 and 0.924, respectively. The obtained statistical parameters of cross-validation test on SVM model were $q^2=0.71$ and SRESS=0.345 which revealed the reliability of this model (Fatemi & Gharaghani, 2007). In another study, QSAR models were developed to predict for CCR5 binding affinity of substituted 1-(3,3-diphenylpropyl)-piperidinyl amides and ureas (Yuan et al., 2009). Studies using multiple linear regression (MLR) and support vector machine (SVM) were performed. Compared with MLR model, the SVM models give better results with the predicted correlation coefficient (R2) of 0.867 and the squared standard error (S^2) of 0.095 for the training set. For the test set R2 was 0.732 and s2 was 0.210. It indicated that the SVMs were more adapted to the set of molecules studied (Yuan et al., 2009).

A more comprehensive study was performed by Oloff *et al.* using various non-linear methods (Oloff et al., 2005). Rigorously validated QSAR models were developed for 48 antagonists of the dopamine D1 receptor. Several QSAR methods have been employed, including CoMFA, SA-PLS, kNN, and SVM. With the exception of CoMFA, these approaches employed 2D topological descriptors generated with the MolConnZ package (EduSoft, LLC. MolconnZ, version 4.05). The original dataset was split into training and test sets to allow for external validation of each training set model. The resulting models were characterized by cross-validated q^2 for the training set and predictive R^2 values for the test set of (q^2/R^2) 0.51/0.47 for CoMFA, 0.7/0.76 for kNN, 0.74/0.71 for SVM, and 0.68/0.63 for SA-PLS.

Baskin et al. also reviewed (Baskin et al., 2008) some important methods being used to build QSAR models with artificial neural networks (ANNs), predominantly the application of multilayer ANNs in the regression analysis of structure-activity data. The review also highlighted the model interpretability, the overfitting and overtraining problems, the learning dynamics as well as the use of neural network ensembles (Baskin et al., 2008). Recently, ANNs were used to build QSAR models for the prediction of biomagnification factor (BMF) based on a dataset of 30 polychlorinated biphenyls and 12 organochlorine pollutants (Fatemi & Baher, 2009). The result demonstrated that the ANN model, which used GA selected descriptors, was superior over other models constructed by other methods. The standard errors for training and test sets of this model are 0.03 and 0.20, respectively, with $q^2 = 0.97$ and SPRESS = 0.084 obtained from cross-validation test.

With lazy learning methods, Helma (2006) recently studied the rodent carcinogenicity and Salmonella mutagenicity studies. However, although the name of the approach he used sounds similar to our method, their approach should be actually regarded as a modified k nearest neighbor method (Helma, 2006). Kulkarni et al. (2004) used locally linear embedding to reduce the nonlinear dimensions and the reduced set was subsequently modeled with robust support vector regressors. Lazy learning technique was also applied to the discovery of toxicological patterns that capture structural regularities among carcinogenic chemical compounds (Armengol & Plaza, 2003). We applied ALL-QSAR method to several experimental chemical datasets that were previously used to develop QSAR models using alternative approaches (Zhang et al., 2006a). These datasets included 48 anticonvulsant agents with known ED_{50} values (Shen et al., 2004), 48 dopamine D_1-receptor antagonists with known competitive binding affinities (Oloff et al., 2005), and a *Tetrahymena pyriformis* dataset containing 250 phenolic compounds with toxicity IGC_{50} values (Cronin et al., 2002). It was demonstrated our results are better than those obtained by other methods applied to these datasets including SVM, kNN, PLS, and CoMFA. In addition to developing predictive training set models, we also define the model applicability domain to avoid excessive extrapolation upon external prediction. This domain is defined as a similarity threshold between the training set compounds and a test set compound (Golbraikh et al., 2003). If the similarity is beyond this threshold, the prediction is considered unreliable.

Machine Learning in Virtual Screening

Virtual screening has been suggested to increase the cost-effectiveness of pharmaceutical discovery programs. Machine learning methods can be used for virtual screening by analyzing the structure characteristics of molecules with known activities. Vert and Jacob (2008) recently summarized the machine learning for *in silico* virtual screening and chemical genomics, and concluded that kernel methods have become prominent in computational biology and chemistry. Chen et al. (2007) evaluated the use of kernel discrimination and naïve Bayesian classifier (NBC) methods on 133,809 compounds from MDDR database and

found that a simple approach based on group fusion would appear to provide superior screening performance, especially for structurally diverse datasets (Kumar et al., 2004). In the same study by Oloff et al. (2005) mentioned above, validated QSAR models with $R^2 > 0.7$, from both kNN and SVM methods, were used to mine three publicly available chemical databases: the National Cancer Institute (NCI) database of ca. 250,000 compounds, the Maybridge Database of ca. 56,000 compounds, and the ChemDiv Database of ca. 450,000 compounds. These searches resulted in only 54 consensus hits, and five of them were confirmed as dopamine D1 ligands. A small fraction of the purported D1 ligands were found as novel structural antagonist leads without containing a catechol ring (Oloff et al., 2005).

With our ALL-QSAR method, models were developed for a series of functionalized amino acids as anticonvulsant agents and then they were applied to the hit identification via chemical database mining (Zhang et al., 2006a). Application of the 10 best individual ALL-QSAR models to 500,357 compounds in the ChemDiv database identified originally 326 compounds. Based on their activity profile ($ED_{50} < 60$ mg/kg) 69 structures were selected. Importantly and interestingly, the compounds have very high structural diversity. Based on the structures, they have been grouped into several families. It turns out that one of the families belongs to semicarbazone analogues which are potent anticonvulsant agents (Dimmock et al., 1996; Dimmock et al., 2000). These anticonvulsant chemical scaffolds do not exist in the training set compounds (functionalized amino acids) and the hits have not ever been identified by any of our previous studies. These initial promising results indicate that the ALL-QSAR method affords robust and externally predictive QSAR models that can be used to discover novel compounds with the desired biological profile. We expect that this novel QSAR approach can be applied to a wide variety of available experimental datasets in combination with the virtual screening using predictive QSAR models leading to the discovery of novel biologically active agents (Zhang et al., 2006a).

With similar concepts, we have used machine learning-based virtual screening methods to identify potent anticancer agents based on natural product derivatives (Zhang et al., 2007). According to a recent review (Newman et al., 2003) on New Chemical Entities (NCE), from 1981 to 2002, approximately 74% of anticancer drugs were either natural products, or natural product-based synthetic compounds, or their mimetics. (+)-(*S*)-Tylophorine and its analogues are phenanthro-indolizidine alkaloids known for their profound cytotoxic activities (Gellert & Rudzats, 1964; Pettit et al., 1984; Rao et al., 1971; Suffness & Cordell, 1985). These discoveries exemplified the great potential of developing tylophorine derivatives as a new class of antitumor drugs. We have advanced a novel series of polar, water-soluble phenanthrene-based tylophorine derivatives (PBTs) with EC_{50} $\square 10^{-7}$ M against the A549 human lung cancer cell line (Wei et al., 2006). However, the biological target of PBTs is unknown, and hence the availability of experimental data on PBT derivatives afforded us an opportunity to apply ligand-based QSAR modeling, towards the discovery of novel

anticancer agents. We applied the *k*NN QSAR method to a dataset of 52 PBTs with known EC_{50} values. The structures were characterized with MolConnZ descriptors. The models developed for the PBT dataset have been extensively validated using several criteria of robustness and accuracy (Golbraikh & Tropsha, 2002). Several validated models with the high predictive power were used to mine the commercially available ChemDiv (2005) database resulting in 34 consensus hits with the moderate to high predicted activities. Ten structurally diverse hits were experimentally tested and eight compounds were confirmed active, with the most potent compound having EC_{50} of 1.8μM. Figure 2 demonstrated the procedures of this virtual screening process. The strong activity and high hit rate (80%) of this study suggested that rigorously validated QSAR models could be successfully used as virtual screening tools for prioritizing untested compounds for experimental biological evaluation (Zhang et al., 2007).

Machine Learning for *in Silico* ADMET Prediction.

Modern drug discovery is focused on the development of effective and safe therapeutics with

Figure 2. Our virtual screening procedures of identifying potent active anticancer agents based on Tylophorine derivatives

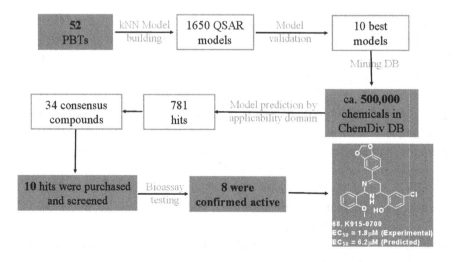

desirable ADMET properties and insignificant adverse drug reactions. Methods for *in silico* ADMET prediction in early discovery stage is critical for drug development and safety evaluation (Clark & Grootenhuis, 2002; Davis & Riley, 2004; Dearden, 2007). Machine learning methods have been explored for performing these tasks to facilitate the drug discovery process (Dearden, 2007; Li et al., 2007; Lobell et al., 2006; Penzotti et al., 2004). Most recently, Sakiyama (2009) studied four different ADMET data sets with six machine learning methods including naive Bayesian classifier, classification and regression tree, random forest, Gaussian process, support vector machine and k nearest neighbors. He demonstrated that the ensemble learning and kernel machine displayed greater accuracy of prediction than classical methods irrespective of the data set size. These results provide insights into the mechanism of machine learning (Sakiyama, 2009).

Machine learning methods can also be used to derive activity level of compounds that exhibit a specific target activity. Combined with the QSAR concept, various automated model building and prediction schemes using machine learning have been proposed (Bolis et al., 1991; King et al., 1992; Mitchell et al., 1989; Sakiyama, 2009; Schneider & Downs, 2003). For instance, Obrezanova et al. (2008) presented an automatic process for building QSAR models using Gaussian Processes, a powerful machine learning modeling method. They tried to ensure that models were built and validated within a rigorous framework: descriptor calculation, data splitting, application of modeling techniques and selection of the best model. They applied this automatic process to data sets of blood-brain barrier penetration and aqueous solubility. The results demonstrated the effectiveness of the automatic modeling process for two types of datasets commonly encountered: small *in vivo* datasets and large physicochemical datasets.

Machine learning techniques are also important to study ADMET properties when the related proteins structures are unknown. Blockade of

the human ether-a-go-go related gene (hERG) potassium channel is regarded as a major cause of drug toxicity and associated with severe cardiac side-effects (Cavalli et al., 2002). However, there is no available 3D structure of this protein. A classification approach was presented for the detection of diverse hERG blockers that combines cluster analysis of training data, feature selection and support vector machine learning (Nisius et al., 2009). Compound learning sets were first divided into clusters of similar molecules. For each cluster, independent support vector machine models were generated utilizing preselected MACCS structural keys as descriptors. These models were combined to predict hERG inhibition of their large compound data set with consistent experimental measurements. Their combined support vector machine model achieved a prediction accuracy of 85% on this data set and performed better than alternative methods used for comparison.

New Applications of Machine Learning

The prediction of the protein-ligand binding affinity is a critical component of *structure-based drug design*. Accurate estimation of *binding affinities*, or at least correct relative ranking of different ligands has proven to be a difficult task due to multiple energetic and entropic factors that must be accounted for (Ajay & Murcko, 1995; Martin, 2001). The limited accuracy and speed of current scoring functions is one of the problems hampering the broad application of docking and virtual screening in lead identification and optimization. Machine learning techniques can be applied to this area with appropriate modifications and integrations.

We have developed a hybrid methodology to predict the binding affinities for a highly diverse dataset of protein-ligand complexes by combining concepts from both *structure-based* and *ligand-based approaches* along with machine learning techniques. It is based on four-body statistical scoring function derived by combined application of the Delaunay tessellation (Carter, Jr. et al., 2001; Sherman et al., 2004; Singh et al., 1996; Tropsha et al., 1996; Zhang et al., 2008) of protein-ligand complexes and the definition of chemical atom types using the fundamental chemical concept of atomic electronegativity. We term them ENTess descriptors. Figure 3 left demonstrates the Delaunay tessellation of the protein-ligand interface, and Figure 4 shows how to calculate the ENTess descriptors for a complex. ENTess descriptors and variable selection kNN machine learning technique were employed to analyze the experimental dataset of 264 diverse protein-ligand complexes with known binding constants. Following the protocols for developing validated and predictive QSAR models established in the course of the previous studies (Golbraikh et al., 2003; Golbraikh & Tropsha, 2002; Tropsha et al., 2003), we divided this datasets into multiple training, test, and independent validation sets. We reported statistically significant quantitative structure-binding affinity relationships models capable of predicting the binding affinities of

ligands in the independent validation set with the R^2 of 0.85. Comparison with other scoring functions has demonstrated that our approach is much more accurate and efficient for the prediction of binding affinities for diverse protein-ligand structures. Therefore, in additional to traditional QSAR modeling, machine learning can also be used develop novel accurate and efficient scoring function which can be applied to molecular docking and structure-based virtual screening.

With the similar concept, we also employed machine learning to predict potential off-target properties of chemical compounds. We implemented a novel chemometric approach to search for Complimentary Ligands Based on Receptor Information (CoLiBRI). It is based on the representation of both receptor binding sites and their respective ligands in a universal chemical descriptor space, as demonstrated in Figure 3. The binding site atoms involved in the interaction with ligands are identified by Delaunay tessellation as described in ENTess. TAE/RECON (Breneman et al., 1995) descriptors were calculated independently for ligands and active site atoms. This allows

Figure 3. Left: Full atom-based protein-ligand interface tessellation for 5HVP. The ribbons are two chains of the protein. The acetylpepstatin ligand is in the spacefill display. Tetrahedra formed by ligand and protein atoms are shown between the ribbons and balls. Right: Transposed chemical descriptor space for both active site and ligand atoms forms anti-symetry correlations established by machine learning techniques

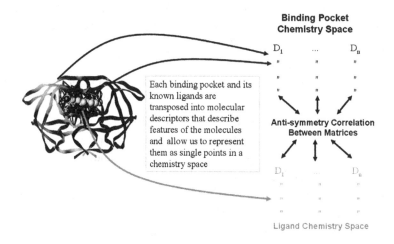

Figure 4. Calculation of ENTess descriptors. The same atom type from the receptor and ligand is treated differently. In the formulas, m is the mth quadruplet composition type; n represents the number of occurrences of this composition type in a given protein-ligand complex, and j is the vertex index within the quadruplet

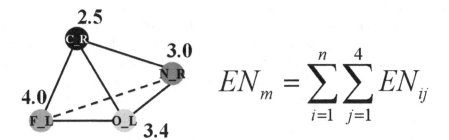

$$EN_m = \sum_{i=1}^{n} \sum_{j=1}^{4} EN_{ij}$$

the application of machine learning techniques in order to correlate chemical similarities between active sites and their respective ligands. From these calculations, we established a protocol to map patterns of nearest neighbor active site vectors in a multidimensional TAE/RECON space onto those of their complementary ligands, and *vice versa*. This protocol affords the prediction of a virtual complementary ligand vector from the position of a known active site vector, and *vice versa*. Consequently, the knowledge of the receptor active site structure affords straightforward and efficient identification of its complementary ligands in large databases of chemical compounds using rapid chemical similarity searches. Conversely, starting from the ligand chemical structure, one may identify possible complementary receptor cavities, and thus predict its potential off target properties or side effects. We have applied the CoLiBRI approach to a dataset of 800 x-ray characterized ligand receptor complexes in the PDBbind database (Wang et al., 2004). Using kNN and variable selection, we have shown that knowledge of the active site structure affords identification of its complimentary ligand among the top 1% of a large chemical database in over 90% of all test active sites when a binding site of the same protein family was present in the training set. The CoLiBRI approach provides an efficient prescreening of large chemical databases for a given protein in lead identification, or of large protein database for a given ligand in *offer-target property* predictions.

CHALLENGES AND FUTURE DIRECTIONS

Although machine learning has been widely used in drug discovery, challenges still exist (Bolis et al., 1991; King et al., 1992; King et al., 1996; Li et al., 2007; Mitchell et al., 1989; Mjolsness & DeCoste, 2001; Schneider & Downs, 2003; Vert & Jacob, 2008). The performance of machine learning methods depends on the diversity of compounds in a training dataset as well as compounds to be predicted. In lead optimization, machine learning can obtain more accurate predictions because the compounds usually have similar scaffolds and thus the similarity level is high. However, in virtual screening the rate of false positives and negatives can be really high if the chemical similarity of the database and the training compounds is very low, and unfortunately in most cases this is true because the training set typically consists of hundreds of compounds, occupying much smaller chemical space compared with millions of compounds in the currently available chemical databases. Another challenge, as we mentioned before, is the representation of chemical compounds with appropriate descriptors, which is equally important in QSAR studies. To date there are more than 3,000 molecular descriptors and selecting the right descriptors is critical to build meaningful models. Furthermore, selecting the appropriate machine learning methods and their parameters also affects the final results in a great deal. For instance, we have found that SVM had the best performance in one case (Oloff et al., 2005) while lazy learning made more accurate prediction in another case (Zhang et al., 2006a).

To tackle the above challenges, sufficiently diverse set of compounds with clean data are needed to develop machine learning models. Thus the curation of such datasets is a key to more extensive exploration of machine learning methods and construction of useful models for prediction and virtual screening. *Databases,* such as PDSP database (http://pdsp.med.unc.edu/indexR.html), DSSTox (http://pdsp.med.unc.edu/indexR.html), and PubChem (http://pubchem.ncbi.nlm.nih.gov/), that provide large experimental data, are particularly useful resources in this regard, and many more are desired. Another solution to the challenges is the rigorous validation of the models. All derived models from machine learning have to

be rigorously evaluated about their robustness and predictivity. To this end, the Y-randomization test can be conducted to exclude chance correlation (Tropsha et al., 2003). Briefly, QSAR calculations should be repeated with the randomized activities of the training sets and the statistical parameters should be compared for actual and random activities of training sets to see if there is any significant difference (Tropsha et al., 2003; Zhang et al., 2006a; Zhang et al., 2006b; Zhang et al., 2007). Predictive power should be validated using test sets based on the certain parameter criteria (e.g., cross validation $q^2 > 05$ and test set correlation coefficient $R^2 > 0.6$). Sometimes a third set is required for the extra external validation. The whole QSAR model procedure, as is illustrated in Figure 5, has been successfully used for many datasets and was described in detail elsewhere (de Cerqueira et al., 2006; Golbraikh et al., 2002; Hoffman et al., 1999; Kovatcheva et al., 2004; Kovatcheva et al., 2005; Zheng & Tropsha, 2000).

As the extrapolation of the models is dangerous and predictions may be unreliable, particularly for diverse datasets, applicability domain needs to be defined. It means the activity of the test set compounds was predicted only if these

Figure 5. Predictive QSAR diagram. Multiple statistical modeling methods, such as SVM, kNN, and ALL-QSAR, will be implemented, and the final resulted models will be used for external prediction or virtual screening

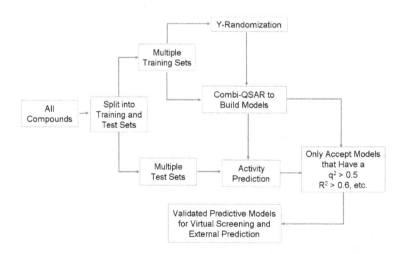

compounds were within the *applicability domain* of the respective training set models. For instance, in our variable selection kNN approach, we define this domain (Golbraikh et al., 2003) as a threshold distance in multidimensional descriptor space between a test set compound and its k nearest neighbors in the training set. If the distance is beyond the threshold, the prediction is considered unreliable. This threshold distance is calculated as $D^2_{cutoff} = <D^2_{nn}> + Z*VAR$, where $<D^2_{nn}>$ is the squared mean distance between each of the training set compound and its k nearest neighbors, VAR is the variance of D_{nn}, and Z is a user-defined parameter (the default value is 0.5). The definition of applicability domain not only controls the prediction reliability limited to similar compounds, but also increases the prediction speed by only predicting certain compounds.

Our experience also suggests that QSAR is a highly experimental area of statistical data modeling where it is impossible to decide *a priori* which particular combination of QSAR modeling method and descriptors will be best. Therefore, to achieve the highest internal and external prediction accuracy, *combinatorial QSAR* approaches (de Cerqueira et al., 2006; Kovatcheva et al., 2004; Kovatcheva et al., 2005) were proposed to explore all possible combinations of various descriptor types and optimization methods along with external model validation. Furthermore, we believe that multiple QSAR models should be developed, as opposed to a single model using some favorite QSAR method, for each data set. Our current approach to QSAR is demonstrated in Figure 4, as published previously (Kovatcheva et al., 2004; Zhang et al., 2006a; Zhang et al., 2006b; Zhang et al., 2007). In order to improve the prediction, consensus predictions should be applied by averaging the activity of each compound predicted by all acceptable models (Perez et al., 1998).

CONCLUSION

Machine learning methods consistently show promising capability in QSAR and ADMET modeling and in making quantitative prediction of the activities if the data are available for a sufficient number of compounds. However, in order to improve the robustness and predictive capability, the models should be subject to rigorous validation. Machine learning methods have also shown comparable virtual screening performance as other receptor-based and ligand-based virtual screening methods. Due to their high computing speed and capability of covering highly diverse spectrum compounds, they can be used to screen very large database efficiently and complement other VS or experimental HTS approaches in hit/lead discovery.

In additional to their application to the traditional QSAR, ADMET, and ligand-based virtual screening, machine learning approaches can also be combined with structure-based approaches to implement hybrid methods which can be used as scoring functions in molecular docking as well as to predict potential off-target properties for any given compound via inverse docking methods This type of new application of machine learning techniques shall open some new avenue for their application in drug discovery and development.

REFERENCES

Aha, D. W. (1997). Lazy learning. *Artificial Intelligence Review*, *11*, 7–10. doi:10.1023/A:1006538427943

Ajay & Murcko, M. A. (1995). Computational methods to predict binding free energy in ligand-receptor complexes. *Journal of Medicinal Chemistry*, *38*, 4953–4967. doi:10.1021/jm00026a001

Armengol, E., & Plaza, E. (2003). Discovery of toxicological patterns with lazy learning. *Knowledge-Based Intellignet Information and Engineering Systems, Pt 2 . Proceedings, 2774*, 919–926.

Atkeson, C. G. (1992). In Casdagli, M., & Eubank, S. G. (Eds.), *Memory-based approaches to approximating continous functions* (pp. 503–521). Redwood City, CA: Addison-Wesley.

Atkeson, C. G., Moore, A. W., & Schaal, S. (1997a). Locally weighted learning. *Artificial Intelligence Review, 11*, 11–73. doi:10.1023/A:1006559212014

Atkeson, C. G., Moore, A. W., & Schaal, S. (1997b). Locally weighted learning for control. *Artificial Intelligence Review, 11*, 75–113. doi:10.1023/A:1006511328852

Baskin, I. I., Palyulin, V. A., & Zefirov, N. S. (2008). Neural networks in building QSAR models. *Methods in Molecular Biology (Clifton, N.J.), 458*, 137–158.

Bolis, G., Dipace, L., & Fabrocini, F. (1991). A Machine Learning Approach to Computer-Aided Molecular Design. *Journal of Computer-Aided Molecular Design, 5*, 617–628. doi:10.1007/BF00135318

Breneman, C. M., Thompson, T. R., Rhem, M., & Dung, M. (1995). Electron Density Modeling of Large Systems using the Transferable Atom Equivalent Method. *Computers & Chemistry, 19*, 161–169. doi:10.1016/0097-8485(94)00052-G

Carter, C. W. Jr, LeFebvre, B. C., Cammer, S. A., Tropsha, A., & Edgell, M. H. (2001). Four-body potentials reveal protein-specific correlations to stability changes caused by hydrophobic core mutations. *Journal of Molecular Biology, 311*, 625–638. doi:10.1006/jmbi.2001.4906

Cavalli, A., Poluzzi, E., De, P. F., & Recanatini, M. (2002). Toward a pharmacophore for drugs inducing the long QT syndrome: insights from a CoMFA study of HERG K(+) channel blockers. *Journal of Medicinal Chemistry, 45*, 3844–3853. doi:10.1021/jm0208875

ChemDiv. (2005). Retrieved from http://www.chemdiv.com

Chen, B., Harrison, R. F., Papadatos, G., Willett, P., Wood, D. J., & Lewell, X. Q. (2007). Evaluation of machine-learning methods for ligand-based virtual screening. *Journal of Computer-Aided Molecular Design, 21*, 53–62. doi:10.1007/s10822-006-9096-5

Clark, D. E., & Grootenhuis, P. D. (2002). Progress in computational methods for the prediction of ADMET properties. *Current Opinion in Drug Discovery & Development, 5*, 382–390.

Cleveland, W. S. (1981). Lowess - A Program for Smoothing Scatterplots by Robust Locally Weighted Regression. *The American Statistician, 35*, 54. doi:10.2307/2683591

Cohen, C., Samuel, G., & Fischel, O. (2004). Computer-assisted drug design, where do we stand? *Chimica Oggi-Chemistry Today, Mar.*, 22-26.

Cramer, R. D. III, Patterson, D. E., & Bunce, J. D. (1988). Comparative molecular field analysis (CoMFA). 1. Effect of shape on binding of steroids to carrier proteins. *Journal of the American Chemical Society, 110*, 5959–5967. doi:10.1021/ja00226a005

Cronin, M. T. D., Aptula, A. O., Duffy, J. C., Netzeva, T. I., Rowe, P. H., & Valkova, I. V. (2002). Comparative assessment of methods to develop QSARs for the prediction of the toxicity of phenols to Tetrahymena pyriformis. *Chemosphere, 49*, 1201–1221. doi:10.1016/S0045-6535(02)00508-8

Davis, A. M., & Riley, R. J. (2004). Predictive ADMET studies, the challenges and the opportunities. *Current Opinion in Chemical Biology, 8,* 378–386. doi:10.1016/j.cbpa.2004.06.005

de Cerqueira, L. P., Golbraikh, A., Oloff, S., Xiao, Y., & Tropsha, A. (2006). Combinatorial QSAR modeling of P-glycoprotein substrates. *Journal of Chemical Information and Modeling, 46,* 1245–1254. doi:10.1021/ci0504317

Dearden, J. C. (2007). In silico prediction of ADMET properties: how far have we come? *Expert Opinion on Drug Metabolism & Toxicology, 3,* 635–639. doi:10.1517/17425255.3.5.635

Dimmock, J. R., Puthucode, R. N., Smith, J. M., Hetherington, M., Quail, J. W., & Pugazhenthi, U. (1996). (Aryloxy)aryl semicarbazones and related compounds: A novel class of anticonvulsant agents possessing high activity in the maximal electroshock screen. *Journal of Medicinal Chemistry, 39,* 3984–3997. doi:10.1021/jm9603025

Dimmock, J. R., Vashishtha, S. C., & Stables, J. P. (2000). Anticonvulsant properties of various acetylhydrazones, oxamoylhydrazones and semicarbazones derived from aromatic and unsaturated carbonyl compounds. *European Journal of Medicinal Chemistry, 35,* 241–248. doi:10.1016/S0223-5234(00)00123-9

Duch, W., Swaminathan, K., & Meller, J. (2007). Artificial intelligence approaches for rational drug design and discovery. *Current Pharmaceutical Design, 13,* 1497–1508. doi:10.2174/138161207780765954

Ducrot, P., Andrianjara, C. R., & Wrigglesworth, R. (2001). CoMFA and CoMSIA 3D-quantitative structure-activity relationship model on benzodiazepine derivatives, inhibitors of phosphodiesterase IV. *Journal of Computer-Aided Molecular Design, 15,* 767–785. doi:10.1023/A:1013104713913

Fatemi, M. H., & Baher, E. (2009). *A novel quantitative structure-activity relationship model for prediction of biomagnification factor of some organochlorine pollutants.* Mol.Divers.

Fatemi, M. H., & Gharaghani, S. (2007). A novel QSAR model for prediction of apoptosis-inducing activity of 4-aryl-4-H-chromenes based on support vector machine. *Bioorganic & Medicinal Chemistry, 15,* 7746–7754. doi:10.1016/j.bmc.2007.08.057

Feng, B. Y., & Shoichet, B. K. (2006). A detergent-based assay for the detection of promiscuous inhibitors. *Nature Protocols, 1,* 550–553. doi:10.1038/nprot.2006.77

Gellert, E., & Rudzats, R. (1964). The antileukemia activity of tylocrebrine. *Journal of Medicinal Chemistry, 15,* 361–362. doi:10.1021/jm00333a029

Golbraikh, A., Bonchev, D., & Tropsha, A. (2002). Novel ZE-isomerism descriptors derived from molecular topology and their application to QSAR analysis. *Journal of Chemical Information and Computer Sciences, 42,* 769–787. doi:10.1021/ci0103469

Golbraikh, A., Shen, M., Xiao, Z. Y., Xiao, Y. D., Lee, K. H., & Tropsha, A. (2003). Rational selection of training and test sets for the development of validated QSAR models. *Journal of Computer-Aided Molecular Design, 17,* 241–253. doi:10.1023/A:1025386326946

Golbraikh, A., & Tropsha, A. (2002). Beware of q(2)! *Journal of Molecular Graphics & Modelling, 20,* 269–276. doi:10.1016/S1093-3263(01)00123-1

Gomeni, R., Bani, M., D'Angeli, C., Corsi, M., & Bye, A. (2001). Computer assisted drug development (CADD): An emerging technology for accelerating drug development. *Clinical Pharmacology and Therapeutics, 69,* 3.

Hansch, C., & Fujita, T. (1995). Status of QSAR at the end of the twentieth century. *Classical and Three-Dimensional Qsar in Agrochemistry, 606,* 1–12. doi:10.1021/bk-1995-0606.ch001

Hansch, C., Muir, R. M., Fujita, T., Maloney, P. P., Geiger, E., & Streich, M. (1963). The Correlation of Biological Activity of Plant Growth Regulators and Chloromycetin Derivatives with Hammett Constants and Partition Coefficients. *Journal of the American Chemical Society, 85,* 2817–2824. doi:10.1021/ja00901a033

Harper, G., Bradshaw, J., Gittins, J. C., Green, D. V., & Leach, A. R. (2001). Prediction of biological activity for high-throughput screening using binary kernel discrimination. *Journal of Chemical Information and Computer Sciences, 41,* 1295–1300. doi:10.1021/ci000397q

Helma, C. (2006). Lazy structure-activity relationships (lazar) for the prediction of rodent carcinogenicity and Salmonella mutagenicity. *Mol.Divers Online First,* 1-12.

Hoffman, B., Cho, S. J., Zheng, W. F., Wyrick, S., Nichols, D. E., & Mailman, R. B. (1999). Quantitative structure-activity relationship modeling of dopamine D-1 antagonists using comparative molecular field analysis, genetic algorithms-partial least-squares, and K nearest neighbor methods. *Journal of Medicinal Chemistry, 42,* 3217–3226. doi:10.1021/jm980415j

Hong, H., Tong, W., Xie, Q., Fang, H., & Perkins, R. (2005). An in silico ensemble method for lead discovery: decision forest. *SAR and QSAR in Environmental Research, 16,* 339–347. doi:10.1080/10659360500203022

Joliffe, I. T. (2002). *Principle Component Analysis* (2nd ed., *Vol. 29*). New York: Springer.

Kellogg, G. E. (2002). *MolConnZ (Version 4.05).* Quincy, MA: Hall Associates Consulting.

King, R. D., Muggleton, S., Lewis, R. A., & Sternberg, M. J. E. (1992). Drug Design by Machine Learning - the Use of Inductive Logic Programming to Model the Structure-Activity-Relationships of Trimethoprim Analogs Binding to Dihydrofolate-Reductase. *Proceedings of the National Academy of Sciences of the United States of America, 89,* 11322–11326. doi:10.1073/pnas.89.23.11322

King, R. D., Muggleton, S. H., Srinivasan, A., & Sternberg, M. J. E. (1996). Structure-activity relationships derived by machine learning: The use of atoms and their bond connectivities to predict mutagenicity by inductive logic programming. *Proceedings of the National Academy of Sciences of the United States of America, 93,* 438–442. doi:10.1073/pnas.93.1.438

Klein, T. E., Huang, C., Ferrin, T. E., Langridge, R., & Hansch, C. (1986). Computer-Assisted Drug Receptor Mapping Analysis. *ACS Symposium Series. American Chemical Society, 306,* 147–158. doi:10.1021/bk-1986-0306.ch013

Kovatcheva, A., Golbraikh, A., Oloff, S., Feng, J., Zheng, W., & Tropsha, A. (2005). QSAR modeling of datasets with enantioselective compounds using chirality sensitive molecular descriptors. *SAR and QSAR in Environmental Research, 16,* 93–102. doi:10.1080/10629360412331319844

Kovatcheva, A., Golbraikh, A., Oloff, S., Xiao, Y. D., Zheng, W. F., & Wolschann, P. (2004). Combinatorial QSAR of ambergris fragrance compounds. *Journal of Chemical Information and Computer Sciences, 44,* 582–595. doi:10.1021/ci034203t

Kubinyi, H. (1986). Quantitative Relationships Between Chemical-Structure and Biological-Activity. *Chemie in Unserer Zeit, 20,* 191–202. doi:10.1002/ciuz.19860200605

Kubinyi, H. (1990). Quantitative Structure-Activity-Relationships (Qsar) and Molecular Modeling in Cancer-Research. *Journal of Cancer Research and Clinical Oncology, 116*, 529–537. doi:10.1007/BF01637071

Kubinyi, H. (1996). Evolutionary variable selection in regression and PLS analyses. *Journal of Chemometrics, 10*, 119–133. doi:10.1002/(SICI)1099-128X(199603)10:2<119::AID-CEM409>3.0.CO;2-4

Kumar, R., Kulkarni, A., Jayaraman, V. K., & Kulkarni, B. D. (2004). Structure-Activity Relationships using Locally Linear Embedding Assisted by Support Vector and Lazy Learning Regressors. *Internet Electron.J.Mol.Des., 3*, 118–133.

Li, H., Yap, C. W., Ung, C. Y., Xue, Y., Cao, Z. W., & Chen, Y. Z. (2005). Effect of selection of molecular descriptors on the prediction of blood-brain barrier penetrating and nonpenetrating agents by statistical learning methods. *Journal of Chemical Information and Modeling, 45*, 1376–1384. doi:10.1021/ci050135u

Li, H., Yap, C. W., Ung, C. Y., Xue, Y., Li, Z. R., & Han, L. Y. (2007). Machine learning approaches for predicting compounds that interact with therapeutic and ADMET related proteins. *Journal of Pharmaceutical Sciences, 96*, 2838–2860. doi:10.1002/jps.20985

Lobell, M., Hendrix, M., Hinzen, B., Keldenich, J., Meier, H., & Schmeck, C. (2006). In silico ADMET traffic lights as a tool for the prioritization of HTS hits. *ChemMedChem, 1*, 1229–1236. doi:10.1002/cmdc.200600168

Mandel, J. (1985). The Regression Analysis of Collinear Data. *Journal of Research of the National Bureau of Standards, 90*, 465–476.

Martin, Y. C. (1991). Computer-Assisted Rational Drug Design. *Methods in Enzymology, 203*, 587–613. doi:10.1016/0076-6879(91)03031-B

Martin, Y. C. (2001). Diverse viewpoints on computational aspects of molecular diversity. *Journal of Combinatorial Chemistry, 3*, 231–250. doi:10.1021/cc000073e

McLachlan, G. J. (2004). *Discriminant Analysis and Statistical Pattern Recognition* (1st ed.). New York: Wiley Interscience.

Mitchell, T., Buchanan, B., Dejong, G., Dietterich, T., Rosenbloom, P., & Waibel, A. (1989). Machine Learning. *Annual Review of Computer Science, 4*, 417–433. doi:10.1146/annurev.cs.04.060190.002221

Mjolsness, E., & DeCoste, D. (2001). Machine learning for science: State of the art and future prospects. *Science, 293*, 2051–2055. doi:10.1126/science.293.5537.2051

Newman, D. J., Cragg, G. M., & Snader, K. M. (2003). Natural products as sources of new drugs over the period 1981-2002. *Journal of Natural Products, 66*, 1022–1037. doi:10.1021/np030096l

Nisius, B., Goller, A. H., & Bajorath, J. (2009). Combining cluster analysis, feature selection and multiple support vector machine models for the identification of human ether-a-go-go related gene channel blocking compounds. *Chemical Biology & Drug Design, 73*, 17–25. doi:10.1111/j.1747-0285.2008.00747.x

Obrezanova, O., Gola, J. M., Champness, E. J., & Segall, M. D. (2008). Automatic QSAR modeling of ADME properties: blood-brain barrier penetration and aqueous solubility. *Journal of Computer-Aided Molecular Design, 22*, 431–440. doi:10.1007/s10822-008-9193-8

Oloff, S., Mailman, R. B., & Tropsha, A. (2005). Application of validated QSAR models of D1 dopaminergic antagonists for database mining. *Journal of Medicinal Chemistry, 48*, 7322–7332. doi:10.1021/jm049116m

Oloff, S., Zhang, S., Sukumar, N., Breneman, C., & Tropsha, A. (2006). Chemometric analysis of ligand receptor complementarity: identifying Complementary Ligands Based on Receptor Information (CoLiBRI). *Journal of Chemical Information and Modeling, 46,* 844–851. doi:10.1021/ci050065r

Penzotti, J. E., Landrum, G. A., & Putta, S. (2004). Building predictive ADMET models for early decisions in drug discovery. *Current Opinion in Drug Discovery & Development, 7,* 49–61.

Perez, C., Pastor, M., Ortiz, A. R., & Gago, F. (1998). Comparative binding energy analysis of HIV-1 protease inhibitors: incorporation of solvent effects and validation as a powerful tool in receptor-based drug design. *Journal of Medicinal Chemistry, 41,* 836–852. doi:10.1021/jm970535b

Pettit, G. R., Goswami, A., Cragg, G. M., Schmidt, J. M., & Zou, J. C. (1984). Antineoplastic agents, 103. The isolation and structure of hypoestestatins 1 and 2 from the East African Hypoestes verticillaris. *Journal of Natural Products, 47,* 913–919. doi:10.1021/np50036a001

Press, W. H., Flannery, B. P., Teukolsky, S. A., & Vetterling, W. T. (1992). *Numerical Recipes in C: The Art of Scientific Computing* (2nd ed.). New York: Cambridge University Press.

Rao, K. V., Wilson, R. A., & Cummings, B. (1971). Alkaloids of tylophora. 3. New alkaloids of Tylophora indica (Burm) Merrill and Tylophora dalzellii Hook f. *Journal of Pharmaceutical Sciences, 60,* 1725–1726. doi:10.1002/jps.2600601133

Rencher, A. C. (2002). *Methods of Multivariate Analysis* (2nd ed.). New York: John Wiley & Sons. doi:10.1002/0471271357

Sakiyama, Y. (2009). The use of machine learning and nonlinear statistical tools for ADME prediction. *Expert Opinion on Drug Metabolism & Toxicology, 5,* 149–169. doi:10.1517/17425250902753261

Schneider, G., & Downs, G. (2003). Machine learning methods in QSAR modelling. *QSAR & Combinatorial Science, 22,* 485–486. doi:10.1002/qsar.200390046

Shen, M., Beguin, C., Golbraikh, A., Stables, J. P., Kohn, H., & Tropsha, A. (2004). Application of predictive QSAR models to database mining: Identification and experimental validation of novel anticonvulsant compounds. *Journal of Medicinal Chemistry, 47,* 2356–2364. doi:10.1021/jm030584q

Sherman, D. B., Zhang, S., Pitner, J. B., & Tropsha, A. (2004). Evaluation of the relative stability of liganded versus ligand-free protein conformations using simplicial neighborhood analysis of protein packing (SNAPP) method. *Proteins, 56,* 828–838. doi:10.1002/prot.20131

Singh, R. K., Tropsha, A., & Vaisman, I. I. (1996). Delaunay tessellation of proteins: four body nearest-neighbor propensities of amino acid residues. *Journal of Computational Biology, 3,* 213–221. doi:10.1089/cmb.1996.3.213

Suffness, M., & Cordell, G. A. (1985). *The Alkaloids, Chemistry and Pharmacology* (*Vol. 25*). New York: Academic Press.

Tetko, I. V., Gasteiger, J., Todeschini, R., Mauri, A., Livingstone, D., & Ertl, P. (2005). Virtual computational chemistry laboratory--design and description. *Journal of Computer-Aided Molecular Design, 19,* 453–463. doi:10.1007/s10822-005-8694-y

Tralau-Stewart, C. J., Wyatt, C. A., Kleyn, D. E., & Ayad, A. (2009). Drug discovery: new models for industry-academic partnerships. *Drug Discovery Today, 14,* 95–101. doi:10.1016/j.drudis.2008.10.003

Tropsha, A. (2005). Application of Predictive QSAR Models to Database Mining . In Oprea, T. (Ed.), *Cheminformatics in Drug Discovery* (pp. 437–455). New York: Wiley-VCH. doi:10.1002/3527603743.ch16

Tropsha, A. (2006). Predictive QSAR (Quantitative Structure Activity Relationships) Modeling . In Martin, Y. C. (Ed.), *Comprehensive Medicinal Chemistry II* (pp. 113–126). New York: Elsevier.

Tropsha, A., & Golbraikh, A. (2007). Predictive QSAR modeling workflow, model applicability domains, and virtual screening. *Current Pharmaceutical Design*, *13*, 3494–3504. doi:10.2174/138161207782794257

Tropsha, A., Gramatica, P., & Gombar, V. K. (2003). The importance of being earnest: Validation is the absolute essential for successful application and interpretation of QSPR models. *QSAR & Combinatorial Science*, *22*, 69–77. doi:10.1002/qsar.200390007

Tropsha, A., Singh, R. K., Vaisman, I. I., & Zheng, W. (1996). Statistical geometry analysis of proteins: implications for inverted structure prediction. *Pacific Symposium on Biocomputing. Pacific Symposium on Biocomputing*, 614–623.

Tropsha, A., & Zheng, W. (2001). Computer Assisted Drug Design . In MacKerell, O. B. B. R. M. W. A. Jr., (Ed.), *Computational Biochemistry and Biophysics* (pp. 351–369). New York: Marcel Dekker, Inc.doi:10.1201/9780203903827.ch16

Vapnik, V. N. (1995). *The Nature of Statistical Learning Theory*. New York: Springer-Verlag.

Vert, J. P., & Jacob, L. (2008). Machine learning for in silico virtual screening and chemical genomics: new strategies. *Combinatorial Chemistry & High Throughput Screening*, *11*, 677–685. doi:10.2174/138620708785739899

Wang, R., Fang, X., Lu, Y., & Wang, S. (2004). The PDBbind database: collection of binding affinities for protein-ligand complexes with known three-dimensional structures. *Journal of Medicinal Chemistry*, *47*, 2977–2980. doi:10.1021/jm0305801

Wei, L., Brossi, A., Kendall, R., Bastow, K. F., Morris-Natschke, S. L., & Shi, Q. (2006). Antitumor agents 251: synthesis, cytotoxic evaluation, and structure-activity relationship studies of phenanthrene-based tylophorine derivatives (PBTs) as a new class of antitumor agents. *Bioorganic & Medicinal Chemistry*, *14*, 6560–6569. doi:10.1016/j.bmc.2006.06.009

Wettschereck, D., Aha, D. W., & Mohri, T. (1997). A review and empirical evaluation of feature weighting methods for a class of lazy learning algorithms. *Artificial Intelligence Review*, *11*, 273–314. doi:10.1023/A:1006593614256

Wold, S., Ruhe, A., Wold, H., & Dunn, W. J. (1984). The Collinearity Problem in Linear-Regression - the Partial Least-Squares (Pls) Approach to Generalized Inverses. *Siam Journal on Scientific and Statistical Computing*, *5*, 735–743. doi:10.1137/0905052

Xue, C. X., Zhang, R. S., Liu, H. X., Yao, X. J., Liu, M. C., & Hu, Z. D. (2004). An accurate QSPR study of O-H bond dissociation energy in substituted phenols based on support vector machines. *Journal of Chemical Information and Computer Sciences*, *44*, 669–677. doi:10.1021/ci034248u

Yao, X. J., Panaye, A., Doucet, J. P., Zhang, R. S., Chen, H. F., & Liu, M. C. (2004). Comparative study of QSAR/QSPR correlations using support vector machines, radial basis function neural networks, and multiple linear regression. *Journal of Chemical Information and Computer Sciences*, *44*, 1257–1266. doi:10.1021/ci049965i

Yap, C. W., Cai, C. Z., Xue, Y., & Chen, Y. Z. (2004). Prediction of torsade-causing potential of drugs by support vector machine approach. *Toxicological Sciences, 79*, 170–177. doi:10.1093/toxsci/kfh082

Yuan, Y., Zhang, R., Hu, R., & Ruan, X. (2009). Prediction of CCR5 receptor binding affinity of substituted 1-(3,3-diphenylpropyl)-piperidinyl amides and ureas based on the heuristic method, support vector machine and projection pursuit regression. *European Journal of Medicinal Chemistry, 44*, 25–34. doi:10.1016/j.ejmech.2008.03.004

Zhang, S., Golbraikh, A., Oloff, S., Kohn, H., & Tropsha, A. (2006a). A Novel Automated Lazy Learning QSAR (ALL-QSAR) Approach: Method Development, Applications, and Virtual Screening of Chemical Databases Using Validated ALL-QSAR Models. *Journal of Chemical Information and Modeling, 46*, 1984–1995. doi:10.1021/ci060132x

Zhang, S., Golbraikh, A., & Tropsha, A. (2006b). Development of quantitative structure-binding affinity relationship models based on novel geometrical chemical descriptors of the protein-ligand interfaces. *Journal of Medicinal Chemistry, 49*, 2713–2724. doi:10.1021/jm050260x

Zhang, S., Kaplan, A. H., & Tropsha, A. (2008). HIV-1 protease function and structure studies with the simplicial neighborhood analysis of protein packing method. *Proteins, 73*, 742–753. doi:10.1002/prot.22094

Zhang, S., Wei, L., Bastow, K., Zheng, W., Brossi, A., & Lee, K. H. (2007). Antitumor agents 252. Application of validated QSAR models to database mining: discovery of novel tylophorine derivatives as potential anticancer agents. *Journal of Computer-Aided Molecular Design, 21*, 97–112. doi:10.1007/s10822-007-9102-6

Zheng, W. F., & Tropsha, A. (2000). Novel variable selection quantitative structure-property relationship approach based on the k-nearest-neighbor principle. *Journal of Chemical Information and Computer Sciences, 40*, 185–194. doi:10.1021/ci980033m

Chapter 13

Learning and Prediction of Complex Molecular Structure– Property Relationships:
Issues and Strategies for Modeling Intestinal Absorption for Drug Discovery

Rahul Singh
San Francisco State University, USA

ABSTRACT

The problem of modeling and predicting complex structure-property relationships, such as the absorption, distribution, metabolism, and excretion of putative drug molecules is a fundamental one in contemporary drug discovery. An accurate model can not only be used to predict the behavior of a molecule and understand how structural variations may influence molecular property, but also to identify regions of molecular space that hold promise in context of a specific investigation. However, a variety of factors contribute to the difficulty of constructing robust structure activity models for such complex properties. These include conceptual issues related to how well the true bio-chemical property is accounted for by formulation of the specific learning strategy, algorithmic issues associated with determining the proper molecular descriptors, access to small quantities of data, possibly on tens of molecules only, due to the high cost and complexity of the experimental process, and the complex nature of bio-chemical phenomena underlying the data. This chapter attempts to address this problem from the rudiments: the authors first identify and discuss the salient computational issues that span (and complicate) structure-property modeling formulations and present a brief review of the state-of-the-art. The authors then consider a specific problem: that of modeling intestinal drug absorption, where many of the aforementioned factors play a role. In addressing them, their solution uses a novel characterization of molecular space based on the notion of surface-based molecular similarity. This is followed by identifying a statistically relevant set of molecular descriptors, which along with an appropriate machine learning technique, is used to build the structure-property model. The authors propose simultaneous use of both ratio and ordinal error-measures for model construction and validation. The applicability of the approach is demonstrated in a real world case study.

DOI: 10.4018/978-1-61520-911-8.ch013

INTRODUCTION

The recent past in human history has been witness to several significant events in the evolution of our understanding at the intersection of biology and medicine. Among others these include, the elucidation of the structure of the DNA, understanding the cell-cycle, cloning of proteins, advances in structure-elucidation techniques, development of rational drug design especially against well identified targets like angiotensin converting enzyme and protein kinases, and most recently, the sequencing of the human genome and mapping of the genomic DNA (Lander, 2001).

Considering the fact that all known commercial drugs today, interact with no more than 500 distinct targets, advances in genomics promise to provide a proliferation of targets that may not only lead to newer or improved therapeutics, but also open exciting avenues like individualized medicine. Somewhat simultaneously, recent developments in industrial robotics, combinatorial chemistry, and high-throughput screening have significantly increased the number of lead compounds that can be synthesized in pharmaceutical drug-discovery settings (Flickinger, 2001; McKinsey Lehman Brothers report 2001). Taken together, these factors may be assumed to point to both advancements in treatment and eradication of diseases as well as a significant reduction in the time-to-market (currently approximately 14 years on average per drug) and cost (currently 100-897 million dollars per drug, depending on the business model) of drug discovery.

Unfortunately, the trends from *pharmaceutical science* and industry differ considerably. A detailed study involving the pharmaceutical sector (McKinsey Lehman Brothers report 2001) accessed the impact of genomics on biopharmaceutical drug development. Broadly speaking, this study found that the cost and number of failures in drug discovery can be expected *to increase* in the immediate future. This startling result can be explained due to two factors. First, once a target

is identified, it needs to be validated to establish its role in a disease. Moreover, its interactions with other genes/targets have to be identified as well, for example, by elucidating the pathways it is involved in. However, validation remains a complex, non-standardized process and the advancements in genomics have, till date, been more effective in increasing our capabilities in identifying new targets, rather than in validating them. This has typically resulted in many insufficiently validated targets being considered for *drug discovery*. Second, newer targets often require that newer classes of molecules be designed to interact with them. However, owing to the structural novelty of such molecules, historical data on their *pharmacokinetics* (influence of the human biological system on the drug molecule), pharmacodynamics (influence of the drug molecule on the human body), or toxicity profiles is scarce. Appropriate pharmacokinetics, pharmacodynamics, and toxicity characteristics are essential for a successful drug. However, these properties are typically tested for, in the later stages of drug discovery due to the associated time and cost. In turn, this leads to the increased possibility of late stage attrition if the pharmacology of a molecule is found to be undesirable.

It is increasingly being recognized that computational approaches can play a significant role in biology and drug discovery, not only at the level of data management, sequence comparison and analysis, and systems biology but also in modeling behavior of molecules and other bio-chemical systems *in-silico* (Palsson, 2000; Singh, 2007). This could, for example be, characterization of the relationship between the structure of a molecule and its properties like binding, localization, or expression. Modeling such relationships constitutes an important research direction of *in-silico* biology called *structure-property modeling* (also known as structure-activity modeling). A structure-property model captures the relationship between the biochemical properties of a molecule and its physicochemical description (Enslein, 1988; Grover,

2000). In such a model the biochemical property Φ of a molecule M_i is envisaged as the function of its "chemical constitution" (Livingstone, 2000):

$$\Phi = (f(M_i)) \tag{1}$$

The basic elements needed for the development of a structure-property model are:

1. A set of parameters describing the molecular structure (Enslein, 1988).
2. Verified assay (experimental) results describing the bio-chemical property of interest (Enslein, 1988).
3. The learning formulation. This can be (a) *Classification*, involving estimation of class decision boundaries, (b) *Regression*, requiring estimation of an unknown continuous function from noisy samples, or (c) *Probability density estimation.*
4. A statistical or *machine learning* technique (e.g. multivariate regression, discriminant analysis, neural networks, or support-vector machine), to establish a relationship between the chemical description and bio-chemical property measurements.

It should be noted that the concept of structure-property modeling significantly predates the idea of *in-silico* biology (Hansch, 1962). The novelty, in context of *in-silico* biology, has been to attempt creating structure-property models for complex biochemical effects that are either too expensive or too time consuming to determine experimentally for large sets of molecules. *In- silico* biology has the potential to be a disruptive technology. For instance, structure-property models that can predict molecular effects that are traditionally determined in the late stages of a drug discovery pipeline can lead to restructuring the conventional cascade of stages in drug discovery by modeling and predicting the pharmacokinetics and pharmacodynamics of a drug molecule early on so as to avoid late stage failure.

In this chapter we identify the salient challenges that are encountered in developing complex structure-property models and present an integrated approach in addressing them. For specificity, we focus on the problem of developing models for intestinal absorption of small (drug) molecules. Notwithstanding this specific focus, the general principles described here are also applicable for designing structure-property models for many other late-stage molecular characteristics. We start this chapter in Section 2, by providing a brief overview of the drug-discovery process and cover the basic biology behind the specific structure-property model we will consider. A brief review of some recent efforts in the computer science community towards solving such problems is also presented. We follow this by enumerating the primary challenges encountered while developing structure-property models in real world situations in Section 3. An approach, investigated by us, that holds promise in context of the issues identified by us is discussed in Section 4. The experimental results from our work on modeling human intestinal absorption using real-life data is presented in Section 5. Finally, the conclusions from this research are presented in Section 6.

THE MODERN DRUG DISCOVERY PIPELINE: INTRODUCTION AND ISSUES

The process of drug discovery is a complex and multi-stage one, with the goal of discovering therapeutically useful synthetic or naturally occurring chemical products called drugs. Generally, but not exclusively (antibody-based therapeutics and gene therapy are some examples of the exception), a drug is a small molecule that interacts with a target, typically a protein or an enzyme, in a therapeutically desirable manner.

The primary stages of a standard drug discovery process are shown in Figure 1. This process begins by determining potentially interesting targets. The identification of a target for drug discovery is done by screening gene databases using homology (sequence similarity) to known targets, co-location (which organ/tissue the gene is expressed in), or other criteria like similar expression profiles that can indicate a biological connection between the gene and targeted symptoms or phenotypes. Once identified, the target is validated using molecular biology techniques like gene-knockouts and/or computational approaches like expression analysis along with pathway elucidation. Design and construction of compound libraries is another early stage in the pipeline. This goal of this stage is to build a collection of compounds to test against the target. The typical criteria used to design compound libraries include structural diversity (to ensure coverage of the molecular space), drug-likeness of the compounds, and inclusion of structural scaffolds (substructure motifs) that are known to support interactions with common classes of targets such as kinases, proteases, and G protein-coupled receptors (GP-CRs). The desired size of the library is another important criterion which influences the tradeoff between the structural diversity of the library and its density, i.e. the number of structural variations of specific scaffolds present in the library. Typical compound library sizes vary between tens of thousands to millions of molecules. Once the compound library is created, a validated target is screened against it to determine which of the molecules in the library interact with the target. These molecules are called *hits*. Subsequently, during *lead-optimization,* the hits are further analyzed and iteratively optimized in terms of

properties like binding potency, pharmacokinetics (PK), pharmacodynamics (PD), and efficacy, to come up with the molecules that are most suited to undergo clinical trials. In typical settings, the number of *hits* against a target represents a two to three order of magnitude reduction in the number of molecules being considered at the start. During the lead optimization stage, the initial focus is on maximizing potency by minimizing the concentration at which the drug-target interaction occurs. This further reduces the candidate list of molecules to (typically) a few hundred. Of these, after testing for PK/PD, only tens of molecules or fewer may remain to be considered for clinical trials.

Late Stage Properties and Prior Research in Structure-Property Modeling

The various stages of drug discovery are essentially filters to weed out unsuitable molecules, with the low-throughput or more expensive experimental stages occurring later in the pipeline. This design is meant to minimize average cost and time. However, late-stage failures in this model are exceedingly disruptive both in monetary terms and even more importantly, in terms of time spent. Poor ADMET (Absorption, Distribution, Metabolism, Excretion, and Toxicity) characteristics of the molecules are amongst the most common causes for late-stage failure. At the state-of-the-art, the cost and time required to run experiments to determine ADMET characteristics precludes early determination on a large number of molecules.

Figure 1. Primary stages of a modern drug discovery pipeline

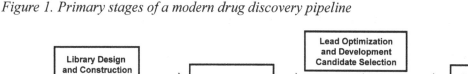

The first of these properties, *Absorption*, refers to the intestinal permeability of a putative drug molecule, and is particularly important for oral-drug entry into the body. Intestinal permeability is influenced by both passive factors like diffusion due to concentration gradient and active factors like monocarboxylic acid carrier (for transporting salicylic acid) and efflux systems like P-glycoprotein (see Figure 2). Other factors that can also play a role include electrostatic interactions between the drug and the lipid surface and partitioning into and across the lipid phase. While a detailed analysis of the biology behind intestinal absorption is beyond the scope of this chapter (the interested reader is referred to (Stenberg, 2000)), it can be inferred from recent research in biology that several transport mechanisms related to intestinal absorption have been identified. However, their relative influence on in-vivo drug absorption is yet to be determined. From the computational perspective, this implies that there is as of now, insufficient knowledge to develop analytical (also known as "first-principle") models (Cherkassky, 2007) for intestinal absorption. The alternative is to develop "soft" models for such properties by using formal or informal fitting techniques to match a mathematical model's behavior to that of the experimental data.

A review of recent literature shows that different research efforts in the computer science community have started to explore various facets of the structure-property modeling problem. For example, based on frequent sub-graph discovery, an algorithm is proposed in (Deshpande, 2003) to identify all topological and geometric substructures that are present in a dataset and distinguish compounds that constitute hits. Another approach based on inductive logic programming, is used in (Srinivasan, 1999) for classifying chemical compounds. A number of researchers have employed string-based molecular representation to discover frequently occurring substrings corresponding to sub-structures. In (Yan, 2003) the idea of graph-based substructure characterization is extended to closed frequent graphs: A graph G is defined to be a closed graph, if it does not consist of any subgraph g having the same support as that of G itself. In this approach, the problem associated with having an exponential number of frequent subgraphs in a graph is avoided. Yu (2003) use a support vector machine to determine important features that define drug activity. Other recent techniques (Berthold, 2002; Eidhammer, 2000) have considered graph-based structure characterization of molecules. It may be noted that graph theoretic approaches towards characterizing molecular properties have a long history of

Figure 2. Schematic illustration of factors involved in intestinal membrane permeability

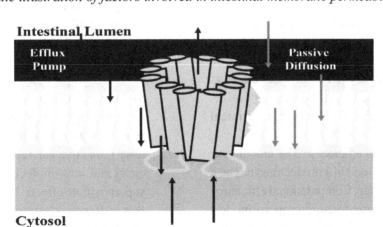

research in computational chemistry. A review of these results can be obtained from (Galvez, 2001) and references therein. The primary inadequacy of such approaches lies in that they can not properly represent the 3D nature of a molecule and associated molecular properties. Other researchers have investigated 3D, volumetric and surface-based characteristics to build structure-activity models (Ghuloum, 1999; Guba, 1998; Kubinyi, 1998; Labute, 2001; Singh, 2007).

COMPUTATIONAL CHALLENGES IN STRUCTURE-ACTIVITY MODELING

A variety of issues typically encountered during real-world drug discovery, complicate the task of building accurate structure-property models. The primary issues and criteria that need to be considered for model design are:

1. *Broad applicability*: The modeling framework should be applicable to various classes of molecules. It should be noted, that this criterion does not necessarily imply that a specific model has to predict biological activity across different structural classes. Rather, the criterion seeks to underline the necessity of developing modeling techniques that are not constrained to specific structural classes.

2. *Interpretability:* The results of a classification or prediction must be interpretable in structural and/or physicochemical terms. This is critical for obtaining a mechanistic understanding of the biochemical effect as well as for structural-driven lead optimization to consolidate the desired bio-chemical behavior.

3. *Computational efficiency:* The calculation of descriptors used in a model must not be a rate-limiting step. Computational efficiency is critical because structure-property models can be used during initial stages of drug dis-covery on millions of molecules. Also, the models can be used on virtual libraries that consist of molecules that have been designed, but not synthesized. The size of such virtual libraries can easily be in the rage of tens to hundreds of millions of compounds.

4. *Training set size:* Many biological properties, especially those related to late-stage effects like intestinal absorption are expensive to determine experimentally. Therefore, the model construction (learning) stage may not have access to large amounts of training data.

5. *Influence of experimental conditions:* Determination of many biological properties depends on experimental protocols involving among others, experimental conditions and choice of reagents. For example, *Verapamil* a P-glycoprotein substrate permeates the human jejunum at higher concentrations (Stenberg, 2000). The permeation rate, in this case, can not be fully explained through molecular structure or derived properties alone. A structure-property model that does not, either explicitly or implicitly, account for such factors is difficult to validate using commonly available experimental data and would be of limited use.

6. *Descriptor selection:* A molecule can be represented using various types of descriptors. For example, it may be represented using simple physical-chemical properties like molecular weight, number of atoms, or its octanol-water partition coefficient. Alternatively, it could be represented by 2D or 3D graph-based descriptors that seek to characterize molecular substructures as sub-graphs. Molecules may also be represented by complex 3D surface-based descriptors that can represent surface properties and intra-molecular interactions like superposition-effects. It is important to note that different descriptors have different representational capabilities (Singh, 2007).

Furthermore, at the current state of the science, there is no agreement on an ideal set of descriptors. Therefore, it is imperative to select descriptors in a manner that takes into account the fundamental nature of the bio-chemical phenomenon being modeled.

7. *Dimensionality of the descriptor set:* Typically molecular descriptors tend to be complex and high-dimensional. For example, the surface based descriptor described in (Jain, 1994) consists of 265 elements, corresponding to the points around a molecule where specific surface properties are measured. Similarly, the sub-structure based descriptor used for representing molecules in the ISIS database system employs close to a thousand relevant features (www.mdl.com). In our context, the high-dimensionality of molecular descriptors leads to two repercussions: First, given the typically small number of molecules for which data is available for model building, issues related to model overfitting can become significant. Second, recent results on the nature of high-dimensional spaces suggest that under reasonable assumptions, for a wide variety of data distributions, the ratio of the distances of the nearest neighbor to the farthest neighbor is almost constant in high-dimensional spaces (Aggarwal, 2001; Beyer, 1999; Lee, 2007). This implies that the concept of nearest-neighbor may be ill-posed in high dimensional spaces since the distinction between distances to different data point does not exist. This can cause fundamental problems in learning formulations and algorithms.

8. *Error interpretation*: Errors in modeling/prediction are either numeric (typical in a regression formulation) or misclassifications (false positives or false negatives). Generally, it should not be assumed that such errors influence the application of the model uniformly. For example, the effect

of misclassifying a promising molecule (a false negative), leading to its rejection is typically considered worse than having a "reasonable" number of false positives. Also, numeric errors that incorrectly change the rank-order of a molecule are undesirable even if their magnitudes are small, since this can change the prioritization of the molecules. Therefore, proper error modeling and handling needs to be an inseparable part of structure-property modeling.

MODELING STRUCTURE-PROPERTY RELATIONSHIPS: PROPOSED APPROACH

The proposed modeling approach consists of the following stages: (i) descriptor design, (ii) determination of the characteristic molecules, (iii) descriptor generation, (iv) feature selection, (v) model construction, and (vi) model application/prediction. In the following we describe each one of these stages in detail.

Descriptor design: The design of descriptors needs to take into account the fundamental biochemistry that we seek to model. This is one of the most crucial steps for ensuring that the computational solution targets the true biological problem and not a computational idealization of it. In our specific case, a recent study on over 1100 drug candidates conducted at GlaxoSmithKline show that the most important molecular properties that influence oral bioavailability are: (i) molecular shape and its conformations (deformations) as defined by the number of rotatable bonds in a molecule and (ii) the polar-surface area of a molecule as defined by the number of H-bond donor and acceptor atoms (Veber, 2002). Molecular shape and its deformation as well as the polar surface area of molecules are examples of three-dimensional and surface-based molecular descriptors. The significance of the results presented in (Veber, 2002) lie in that they show, through in-vivo studies, a direct relationship between three dimensional and surface-based descriptors and observed biological effects.

Most methods till date have tried to incorporate the effects of molecular shape using one of two general approaches. The first of these involves finding similar substructures or similar arrangements of groups of atoms between molecules (Berthold, 2002; Deshpande 2003; King 1996). The second approach is based on the fact that two three dimensional molecular structures can be related in terms of translation and/or rotation along the three axes. The goal of methods falling in this category is therefore to determine the optimal Euclidean transformation in the ensuing 6-degree of freedom (DOF) space such that the two structures are aligned with minimal error (Ghuloum, 1999; Jain, 1994).

Since molecular interactions occur at the molecular interface (i.e. the molecular surface), recent research has started to focus on the problem of determining the surface similarity between molecules, both in the context of small and large molecules (Binkowski, 2003; Bock, 2007; Huang, 2006; Labute, 2001; Pawlowski, 2001; Singh, 2007). Such approaches are arguably more effective in capturing the physics of molecular interactions. However, the complex nature of molecular surfaces requires solving significant modeling and computational challenges. It may be noted at this point that molecular surface generation is, in itself, a problem that has been reasonably well solved. Typically, such surfaces are generated by rolling a probe-atom over the molecule and the actual molecular surface defined as the set of points where the surface of the probe atom touches the *van der Walls* surfaces of the atoms constituting the molecule. In spite of its seeming complexity, efficient algorithms, for example (Halperin, 1994), requiring $O(nlogn)$ deterministic time and using $O(n)$ space exist for computing surface-based representations.

As mentioned above, molecular surfaces tend to be highly complex. This complicates the problem of determining the similarity of features between such surfaces – a step which is critical in creating structure-activity models. To deal with these issues, we define surface-based descriptors which are determined by placing a molecule inside a tessellated sphere. Subsequently, a one-to-one mapping is established between points on the molecular surface and points on the sphere. Based on this mapping, specific properties computable at the molecular surface, such as molecular geometry or donor/acceptor fields can be mapped to the corresponding point on the sphere. The advantage of this design is that arbitrary molecular surfaces can be mapped to a standard (spherical) coordinate system. This facilitates comparison and descriptor generation.

A central challenge in mapping arbitrary molecular surfaces to standard coordinate systems lies in dealing with the non-convexity of molecular surfaces. Motivated by research in computer vision on extended Gaussian images (Hebert, 1995), we use the idea of deformable models (Kass, 1987; McInerney, 1995) to solve this problem. Our idea lies in continuously deforming the enclosing sphere until it settles on the molecular surface. The deformation is accomplished by minimizing the energy of the surface of the sphere until convergence. The energy functional is formulated as shown in Equation (2). In it, the integral gives the total energy of the model and the terms $E_{internal}$ and $E_{external}$ denote the internal and external energy components of the model.

$$E_{model} = \int E_{internal} + E_{external} d\gamma \qquad (2)$$

The sphere is deformed iteratively. Each iteration consists of two components. First, each node of the tessellated sphere is moved based on the sum of acting forces. Once the position of each node has been computed, in the second step, the geometry of the deformed sphere is checked and corrected, if necessary, for self intersections. The motion of the nodes themselves is driven by a combination of internal and external forces. We use a physics-based model where the internal force is computed as approximate spring force minus the node normal vector (Equation 3). In this equation, NN is used to denote the number of neighbors of a node which is being moved. Further, v_n denotes the nth neighbor of node v_i and $\vec{N}_{v_i}$ denotes the surface normal vector at node v_i.

$$\vec{F}_{internal} = \left(\sum_{n=1}^{NN} \left(v_n - v_i \right) \right) - \vec{N}_{v_i} \qquad (3)$$

The formulation of the external force is described in Equation (4). This force is approximated as the pressure force, where a pressure drop is assumed from exterior of the sphere to the interior. In Equation (4), p denotes as positive constant.

$$\vec{F}_{external} = -p\vec{N}_{v_i} \qquad (4)$$

In the second stage of each iteration, the new position of each node is checked to ensure that the node has not penetrated the molecular surface. Otherwise, from the calculated point of intersection of the node and the surface, the vector component responsible for penetration into the molecular surface is subtracted. This effectively simulates the sliding of the node on the molecular surface. Empirical results indicate that on convergence this strategy also ensures better uniformity of the node distribution on the molecular surface. An example of the convergence process of the sphere on a CDK2 inhibitor is shown in Figure 3. For further details of this method we refer the reader to (Postarnakevich, 2009).

We use three surface-based features to describe a molecule. The first of these is the surface curvature, which described the geometry of the molecular surface at each surface point. We also defined the donor and acceptor fields at each surface point P_j. The donor (or acceptor) field at point P_j due to an appropriate atom at position X_i having van der Walls radii r_i is defined as:

Figure 3. Mapping of the molecular surface onto the spherical coordinate system. Top row left: the CDK2 inhibitor, middle: encapsulation of the molecular surface inside the tessellated sphere, right: early snapshot of the deformation of the sphere. Bottom row left and middle: the deforming sphere converges on the molecular surface providing a one-to-one mapping of the surface points to the sphere, right: the surface curvature mapped to the sphere. The curvature values are presented using a grey-scale with high curvature points in dark shades and low curvature points in light shades

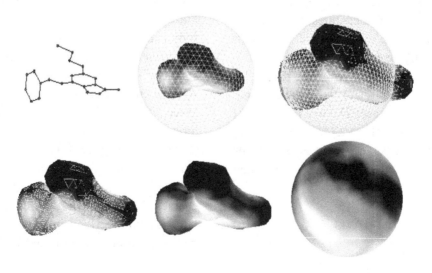

$$f(P_j, X_i) = \left(\frac{a^2}{2\pi r_i^2}\right)^{\frac{3}{2}} \exp(\frac{-a^2}{2r_i^2} \mid X_i - P_j \mid^2)$$

$$(5)$$

In addition to the above three descriptors, we also use the computed octanol-water partition coefficient (clogP), which indicates the lipophilicity of a molecule, as an additional descriptor.

Determination of characteristic molecules: One of the primary challenges induced by the aforementioned descriptor is its high dimensionality, as the sphere encapsulating the molecule needs to be densely sampled to capture the relevant molecular properties. In our method the number of tessellate points on the sphere is lower bounded by the number of points on the Connolly surface and can be substantial depending on the size of the molecule and complexity of the molecular surface. Other 3D molecular similarity techniques like (Jain, 1994) have used a lower tessellation frequency (265 points). However, this can lead to loss of resolution and may yet be insufficient to reduce the dimensionality of the descriptors significantly. Our strategy to address this problem is based on a form of dimensionality reduction that leads to chemically meaningful results. This strategy consists of two steps. Central to the first step is the selection of a predefined set of molecules which we shall henceforth refer to as the characteristic molecules. These molecules are selected in a manner that captures the structural and physicochemical variability of the set of compounds being investigated.

The idea behind the characteristic molecules is to tessellate and quantize the d-dimensional molecular descriptor space D into a finite subset C of the d-dimensional space. We assume here that d is a sufficiently large number. Formally, this process can be denoted as a mapping Q, which we defined as:

$$Q:D^d \rightarrow C, \ C=\{c_1, c_2, ...c_m\} \wedge \forall j, C_j \in D^d \qquad (6)$$

In this mapping, each characteristic molecule c_j is itself defined in the high-dimensional descriptor space. As mentioned above, the set C is selected to represent the structural and physicochemical diversity of the molecular space of interest. It is thus a 'representative subset' of the set of molecules being considered. The problem of representative subset selection is a mature area in clustering research and a variety of methods are available to solve it (Jain, 1988). Additionally, whenever applicable, selecting the set C can also be done based on domain knowledge. The first step concludes by selecting m characteristic molecules ($m \ll d$). The reader may note that this step is closely related to the concepts behind vector quantization (Cherkassky, 2007).

A transform coding is performed in the second step. In it, each molecule M_k which is involved in the structure-property relationship modeling is compared to the set C of the characteristic molecules. This comparison yields an m-dimensional vector where the ith element of the vector represents the similarity of M_k to c_i, the ith characteristic molecule (the similarity computation is described in the following paragraph). As a result of this transform coding, each molecule participating in the structure-activity model is represented by its m-dimensional description vector. Since m is much smaller than d, the dimensionality of the original descriptor space, the transform coding leads to a significant dimensionality reduction. Furthermore, the process of computing the transform coding implies that the original descriptors are taken into account. Finally, the selection of characteristic molecules can be done in a manner so as to ensure that the regions of the chemical space which are of physicochemical interest are directly be taken into account.

Descriptor generation: The descriptor generation step involves computing the similarity score of each input molecule with respect to the characteristic molecules. In this work, this similarity is computed by using the surface similarity matching technique proposed by us in (Singh,

2007), where each molecular surface is mapped to the standard spherical coordinate system as described earlier. Next, the idea of histogram intersection (Swain, 1991) is employed to compare the distribution of the molecular surface property features on the sphere. The computation of the histogram intersection score is augmented by topological constraints between patches having similar property distributions on the molecular surface. Essentially, the similarity score between two molecules reflects the relative similarity of the distribution of surface properties (in our case curvature, donor field, and acceptor field). An important attribute of this matching strategy is that it directly matches the molecular surfaces and does not require pose optimization in 6-degree of freedom (DOF)-space. This allows the similarity computation to be extremely fast. At the same time, this matching method leads to very accurate matching as indicated by results in (Singh, 2007).

Descriptor selection: In the feature selection step, the pair-wise correlation between all descriptors is iteratively used in a greedy algorithm, to determine the descriptor set that is least intra-correlated. This step removes redundant information in the descriptors and further reduces the descriptor dimensionality. In the following, we denote the set of all available descriptors by Ω and the set of least intra-correlated descriptors by Δ and explain the feature selection process in detail.

The selection process starts by determining the complete cross-correlation matrix of the descriptors. Next, for each descriptor in Ω, the sum of correlation values across all the other descriptors is determined. The descriptor corresponding to the smallest sum is least correlated with the rest of the descriptors and is selected as the first element of the set Δ. Subsequently this specific descriptor is removed from Ω. Next, the process iterates by selecting another descriptor from (the now truncated set) Ω that is least correlated, in terms of the aforementioned sum of correlation scores, from the elements of Δ. Like in the previous iteration, this descriptor is added to the set Δ

and removed from Ω. The process repeats, till a predetermined number of descriptors are included in Δ (in these investigations eight minimally correlated descriptors are used). The reader may note that the descriptor selection process reduces redundancy as well as the dimensionality of the descriptor space. In this context, the method used by us contrasts with classical dimensionality reduction techniques from statistics, such as principle component analysis (PCA) where the mean squared error of approximating the data is minimized. One of the problems of PCA is the physical interpretation of the resultant descriptors, which often are linear combinations of the input descriptors. Unlike PCA, one of the advantages of our method is its interpretability: descriptors are selected based on the extent of their correlation with the other descriptors and are either included in the final set or are excluded from it.

Model construction/selection: In the model construction phase, a learning algorithm (back-propagation-based neural network with a single hidden layer), is used to build a structure-property model using the descriptors made available from the previous step for the training molecules. Model selection is done using leave-one-out cross-validation (LOOCV) on the training set. During the LOOCV process, each compound is systematically excluded from the training set and the remaining compounds are used to build a model and predict the permeability of the missing compound. The mean error (in terms of the two measures described below) is reported once the entire set is processed. The evaluation of the model is done using two measures. The first is a ratio-measure called cross-validated r^2 and shows how well the model predicts data not used during model construction. It is defined as:

$$r^2 = 1 - \frac{\sum_i (V_i - P_i)^2}{\sum_i (V_i - \bar{V})^2} \qquad (7)$$

where V_i is the experimentally determined property of the molecule i, P_i is its predicted molecular property, and $\bar{V}$ is the mean experimental property value. The second measure is an ordinal measure called Kendall's τ that shows how well the *ordering* of the data is preserved during prediction by the model. A value of $\tau=1$ is obtained when the predicted order coincides with the order as determined by actual experimental property values. This measure is computed, for n molecules as:

$$\tau = \frac{correct\_ordering - incorrect\_ordering}{n(n-1)/2}$$

(8)

Using a combination of the above measures allows evaluation of a model both in terms of its numeric predictive accuracy, and in terms of how well it can maintain prioritization of molecules.

It should be noted, that leave-one-out cross-validation is one of many different ways to assess how a predictor will generalize to a different data set, that is, how well it will perform in practice. Other forms of cross-validation include random sub-sampling validation (where the data set is randomly split into training and test sets) and K-fold cross validation (data set is split into K subsamples and one of the subsamples is used for testing and the other K-1 subsamples are used for training). We refer the reader to (Singh, 2007), where we have analyzed model selection using K-fold cross validation. Well studied alternatives to cross-validation-based model selection include the Akaike information criterion, the Bayesian information criterion, and structural risk minimization using the VC (Vapnik-Chervonenkis) dimension (Cherkassky, 2007). Descriptor selection, to best fit the given data, can be another aspect of model construction. In the current research descriptor selection involved selecting the least redundant subset and was independent of assessing the model. However, the learning formulation can

be expanded to consider this possibility. In such cases cross-validation can also be used to find a useful subset of descriptors and techniques such as Y-randomization (Rucker, 2007) used to determine if chance correlations occur.

Prediction: In the predictive setting, first the similarity scores for the input molecules are computed with respect to the characteristic molecule set. Additionally the clogP descriptor is calculated. The final feature set defining the input molecules is obtained by selecting the least correlated descriptors. This information is entered in the model to obtain the predicted permeation values.

EXPERIMENTAL INVESTIGATION AND CASE STUDY

In this section we present experimental evaluation of the proposed approach using a dataset of 30 compounds which were tested for human intestinal permeation using the Caco-2 assay. Some of the compounds used in the study are shown in Figure 4. The assay protocol was designed to measure uni-directional flux and all compounds were analyzed at identical initial concentrations. The range of measured values was between 0.0% (no permeation) to 2.8% (maximum permeation) flux units. The data for the experiment was provided in two stages. In the first stage, 20 molecules were selected by hand (by a team of medicinal chemists who were not involved in the computational investigations) and provided for analysis and training of the predictor. The data for these molecules consisted of their chemical structures and their permeation values as defined by the Caco-2 assay. The remaining 10 molecules constituted the test data and were not made available during the descriptor selection and training process. Two observations should be noted here. First, the selection of the training set was done so as to ensure that it was representative of the entire set of 30 molecules. This is a fundamental requirement in supervised learning without which a machine learning strategy cannot be expected to perform well. Second, due to the fact that the data came from advanced stages of lead-optimization, only a small number of molecules were actually available. This is archetypical of real-world drug discovery scenarios where late-stage properties need to be modeled and data is scarce.

Model construction was done using the 20 training molecules. For each molecule in the training set, its corresponding structure was used to obtain its Connolly surface. Subsequently, the local curvature, donor-field, and acceptor field were calculated at each surface point. Next, the surfaces were mapped to a standard spherical coordinate system and the surface properties transferred to the corresponding points on the sphere. Following this, the surface-based similarity of these molecules was computed with respect to each of the 30 characteristic molecules using the method from (Singh, 2007). This information along with the computed octanol-water partition coefficient (clogP) constituted the (31 dimensional) descriptors. As part of the descriptor selection step, the complete cross-correlation matrix

of the descriptors was computed and the top eight least correlated descriptors selected. A back-propagation network with a single hidden layer was used to learn the empirical mapping between the molecules as defined by the 8-dimensional feature vector and their permeability values. Learning was stopped when the cross-validated error became lower than a predefined threshold. Figure 5 shows the performance of the model in a leave-one-out cross-validation setting for the training set. In this setting, one compound was randomly excluded from the training set and the remaining compounds used to learn a model that predicted the permeability for the excluded compound. For this model, the cross-validated r^2 equaled 0.97 and the value for Kendall's τ was 0.65. We note, that our experience across experiments, suggests that values for Kendall's τ typically tended to be lower than those for cross-validated r^2, underscoring thereby the necessity to use *both* these metrics in conjunction.

For the test set, the descriptors were computed as described above for the training data. The predicted values for the test molecules along

Figure 4. Examples of compounds used in the experiments. The molecules in the first and the second rows are from the training set. The molecules in the third row are from the test set

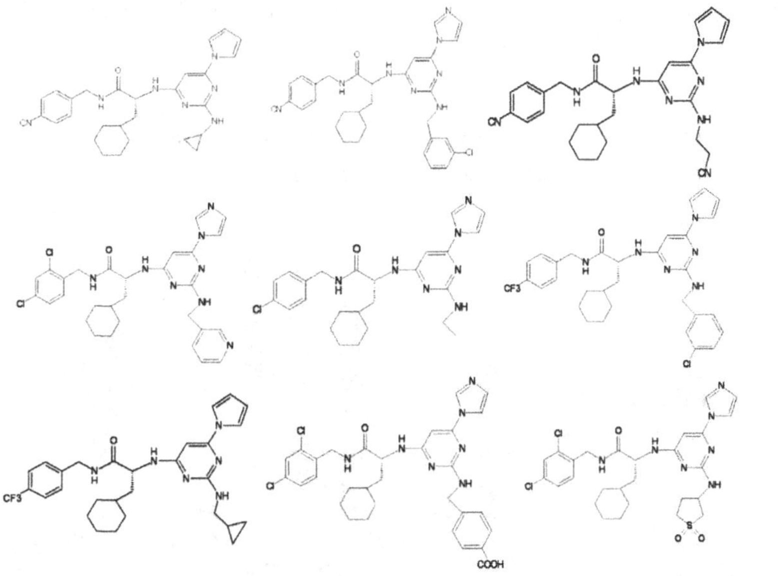

Figure 5. Leave-one-out cross-validation results on the training set using the model. Predicted values shown as lighter shaded bars on the left. Actual (measured) values are darker bars on the right

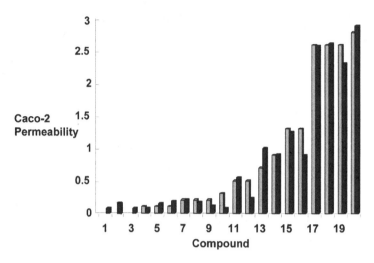

with the experimentally determined values are graphically shown in Figure 6. It may be noted, that in spite of the various computational challenges that accompanied the given problem, the proposed approach resulted in highly accurate structure-property models.

CONCLUSION AND DISCUSSION

In this chapter we considered the problem of building structure-property models for pharmaceutically relevant properties that are typically assessed in late-stages of drug discovery. We have described various factor that cause complications for direct application of standard machine-learning or data mining algorithms to such problems. Based on these observations, a general approach towards designing such models is proposed. This method ameliorates challenges associated with factors like lack of large quantities of data, high-dimensional descriptor spaces, and interpretability of resultant models. The efficacy of our approach is demonstrated by building structure-property models that demonstrate high predictability on

Figure 6. Prediction performance on the test set. Predicted values shown as lighter shaded bars on the left. Actual (measured) values are darker bars on the right

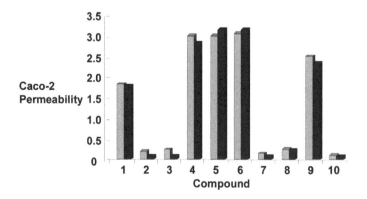

real-world experimental data for human-intestinal absorption using Caco-2 cell lines.

ACKNOWLEDGMENT

The author thanks the anonymous reviewers whose comments were instrumental in improving the presentation of the chapter. The author also thanks N. Postarnakevich for generating the visualization of the molecular surface mapping. This research was funded by the National Science Foundation through the CAREER award IIS-0644418.

REFERENCES

Aggarwal, C. C., Hinneburg, A., & Keim, D. A. (2001). On the Surprising Behavior of Distance Metrics in High Dimensional Space. *International Conference on Database Theory* (pp. 420-434). New York: Springer.

Berthold, M., & Borgelt, C. (2002). Mining Molecular Fragments: Finding Relevant Substructures of Molecules. *International Conference on Data Mining* (pp. 51-58). Washington, DC: IEEE Press.

Beyer, K. S., Goldstein, J., Ramakrishnan, R., & Shaft, U. (1999). When is nearest neighbor meaningful. *International Conference on Database Theory*, (pp. 217-235). Washington, DC: IEEE Press.

Binkowski, A. T., Adamian, L., & Liang, J. (2003). Inferring functional relationships of proteins from local sequence and spatial surface patterns. *Journal of Molecular Biology*, *332*, 505–526. doi:10.1016/S0022-2836(03)00882-9

Bock, M. E., Garutti, C., & Guerra, C. (2007). Discovery of Similar Regions on Protein Surfaces. *Journal of Computational Biology*, *14*(3), 285–299. doi:10.1089/cmb.2006.0145

Cherkassky, V., & Mulier, F. (2007). *Learning From Data*. New York, NY: Wiley Inter-Science. doi:10.1002/9780470140529

Deshpande, M. Kuramochi, M. & Karypis, G. (2003) Frequent Sub-Structure-Based Approaches for Classifying Chemical Compounds. *International Conference on Data Mining* (pp. 35-42). Washington, DC: IEEE Press.

Eidhammer, I., Jonasses, I., & Taylor, W. R. (2000). Structure Comparison and Structure Patterns. *Journal of Computational Biology*, *7*(5), 685–716. doi:10.1089/106652701446152

Enslein, K. (1988). An overview of structure-activity relationships as an alternative to testing in animals for carcinogenicity, mutagenicity, dermal and eye irritation, and acute oral toxicity. *Toxicology and Industrial Health*, *4*(4), 479–497.

Flickinger, B. (2001) Using metabolism data in early development. *Drug Discovery and Development*, September 2001.

Galvez, J., Julian-Ortiz, J. V., & Garcia-Domenech, R. (2001). General Topological Patterns of Known Drugs. *Journal of Molecular Graphics & Modelling*, *20*(1), 84–94. doi:10.1016/S1093-3263(01)00103-6

Ghuloum, A., Sage, C., & Jain, A. (1999). Molecular hashkeys: a novel method for molecular characterization and its application for predicting important pharmaceutical properties of molecules. *Journal of Medicinal Chemistry*, *42*(10), 1739–1748. doi:10.1021/jm980527a

Grover, M., Singh, B., Bakshi, M., & Singh, S. (2000). Quantitative structure-property relationships in pharmaceutical research – part 1. *Pharmaceutical Science & Technology Today*, *3*(1), 28–35. doi:10.1016/S1461-5347(99)00214-X

Guba, W., & Cruciani, G. (2000). Molecular field-derived descriptors for the multivariate modeling of pharmacokinetic data . In Gundertofte, K., & Jorgensen, F. (Eds.), *Molecular modeling and prediction of bioactivity* (pp. 89–94). New York: Kluwer Academic/Plenum Publishers.

Halperin, D., & Overmars, M. (1994). Spheres, Molecules, and Hidden Surface Removal. *10th Annual Symposium on Computational Geometry* (pp. 113-122). New York: ACM Press

Hansch, C., Maloney, P. P., Fujita, T., & Muir, R. M. (1962). Correlation of biological activity of phenoxyacetic acids with hammett substituent constants and partition coefficients. *Nature, 194*, 178–180. doi:10.1038/194178b0

Hebert, M., Ikeuchi, K., & Delingette, H. (1995). A spherical representation for recognition of free-form surfaces. *IEEE Transactions on Pattern Analysis and Machine Intelligence, 17*(7), 681–689. doi:10.1109/34.391410

Huang, B., & Schroeder, M., M. (2006). Ligsite: predicting ligand binding sites using the connolly surface and degree of conservation. *BMC Structural Biology, 6*(19).

Jain, A., & Dubes, R. (1988). *Algorithms for Clustering Data*. Upper Saddle River, NJ: Prentice Hall.

Jain, A., Koile, K., & Chapman, D. (1994). Compass: predicting biological activities from molecular surface properties. performance comparisons on a steroid benchmark. *Journal of Medicinal Chemistry, 37*(15), 2315–2327. doi:10.1021/jm00041a010

Kass, M., Witkin, A., & Terzopoulos, D. (1987). Snakes - active contour models. *International Journal of Computer Vision, 1*(4), 321–331. doi:10.1007/BF00133570

King, R. D., Muggleton, S., Srinivasan, A., & Sternberg, M. (1996). Structure-activity relationships derived by machine learning: The use of atoms and their bond connectivities to predict mutagenicity by inductive logic programming. *Proceedings of the National Academy of Sciences of the United States of America, 93*(1), 438–442. doi:10.1073/pnas.93.1.438

Kubinyi, H., Folkers, G., & Martin, Y. (1998). *3D QSAR in Drug Design*. New York: Kluwer.

Labute, P., & Williams, C. (2001). Flexible alignment of small molecules. *Journal of Medicinal Chemistry, 44*(10), 1483–1490. doi:10.1021/jm0002634

Lander, E. (2001). Initial Sequencing and Analysis of the Human Genome. *Nature, 409*(6822), 860–921. doi:10.1038/35057062

Lee, J. A., & Verleysen, M. (2007). *Nonlinear Dimensionality Reduction*. New York: Springer. doi:10.1007/978-0-387-39351-3

Lehman Brothers and McKinsey&Company. (2001). The Fruits of Genomics.

Livingstone, D. J. (2000). The characterization of chemical structures using molecular properties: A survey. *Journal of Chemical Information and Computer Sciences, 40*(2), 195–209. doi:10.1021/ci990162i

McInerney, T., & Terzopolous, D. (1999). Topology adaptive deformable surfaces for medical image volume segmentation. *IEEE Transactions on Medical Imaging, 18*(10), 840–850. doi:10.1109/42.811261

Palsson, B. (2000). The challenges of in-silico biology. *Nature Biotechnology, 18*(11), 1147–1150. doi:10.1038/81125

Pawlowski, K., & Godzik, A. (2001). Surface map comparison: studying function diversity of homologous proteins. *Journal of Molecular Biology, 309*, 793–806. doi:10.1006/jmbi.2001.4630

Postarnakevich, N., & Singh, R. (2009). Global-to-local representation and visualization of molecular surfaces using deformable models. *ACM Symposium on Applied Computing* (pp. 782-787). New York: ACM Press.

Rucker, C., Rucker, G., & Meringer, M. (2007). Y-Randomization and its Variants in QSPR/QSAR. *Journal of Chemical Information and Modeling, 47*(6), 2345–2357. doi:10.1021/ci700157b

Singh, R. (2007). Surface similarity-based molecular query-retrival. *BMC Cell Biology, 8*(supplement 1), S6. doi:10.1186/1471-2121-8-S1-S6

Srinivasan, A., & King, R. (1999). Feature construction with inductive logic programming: A study of quantitative oredictions of biological activity aided by structural attributes. *Knowledge Discovery and Data Mining Journal, 3*(1), 37–57. doi:10.1023/A:1009815821645

Stenberg, P., Luthman, K., & Artursson, P. (2000). Virtual screening of intestinal drug permeability. *Journal of Controlled Release, 65*(1-2), 231–243. doi:10.1016/S0168-3659(99)00239-4

Swain, M., & Ballard, D. (1991). Color indexing. *International Journal of Computer Vision, 7*(1), 11–32. doi:10.1007/BF00130487

Veber, D., Johnson, S., Cheng, H.-Y., Smith, B., Ward, K., & Kopple, K. (2002). Molecular properties that influence the oral bioavailability of drug candidates. *Journal of Medicinal Chemistry, 45*(12), 2615–2623. doi:10.1021/jm020017n

Yan, X., & Han, J. (2003). CloseGraph: mining closed frequent graph patterns. *ACM International Conference on Knowledge Discovery and Data Mining* (pp. 286-295). New York: ACM Press.

Yu, H. Yang, J., Wang, W., & Han, J. (2003). Discovering compact and highly discriminative features or feature combinations of drug activities using support vector machines, *IEEE Computer Society Bioinformatics Conference* (pp. 220-228). Washington, DC: IEEE Press.

Chapter 14
Learning Methodologies for Detection and Classification of Mutagens

Huma Lodhi
Imperial College London, UK

ABSTRACT

Predicting mutagenicity is a complex and challenging problem in chemoinformatics. Ames test is a biological method to assess mutagenicity of molecules. The dynamic growth in the repositories of molecules establishes a need to develop and apply effective and efficient computational techniques to solving chemoinformatics problems such as identification and classification of mutagens. Machine learning methods provide effective solutions to chemoinformatics problems. This chapter presents an overview of the learning techniques that have been developed and applied to the problem of identification and classification of mutagens.

INTRODUCTION

Mutagenicity is an unfavorable characteristic of drugs that can cause adverse effects. In chemoinformatics, it is crucial to develop and design effective and efficient computational tools to identify toxic and mutagenic molecules. Accurate prediction of mutagenicity will not only accelerate the process of finding quality lead molecules but will also decrease the potential drug attrition. During recent years considerable efforts have been devoted to developing, analyzing and applying

statistical and relational learning techniques to identify undesirable biological effects such as mutagenicity.

Mutagens produce mutations to DNA and may/ may not cause cancers. However the use of drugs that are characterized by mutagenicity but not carcinogenicity is not recommended (Debnath, Compadre, Debnath, Schusterman, & Hansch, 1991). The *Ames test* (Ames, Lee, & Durston, 1973) is viewed a biological means to identify mutagenic molecules. In this test, a bacterium, generally Salmonella typhimurium, is used to categorize mutagens and non-mutagens. The novel molecules are exposed to the bacterium that lacks

DOI: 10.4018/978-1-61520-911-8.ch014

the ability to produce amino acid, histidine. The growth of the bacterial culture demonstrates the mutations in DNA, hence the molecule is classified mutagen. Figure 1 shows a mutagenic molecule. Machine learning methods and techniques provides an accurate, useful and efficient means to classify mutagens. In this chapter we present an overview of a number of techniques that have been developed and applied to the problem of predicting mutagenicity. The review, presented in the chapter, is not exhaustive and recent research and seminal work has been outlined.

BACKGROUND

In machine learning the problem of recognition and identification of mutagens is generally solved by viewing it as a classification problems. Methods ranging from Inductive Logic Programming

Figure 1. An example of mutagenic molecule

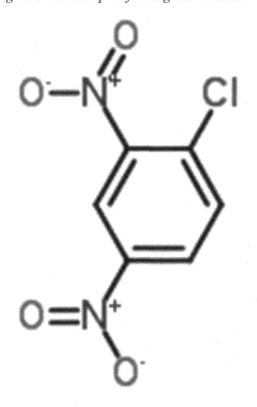

(ILP) techniques to kernel based methods (KMs) have been developed and applied to *mutagenicity classification*. Mutagenesis dataset presented by Debnath et al. (1991) is a benchmark dataset on which the efficacy of learning methods has been evaluated. We, therefore, present an overview of the techniques that have been applied to the dataset.

Mutagenesis dataset comprises 230 molecules trialled for mutagenicity on Salmonella typhimurium. Debnath et al. (1991) showed that a subset of 188 molecules are learnable using linear regression. This subset was later termed the "regression friendly" dataset (hereafter referred to as mutagenesis dataset). The remaining 42 molecules are named the "regression unfriendly" subset. Of the 188 molecules 125 have positive log mutagenicity whereas 63 molecules have zero or negative log mutagenicity. Debnath et al. identified two chemical features, C, and two structural (indicator) variables, I, to predicting mutagenicity. The chemical features are lowest unoccupied molecule orbital (LUMO) and water/octanol partition coefficient (LOGP). The two indicator variables are number of fused rings (fused rings count), I_{N1}, and examples of acenthrylenes, I_{N2}. These are structural binary variables where I_{N1} is assigned value "1" if a molecule has 3 or more fused rigs and I_{N1} is set to "0" for all the molecules that have less than 3 fused rings. Similarly the value of I_{N2} is set to 1 for 5 examples of acenthrylenes and alternatively 0. On the basis of linear regression based quantitative structure activity relation analysis, Debanth et al. suggested that *mutagenicity* of molecules that are aromatic nitro compounds is characterized by hydrophobicity, nitro groups in conjunction with electron attracting elements and 3 or more fused rings.

Srinivasan, Muggleton, King, and Sternberg (1996) introduced more features for the mutagenesis dataset by exploiting atom bond connectivities and using first order logic. The key information is given in the form of atom and bond, *AB*, description. Furthermore, atom and bond description is used to define functional groups, *FG*, including

Table 1. Features for the Mutagenesis dataset

Feature	Abbreviation	Name
Chemical	C	LUMO, LOGP
Structural (Indicator)	I	number of fused rings, examples of acenthrylenes
Structural	AB	atom, bond
Functional groups	FG	methyl groups, nitro groups, aromatic rings,
		hetero-aromatic rings, connected rings,
		ring length, benzene rings

methyl groups, nitro groups, aromatic rings, hetero-aromatic rings, connected rings, ring length and benzene rings. Table 1 summarizes the features identified by Debnath et al. (1991) and Srinivasan et al. (1996). All or some of these feature have been utilized in machine learning based mutagenesis studies. The dataset is viewed as a classification dataset by considering the molecules with positive log mutagenicity as active (positive) where the remaining are labeled inactive (negative).

MUTAGENICITY CLASSIFICATION

In machine learning classification of mutagens has been a focus of researchers. We can divide learning methods applied to mutagenesis dataset into different categories that are described below. Table 3 presents a comparison of these methods. The categories are as follows:

- Inductive Logic Programming (ILP) techniques
- Propositional methods within ILP (PMILP)
- ILP based ensemble methods (EMILP)
- Naive Bayes in conjunction with relational methods (NBR)
- Kernel based methods (KMs)

We now present an overview of these novel techniques.

Inductive Logic Programming

Inductive Logic Programming (ILP) systems have been successfully applied to solving chemoinformatics problems including *mutagenicity classification.* ILP techniques are well known for their expressive language formalism and the use of background (prior) knowledge. These methods can capture relations in data and can generate comprehensible rules. In this chapter we use "clause" and "rule" interchangeably.

In ILP, data, background knowledge and the induced theory are generally, expressed, in first order logic. The terms of first order logic are given by: a) a constant b) a variable c) a function symbol applied to terms. Atomic formulae are defined by predicate symbols applied to terms and a literal is an atomic formula or negated atomic formula. For example, $atom(m1,m2\_1,c)$ is an example of positive literal that states that atom $m1\_1$ is a carbon atom in molecule $m1$. A clause is a disjunction of literals. A Horn clause, given

Table 2. References for Table 3

Reference	Number
(Srinivasan et al., 1996)	1
(Quinlan, 1996a)	2
(Sebag & Rouveirol, 1997)	3
(Kramer & Raedt, 2001)	4
(Krogel et al., 2003)	5
(Kuzelka & Zelenzy, 2009)	6
(Hoche & Wrobel, 2001)	7
(Hoche & Wrobel, 2002)	8
(Lodhi & Muggleton, 2005)	9
(Landwehr, Kersting, & Raedt, 2005)	10
(Flach & Lachiche, 2004)	11
(Kashima, Tsuda, & Inokuchi, 2003)	12
(Mahe, Ueda, Akutsu, Perret, & Vert, 2004)	13
(Ralaivola, Swamidass, Saigo, & Baldi, 2005)	14
(Lodhi, Muggleton, & Sternberg, 2009)	15
(Landwehr et al., 2006)	16
(Ruckert & Kramer, 2008)	17

Table 3. The performance of different learning methods for the mutagenesis dataset. Accuracy has been reported for all the methods. The background knowledge has been reported as n.a (not available) for some of the methods that do not provide this information. Best accuracy values are reported in bold and second best in italics

Category	Method	Background Knowledge	Reference	Evaluation	Value
ILP					
	P-Progol	C+AB	1	10-fold	82.0
	P-Progol	C+AB+FG	1	10-fold	88.0
	FFOIL	n.a	2	10-fold	86.7
	Aleph	AB	15	10-fold	73.4
	STILL	n.a	3	train-test	93.6
				90%-10%	
PMILP					
	MFLOG	C+AB	4	10-fold	*95.7*
	MFSVM	C+AB	4	10-fold	94.7
	MFPART	C+AB	4	10-fold	93.1
	MFC4.5	C+AB	4	10-fold	90.4
	RSD	C+AB+FG	5	10-fold	92.6
	SINUS	C+AB+FG	5	10-fold	84.5
	RELAGGS	C+AB+FG	5	10-fold	88.0
	RELF	C+AB	6	10-fold	89.8
EMILP					
	C²RIB	C+AB+FG	7	10-fold	88.0
	C²RIB[D]	C+AB+FG	8	10-fold	88.8
	RS	AB	9	10-fold	88.9
	RS	AB+FG	9	10-fold	89.9
	RS	C+AB+FG+I	9	10-fold	**95.8**
	boosted-FFOIL	n.a	2	10-fold	88.3
NBR					
	nFOIL	AB	10	10-fold	78.3
	1BC2	AB	11	10-fold	82.4
	1BC2	AB+C	11	10-fold	80.3
	1BC2	AB+C+I	11	10-fold	84.6
	1BC2	C+I	11	10-fold	91.5
KMs					
	GK[1]	AB	12	leave-one-out	85.1
	GK[2]	AB	13	leave-one-out	91.0
	GK[3]	AB	14	leave-one-out	87.8
	SVILP	AB	15	10-fold	87.2
	kFOIL	AB	16	10-fold	81.3
	RUMBLE	AB	17	10-fold	84.0

continues on following page

Table 3. continued

Category	Method	Background Knowledge	Reference	Evaluation	Value
Others					
	LR	C+I	1	10-fold	89.0
	NN	C+I	1	10-fold	89.0
	IndCART	C+I	1	10-fold	88.0

by the expression $Hd \leftarrow B_1,...,B_n$, contains at most one positive literal and zero or more negative literals. In the expression, the negative literals $B_1,...,B_n$ are body of the clause and the positive literal Hd is head of the clause. Definite clauses are the Horn clauses which contain exactly one positive literal and zero or more negative literals. The Horn clause with an empty head is termed as goal while the Horn clause with empty body is known as fact.

Learning from entailment is viewed a standard framework within which ILP algorithms are designed. In this setting, a logic based learning method learns a theory (set of clauses, set of rules), H, from background knowledge, B, and a set of examples, D, where the set comprises positive examples, D^+, and negative examples, D^-. The aim of the learning algorithm is to induce a theory that does not imply any negative example and all the positive examples satisfy the conditions of the theory. The theory construction process is as follows. An *ILP* algorithm randomly selects a positive example (seed) and constructs most specific clause. The constructed clause is then generalized. All the training examples that fulfil its conditions are removed from the training set and the clause is added to the theory. This process is repeated till some stopping criterion is met.

Srinivasan et al. (1996) investigated the efficacy of ILP techniques for the detection and classification of mutagens. The study compared performance of logic based methods with attribute value learners. ILP system P-Progol (Muggleton, 1995), developed within the framework of learning from entailment, was applied to induce theories for mutagenicity. Neural Network, linear regression

and Classification and Regression Trees (CART) (Breiman, Friedman, Olshen, & Stone, 1984) were used to learn hypotheses by using attribute value representation of molecules. Classification trees were constructed by using Ind software systems (Buntine & Caruana, 1992). The performance of P-Progol was analyzed by varying the background knowledge. The study showed that the performances of P-Progol, IndCART, Neural Networks and linear regression were comparable and confirmed that ILP methods could be successfully applied to chemoinformatics problems. Furthermore, P-Progol discovered novel concepts (structural properties) of mutagenesis.

In another study Quinlan (1996a) applied FFOIL (Quinlan, 1996b), a first-order relational learning system, to mutagenicity problem with encouraging results. FFOIL is a variation of FOIL (Quinlan, 1990) specialized on learning functional relations expressed as Horn clauses and it employs the same search strategy and the selection criterion for literals that FOIL does. The learning process of FOIL is based on divide and conquer strategy and it iteratively constructs a clause by including literals to the body of the clause and removing examples that have been implied by it. FFOIL and FOIL share algorithmic similarities and the significant difference between the two approaches is FFOIL's handling of partial clauses. The application of FFOIL to mutagenesis dataset produced an accuracy of 86.7%.

Sebag and Rouveirol (1997) also investigated the problem of predicting mutagenic activity. They introduced a novel learning algorithm, Stochastic Inductive Learning (STILL), to induce first order logic clauses. The technique is based on the use of stochastic matching and is characterized by learning theories in polynomial time as compared to the earlier ILP techniques that induce theories in exponential time. Let $D(P,N)$ represents a set of hypotheses (definite clauses), that includes P clauses whose conditions are satisfied by positive example and N clauses which discriminate negative examples. STILL constructs a set of hypotheses $D_a(P,N)$ by using stochastic sampling where $D_a(P,N)$ is an approximation of $D(P,N)$ and $\lim_{a \to \infty}(D_a(P,N)) = D(P,N)$. Once the clauses are constructed, STILL applies stochastic matching to classify example in polynomial time. Experimental validation of the method demonstrated its efficacy to classifying mutagens. Table 3 shows that the predictive accuracy of STILL is higher than other logic based algorithms. In a related study, Costa et al. (2003) presented a methodology based on clause transformation to speed up ILP algorithms. The experimental results showed that ILP system Aleph obtained efficiency gain for mutagenesis dataset. In (Landwehr, Passerini, Raedt, & Frasconi, 2006) the reported accuracy of Aleph is 73.4% for mutagenesis dataset.

Propositionalization Methods

Propositionalization (Kramer, Lavrac, & Flach, 2001, Kramer, 2000) may be viewed as a learning process that transforms relational data into a form that is amenable to attribute value learning algorithms. In propositionalization, structural features are extracted from relational data and background knowledge that are generally, encoded in first order logic. Learning systems based on propositionalization consist of two modules where one module performs propositional feature extraction and the second module involves induction of a hypothesis by utilizing a feature based learner. In propositionalization stage the system is aimed at constructing all or a subset of features from relational data. The extraction of all possible features has been termed as complete propositionalization and it is characterized by zero loss of information. In the other case the system performs partial propositionalization by extracting k relevant features. There may be loss of information in partial propositionalization. The k features obtained during partial propositionalization are given by the clauses of the form as follows:

$$Clause_1(M) - Literat_{1,1}, \cdots, Literal_{1,r_1}$$
$$Clause_2(M) - Literat_{2,1}, \cdots, Literal_{2,r_2}$$
$$\cdots$$
$$Clause_k(M) - Literat_{k,1}, \cdots, Literal_{k,r_k}$$

In this setting the values of the features can be binary or non-binary. In order to obtain Boolean features for an example M, a clause i is assigned a value 1 (true) if the example satisfies the conditions of the clause otherwise the value is set 0 (false). For example a first order feature for mutagenic molecules is of the form: $C_1(M)$: $-atom(M,B,c,22,C), gteq(C,0.05)$ where the feature C_1 states that molecule M has an aromatic carbon atom with charge ≥ 0.05. The value of the feature C_1 will be "1" for all the molecules that satisfy the conditions of the clause otherwise it will be "0".

RSD (Lavrac, Zelezny, & Flach, 200), SINUS (Lavrac & Dzeroski, 1994) and RELAGGS (Krogel & Wrobel, 200) are well known relational systems that are based on partial propositionalization. RSD generates a set of first order features (clauses). During the feature construction process conjunction of literals are identified, variables are instantiated, irrelevant clauses are removed and finally the binary values of the features are computed according to examples. SINUS is a similar system that performs partial propositionalization and constructs first order features. Unlike RSD, SINUS may generate redundant features and can transform the hypothesis generated by propositional learner into relational form. In contrast to RSD and SINUS, RELAGGS (relational aggregations) is a database oriented partial propositionalization system. It has been developed by implementing techniques from databases including aggregation, database schema information and optimization techniques. In (Krogel et al., 2003) a comparative analysis of the three systems was presented for mutagenesis dataset. Molecules, encoded in first order logic, were proportionalized by using RSD, SINUS and RELAGGS. Once the data was processed, it was used in conjunction with a decision tree learner, namely J48 (reimplementation of C4.5 (Quinlan, 1993)). Experiments were performed by setting the parameters of J48 learner to their default values. In the propositionalization stage different subsets of features, with varying sizes, were obtained. The predictive accuracies of the three systems were compared with respect to the number of features. The paper reported an accuracy of 92.6% for RSD where the method achieved this accuracy by using 25 features.

Kramer and Raedt (2001) presented a propositionalization technique to generate features for molecular data. The method takes a two dimensional representation of molecules, i.e., an atom bond representation as input and generates molecular fragments as features. A molecular fragment is a string of connected atoms and within graph theoretic terms a fragment of a molecule may be viewed as the subgraph of a graph. Fragments are generated in a way so that they satisfy a number of conditions including partial ordering. The method allows the specification of interesting features by the use of query. The query is constructed by conjunction of constraints where the constrains controls the generality and relative frequency of fragments. Features that are very rare or common, in the dataset, are not selected by constructing a query of the form, $(freq(V,D) \leq \delta) \wedge (freq(V,D) \geq \varepsilon)$ for a fragment V with respect to dataset D. Here $\varepsilon \in [0,1]$, $\delta \in [0,1]$ and they provide threshold values for maximum and minimum frequency respectively. Category specific fragments can be generated by introducing class (label) information in queries. The solution of the query defines a version space (Micthell,

1982) that is given by all the generated features. A subset of features can be selected by choosing most specific and most general fragments. The authors used the constructed features in conjunction with a number of propositional techniques to classify mutagens. C4.5, PART (Frank & Witten, 1998), logistic regression and SVMs with linear kernels were used as classifiers in the study. Weka machine learning system was applied to learn the classifiers. The experiments were performed by varying the frequency thresholds and incorporating class information during feature construction. The best predictive accuracies of the algorithms are given in the Table 3 where MFLOG, MFSVM, MFPART and MFC4.5 represent logistic regression, SVMs, PART and C4.5 in conjunction with fragments respectively. Comparison among these methods, for the detection and identification of mutagens, showed that MFLOG obtained accuracy (95.7%) that is higher than the best accuracies of MFSVM, MFPART and MFC4.5. The accuracy, 95.7%, was achieved by selecting class sensitive most specific and most general fragments.

Recently Kuzelka and Zelenzy (2009) presented a propositionalization method, namely RELF. The algorithm is provided with a set of training examples, represented as first order logic interpretations, and a feature template. It filters irrelevant features from the constructed set and hence finds the relevant features for the task at hand. The authors applied random forest algorithm (Breiman, 2001) in conjunction with the computed propositional features to perform classification tasks. The application of RELF to mutagenesis dataset showed its efficacy in solving the problem of classifying mutagens.

Naive Bayes in Conjunction with Relational Methods

Recently methods have been designed to apply *probabilistic techniques* to structured data. 1BC2 (Flach & Lachiche, 2004) is an example of these techniques. It upgrades naive Bayes to perform classification of structured examples. 1BC2 views examples as bag of parts where the probability distribution over examples is computed by the probability of the parts. For example, in this setting a molecule is viewed as a bag or set of atoms and bonds. The conditional probability of an atom of particular type is computed by counting the number of atoms of the type that satisfy a given condition. Finally class conditional probability of the bag is computed from the probability of the parts. IBC2 correctly classified 91.5% molecules in mutagenesis dataset.

nFOIL (Landwehr et al., 2005) is another related approach that integrates naive Bayes and FOIL. FOIL, as described in preceding paragraphs, is a well known ILP system that has been developed within the framework of learning from entailment and is based on separate-and-conquer strategy. The algorithm employs a general-to-specific hill climbing search criterion iteratively to find good rules and the selected rules are added to the current hypothesis (set of rules). The goodness of the hypothesis is generally measured by the number of correct classification of training examples. During learning the training set is updated by removing the examples that satisfy the conditions of the current hypothesis. nFOIL is not based on separate-and-conquer strategy and does not update training set by removing the training examples that satisfy the conditions of the current hypothesis. Furthermore the goodness of each rule is measured by the conditional likelihood of data. The successful application of nFOIL to mutagenesis dataset demonstrated the usefulness of the development of probabilistic methods for structured data.

Ensemble Methods in ILP

Ensemble methods (voting methods) are a popular class of machine learning algorithms. Boosting (Schapire, 1999a, 1999c, 1999b, Freund & Schapire, 1999b) and bagging (Breiman, 1998)) are the most popular ensemble methods. The underlying

aim of these techniques is to construct a classifier that performs classification tasks with low error probability. Ensemble methods achieve this aim by constructing a series of classifiers. The final classifier performs classification tasks by voting of individual classifiers.

Quinlan (1996a) was the first to apply the boosting algorithm AdaBoost (Freund & Schapire, 1995) with FFOIL (Quinlan, 1996b), a first-order relational learning system, to mutagenesis dataset with encouraging results. AdaBoost (Freund & Schapire, 1995), is a boosting algorithm that has been extensively used and analyzed in the machine learning community over the last years. Empirical results have demonstrated AdaBoost's ability to generate highly accurate classifiers irrespective to the complexity of the base classifiers. It trains a classifier by using an iterative learning process and at each boosting iteration it adapts to the error of the base hypothesis.

The boosting FFOIL works by taking, as input, a set of relational training examples, S, a base learning algorithm (FFOIL), L, and returns a function (sequence of clauses), $h=BL(S,P)$. A distribution P that is a measure of the importance of the training examples is maintained on the training dataset. Initially the training dataset has a uniform distribution, giving equal importance or weights to all examples. Boosting FFOIL learns a classifier in a three stage learning process that is repeated T times to create an ensemble. The three stages are outlined as follows. In the first stage the algorithm calls FFOIL. As FFOIL cannot use a weighted training dataset, boosting by resampling is applied. In boosting by resampling the weak learner (FFOIL) is provided with a sample of the size of the training dataset which is selected according to the distribution over the training dataset. The base learner generates a base classifier (ordered sequence of clauses) h, $h=BL(S,P)$. The second stage consists of calculating the error ε of the hypothesis. Error, ε, is measured by the sum of the weights of all misclassified examples, $\varepsilon = \sum_{i:h(m_i) \neq c_i} P(i)$, where c_i are the labels of the training examples m_i. In third stage, boosting FFOIL updates the weights of the training examples so that the weights of misclassified examples are increased and the weights of correctly classified examples are decreased. An ensemble is constructed by repeating the three stages for T times. After completion of this process boost FFOIL solves the problem at hand, for example classification of molecules according to their mutagenicity, by taking a majority vote of the individual classifiers. Quinlan applied boosting FFOIL to mutagenesis dataset. The experiments were performed by setting the parameter T to 10. The results demonstrated that boosting improved over the performance of base learner, FFOIL.

Successful experimental studies on boosting led Schapire, Freund, Barlett, and Lee (1998) to analyse boosting in terms of another quantity, the margin. It is defined as follows: Let $S=\{(m_1 c_1),\ldots,(m_n c_n)\}$ be a set of n labeled training examples. For a real valued function f, the margin of an example (m,c) is $m(f,(m,c))=cf(m)$. *Classification* of examples may be characterized by the margin, where a positive margin suggests correct classification and negative margin tells incorrect classification. The minimum of this quantity over the whole training set gives the margin of the training set (for details, see (Lodhi, Karakoulas, & Shawe-Taylor, 2002)). The paper (Schapire et al., 1998) showed that the success of boosting methods is due to their ability of producing large margin classifiers. At each boosting iteration, the boosting algorithm, Adaboost, concentrates on the examples that have small or negative margin and forces the base learning algorithm to generate a classifier that increases their margin. At the same time, the margin of correctly classified examples may be decreased by these classifiers. In summary, the margin of an ensemble classifier is increased continuously by combining the base classifiers.

The analysis of boosting within margin theory has led to the development of effective methods to solving challenging classification problems.

The techniques presented by Hoche and Wrobel (2001), 2002) are examples of such efforts. The methods are based on the integration of boosting with ILP. The algorithm presented by Hoche and Wrobel (2001) is an instance of confidence-rated boosting (Cohen & Singer, 1999). In this approach the underlying relational base learner generates a a hypothesis, in the form of a single Horn clause, with an associated coefficient, α (real number). The underlying base hypothesis can be of two forms: a hypothesis with a positive coefficient or a default hypothesis with a negative coefficient. Given that the associated coefficient is positive, all the examples that satisfy the conditions of the hypothesis are classified positive. In the same way the examples that fulfil the conditions of the hypothesis with negative coefficient are classified negative. Confidence-rated ILP-boosting algorithm, C^2RIB, takes relationally encoded set of examples and background knowledge as input. At each boosting iteration, the algorithm calls relational base learner and obtains a hypothesis, $h(m_i)$ where $h(m_i)=\alpha$ if h implies an example m_i and alteratively $h(m_i)=0$. The algorithm then updates the weights on the examples that are maintained as a distribution, P, on the examples, $\dfrac{P(i)\exp(-c_i h(m_i))}{Z}$, where Z is normalization constant and c_i are the labels. The final classification is made by taking the sum of the individual predictions. Hoche and Wrobel (2002) extended their boosting approach and speeded up the learning process by exploiting the large margin characteristics of boosting. The proposed algorithm, C^2RIB^D starts with a few features and incrementally selects the features that are highly informative to solving the task at hand. The application of methods to the mutagenesis dataset demonstrated that their performance, in terms of accuracy, was comparable to standard ILP systems whereas the learning time was significantly less.

The best results on mutagenesis dataset were obtained by the application of an ensemble method and were reported in (Lodhi & Muggleton, 2005).

In the work an ILP based ensemble method, random seeds (RS) (Dutra, Page, & Shavilk, 2002) was used. RS may be viewed as a variant of bagging.

Bagging is based on the idea of resampling and combining. It obtains a classifier from the base learner by providing a bootstrap replicate (Efron & Tibshirani, 1993) of the training data. A bootstrap replicate is constructed by randomly drawing, with replacement, n examples from the training data of size n. The process of drawing bootstrap samples and obtaining base classifiers is repeated for T times. The final bagged classifier classifies an example by majority vote.

In RS an ILP algorithm is used as base learner. The ILP algorithm is provided with training data and background knowledge. The algorithm randomly selects a seed example and generates a theory. The process of obtaining randomized theories is repeated for T times and at each iteration the same training data is fed to the ILP learner. The final classifier is obtained by voting. In the study Aleph (Srinivasan, 2001) was used to induce theories. In this way Aleph is applied iteratively on the same training data with random choice of seeds. At each iteration Aleph induces a diverse theory by the use of a new random seed. The final combined classifier is obtained by using a voting threshold scheme. In the voting threshold scheme a molecule is classified positive if it is implied by the number of theories greater or equal to the threshold and otherwise negative. In (Lodhi & Muggleton, 2005) a simplified experimental methodology was used as compared to the scheme described in (Dutra et al., 2002). The scheme described in (Dutra et al., 2002) may pose challenges for ILP based ensemble methods due to its computational limitations and requirement of specialized tools such as Condor (Nasney & Livny, 2000). Lodhi and Muggleton (2005) conducted experiments by keeping default setting for Aleph parameters except for the number of nodes and *minacc*. Nodes were set to 20000 and *minacc* and voting threshold were tuned by applying 3-fold

cross validation in the training set of the first fold. The method obtained 88.9% accuracy by using atom bond description. There was an increase in accuracy by adding background knowledge. The accuracy of RS was 89.9% by the addition of functional groups. Finally an ensemble of 25 theories obtained an accuracy of 95.8% by using atom, bond, functional groups, indicator and chemical features. It is worth noting the results of random seed in conjunction with a simple experimental methodology are better than all the reported results in the literature.

Kernel Methods

Kernel based methods comprise a class of machine learning algorithms that is well known for strong mathematical foundations and high generalization ability. Support Vector Machine (Cortes & Vapnik, 1995) is a well known example of KMs. The building block of kernel methods is an entity known as the kernel. The non-dependence of KMs on the dimensionality of the feature space and flexibility of using any *kernel function* make them an optimal choice for different prediction tasks. KMs work by embedding the input data into a higher dimensional space through a mapping ϕ and constructing a linear function in this space. Mapping may not be known explicitly but be accessed via the kernel function that is a similarity measure and returns the inner product between the mapped examples in a higher dimensional space. A *kernel function* is a symmetric function and satisfies positive semi-definiteness. Successful construction and application of kernels for discrete objects have set new directions (Haussler, 1999, Lodhi, Saunders, Shawe-Taylor, Cristianini, & Watkins, 2002, Collins & Duffy, 2002).

Recently a number of kernels for graph based objects have been presented and evaluated on the mutagenesis dataset. We refer to the graph kernels appearing in (Kashima et al., 2003) as GK^1 and the kernels presented in (Mahe et al., 2004, Mahe, Ueda, Akutsu, Perret, & Vert, 2005) as GK^2 respectively. The kernel function defined in (Kashima et al., 2003) maps graph data into infinite dimensional feature space. In this setting a graph is implicitly represented by a vector indexed by all possible label paths. The kernel between two graphs m_1 and m_2 is given by the expression

$$k(m1, m2) = \sum_{n1 \in N1} \sum_{n2 \in N2} P_1(n1) P_2(n2) k_d(l(n_1), l(n_2)),$$

where $l(n_1)$ and $l(n_2)$ are path labels and P_1 and P_2 are probability distribution on set of paths with respect to graphs m_1 and m_2 respectively. The kernel k_d is the dirac function. Mahe et al. (2004, 2005) further extended the graph kernels presented by Kashima et al. (2003). They encoded more information in kernel matrix by incorporating a vertex neighbourhood information (in its label) and reducing the noise by preventing the successive revisiting of a vertex. In (Kashima et al., 2003) the classification of molecules was performed by using voted kernel perceptron (Freund & Schapire, 1999a) with GK[1] where in (Mahe et al., 2004, 2005) the experiments were conducted by using SVMs with GK[2]. Table 3 shows that classification accuracy of GK[2] is higher than GK[1] for mutagenesis dataset.

Other related graph based kernels were presented by Ralaivola et al. (2005). The kernels are based on the use of molecular fingerprints and label path counts. Fingerprints based representation maps a molecule to a vector indexed by substructures (labeled paths) comprising the molecule. In other words, each entry of the vector represents the occurrence or non-occurrence of a label path of given length l. The inner product between two such vectors is a kernel given by

$$k_l(m1, m2) = \sum_{l \in \Sigma^*} \phi_l(m_1) \phi_l(m_2),$$ where $\sum^*$ is the

set of all label paths of a given length l. The authors computed a number of kernels including Tanimoto kernel by employing k_l. Tanimoto kernel is given by expression

$$k_t(m_1, m_2) = \frac{k_l(m_1, m_2)}{k_l(m_1, m_1) + k_l(m_2, m_2) - k_l(m_1, m_2)}.$$

We refer to Tanimoto kernel as GK[3]. The applica-

tion of GK[3] to mutagenesis dataset showed that the proposed method successfully performs the classification task.

We now present an overview of recent approaches that integrates ILP and KMs. Support Vector Inductive Logic Programming (SVILP) (Muggleton, Lodhi, Amini, & Sternberg, 2005, Lodhi et al., 2009) is an example of such techniques. It is at the intersection of SVMs and ILP and has been applied to identify and classify mutagens. SVILP compares two molecules by means of structural and relational features they contain; the more features in common the more similar the molecules are. The induction process of SVILP can be viewed as a multi-stage learning algorithm where the four stages of SVILP learning are as follows. In the first stage a set of rules H is obtained from an ILP system. This stage maps the examples into a logic based relational space. A first order rule, $h \in H$, can be viewed as a boolean function of the form, $h:D \rightarrow \{0,1\}$. In the next stage a subset $H \subseteq H$ is selected by using a statistical measure. The third stage computes a kernel function by using the selected set of rules that can be weighted/unweighted. The kernel for examples m_1 and m_2 is given by,

$$k(m_1, m_2) = \langle \phi(m_1), \phi(m_2) \rangle = \sum_{l=1}^{t} \sqrt{\pi(h_l(m_1))} \sqrt{\pi(h_l(m_2))}.$$

The mapping ϕ for an example m is given by

$$\phi : m \rightarrow \left(\sqrt{\pi(h_1(m))}, \sqrt{\pi(h_2(m))}, \ldots, \sqrt{\pi(h_t(m))} \right)',$$

where π is the weight assigned to each rule. The construction maps the data into a feature space, where dimensionality of the space is the same as the cardinality of the set of rules. An RBF kernel using kernel k is computed by,

$$k_{RBF}(m_1, m_2) = \exp \left(\frac{-\|\varphi(m_1) - \varphi(m_2)\|^2}{2\sigma^2} \right),$$

where

$$\|\varphi(m_1) - \varphi(m_2)\|^2 = \sqrt{k(m_1, m_1) - 2k(m_1, m_2) + k(m_2, m_2)}.$$

In the final stage SVILP trains a classifier by using an SVM. Another recent technique, kFOIL (Landwehr et al., 2006), combines ILP system

FOIL with kernel based methods. kFOIL has been developed within the same framework in which nFIOL has been designed. However, In kFIOIL ILP rules are selected dynamically and the goodness of a rule is measured by the use of SVMs. RUMBLE (Ruckert & Kramer, 2008) is another related technique. All these methods have been applied to mutagenesis dataset with positive results. The results presented in Table 3 show that the SVILP outperforms related approaches kFOIL and RUMBLE in classifying mutagens. It is to be noted that some of the methods that belong to category NBR and KMs also belong to category PMILP.

CONCLUSION

In this chapter we have presented an overview of machine learning methods for the problem of predicting mutagenicity in chemoinformatics. We are witnessing a rapid development in the field of chemoinformatics that will provide solutions to challenging chemical problems. We believe that the growing popularity of machine learning methods to studying and analyzing problems such as identification and classification of mutagens will foster collaboration between researchers from diverse disciplines including chemistry, biology and computer science and will lead to significant development and progress in the field of chemoinformatics.

REFERENCES

Ames, B. N., Lee, F. D., & Durston, W. E. (1973). An improved bacterial test system for the detection and classification of mutagens and carcinogens. *Proceedings of the National Academy of Sciences of the United States of America, 70*(3), 782–786. doi:10.1073/pnas.70.3.782

Breiman, L. (1998). Arcing classifiers. *Annals of Statistics, 3*(26), 801–849.

Breiman, L. (2001). Random forests. *Machine Learning, 45*, 5–32. doi:10.1023/A:1010933404324

Breiman, L., Friedman, J., Olshen, R. A., & Stone, C. (1984). *Classification and regression trees.* Belmont, CA: Wadsworth International Group.

Buntine, W., & Caruana, R. (1992). *Introduction to IND version 2.1 and recursive partitioning.* CA: Moffet Field.

Cohen, W., & Singer, Y. (1999). A simple, fast and effective rule learner. In *Proceedings of the 16th national conference on artificial intelligence.*

Collins, M., & Duffy, N. (2002). Convolution kernels for natural language . In *Advances in neural information processing system (NIPS-14).* Cambridge, MA: MIT Press.

Cortes, C., & Vapnik, V. (1995). Support vector networks. *Machine Learning, 20*, 273–297. doi:10.1007/BF00994018

Costa, V. S., Srinivasan, A., Camacho, R., Blockeel, H., Demoae, B., & Janssens, G. (2003). Query transformations for improving the efficiency of ILP systems. *Journal of Machine Learning Research, 4*, 465–491. doi:10.1162/153244304773936027

Debnath, A. K., de Compadre, R. L. L., Debnath, G., Schusterman, A. J., & Hansch, C. (1991). Structure-activity relationship of mutagenic aromatic and heteroaromatics nitro compounds. correlation with molecular orbital energies and hydrophobicity. *Journal of Medicinal Chemistry, 34*(2), 786–797. doi:10.1021/jm00106a046

Dutra, I. C., Page, D., & Shavilk, J. (2002). An emperical evaluation of bagging in inductive logic programming. In *Proceedings of the international conference on inductive logic programming.*

Efron, B., & Tibshirani, R. (1993). *An introduction to bootstrap*. New York: Chapman and Hall.

Flach, P., & Lachiche, N. (2004). Naive bayesian classifcation of structured data. *Machine Learning, 57*(3), 233–269. doi:10.1023/B:MACH.0000039778.69032.ab

Frank, E., & Witten, I. (1998). Generating accurate rule sets without global optimization. In *Proceedings of the fifteenth international conference on machine learning (icml-98)* (pp. 144–151).

Freund, Y., & Schapire, R. (1999a). Large margin classification using the perceptron algorithm. *Machine Learning, 37*, 277–296. doi:10.1023/A:1007662407062

Freund, Y., & Schapire, R. E. (1995). A decision-theoretic generalization of on-line learning and an application to boosting . In *Computational learning theory: Eurocolt '95* (pp. 23–37). New York: Springer-Verlag.

Freund, Y., & Schapire, R. E. (1999b). A short introduction to boosting. *Journal of Japanese Society for Artificial Intelligence, 5*(14), 771–780.

Haussler, D. (1999, July). *Convolution kernels on discrete structures* (Tech. Rep. No. UCSC-CRL-99-10).: University of California in Santa Cruz, Computer Science Department.

Hoche, S., & Wrobel, S. (2001). Relational learning using constrained confidence-rated boosting. In C. Rouveirol & M. Sebag (Eds.), *Proceedings of the eleventh international conference on inductive logic programming ILP* (pp. 51–64). New York: Springer-Verlag.

Hoche, S., & Wrobel, S. (2002). Scaling boosting by margin-based inclusion of features and relations. In *Proceedings of the 13th european conference on machine learning (ECML)* (pp. 148–160).

Kashima, H., Tsuda, K., & Inokuchi, A. (2003). Marginalized kernels between labeled graphs. In T. Faucett & N. Mishra (Eds.), *Proceedings of the twentieth international conference on machine learning (ICML-2003)* (pp. 321–328).

Kramer, S. (2000). Relational learning vs. propositionalization. *AI Communications, 13*(4), 275–276.

Kramer, S., Lavrac, N., & Flach, P. (2001). Propositionalisation approaches to Relational Data Mining. In Dzeroski, S., & Larac, N. (Eds.), *Relational data mining* (pp. 262–291). Berlin: Springer.

Kramer, S., & Raedt, L. D. (2001). Feature construction with version spaces for biochemical applications. In *Proceedings of the eighteenth international conference on machine learning (ICML-2001)*.

Krogel, M. A., Rawles, S., Zelezny, F., Flach, P. A., Lavrac, N., & Wrobel, S. (2003). Comparative evaluation of approaches to propositionalization. In *Proceedings of the 13th international conference on inductive logic programming*. New York: Springer-Verlag.

Krogel, M. A., & Wrobel, S. (200). Transformation-based learning using multirelational aggregation. In S. Rouveirol & M. Sebag (Eds.), *Proceedings of the eleventh international conference on inductive logic programming (ILP)*. New York: Springer.

Kuzelka, O., & Zelenzy, F. (2009). Block-wise construction of acyclic relational features with monotone irreducibility and relevanve properties. In *Proceedings of the 26th international conference on machine learning* (pp. 569–575).

Landwehr, N., Kersting, K., & Raedt, L. D. (2005). nFOIL:integrating naive bayes and foil. In *Proceedings of the twentieth national conference on artificial intelligence (AAAI-05)*.

Landwehr, N., Passerini, A., Raedt, L., & Frasconi, P. (2006). kFOIL: Learning simple relational kernels. In *Proceedings of the national conference on artificial intelligence (AAAI)* (Vol. 21, pp. 389–394).

Lavrac, N., & Dzeroski, S. (1994). *Inductive logic programming*. Ellis Horwood. Lavrac, N., Zelezny, F., & Flach, P. A. (200). RSD: Relational subgroup discovery through first-order feature construction. In S. Matwin & C. Sammut (Eds.), *Proceedings of the twelfth international conference on inductive logic programming (ILP)*. New York: Springer.

Lodhi, H., Karakoulas, G., & Shawe-Taylor, J. (2002). Boosting strategy for classification. *Intelligent Data Analysis, 6*(2), 149–174.

Lodhi, H., & Muggleton, S. (2005). Is mutagenesis still challenging. In *International conference on inductive logic programming, (ILP - late-breaking papers)* (pp. 35–40).

Lodhi, H., Muggleton, S., & Sternberg, M. J. E. (2009). Learning large margin first order decision lists for multi-class classification. In *Proceedings of the twelfth international conference on discovery science (DS-09)* (Vol. LNAI).

Lodhi, H., Saunders, C., Shawe-Taylor, J., Cristianini, N., & Watkins, C. (2002). Text classification using string kernels. *Journal of Machine Learning Research, 2*, 419–444. doi:10.1162/153244302760200687

Mahe, P., Ueda, N., Akutsu, T., Perret, J. L., & Vert, J. P. (2004). Extensions of marginalized graph kernels. In R. Greiner & D. Schuurmans (Eds.), *Proceedings of the twenty-first international conference on machine learning (ICML-2004)* (pp. 552–559). New York: ACM Press.

Mahe, P., Ueda, N., Akutsu, T., Perret, J. L., & Vert, J. P. (2005). Graph kernels for molecular structure-activity relationship analysis with support vector machines. *Journal of Chemical Information and Modeling, 45*(4), 939–951. doi:10.1021/ci050039t

Micthell, T. (1982). Generalization as search. *Artificial Intelligence, 18*, 203–226. doi:10.1016/0004-3702(82)90040-6

Muggleton, S. (1995). Inverse entailment and progol. *New Generation Computing, 13*, 245–286. doi:10.1007/BF03037227

Muggleton, S., Lodhi, H., Amini, A., & Sternberg, M. J. E. (2005). Support Vector Inductive Logic Programming. In *Proceedings of the eighth international conference on discovery science* (Vol. LNCS, pp. 163–175). New York: Springer Verlag.

Nasney, J., & Livny, M. (2000). Managing network resources in Condor. In *Proceedings of the ninth ieee symposium on high performance distributed computing (hpdc9)* (pp. 298–299).

Quinlan, J. R. (1990). Learning logical definitions from relations. *Machine Learning*, 239–266. doi:10.1007/BF00117105

Quinlan, J. R. (1993). *C4.5: Programs for machine learning*. New York: Morgan Kaufmann.

Quinlan, J. R. (1996a). Boosting first-order learning. In S. Arikawa & A. Sharma (Eds.), *Proceedings of the 7th international workshop on algorithmic learning theory* (Vol. LNAI, pp. 143–155). New York: Springer.

Quinlan, J. R. (1996b). Learning first-order definitions of functions. *Journal of Artificial Intelligence Research, 5*.

Ralaivola, L., Swamidass, S., Saigo, H., & Baldi, P. (2005). Graph kernels for chemical informatics. *Neural Networks, 18*(8), 1093–1110. doi:10.1016/j.neunet.2005.07.009

Ruckert, U., & Kramer, S. (2008). Margin-base first-order rule learning. *Machine Learning, 70*(2-3), 189–206. doi:10.1007/s10994-007-5034-6

Schapire, R. E. (1999a). A brief introduction to boosting. In *Proceedings of the sixteenth international conference on artificial intelligence* (pp. 1401–1406).

Schapire, R. E. (1999b). Theoretical views of boosting. In *European conference on computational learning theory* (pp. 1–10).

Schapire, R. E. (1999c). Theoretical views of boosting and applications . In *Tenth international conference on algorithmic learning theory* (pp. 13–25). New York: Springer-Verlag. doi:10.1007/3-540-46769-6_2

Schapire, R. E., Freund, Y., Barlett, P., & Lee, W. S. (1998). Boosting the margin: A new explanation for the effectiveness of voting methods. *Annals of Statistics, 5*(26), 1651–1686.

Sebag, M., & Rouveirol, C. (1997). Tractable induction and classification in FOL via stochastic matching. In . *Proceedings, IJCAI-97,* 888–892.

Srinivasan, A. (2001). *The aleph manual.*

Srinivasan, A., Muggleton, S. H., King, R., & Sternberg, M. (1996). Theories for mutagenicity: a study of first-order and feature based induction. *Artifical Intelligence, 85*(1,2), 277–299.

Chapter 15
Brain–like Processing and Classification of Chemical Data:
An Approach Inspired by the Sense of Smell

Michael Schmuker
Freie Universität Berlin, Germany

Gisbert Schneider
Johann-Wolfgang-Goethe Universität, Germany

ABSTRACT

The purpose of the olfactory system is to encode and classify odorants. Hence, its circuits have likely evolved to cope with this task in an efficient, quasi-optimal manner. In this chapter the authors present a three-step approach that emulate neurocomputational principles of the olfactory system to encode, transform and classify chemical data. In the first step, the original chemical stimulus space is encoded by virtual receptors. In the second step, the signals from these receptors are decorrelated by correlation-dependent lateral inhibition. The third step mimics olfactory scent perception by a machine learning classifier. The authors observed that the accuracy of scent prediction is significantly improved by decorrelation in the second stage. Moreover, they found that although the data transformation they propose is suited for dimensionality reduction, it is more robust against overdetermined data than principal component scores. The authors successfully used our method to predict bioactivity of drug-like compounds, demonstrating that it can provide an effective means to connect chemical space with biological activity.

INTRODUCTION

In many application domains, our senses are more efficient in analyzing information than most computational implementations. In addition, it appears that when a specific kind of information must be encoded, engineers often find solutions which are similar to information processing strategies that have evolved in nature. For example, various image compression algorithms use a wavelet-like encoding strategy, which is comparable to what happens in the retina and subsequent stages of the visual system (Mallat, 1989). Similarly, the cochlea in the inner ear decomposes the auditory signal into its frequency spectrum. An analysis of

DOI: 10.4018/978-1-61520-911-8.ch015

the cochlea's coding properties has contributed to improved coding of audio signals (Baumgarte, 2002).

The task of the olfactory sense is to efficiently encode, process and classify chemical information. The physiological architecture of the chemical sense has striking similarities in insects and mammals, although it has evolved independently (Hildebrand & Shepherd, 1997, Firestein, 2001). This fact may indicate that there is an optimal way for coping with chemical information, that is, identifying specific odorants out of a rich assortment of olfactory agents.

In this study we summarize our computational concept of employing processing principles from the olfactory system to a chemical classification task.

BACKGROUND

Figure 1 outlines the basic architecture of an *olfactory system*. "Chemical space" consists of the multitude of odorant molecules which float around us in the air. These molecules activate an array of receptor neurons. Subgroups of receptor neurons are distinguished into classes according to the particular olfactory receptor protein they present on the membrane (indicated by different shades

of gray in Figure 1). Since this membrane-bound receptor protein determines the ligand selectivity of the receptor neuron, neurons of one class respond to the same *odorants.*

The number of receptor neuron classes varies greatly between species and exhibits weak correlation with olfactory capability. The fruit fly *Drosophila melanogaster* possesses about 60 different functional receptor genes (Vosshall, Wong, & Axel, 2000), the honeybee *Apis mellifera* has about 160 (Robertson & Wanner, 2006). Humans are believed to possess approximately 250 different receptor classes (Glusman, Yanai, Rubin, & Lancet, 2001, Zozulya, Echeverri, & Nguyen, 2001), while mice have about 1000 (Zhang & Firestein, 2002) and dogs around 1200 (Olender et al., 2004).

Olfactory receptor neurons typically exhibit broad ligand selectivity (Araneda, Kini, & Firestein, 2000, Bruyne, Foster, & Carlson, 2001, Mori, Takahashi, Igarashi, & Yamaguchi, 2006, Hallem & Carlson, 2006). A receptor neuron of a given class is therefore activated by many different odorants, as well as an odorant may activate many different classes of receptor neurons. As a consequence, the identity of the odor stimulus is encoded in the combinatorial activity pattern of all receptor neurons rather than having a dedicated receptor for each odorant.

Figure 1. Schematic of the basic architecture of the olfactory system in insects and mammals

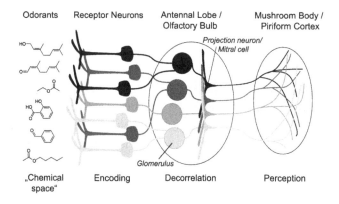

Receptor neurons form the first neuronal layer in the olfactory system. They project to a secondary layer, the antennal lobe in insects, or the olfactory bulb in mammals, respectively. This layer is organized into so-called *glomeruli* small compartments where receptor neurons make contacts with their downstream partners (projection neurons in insects; mitral cells in mammals). Receptor neurons from one class (i.e. with the same ligand selectivity) converge onto the same glomerulus. Hence, an odorant evokes a specific pattern of activity on the glomeruli.

Several lines of evidence suggest that the glomeruli exhibit *chemotopy* in their spatial arrangement, meaning that neighboring glomeruli are activated by structurally similar odorant molecules (Friedrich & Korsching, 1997, Uchida, Takahashi, Tanifuji, & Mori, 2000, Meister & Bonhoeffer, 2001, Couto, Alenius, & Dickson, 2005, Johnson, Farahbod, Saber, & Leon, 2005). A similar organization has been described for other sensory modalities: We have *tonotopy* in the auditory system, where neighboring neurons respond to similar frequencies, and *retinotopy* the visual system, where neighboring points on the retina are represented in neighboring neurons in the first stages of the visual system.

Neurons in the secondary layer exhibit pronounced lateral interconnectivity, indicating that they interact during signal processing before relaying the information to higher brain centers. Several studies pointed out that this kind of processing may serve to decorrelate the signals (for review see Cleland and Linster (2005)).

The output neurons of the secondary neuron layer project to the piriform cortex in mammals, respectively the mushroom bodies in insects (Firestein, 2001). These regions represent parts of the brain that are thought to integrate over different sensory modalities. This likely is the level of perception where an olfactory quality is assigned to the stimulus.

The architecture of the olfactory system can be regarded as an implementation of a workflow for processing and classification of chemical data. The receptor neurons translate molecular properties into neuronal activity ("encoding"). Then, the data is preprocessed to facilitate further analysis in higher regions. Finally, in higher brain centers the stimulus is assigned some olfactory *meaning*, that is, a scent quality. Note that this scheme is roughly equivalent to a typical workflow for predicting bioactivity of molecules: Molecular structures are encoded by calculating physico-chemical descriptors, which are then preprocessed in order to reduce the dimensionality of the data. This data is then fed into a classifier which is trained to assign the proper label to the molecule (G. Schneider & Baringhaus, 2008).

PROCESSING AND CLASSIFICATION OF CHEMICAL DATA INSPIRED BY INSECT OLFACTION

Our goal was to model information processing and classification by the olfactory system in a computational framework. The concept has been published by us previously, and details can be found in the original publication (Schmuker & Schneider, 2007). In a first approach and as a proof of concept we applied the olfactory processing scheme to predicting the scent of odorant molecules. In a second study we tested the model's capability to predict the pharmacological targets of drugs and other bioactive compounds.

The odorant data set was provided by the 2004 Sigma-Aldrich *Flavors & Fragrances* catalog (Sigma Aldrich, 2004). We extracted the molecular structures of 836 *odorants* together with their annotated scent quality. A total of 66 scent qualities were defined, like *fruity, banana, floral* or *nutty*, for example. An odorant could have one or more annotated scents.

Encoding Molecules by "Virtual Receptors"

In the first stage of our model, molecules are encoded into a numerical representation, in analogy to the role of olfactory receptor neurons. In real olfactory systems, an *olfactory receptor* is activated if a molecule possesses a certain combination of features (Schmuker, Bruyne, Hähnel, & Schneider, 2007, Araneda et al., 2000). We mimicked this behavior in "virtual receptors": First, we computed 184 physico-chemical properties of the odorant molecules using the software MOE version 2005.06 (Chemical Computing Group, Montreal, Canada). A virtual receptor is simply a point in this multidimensional property space, as it represents the combination of chemical features an olfactory receptor is tuned to (Figure 2A). The strength of receptor response to a molecule depends on how well the ligand features (stimulus) match the receptor's preferred ones. This proportionality was computed as the distance of the stimulus and the receptor in physicochemical property space.

In analogy to the olfactory system our model contains several receptors, each with a different set of features. Considering an array of n virtual receptors, each receptor has a position described by a coordinate vector p in m-dimensional descriptor space. Formally, the response r_i of the ith receptor ($i=1,2,\ldots,n$)) to an odorant s is described by Equation (1)

$$r_i = 1 - \frac{d(\mathbf{s}, \mathbf{p}_i) - d_{min}}{d_{max} - d_{min}}, \qquad (1)$$

with $\mathbf{p}_i$ the coordinates of the ith receptor, $d(\mathbf{s},\mathbf{p}_i)$ the Manhattan distance (sum of absolute coordinate differences) between $\mathbf{s}$ and $\mathbf{p}_i$, d_{min} and d_{max} the minimal and maximal distance between any $\mathbf{s}$ and $\mathbf{p}_i$. Thus, $r_i=0$ if $d(\mathbf{s},\mathbf{p}_i)$ is maximal and $r_i=1$ if $d(\mathbf{s},\mathbf{p}_i)$ is minimal.

We used Manhattan distance in this study for reasons of computational efficiency, but in principle any other suitable distance measure should be applicable. Manhattan distance has been shown to be particularly useful for similarity searching in drug design (Fechner, Franke, Renner, Schneider, & Schneider, 2003).

The virtual receptors should cover the property space evenly to avoid over- or undersampling of specific chemical classes. We achieved this by training a *Self-Organizing Map (SOM)* on the data set. The SOM has originally been described

Figure 2. Encoding molecules by "virtual receptors". A) Molecules (black squares) are represented as points in a multidimensional property space. The strength of response of a virtual receptor (gray circle) depends on its distance to the molecule in property space (arrows). B) An array of virtual receptors is placed in property space by a self-organizing map (SOM); the distance-dependent response is computed for every receptor. C) Simplified two-dimensional activation pattern received by the second layer (the glomeruli) of the virtual olfactory system; each square represents the activity of one virtual glomerulus

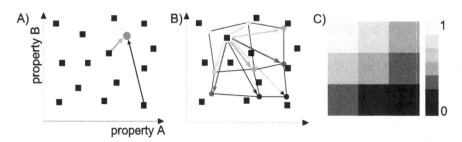

by Kohonen (1982) and can be thought of as a set of points (or *units*) in high-dimensional space which are connected through a lower-dimensional topological grid, like nodes in a two-dimensional fishing net that floats in the three-dimensional water. An SOM can be trained to adapt its nodes to the distribution of training vectors in high-dimensional space. The inherent topology allows for a low-dimensional mapping of the input space. Kohonen (1982) has shown that the SOM preserves local topology in low-dimensional projections.

In our framework the final positions of SOM units determined the positions of virtual receptors. We used the software SOMMER (Schmuker, Schwarte, et al., 2007) for SOM training, employing a toroidal SOM architecture. The preservation of topology guarantees a *chemotopic* arrangement of virtual receptors: Receptors which are close on the two-dimensional SOM grid are also neighbors in the original, high-dimensional space.

When presenting an odorant, distance-dependent activation was computed for each receptor (Figure 2B). The toroidal SOM topology enabled us to project activation patterns of receptors onto a two-dimensional plane (Figure 2C). The resulting activation pattern is analoguous to the activation pattern of glomeruli in the secondary layer of the olfactory system: Each glomerulus receives input

from one receptor type, and their arrangement on the surface of the olfactory bulb (resp. the antennal lobe) evokes a specific two-dimensional pattern of activity.

Decorrelation by Lateral Inhibition

Figure 3 A and B present two odorants, butyl levulinate which has a rose-like, balsamic smell, and butyl phenylacetate which has a fruity scent. Both molecules have in common a pentyl chain that is connected to an ester group, but they also differ in some structural features. This is reflected in the activity patterns that we obtained from our virtual receptor array, which exhibited some overlap but also some differences (Figure 3C and D). These activity patterns are qualitatively similar to what is observed in animals: One odorant activates many glomeruli (indicated by the large patches of light gray in the activation patterns), as well as one glomerulus is activated by several odorants (visibile through the partial overlap of activation between both patterns).

The olfactory bulb/antennal lobe transforms information coming from the receptor neurons to facilitate processing in higher brain areas. A common feature in most olfactory systems is the pronounced inhibitory connectivity between

Figure 3. Response patterns from virtual glomeruli before and after decorrelation. A, B) The chemical structure of two odorants along with their annotated scent properties. C, D) Glomerular activation patterns at the input of the virtual antennal lobe for the two odorants. White indicates high, black low activity. E, F) Patterns after transformation by correlation-based lateral inhibition

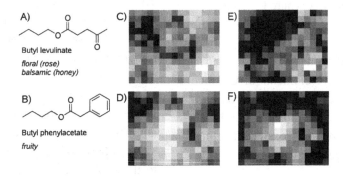

glomeruli in the olfactory bulb resp. the antennal lobe. This leads to a *winner-take-most* situation: If two adjacent glomeruli are activated, the one with the higher activation will reduce activity in the weaker one.

The strength of lateral inhibition is not uniform. Linster, Sachse, and Galizia (2005) have shown in the honeybee that it depends on the correlation between the odorant response spectra of the glomeruli: If two glomeruli have a very similar response profile their inhibitory connection tends to be stronger. A similar effect has been demonstrated in the olfactory bulb of mice (Arevian, Kapoor, & Urban, 2008).

We can easily compute the correlation between any two glomeruli from the activation patterns. The effect of lateral inhibition is then a reduction in activity of each glomerulus. The amount of reduction is proportional to the sum of the activity of all other glomeruli, weighted by their correlation to the glomerulus under consideration. Thus, in principle the processing scheme in the second layer of our model consisted of applying a correlation-dependent filter on the input patterns.

Formally, the transformed response pattern $\mathbf{r}'$ was computed from from the input pattern $\mathbf{r}$ (cf. Equation 1) by Equation (2)

$$\mathbf{r}' = \mathbf{r} - q\left(\frac{\mathbf{C}\cdot \mathbf{r}^T}{n}\right), \qquad (2)$$

with n the number of virtual receptors, q an arbitrary weight and $\mathbf{C}$ a matrix where $\mathbf{C}_{i,j}$ contained Pearson's correlation coefficient between the responses of the ith and jth receptor. In addition, all elements on the diagonal of $\mathbf{C}$ were set to zero, as were all negative elements of $\mathbf{C}$, since it is not clear which effect a negative correlation might have in a neuronal context. Similarly, all elements of $\mathbf{r}'<0$ were set to zero, since neuronal activity cannot take negative values.

Note the factor q in Equation 2, which is an arbitrary scaling factor that allowed us to adjust the strength of the filter. Setting $q=0$ effectively switches off processing, while increasing the value of q also increases the strength of the filter.

Figure 3E and F present response patterns after signal processing by the virtual glomerular layer. Two effects are visible: First, overall activity is sparser. Second, the patterns become less similar.

Mimicking Perception by a Machine Learning Classifier

Perception means that a scent quality is assigned to an odorant stimulus. In the scope of our model this is equivalent to assigning the correct labels (fruity, floral, balsamic, etc.) to the activation patterns obtained in the previous stages. We mimicked the process of perception by training a *naive Bayes classifier* (Witten & Frank, 2005). One aim of this study was to analyze whether our olfaction-inspired processing scheme improves the classifyability of the data. To address this question, we compared the performance of the classifier without previous processing of the activation patterns through lateral inhibition ($q=0$) to a situation with such a signal processing ($q>0$). We also checked whether stronger filtering would improve or impair classifyability of the input data.

The majority of ROC curves for $q=2$ are above those for $q=1$, indicating superior training succes for $q=2$. When turning lateral inhibition off ($q=0$) the classifier was generally performing worse. This is true even when considering the best (Figure 4B) and the worst (Figure 4C) crossvalidation repetition for each setting of q.

We assessed classifier performance on the remaining 65 scent annotations. Performance was quantified by the area under the ROC curve (AUC, *cf.* shaded regions in Figure 4). The median AUC values over all cross-validation runs and scent annotations are presented in (Figure 5). AUC values obtained with $q=2$ were always superior to those obtained with $q=1$ or $q=0$, confirming

Figure 4. Impact of q on classifier performance. Shown are ROC curves from training the classifier on the fruity scent. A) 50 times crossvalidated training when setting q to 0, 1 or 2. The filled area denotes the area-under-curve (AUC). B) Best and C) worst cases from 50 cross-validation runs

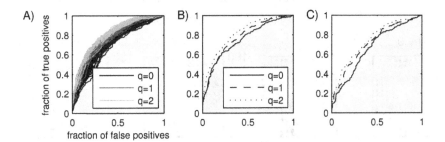

the results for *fruity* as a general trend that applies to the entire data set. In addition, we analyzed the role of the number of receptors and reduced it in several steps from 180 (12×15) to two (1×2). Classification performance was best when setting $q=2$ for all receptor counts analyzed, followed by $q=1$. Moreover, median classification performance did not significantly decrease when reducing the number of virtual receptors from 180 (12×15) to 35 (5×7), which equals more than five-fold *reduction in dimensionality*.

Effect of Lateral Inhibition on Data Classification

Lateral inhibition is most pronounced between strongly correlated glomeruli. This observation suggests that it acts to reduce the correlation between these input dimensions. To test this hypothesis, we calculated correlation matrices for the input dimensions (glomeruli) with different strengths of lateral inhibition. A glomerulus was hereby represented by a vector containing its responses to all 836 odorants in the data set. The pairwise correlations between all glomerulus

Figure 5. Median AUC values after crossvalidated classifier training for all scent annotations in the dataset and for different numbers of virtual receptors. n.s.: Non-significant difference (Wilcoxon rank-sum test, p<0.05

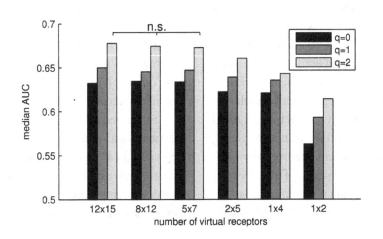

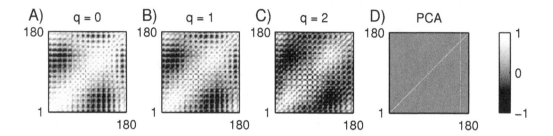

Figure 6. Effect of lateral inhibition on residual correlation between input dimensions. Shown are correlation matrices for the data after transformation with 180 virtual receptors. A) Without processing by lateral inhibition (q=0). B) With processing by lateral inhibition (q=1). C) With processing by stronger lateral inhibition (q=2). D) Correlation between principal component scores of the original descriptors

reponse vectors were used to construct the correlation matrices, shown in Figure 6.

When lateral inhibition was switched off by setting $q = 0$, there was high correlation between dimensions of the data (Figure 6A). With lateral inhibition turned on ($q = 1$), the residual correlation decreased (Figure 6B). When increasing the strength of lateral inhibition by setting $q = 2$ residual correlation was further reduced (Figure 6C). We conclude that correlation-based lateral inhibition actually had a decorrelating effect on the data.

Figure 5 indicates that the classifyability of the data increased with *decorrelation*. But does maximal decorrelation also lead to best classifyability? To answer this question we applied *Principal Component Analysis (PCA)* to the original set of physicochemical descriptors and applied cross-validated classifier training using the principal component scores as input. PCs were directly fed into the classifier, without creating an SOM and deriving distance-dependent activation of virtual receptors in order to mimick a typical workflow for *virtual screening*. PCA removes all correlation between dimensions, which results in a diagonal correlation matrix (Figure 6D). Hence, applying PCA corresponds to maximal decorrelation.

We compared the performance of the classifier when decorrelating the data with our correlation-based method to using PCA on the original data

set (Figure 7). For a comparison of the results obtained with different numbers of virtual receptors (i.e. different dimensionalities) we used the same number of principal components, first considering those which explained the most variance in the data. The results were intriguing: PCA yielded best results when using four components. In this dimensionality it also outperformed the transformation obtained from lateral inhibition when using four virtual receptors. But when considering higher dimensionalities, the classifier performance quickly declined for PCA, and was not much better than random when using 180 dimensions. This behaviour is not surprising: Including components which explain little variance does not add much information for the classifier to learn, but rather tends to introduce noise into the data which inevitably leads to a decline in classifier performance.

In contrast, classification performance on the data that was transformed with lateral inhibition increased with increasing dimensionality, saturating on a high level. Adding more dimensions had no negative effect on classifier performance.

Sustained classifier performance for high-dimensional data representations has particular benefits in a highly parallel computing context, where data dimensionality may have less impact on computing time than in serial computing. The brain has a highly parallel architecture, and thus

Figure 7. Crossvalidated classifier performance after transforming the data with principal component analysis (PCA) compared to processing by lateral inhibition (q=2). Dimensionality denotes number of principal components or number of virtual receptors, respectively

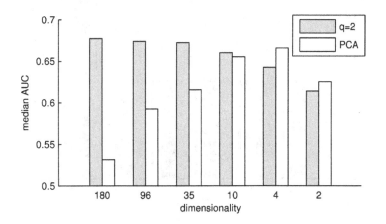

its processing strategies may have evolved to take full advantage of parallelism. If including more dimensions does not decrease performance, one can benefit from redundancy and increased robustness against noise. These properties are of vital importance if input data is noisy and not all channels are reliable.

Application to Predicting the Bioactivity of Compounds

Our olfaction-inspired processing method yielded good classification results for odorant data. But how is the performance when dealing with non-olfactory chemical data? To answer this question we performed a retrospective *virtual screening* study using the COBRA collection of drugs and lead compounds, version 6.1 (P. Schneider & Schneider, 2003). This database was manually compiled from the recent scientific literature and contained the chemical structures of 8374 bioactive molecules together with their pharmaceutical target, e.g. "cyclooxygenase-2" or "thrombin". We transformed the data using the exact same steps as described above: Calculation of physico-chemical properties, encoding the data by virtual receptors and transforming it using lateral inhibition. Instead

of assigning scent quality, we trained the classifier to predict the macromolecular target (receptor) of the compounds. Figure 8 presents the results.

The picture is different than for the odorant data. First, overall performance is better than for odorant data. Since the software we used to calculate the descriptors is targeted at drug development, the selection of descriptors it is offering was likely optimized to quantify properties of drug-like molecules. Odorant molecules are considerably smaller (typically ≤ 350 u) than drugs, and some of the descriptors may be better suited for molecules with higher molecular mass. In addition, the *Flavors & Fragrances* catalog does not state how the scent annotations had been obtained, so possibly different protocols for defining the scent of an odorant molecule have been used. As a consequence, the odorant data may be more "noisy" than the COBRA database, which also would explain why overall classification performance is better for the COBRA data.

Second, $q=2$ did not always deliver the best results. This indicates that a higher value is not always better, and that the value of q can be optimized to fit the data.

Third, our olfaction-inspired method performed better than PCA in the 180-dimensional

Figure 8. Classifier performance on the COBRA database when using virtual receptors only (q=0), virtual receptors with lateral inhibition (q=1, q=2), and using PCA to preprocess the data

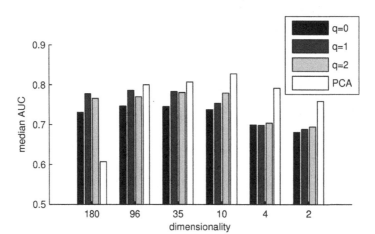

case only. Lower-dimensional representations yielded higher classification accuracy when principal components were used. The reason for this difference is not clear; it may indicate that for this data set more dimensions are required to be appropriately learnt by the classifier.

Taken together, we have shown that our olfaction-inspired method is not only suitable for odorant data, but can also be applied to predicting pharmacological activity. It also turned out that there is still some potential for improving the method.

FUTURE RESEARCH DIRECTIONS

In the processing scheme we presented, chemical structures were encoded by "virtual receptors". This concept corresponds to resampling the data using sample points obtained with an SOM. SOMs were previously applied in cheminformatics in numerous studies, and a comprehensive review on this topic is also available (P. Schneider, Tanrikulu, & Schneider, 2009). However, all these studies have used SOMs to directly derive descriptors from the molecular structure. This is different from the approach we presented here: We used

SOMs to resample the high dimensional space of physicochemical properties instead of using the SOM to derive these properties.

We have shown that the resampling step enables stable classifier performance even for high-dimensional representations of the data. But how this robustness emerges is an open question. Intuitively, data resampling should come with a trade-off in accuracy, but this does not seem to be the case. The odorant data apparently benefits from resampling, maybe because it is inherently more noisy than the bioactivity data. The resampling process might "smooth out" some of the noise and lead to improved classifyability of the data.

Support for this hypothesis comes from a recent study in which SOM-transformed molecular descriptors outperformed descriptors without the preprocessing step (G. Schneider, Tanrikulu, & Schneider, 2009). It was demonstrated that the SOM approach smoothes the descriptor data and eliminates noise. This is most relevant for descriptor-based similarity searching, virtual screening and clustering, as it can help avoid artifacts resulting from raw data (Weisel, Proschak, Kriegl, & Schneider, 2009). In molecular design studies, noise can result from poor conformer sampling, erroneous atom charge and pharmacophoric feature assignment, or heterogeneous sets of ligand chemotypes, just to name major sources of error.

Another direction for future research is the optimization of q. The results for the COBRA database demonstrated that there is no generally best setting for q that is optimal for all applications. Instead, q should be optimized on the data that is to be classified. Possibly, q can also be estimated from some statistical properties of the data, e.g. the average covariance between dimensions.

The largest advantage of our method might be its robustness when using high-dimensional data representations. Certainly, this feature can only be fully exploited if parallel, "brain-like" classification algorithms are available. A tantalizing prospect is the application of our processing scheme on neuronal hardware which implements the physical properties of real neurons on a chip. Such devices exist and neuronal processing schemes have already been realized with them (Schemmel, Brüderle, Meier, & Ostendorf, 2007, Brüderle, Meier, Mueller, & Schemmel, 2007, Mitra, Fusi, & Indiveri, 2009). It will be most interesting to see how our method performs when implemented in neuronal hardware. As a consequence of miniaturization neuronal hardware can run at up to 10,000 times the speed of real neurons (Schemmel et al., 2007). This speedup could be sufficient for our method to compete with current on-line algorithms for decorrelating and denoising input data, for example from sensor arrays.

CONCLUSION

We have presented a method for processing and classification of chemical data that is inspired by the olfactory system. We have shown that this method is capable of improving classifyability of chemical data. The main advantage of the method is its robustness in a high-dimensional context, favoring highly parallel, "brain-like" classifiers. We also demonstrated that this method is suited

for the prediction of the bioactivity of drug-like compounds.

ACKNOWLEDGMENT

Volker Majczan is thanked for assistance in the preparation of the odorant database from the Sigma-Aldrich catalog. Natalie Jäger and Joanna Wisniewska have done preliminary research on odorant classification that was helpful in the design of the experiments described here. Petra Schneider kindly provided the COBRA database for this study. This research was supported by the Beilstein-Institut zur Förderung der Chemischen Wissenschaften.

REFERENCES

Araneda, R. C., Kini, A. D., & Firestein, S. (2000, Dec). The molecular receptive range of an odorant receptor. *Nature Neuroscience*, *3*(12), 1248–1255. Available from http://dx.doi.org/10.1038/81774. doi:10.1038/81774

Arevian, A. C., Kapoor, V., & Urban, N. N. (2008, Jan). Activity-dependent gating of lateral inhibition in the mouse olfactory bulb. *Nature Neuroscience*, *11*(1), 80–87. Available from http://dx.doi.org/10.1038/nn2030. doi:10.1038/nn2030

Baumgarte, F. (2002, Oct.). Improved audio coding using a psychoacoustic model based on a cochlear filter bank. *IEEE Transactions on Speech and Audio Processing*, *10*(7), 495–503. doi:10.1109/TSA.2002.804536

Br¨uderle, D., Meier, K., Mueller, E., & Schemmel, J. (2007). Verifying the biological relevance of a neuromorphic hardware device. *BMC Neuroscience, 8*(Suppl 2), P10.

Cleland, T. A., & Linster, C. (2005, Nov). Computation in the olfactory system. *Chemical Senses, 30*(9), 801–813. Available from http://dx.doi.org/10.1093/chemse/bji072. doi:10.1093/chemse/bji072

Couto, A., Alenius, M., & Dickson, B. J. (2005, Sep). Molecular, anatomical, and functional organization of the *Drosophila* olfactory system. *Current Biology, 15*(17), 1535–1547. Available from http://dx.doi.org/10.1016/j.cub.2005.07.034. doi:10.1016/j.cub.2005.07.034

de Bruyne, M., Foster, K., & Carlson, J. R. (2001, May). Odor coding in the *Drosophila* antenna. *Neuron, 30*(2), 537–552. doi:10.1016/S0896-6273(01)00289-6

Fechner, U., Franke, L., Renner, S., Schneider, P., & Schneider, G. (2003, Oct). Comparison of correlation vector methods for ligand-based similarity searching. *Journal of Computer-Aided Molecular Design, 17*(10), 687–698. doi:10.1023/B:JCAM.0000017375.61558.ad

Firestein, S. (2001, Sep). How the olfactory system makes sense of scents. *Nature, 413*(6852), 211–218. Available from http://dx.doi.org/10.1038/35093026. doi:10.1038/35093026

Friedrich, R. W., & Korsching, S. I. (1997, May). Combinatorial and chemotopic odorant coding in the zebrafish olfactory bulb visualized by optical imaging. *Neuron, 18*(5), 737–752. doi:10.1016/S0896-6273(00)80314-1

Glusman, G., Yanai, I., Rubin, I., & Lancet, D. (2001, May). The complete human olfactory subgenome. *Genome Research, 11*(5), 685–702. Available from http://dx.doi.org/10.1101/gr.171001. doi:10.1101/gr.171001

Hallem, E. A., & Carlson, J. R. (2006, Apr). Coding of odors by a receptor repertoire. *Cell, 125*(1), 143–160. Available from http://dx.doi.org/10.1016/j.cell.2006.01.050. doi:10.1016/j.cell.2006.01.050

Hildebrand, J. G., & Shepherd, G. M. (1997). Mechanisms of olfactory discrimination: converging evidence for common principles across phyla. *Annual Review of Neuroscience, 20*, 595–631. Available from http://dx.doi.org/10.1146/annurev.neuro.20.1.595. doi:10.1146/annurev.neuro.20.1.595

Johnson, B. A., Farahbod, H., Saber, S., & Leon, M. (2005, Mar). Effects of functional group position on spatial representations of aliphatic odorants in the rat olfactory bulb. *The Journal of Comparative Neurology, 483*(2), 192–204. Available from http://dx.doi.org/10.1002/cne.20415. doi:10.1002/cne.20415

Kohonen, T. (1982, January). Self-organized formation of topologically correct feature maps. *Biological Cybernetics, V43*(1), 59–69. Available from http://dx.doi.org/10.1007/BF00337288. doi:10.1007/BF00337288

Linster, C., Sachse, S., & Galizia, C. G. (2005, Jun). Computational modeling suggests that response properties rather than spatial position determine connectivity between olfactory glomeruli. *Journal of Neurophysiology, 93*(6), 3410–3417. Available from http://dx.doi.org/10.1152/jn.01285.2004. doi:10.1152/jn.01285.2004

Mallat, S. G. (1989). Multifrequency channel decompositions of images and wavelet models. *IEEE Transactions on Acoustics, Speech, and Signal Processing, 37*, 2091–2110. doi:10.1109/29.45554

Meister, M., & Bonhoeffer, T. (2001, Feb). Tuning and topography in an odor map on the rat olfactory bulb. *The Journal of Neuroscience, 21*(4), 1351–1360.

Mitra, S., Fusi, S., & Indiveri, G. (2009, Feb.). Real-time classification of complex patterns using spike-based learning in neuromorphic VLSI. *IEEE Transactions on Biomedical Circuits and Systems, 3*(1), 32–42. doi:10.1109/TBCAS.2008.2005781

Mori, K., Takahashi, Y. K., Igarashi, K. M., & Yamaguchi, M. (2006, Apr). Maps of odorant molecular features in the mammalian olfactory bulb. *Physiological Reviews, 86*(2), 409–433. Available from http://dx.doi.org/10.1152/physrev.00021.2005. doi:10.1152/physrev.00021.2005

Olender, T., Fuchs, T., Linhart, C., Shamir, R., Adams, M., & Kalush, F. (2004, Mar). The canine olfactory subgenome. *Genomics, 83*(3), 361–372. Available from http://dx.doi.org/10.1016/j.ygeno.2003.08.009. doi:10.1016/j.ygeno.2003.08.009

Robertson, H. M., & Wanner, K. W. (2006, Nov). The chemoreceptor superfamily in the honey bee, apis mellifera: expansion of the odorant, but not gustatory, receptor family. *Genome Research, 16*(11), 1395–1403. Available from http://dx.doi.org/10.1101/gr.5057506. doi:10.1101/gr.5057506

Schemmel, J. Br¨uderle, D., Meier, K., & Ostendorf, B. (2007). Modeling synaptic plasticity within networks of highly accelerated I&F neurons. In *Proceedings of the 2007 IEEE International Symposium on Circuits and Systems (iscas'07)*. Washington, DC: IEEE Press.

Schmuker, M., de Bruyne, M., H¨ahnel, M., & Schneider, G. (2007). Predicting olfactory receptor neuron responses from odorant structure. *Chemistry Central Journal, 1*, 11. Available from http://dx.doi.org/10.1186/1752-153X-1-11. doi:10.1186/1752-153X-1-11

Schmuker, M., & Schneider, G. (2007, Dec). Processing and classification of chemical data inspired by insect olfaction. *Proceedings of the National Academy of Sciences of the United States of America, 104*(51), 20285–20289. Available from http://dx.doi.org/10.1073/pnas.0705683104. doi:10.1073/pnas.0705683104

Schmuker, M., & Schwarte, F. Br¨uck, A., Proschak, E., Tanrikulu, Y., Givehchi, A., et al. (2007, January). SOMMER: self-organising maps for education and research. *Journal of Molecular Modeling, 13*(1), 225–228. Available from http://dx.doi.org/10.1007/s00894-006-0140-0

Schneider, G., & Baringhaus, K.-H. (2008). *Molecular design*. Weinheim: Wiley-VCH.

Schneider, G., Tanrikulu, Y., & Schneider, P. (2009). (in press). Self-organizing molecular fingerprints: a ligand-based view on druglike chemical space and off-target prediction. *Future Medicinal Chemistry*. doi:10.4155/fmc.09.11

Schneider, P., & Schneider, G. (2003). Collection of bioactive reference compounds for focused library design. *QSAR & Combinatorial Science, 22*(7), 713–718. Available from http://dx.doi.org/10.1002/qsar.20033082514. doi:10.1002/qsar.200330825

Schneider, P., Tanrikulu, Y., & Schneider, G. (2009). Self-organizing maps in drug discovery: compound library design, scaffold-hopping, repurposing. *Current Medicinal Chemistry, 16*(3), 258–266. doi:10.2174/092986709787002655

Sigma Aldrich. (2004). *Flavors and fragrances catalog*. Milwaukee, WI.

Uchida, N., Takahashi, Y. K., Tanifuji, M., & Mori, K. (2000). Odor maps in the mammalian olfactory bulb: domain organization and odorant structural features. *Nature Neuroscience, 3*, 1035–1043. doi:10.1038/79857

Vosshall, L. B., Wong, A. M., & Axel, R. (2000, Jul). An olfactory sensory map in the fly brain. *Cell, 102*(2), 147–159. doi:10.1016/S0092-8674(00)00021-0

Weisel, M., Proschak, E., Kriegl, J. M., & Schneider, G. (2009, Jan). Form follows function: shape analysis of protein cavities for receptor-based drug design. *Proteomics*, 9(2), 451–459. Available from http://dx.doi.org/10.1002/pmic.200800092. doi:10.1002/pmic.200800092

Witten, I. H., & Frank, E. (2005). *Data mining: Practical machine learning tools and techniques* (2nd ed.). San Francisco: Morgan Kaufmann.

Zhang, X., & Firestein, S. (2002, Feb). The olfactory receptor gene superfamily of the mouse. *Nature Neuroscience*, 5(2), 124–133. Available from http://dx.doi.org/10.1038/nn800.

Zozulya, S., Echeverri, F., & Nguyen, T. (2001). The human olfactory receptor repertoire. *Genome Biology*, 2(6), RESEARCH0018.

Section 5
Machine Learning Approaches for Chemical Genomics

Chapter 16
Prediction of Compound–Protein Interactions with Machine Learning Methods

Yoshihiro Yamanishi
Mines ParisTech, Institut Curie, INSERM U900, France

Hisashi Kashima
University of Tokyo, Japan

ABSTRACT

In silico prediction of compound-protein interactions from heterogeneous biological data is critical in the process of drug development. In this chapter the authors review several supervised machine learning methods to predict unknown compound-protein interactions from chemical structure and genomic sequence information simultaneously. The authors review several kernel-based algorithms from two different viewpoints: binary classification and dimension reduction. In the results, they demonstrate the usefulness of the methods on the prediction of drug-target interactions and ligand-protein interactions from chemical structure data and genomic sequence data.

INTRODUCTION

Most drugs are small compounds which interact with their target proteins and inhibit or activate the biological behavior of the proteins. Therefore, the identification of interactions between compounds (ligands, small molecules, drugs) and proteins (targets) is an important part of genomic drug discovery. Examples of pharmaceutically useful target proteins are enzymes, ion channels, G protein-coupled receptors (GPCRs) and nuclear receptors. Owing to the completion of the human

genome sequencing projects, we are beginning to understand the genomic spaces populated by these protein classes. At the same time, the high-throughput screening of large-scale chemical compound libraries with various biological assays is enabling us to explore the chemical space of possible compounds (Kanehisa et al., 2006, Stockwell, 2000, Dobson, 2004). However, our knowledge about the relationship between the chemical and genomic spaces is very limited.

In 2003 the U.S. National Institutes of Health announced the Roadmap, which contained new chemical genomics initiatives. The aim of chemical genomics research is to relate this chemical

DOI: 10.4018/978-1-61520-911-8.ch016

space with the genomic space in order to identify potentially useful compounds such as imaging probes and drug leads. Toward the goal, the Pub-Chem database was established at NCBI (Wheeler et al., 2006) in order to store various chemical information about millions of compounds, but the number of compounds with information on their target protein is very limited. This implies that many potential interactions between the chemical and genomic spaces remain undiscovered. There is therefore a strong incentive to develop new methods capable of detecting these potential compound-protein interactions efficiently.

Although some bio-technologies such as binding assays are becoming available, experimental determination of compound-protein interactions remains very challenging and expensive even nowadays. It is therefore of great practical interest to develop effective *in-silico* prediction methods which can both provide new predictions to experimentalists and provide supporting evidence to experimental studies. The computational prediction is expected to increase research productivity toward genomic drug discovery.

In this chapter we review various computational approaches to predict compound-protein interactions from chemical structures and protein sequences. From the viewpoint of machine learning, we formulate the problem of predicting compound-protein interactions, and introduce several supervised machine learning methods which have been recently developed from two different viewpoints: binary classification and dimension reduction. In the results, we show the usefulness of the methods on the predictions of compound-protein interactions from chemical structure data and genomic sequence data. We also discuss the characteristics of the methods, and show some perspectives toward future work.

BACKGROUND

A variety of computational approaches have been developed to analyze and predict compound-protein interactions. One of the most commonly used is docking simulations (Rarey, Kramer, Lengauer, & Klebe, 1996, Cheng et al., 2007). However, the docking cannot be applied to proteins whose 3D structures are unknown, so this limitation is serious for membrane proteins. For example, there are only two GPCRs with 3D structure information (bovine rhodopsin and human β_2-adrenergic receptor) as of writing. Therefore it is difficult to use the docking simulations on a large scale. Another unique approach is text mining which are usually based on key-word searching in a huge number of literatures (Zhu, Okuno, Tsujimoto, & Mamitsuka, 2005), but it suffers from an inability to detect new biological findings and the problem of redundancy in the compound names and protein names in the literature. Recently, a classification of target proteins based on their ligand structures has been performed (Keiser et al., 2007) and an analysis of the drug-target network has revealed characteristic features of its network topology (Yildirim, Goh, Cusick, Barabasi, & Vidal, 2007). However, neither protein sequence information nor chemical structure information were taken into consideration simultaneously.

The current state-of-the-art involves more integrative methods that simultaneously take into account compound chemical structures, protein sequences, and the currently known compound-protein interactions. A straightforward supervised approach for predicting compound-protein interactions is to use binary classiþcation methods where they take compound-protein pairs as an input for machine learning classifiers such as neural network and support vector machine (SVM) (Bock & Gough, 2005, Erhan, LÕheureux, Yue, & Bengio, 2006, Nagamine & Sakakibara, 2007, Faulon, Misra, Martin, Sale, & Sapra, 2008, Jacob & Vert, 2008). One serious problem of the pair-wise SVM is that the complexity of the "training"

Figure 1. An illustration of the problem of compound-protein interaction prediction

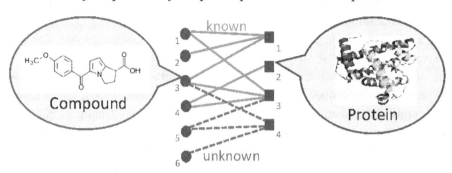

phase scales with the *square* of the "number of training compounds *times* the number of training proteins," leading to computational difficulties for large-scale problems and requiring prohibitive computational cost.

Another supervised approach for predicting compound-protein interactions is the dimension reduction (Yamanishi, Araki, Gutteridge, Honda, & Kanehisa, 2008, Yamanishi, 2009). The basic idea is to map compound and protein nodes into a unified feature space in which known interacting compounds and proteins are "close" to each other, then to infer potentially new compound-protein interactions between other pairs of compounds and proteins that were mapped close to each other in the unified space. In this chapter, we introduce the SVM-based classification approach (Erhan et al., 2006, Nagamine & Sakakibara, 2007, Faulon et al., 2008, Jacob & Vert, 2008) and the dimension reduction approach (Yamanishi et al., 2008, Yamanishi, 2009).

FORMALISM OF THE COMPOUND-PROTEIN INTERACTION PREDICTION BY SUPERVISED BIPARTITE GRAPH INFERENCE

The compound-protein interaction network can be regarded as a bipartite graph with compounds and proteins as heterogeneous nodes and their interactions as edges. From the viewpoint of machine learning, the prediction of compound-protein interactions can be formulated as the problem of supervised bipartite graph inference. The question is to predict the presence or absence of edges between heterogeneous objects known to form the nodes of the bipartite graph, based on the observation about the heterogeneous objects. Let us formally define the supervised bipartite graph inference problem in the context of compound-protein interaction prediction below.

Suppose that we are given an undirected bipartite graph $G=(U+V,E)$, where $U = (x_1,\cdots,x_{n_c})$ is a set of ligand compound nodes, $V = (y_1,\cdots,y_{n_p})$ is a set of target protein nodes and $E \subset (U{\times}V) \cup (V{\times}U)$ is a set of compound-protein interaction edges. The problem is, given additional sets of ligand candidate compound nodes $U' = (x'_1,\cdots,x'_{m_c})$ and target candidate protein nodes $V' = (y'_1,\cdots,y'_{m_p})$, to infer a set of new interaction edges $E'{\subset}\,U' \times (V+V'){\cup}V'{\times}(U+U'){\cup}(U+U'){\times}V'{\cup}(V+V'){\times}U$ involving the additional nodes in U' and V'. Sets of U and V are referred to as training sets below.

The prediction is performed based on available observations about the compounds and proteins. For example, compounds are represented by chemical graph structures and proteins are represented by amino acid sequences. The question is how to predict unknown compound-protein interactions from compound structures and protein sequences using prior knowledge about known compound-protein interactions. Figure 1 shows an illustration of this problem, where solid lines indicate known interactions and dot lines indicate unknown interactions.

BINARY CLASSIFICATION APPROACH

SVM with Pairwise Kernels (Pairwise SVM)

A straightforward approach is to use binary classification methods by regarding each compound-protein pair as an object. The problem of binary classification is a typical issue in the field of statistics and machine learning. A variety of methods have been developed, for example, decision tree, neural networks, and SVM. Especially, the SVM has been receiving a lot of attentions in the recent years because of its high-performance classification ability and applicability to structured data. For example, the SVM is gaining popularity in bioinformatics (Schölkopf, Tsuda, & Vert, 2004) and in chemoinformatics (Ivanciuc, 2007).

An SVM basically learns how to classify an object z into two classes $\{-1, +1\}$ from a set of labelled samples $\{z_1, z_2, L, z_n\}$. The resulting classifier is formulated as

$$f(z) = \sum_{i=1}^{n} \tau_i k(z_i, z), \qquad (1)$$

where z is any new object to be classified, n is the number of training samples, $k(\cdot, \cdot)$ is a positive definite kernel, that is, a symmetric function k: $Z \times Z \rightarrow R$ satisfying $\sum_{i,j=1}^{n} a_i a_j k(z_i, z_j) \geq 0$ for any $a_i, a_j \in N$, and $\{\tau_1, \tau_2, L, \tau_n\}$ are the parameters learned. If $f(z)$ is positive, z is classified into class $+1$. On the contrary, if $f(z)$ is negative, z is classified into class -1.

In the context of compound-protein interaction prediction, we can directly use the SVM by assuming that we have a set of compound-protein pairs $\{(x_1, y_1), (x_2, y_2), L, (x_n, y_n)\}$, class $+1$ corresponds to interaction, and class -1 corresponds to non-interaction. Note that n is the number of training compounds *times* the number of training proteins. Therefore, the essential question here is how to design the kernel function for compound-protein pairs using chemical structure information of compounds and sequence information of proteins.

Vector Representation of Compound-Protein Pairs

We consider a vector representation of a compound-protein pair (x, y). Suppose that a compound x is represented by a vector $\Phi_c(x) \in \mathbf{R}^{d_c}$, which corresponds to molecular descriptors encoding several features related to the physico-chemical property or substructure fingerprint (Todeschini & Consonni, 2002). Likewise, suppose that a protein y is represented by a vector $\Phi_p(y) \in \mathbf{R}^{d_p}$, , which corresponds to the features related with amino acid composition or functional motif profiles, for example.

We then consider representing a compound-protein pair (x, y) by a vector of $\Phi(x, y)$. A simple vector representation is to concatenate $\Phi_c(x)$ and $\Phi_p(y)$ as $\Phi(x,y) = (\Phi_c(x)^T, \Phi_p(y)^T)^T$Erhan et al., 2006, Nagamine & Sakakibara, 2007). Note that the size of the vector is $(d_c + d_p)$ in this case.

Another vector representation approach is to use the set of all possible products of features of x and y by the tensor product as follows (Faulon et al., 2008, Jacob & Vert, 2008):

$$\Phi(x,y) = \Phi_c(x) \otimes \Phi_p(y). \qquad (2)$$

Note that the tensor product is a vector of size $(d_c \times d_p)$, where the (i, j)-th entry is the product of the i-th entry of $\Phi_c(x)$ and the j-th entry of $\Phi_p(y)$. This operation enables us to combine any vector representation of compounds and any vector representation of proteins into a single vector representation, but it requires considerable computationa burden.

Kernels for Compound-Protein Pairs

To avoid such a problem, a computational technique has been proposed by using a property of tensor product (Faulon et al., 2008, Jacob & Vert, 2008). A classical property of tensor products allows us to factorize the inner product between two tensor product vectors as follows:

$$(\Phi_c(x) \otimes \Phi_p(y))^T (\Phi_c(x') \otimes \Phi_p(y')) = \Phi_c(x') \times \Phi_p(y')^T \Phi_p(y') \tag{3}$$

The factorization has dramatical effect of reducing the computational burden of working with tensor products in large dimensions. Let us define the kernels for compounds and proteins as follows:

$$k_c(x,x') = \Phi_c(x)^T \Phi_c(x'), \, k_p(y,y') = \Phi_p(y)^T \Phi_p(y'). \tag{4}$$

Then, the inner product between tensor products can be computed by

$$k((x,y), (x', y')) = k_c(x,x') \times k_p(y, y'). \tag{5}$$

This implies that, as soon as we obtain the compound kernel $k_c(x,x')$ and the protein kernel $k_p(y, y')$, we can compute the kernel for the corresponding compound-protein pair. A variety of kernel functions for structured data are available (Schölkopf et al., 2004), so a graph kernel and a string kernel would be appropriate candidates for representing compounds and proteins, respectively. Finally, the pairwise kernel is used as an input in the framework of SVM in order to predict whether compound-protein pairs are likely to interact or not.

DIMENSION REDUCTION APPROACH WITH REGRESSION MODELS

Here we introduce a dimension reduction method with regression models (Yamanishi et al., 2008), which is based on the modelling the correlation between the data similarity space and the interaction space that we call "interaction space". The proposed procedure is as follows:

1. Embed compounds and proteins on the interaction network into a unified feature space that we call "interaction space".
2. Learn a regression model between the chemical structure similarity space (resp. genomic sequence similarity space) and the interaction space, and map any compounds (resp. proteins) onto the interaction space.
3. Predict interacting compound-protein pairs by connecting compounds and proteins which are closer than a threshold δ in the interaction space.

The details of each step are explained below.

First, we represent the bipartite graph structure of a compound-protein interaction network by an Euclidian space such that both compounds and proteins are represented by sets of d-dimensional feature vectors $\{\mathbf{u}_{x_i}\}_{i=1}^{n_c}$ and $\{\mathbf{u}_{y_j}\}_{j=1}^{n_p}$, respectively. To do so, based on the shortest distance between compounds and proteins on the bipartite graph, we first construct a graph-based similarity matrix $L = \begin{pmatrix} L_{cc} & L_{cp} \\ L_{cp}^T & K_{pp} \end{pmatrix}$, where the elements of L_{cc}, L_{pp} and L_{cp} are computed by using Gaussian functions as follows: $(L_{cc})_{ij} = \exp(-d(x_i, x_j)^2/h^2)$ for i, j = 1,L, n_c, $(L_{pp})_{ij} = \exp(-d(y_i, y_j)^2/h^2)$ for i, j = n_p, and $\exp(-d(x_i, y_j)^2/h^2)$ and for i = 1,L, n_c, for j = 1, L, n_p, where $d(\cdot, \cdot)$ is the shortest distance between all possible objects (including compounds and proteins) on the interaction network with a bipartite graph structure, h is a width parameter, and the distance between unreachable object pairs is treated as infinity. Note that the size of the resulting matrix L is $(n_c + n_p) \times (n_c + n_p)$. The matrix L is not always positive definite, so an appropriate identity matrix is added to the L such that the matrix L meets the positive definite property. Another possibility of constructing the graph-based similarity matrix L is to use the diffusion kernel (Kondor & Lafferty, 2002).

Borrowing a similar idea with kernel principal component analysis (Scholkopf, Smola, & Muller, 1998), we apply the eigen-value decomposition of L as

$$L = \Gamma \Lambda^{1/2} \Lambda^{1/2} \Gamma^T \qquad (6)$$

where the diagonal elements of matrix Λ are eigenvalues and columns of matrix Γ are eigenvectors, and we construct a $(n_c + n_p) \times d$ feature matrix U as $U = \Gamma_d \Lambda_d^{1/2}$ by using the d largest eigenvalues and associated eigenvectors. Then, we represent all compounds and proteins by using the row vectors of the feature matrix $U = (\mathbf{u}_{x_1}, \cdots, \mathbf{u}_{x_{n_c}}, \mathbf{u}_{y_1}, \cdots, \mathbf{u}_{y_{n_p}})^T$. The space spanned by features $\mathbf{u}_x$ and $\mathbf{u}_y$ is referred to as "interaction feature space".

Second, we consider a model representing the correlation between the data similarity space and the interaction feature space. To do so, we propose to apply a variant of the kernel regression model as follows:

$$\mathbf{u} = f(z, z_i) = \sum_{i=1}^{n} k(z, z_i) \mathbf{w}_i + \varepsilon, \qquad (7)$$

where z is an object, n is the number of training samples, f is the projection from a similarity space to an Euclidean space, $k(\cdot, \cdot)$ is a kernel similarity function, $\mathbf{w}_i$ is a weight vector of size d, and ε is a noise vector. For simplicity, we assume that all the feature values are centered. The optimization can be done by finding $\mathbf{w}_i$ which minimizes the following loss function:

$$R(f) = \| UU^T - KWW^T K^T \|_F^2, \qquad (8)$$

where K is an $n \times n$ similarity matrix, $W = (\mathbf{w}_1, L, \mathbf{w}_n)^T$ and $\|\cdot\|_F$ is Frobenius norm. In this study, we learn two models: f_c for the correlation between the chemical structure similarity space and the interaction feature space with respect to compounds, and f_p for the correlation between the genomic sequence similarity space and the interaction feature space with respect to proteins, respectively.

Given a new compound x' and a new protein y', we apply the above regression models to the compound and the protein separately. Applying the model f_c, we map the new compound x' onto the interaction feature space as

$$\mathbf{u}_{x'} = f_c(x', x_i) = \sum_{i=1}^{n_c} k_c(x', x_i) \mathbf{w}_{x_i}, \qquad (9)$$

where $\mathbf{w}_{x_i}$ is a weight vector and $k_c(\cdot, \cdot)$ is a chemical structure kernel function. Applying the model f_g, we map the new protein y' onto the interaction feature space as

$$\mathbf{u}_{y'} = f_p(y', y_j) = \sum_{j=1}^{n_p} k_p(y', y_j) \mathbf{w}_{y_j}, \qquad (10)$$

where $\mathbf{w}_{y_j}$ is a weight vector and $k_p(\cdot, \cdot)$ is a sequence kernel function.

Finally, we compute the closeness between compounds and proteins by the inner product of the features between the corresponding compounds and proteins in the interaction feature space. Then, compound-protein pairs whose closeness is larger than a threshold δ are predicted to interact with each other.

DIMENSION REDUCTION APPROACH WITH DISTANCE LEARNING

Here we introduce a different dimension reduction method based on distance learning (Yamanishi, 2009), which proposes the following two step procedure:

1. learn two mappings f and g in order to implicitly embed compounds and proteins into a unified Euclidean space representing the network topology, where interacting compounds and proteins are close to each other

2. apply the mappings f and g to any compounds and proteins respectively, and predict new interactions between compounds and proteins if the distance between mapped compounds and proteins is smaller than a fixed threshold δ.

The difference with the regression-based dimension reduction approach in the previous section is that this method *implicitly* embed compounds and proteins in a unified feature space. The details of the procedure is described below.

Given functions $f:U\rightarrow\mathbb{R}$ and $g:V\rightarrow\mathbb{R}$, a possible criterion to assess whether connected (resp. disconnected) compound-protein pairs are mapped onto similar (resp. dissimilar) points in $\mathbb{R}$ is the following:

$$R(f,g) = \frac{\sum_{(x_i,y_j)\in E}(f(x_i)-g(y_j))^2 - \sum_{(x_i,y_j)\notin E}(f(x_i)-g(y_j))^2}{\sum_{(x_i,y_j)\in U\times V}(f(x_i)-g(y_j))^2},$$

(11)

where $E\subset(U\times V)\cup(V\times U)$ is a set of compound-protein interaction edges on the graph. A small value of $R(f,g)$ ensures that connected compound-protein pairs tend to be closer than disconnected compound-protein pairs in the sense of quadratic error.

To represent the connectivity between compound nodes and protein nodes on the interaction network, we define a kind of the adjacency matrix A_{cp}, where element $(A_{cp})_{ij}$ is equal to 1 (resp. 0) if compound x_i and protein y_j are connected (resp. disconnected). Note that the size of the matrix A_{cp} is $n_c \times n_p$. We also define degree matrices D_c and D_p for compounds and proteins respectively, where diagonal elements $(D_c)_{ii}$ and $(D_p)_{jj}$ are the degrees of compound x_i and protein y_j (the numbers of edges involving compound x_i and protein y_j), respectively. Note that all non-diagonal elements in D_c and D_p are zero, and the sizes of the matrices are $n_c \times n_c$ and $n_p \times n_p$, respectively.

Let us denote by $f_U = (f(x_1),\cdots,f(x_{n_c}))^T \in \mathbf{R}^{n_c}$ and $g_V = (g(y_1),\cdots,g(y_{n_p}))^T \in \mathbf{R}^{n_p}$ the values taken by f and g on the training set. If the feature values f_U and g_V are centerized as $\sum_{i=1}^{n_c}f(x_i) = 0$ and $\sum_{i=1}^{n_p}g(y_i) = 0$, then the criterion (11) can be rewritten as follows:

$$R(f,g) = 4\frac{\begin{pmatrix}f_U\\g_V\end{pmatrix}^T\begin{pmatrix}D_c & -A_{cp}\\-A_{cp}^T & D_p\end{pmatrix}\begin{pmatrix}f_U\\g_V\end{pmatrix}}{\begin{pmatrix}f_U\\g_V\end{pmatrix}^T\begin{pmatrix}f_U\\g_V\end{pmatrix}} - 2$$

(12)

To avoid the over-fitting problem and obtain meaningful solutions, we regularize the criterion (12) by a smoothness functional on f and g based on a classical approach in statistical learning (Wahba, 1990, Girosi, Jones, & Poggio, 1995). We assume that f and g belong to the reproducing kernel Hilbert space (r.k.h.s.) H_U and H_V defined by the kernels k_c for compounds and k_p for proteins, and to use the norms of f and g as regularization operators. Let us define by $\|f\|$ and $\|g\|$ the norms of f and g in H_U and H_V. Then, the regularized criterion to be minimized becomes:

$$R(f,g) = \frac{\begin{pmatrix} f_U \\ g_V \end{pmatrix}^T \begin{pmatrix} D_c & -A_{cp} \\ -A_{cp}^T & D_p \end{pmatrix} \begin{pmatrix} f_U \\ g_V \end{pmatrix} + \lambda_1 \| f \|^2 + \lambda_2 \| g \|^2}{\begin{pmatrix} f_U \\ g_V \end{pmatrix}^T \begin{pmatrix} f_U \\ g_V \end{pmatrix}},$$

(13)

where λ_1 and λ_2 are regularization parameters which control the trade-off between minimizing the original criterion (11) and smoothing the functions, and the norms are set as $\|f\|=\|g\|=1$.

The above criterion can be used for extracting one-dimentional feature of the compounds and proteins. In order to obtain a d-dimensional feature representation of the objects, we iterate the minimization of the regularized criterion (13) under orthogonality constraints in the r.k.h.s., that is, we recursively define the l-th features f_l and g_l for $l = 1, L, d$ as follows:

$$(f_l, g_l) = \arg\min \frac{\begin{pmatrix} f_U \\ g_V \end{pmatrix}^T \begin{pmatrix} D_c & -A_{cp} \\ -A_{cp}^T & D_p \end{pmatrix} \begin{pmatrix} f_U \\ g_V \end{pmatrix} + \lambda_1 \| f \|^2 + \lambda_2 \| g \|^2}{\begin{pmatrix} f_U \\ g_V \end{pmatrix}^T \begin{pmatrix} f_U \\ g_V \end{pmatrix}}$$

(14)

under the orthogonality constraints: $f \perp f_1, ..., f_{l-1}$ and $g \perp g_1, ..., g_{l-1}$

According to the representer theorem (Shawe-Taylor & Cristianini, 2004), for any $l = 1, L, d$, the solution to Equation (14) has the following expansions: $f_l(x) = \sum_{j=1}^{n_c} \alpha_{l,j} k_c(x_j, x)$ and $g_l(y) = \sum_{j=1}^{n_p} \beta_{l,j} k_p(y_j, y)$ for some vectors $\boldsymbol{\alpha}_l = (\alpha_{l,1}, \cdots, \alpha_{l,n_c})^T \in \mathbf{R}^{n_c}$ and $\boldsymbol{\beta}_l = (\beta_{l,1}, \cdots, \beta_{l,n_p})^T \in \mathbf{R}^{n_p}$. Similarly as kernel

principal component analysis and kernel canonical correlation analysis, the optimization problem can be reduced to the generalized eigen-value problem with respect to α and β. More details of the algorithm can be found in the original paper (Yamanishi, 2009). Note that if the denominator and numerator in Equation (13) are interchanged, the above minimization problem can be thought of as the maximization problem. In the implementation the solution of the maximization problem is more stable than that of the minimization problem in solving the corresponding generalized eigenvalue problem from a numerical viewpoint.

Finally, we compute the closeness between compounds and proteins by the cosine correlation coefficient of the features between the corresponding compounds and proteins in the interaction feature space. Then, compound-protein pairs whose closeness is larger than a threshold δ are predicted to interact with each other.

EXPERIMENT

Data

The information about compound-protein interactions were obtained from the KEGG BRITE (Kanehisa et al., 2008), BRENDA (Schomburg et al., 2004), SuperTarget (Gunther, Guenther, Kuhn, Dunkel, & al., 2007) and DrugBank databases (Wishart et al., 2007). In this study we focus on compound-protein interactions involving four pharmaceutically useful protein classes: enzymes, ion channels, G protein-coupled receptors (GPCRs), and nuclear receptors.

We constructed the interaction data in two ways. First, we constructed a set of drug-target interactions, where the number of known interactions involving enzymes, ion channels, GPCRs, and nuclear receptors is 2926, 1476, 635, and 90, respectively. The number of known drugs targeting enzymes, ion channels, GPCRs, and nuclear receptors are 445, 210, 223, and 54 respectively, and the number of target proteins in these classes is 664, 204, 95, and 26, respectively. This is the same data used in (Yamanishi et al., 2008).

Second, we constructed a set of ligand-protein interactions, where the number of known interactions involving enzymes, ion channels, GPCRs, and nuclear receptors is 5449, 3020, 1124, and 199, respectively. The number of proteins involving the interactions is 1062, 242, 134, and 38, respectively, and the number of compounds involving the interactions is 1123, 475, 417, and 115, respectively. Note that the ligand compound set includes not only drugs but also experimentally confirmed ligand compounds. This is the same data used in (Yamanishi, 2009). These data sets are used as gold standard data to evaluate the prediction performance below.

Chemical structures of the compounds and amino acid sequences of the human proteins were obtained from the KEGG database (Kanehisa et al., 2008). We computed the kernel similarity value of chemical structures between compounds using the SIMCOMP algorithm (Hattori, Okuno, Goto, & Kanehisa, 2003), where the similarity value between two compounds is computed by Tanimoto coefficient defined as the ratio of common substructures between two compounds based on the chemical graph alignment. We computed the sequence similarities between the proteins using Smith-Waterman scores based on the local alignment between two amino acid sequences (Smith & Waterman, 1981).

In this study we used the above similarity measures as kernel functions, because these measures are very popular, efficient and widely used in the field of chemistry and genomics. However, the graph-based Tanimoto coefficients and the Smith-Waterman scores are not always positive definite, so we added an appropriate identity matrix such that the corresponding kernel Gram matrix is positive definite. All the kernel matrices are normalized such that all diagonals are ones.

Performance Evaluation

As a baseline method, we used the nearest neighbor method, because this idea has been used in traditional molecular screening so far. Given a new ligand candidate compound, we find a known ligand compound (in the training set) sharing the highest structure similarity with the new compound, and predict the new compound to interact with proteins known to interact with the nearest ligand compound. Likewise, given a new target candidate protein, we find a known target protein (in the training set) sharing the highest sequence similarity with the new protein, and predict the new protein to interact with ligand compounds known to interact with the nearest target protein. Newly predicted compound-protein interaction pairs are assigned prediction scores with the highest structure or sequence similarity values involving new compounds or new proteins in order to draw a receiver operating curve (ROC) below.

We tested the four different methods: i) nearest neighbor (NN), ii) pairwise SVM (P-SVM), iii) dimension reduction with regression (DR-R), and iv) dimension reduction with distance learning (DR-D) on their abilities to predict the compound-protein interactions. In the application of the P-SVM, it is impossible to apply standard SVM implementations such as SVM^{light} (Joachims, 2003), because the size of the kernel matrix for all possible compound-protein pairs is too huge to construct explicitly in the memory. In fact, the space complexity is $O(n_c^2 \times n_p^2)$ which is just for storing the kernel matrix, where n_c and n_p are the numbers of compounds and proteins, respectively. Therefore, we trained the models of the P-SVM by using an on-line learning algorithm which processes one training example at each training step, so it is computationally and spatially efﬁcient. In this study, we used PUMMA (Ishibashi, Hatano, & Takeda, 2008) whose solutions asymptotically converge to those by the SVM with the squared hinge loss. In the experi-

Table 1. AUC (ROC scores) for drug-target interaction data

Data	Method	AUC ± S.D.
Enzyme	Nearest neighbor	0.752 ± 0.009
	Pairwise SVM	0.859 ± 0.006
	Dimension reduction with regression model	0.899 ± 0.007
	Dimension reduction with distance learning	0.880 ± 0.006
Ion	Nearest neighbor	0.736 ± 0.010
channel	Pairwise SVM	0.812 ± 0.013
	Dimension reduction with regression model	0.838 ± 0.006
	Dimension reduction with distance learning	0.841 ± 0.008
GPCR	Nearest neighbor	0.711 ± 0.009
	Pairwise SVM	0.827 ± 0.003
	Dimension reduction with regression model	0.885 ± 0.009
	Dimension reduction with distance learning	0.870 ± 0.007
Nuclear	Nearest neighbor	0.722 ± 0.015
Receptor	Pairwise SVM	0.743 ± 0.038
	Dimension reduction with regression model	0.828 ± 0.015
	Dimension reduction with distance learning	0.771 ± 0.014

Table 2. AUC (ROC scores) for ligand-protein interaction data

Data	Method	AUC ± S.D.
Enzyme	Nearest neighbor	0.691 ± 0.009
	Pairwise SVM	0.837 ± 0.008
	Dimension reduction with regression model	0.870 ± 0.007
	Dimension reduction with distance learning	0.855 ± 0.006
Ion	Nearest neighbor	0.770 ± 0.008
channel	Pairwise SVM	0.849 ± 0.007
	Dimension reduction with regression model	0.879 ± 0.009
	Dimension reduction with distance learning	0.853 ± 0.008
GPCR	Nearest neighbor	0.734 ± 0.010
	Pairwise SVM	0.850 ± 0.005
	Dimension reduction with regression model	0.919 ± 0.008
	Dimension reduction with distance learning	0.900 ± 0.006
Nuclear	Nearest neighbor	0.660 ± 0.014
Receptor	Pairwise SVM	0.701 ± 0.035
	Dimension reduction with regression model	0.832 ± 0.010
	Dimension reduction with distance learning	0.807 ± 0.009

ment the hyperparameter for regularization in the P-SVM is set to $C = 1$, and all of the training data were processed one time in the training phase. In the application of DR-R, the width parameter in the Gaussian function is set to $h = 2$. In the application of DR-D, the regularization parameter is set to $\lambda = 2$ and the number of features is set to $d = 100$, but $d = 20$ is set for the nuclear receptor classes because of the data size.

For the performance evaluation, we performed the following 5-fold cross-validation procedure:

the gold standard set was split into 5 subsets of roughly equal size by compounds and proteins, each subset was then taken in turn as a test set, and the training is performed on the remaining 4 sets. We draw the ROC curve, the plot of true positives as a function of false positives based on various thresholds δ, where true positives are correctly predicted interactions and false positives are predicted interactions that are not present in the gold standard interactions. The performance was evaluated by AUC (area under the ROC curve)

Figure 2. Execution time for enzyme data: drug-target interaction (left) and ligand-protein interaction (right). NN, P-SVM, DR-R, and DR-D indicate nearest neighbor, pairwise SVM, dimension reduction with regression, and dimension reduction with distance learning, respectively.

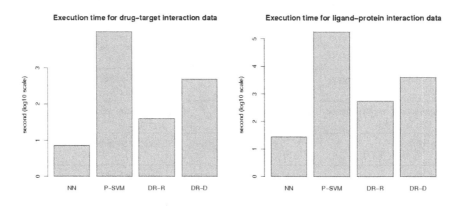

score. To obtain robust results, we repeated the above cross-validation experiment five times, and computed the average and standard deviation of the resulting AUC scores.

Table 1 and Table 2 show the resulting AUC scores for drug-target interaction data and ligand-protein interaction data, respectively. Comparing the four different methods, DR-R method seems to have the best performance for all the four protein families, and consistently outperform the other methods in both drug-target data and ligand-protein data. One explanation of the worst performance of NN is that that raw compound structure or protein sequence similarities do not always reflect the tendency of interaction partners in true compound-protein interactions.

Finally, we investigated the execution times for conducting each method. Figure 2 shows the comparison of the execution times between the four different methods, where the total execution times for training and test processes are shown in the scale of base 10 logarithm. Here we focused on the application to enzyme data. All of the algorithms were implemented in R on a Macbook(TM) with 2.16GHz Intel Core 2 Duo processor and 2 GB RAM. It seems that the dimension reduction approaches (DR-R and DR-D) are consistently

and significantly faster than the binary classification approach (P-SVM). Note that the space complexity of the binary classification approach is $O(n_c^2 \times n_p^2)$, while that of the dimension reduction is $O(n_c^2 + n_p^2)$. A strong advantage of the dimension reduction approach over the binary classification approach is computational efficiency.

FUTURE RESEARCH DIRECTION

There are several possible research directions for compound-protein interaction prediction. In recent years, in the context of protein interaction prediction, new network inference methods have been proposed, such as local model approach (Bleakley, Biau, & Vert, 2007), semi-supervised approach (Kashima, Kato, Yamanishi, Sugiyama, & Tsuda, 2009), and severals more. It would be interesting to investigate the possibility of extending these recent algorithms to the compound-protein interaction prediction problem.

All the methods introduced in this chapter belong to a class of kernel methods (Schölkopf et al., 2004), so the performance could be improved by using more sophisticated kernel similarity functions designed for genomic sequences and chemical structures. Considering how small compounds interact with their biding pockets of the target proteins, the incorporation of prior information about pharmacophores (Mahe, Ralaivola, Stoven, & Vert, 2006) and biding pockets (Kratochwil et al., 2005) into the design of compound similarity and protein similarity is an important future work. At the same time, we need to take into account the efficiency of computing the compound and protein similarities. In practical applications we will be faced with the necessity of handling with million of compounds and thousands of proteins, so the scalability is also an important issue for large-scale screening.

Another challenging issue is the interpretability of the prediction models. From a biological viewpoint, the interactions are determined by the combination of compound substructures and protein functional domains corresponding the biding pockets. In this sense, it is interesting to extract pairs of compound substructures and protein domains which are involved in the interactions. Recently, several graph mining techniques for structured data have been proposed in order to extract biologically meaningful features (Saigo, Nowozin, Kadowaki, Kudo, & Tsuda, 2008, Saigo, Uno, & Tsuda, 2007). An extension of such mining techniques for extracting compound substructures and protein sub-sequences is an promising research direction.

CONCLUSION

In this chapter, the authors reviewed several machine learning methods which have been recently developed for predicting compound-protein interactions. They reviewed several kernel-based learning algorithms from two different viewpoints: binary classification and dimension reduction. Among the three supervised machine learning methods, the DR-R method (dimension reduction approach with regression model) achieves the best performance in terms of prediction accuracy and computational efficiency. These methods are expected to be useful for virtual screening of a huge number of ligand candidate compounds being generated with various biological assays and for identification of drug-targets in the drug development process.

REFERENCES

Bleakley, K., Biau, G., & Vert, J.-P. (2007). Supervised reconstruction of biological networks with local models. *Bioinformatics (Oxford, England)*, *23*, i57–i65. doi:10.1093/bioinformatics/btm204

Bock, J. R., & Gough, D. A. (2005). Virtual screen for ligands of orphan g protein-coupled receptors. *Journal of Chemical Information and Modeling*, *45*(5), 1402–1414. doi:10.1021/ci050006d

Cheng, A., Coleman, R., Smyth, K., Cao, Q., Soulard, P., & Caffrey, D. (2007). Structure-based maximal affinity model predicts small-molecule druggability. *Nature Biotechnology*, *25*, 71–75. doi:10.1038/nbt1273

Dobson, C. (2004). Chemical space and biology. *Nature*, *432*, 824–828. doi:10.1038/nature03192

Erhan, D., LÕheureux, P.-J., Yue, S. Y., & Bengio, Y. (2006). Collaborative Þltering on a family of biological targets. *Journal of Chemical Information and Modeling*, *46*(2), 626–635. doi:10.1021/ci050367t

Faulon, J., Misra, M., Martin, S., Sale, K., & Sapra, R. (2008). Genome scale enzymeÐmetabolite and drugÐtarget interaction predictions using the signature molecular descriptor. *Bioinformatics (Oxford, England)*, *24*, 225–233. doi:10.1093/bioinformatics/btm580

Girosi, F., Jones, M., & Poggio, T. (1995). Regularization theory and neural networks architectures. *Neural Computation, 7,* 219–269. doi:10.1162/neco.1995.7.2.219

Gunther, S., Guenther, S., Kuhn, M., Dunkel, M., & al. et. (2007). Supertarget and matador: resources for exploring drug-target relationships. *Nucleic Acids Res.*

Hattori, M., Okuno, Y., Goto, S., & Kanehisa, M. (2003). Development of a chemical structure comparison method for integrated analysis of chemical and genomic information in the metabolic pathways. *Journal of the American Chemical Society, 125,* 11853–11865. doi:10.1021/ja036030u

Ishibashi, K., Hatano, K., & Takeda, M. (2008). Online learning of approximate maximum p-norm margin classiÞers with biases. In *Proceedings of the 21st annual conference on learning theory (colt2008).*

Ivanciuc, O. (2007). *Applications of support vector machines in chemistry.* Weiheim: Wiley-VCH.

Jacob, L., & Vert, J.-P. (2008). Protein-ligand interaction prediction: an improved chemogenomics approach. *Bioinformatics (Oxford, England), 24,* 2149–2156. doi:10.1093/bioinformatics/btn409

Joachims, T. (2003). *Learning to classify text using support vector machines: Methods, theory and algorithms.* New York: Kluwer Academic Publishers.

Kanehisa, M., Araki, M., Goto, S., Hattori, M., Hirakawa, M., & Itoh, M. (2008). Kegg for linking genomes to life and the environment. *Nucleic Acids Research, 36*(Database issue), D480–D485. doi:10.1093/nar/gkm882

Kanehisa, M., Goto, S., Hattori, M., Aoki-Kinoshita, K., Itoh, M., & Kawashima, S. (2006, Jan). From genomics to chemical genomics: new developments in kegg. *Nucleic Acids Research, 34*(Database issue), D354–D357. doi:10.1093/nar/gkj102

Kashima, H., Kato, T., Yamanishi, Y., Sugiyama, M., & Tsuda, K. (2009). Link propagation: A fast semi-supervised learning algorithm for link prediction. In *Proceedings of the 2009 siam conference on data mining* (p. 1099-1110).

Keiser, M., Roth, B., Armbruster, B., Ernsberger, P., Irwin, J., & Shoichet, B. (2007). Relating protein pharmacology by ligand chemistry. *Nature Biotechnology, 25,* 197–206. doi:10.1038/nbt1284

Kondor, R., & Lafferty, J. (2002). Diffusion kernels on graphs and other discrete input spaces. In T. Faucett & N. Mishra (Eds.), *Proceedings of the twentieth international conference on machine learning* (p. 321-328). Boston: AAAI Press.

Kratochwil, N., Malherbe, P., Lindemann, L., Ebeling, M., Hoener, M., & Muhlemann, A. (2005). An automated system for the analysis of g protein-coupled receptor transmembrane binding pockets: Alignment, receptor-based pharmacophores, and their application. *Journal of Chemical Information and Modeling, 45*(5), 1324–1336. doi:10.1021/ci050221u

Mahe, P., Ralaivola, L., Stoven, V., & Vert, J. (2006). The pharmacophore kernel for virtual screening with support vector machines. *Journal of Chemical Information and Modeling, 46,* 2003–2014. doi:10.1021/ci060138m

Nagamine, N., & Sakakibara, Y. (2007). Statistical prediction of proteinÐchemical interactions based on chemical structure and mass spectrometry data. *Bioinformatics (Oxford, England), 23,* 2004–2012. doi:10.1093/bioinformatics/btm266

Rarey, M., Kramer, B., Lengauer, T., & Klebe, G. (1996). A fast flexible docking method using an incremental construction algorithm. *Journal of Molecular Biology, 261,* 470–489. doi:10.1006/jmbi.1996.0477

Saigo, H., Nowozin, S., Kadowaki, T., Kudo, T., & Tsuda, K. (2008). gboost: A mathematical programming approach to graph classification and regression. *Machine Learning, 75*(1), 69–89. doi:10.1007/s10994-008-5089-z

Saigo, H., Uno, T., & Tsuda, K. (2007). Mining complex genotypic features for predicting hiv-1 drug resistance. *Bioinformatics (Oxford, England), 23*(18), 2455–2462. doi:10.1093/bioinformatics/btm353

Scholkopf, B., Smola, A., & Muller, K.-R. (1998). Nonlinear component analysis as a kernel eigenvalue problem. *Neural Computation, 10,* 1299–1319. doi:10.1162/089976698300017467

Schölkopf, B., Tsuda, K., & Vert, J. (2004). *Kernel methods in computational biology.* Cambridge, MA: MIT Press.

Schomburg, I., Chang, A., Ebeling, C., Gremse, M., Heldt, C., & Huhn, G. (2004). Brenda, the enzyme database: updates and major new developments. *Nucleic Acids Research, 32,* D431–D433. doi:10.1093/nar/gkh081

Shawe-Taylor, J., & Cristianini, N. (2004). *Kernel methods for pattern analysis.* Cambridge, UK: Cambridge University Press.

Smith, T., & Waterman, M. (1981). Identification of common molecular subsequences. *Journal of Molecular Biology, 147,* 195–197. doi:10.1016/0022-2836(81)90087-5

Stockwell, B. (2000). Chemical genetics: ligand-based discovery of gene function. *Nature Reviews. Genetics, 1,* 116–125. doi:10.1038/35038557

Todeschini, R., & Consonni, V. (2002). *Handbook of molecular descriptors.* New York: Wiley-VCH.

Wahba, G. (1990). *Splines models for observational data: Series in applied mathematics.* Philadelphia: SIAM.

Wheeler, D., Barrett, T., Benson, D., Bryant, S., Canese, K., & Chetvernin, V. (2006). Database resources of the national center for biotechnology information. *Nucleic Acids Research, 34,* D173–D180. doi:10.1093/nar/gkj158

Wishart, D., Knox, C., Guo, A., Cheng, D., Shrivastava, S., & Tzur, D. (2007). *Drugbank: A knowledgebase for drugs, drug actions and drug targets.* Nucleic Acids Res.

Yamanishi, Y. (2009). Supervised bipartite graph inference . In Koller, D., Schuurmans, D., Bengio, Y., & Bottou, L. (Eds.), *Adv. Neural Inform. Process. Syst. 21* (pp. 1841–1848). Cambridge, MA: MIT Press.

Yamanishi, Y., Araki, M., Gutteridge, A., Honda, W., & Kanehisa, M. (2008). Prediction of drug-target interaction networks from the integration of chemical and genomic spaces. *Bioinformatics (Oxford, England), 24,* i232–i240. doi:10.1093/bioinformatics/btn162

Yildirim, M., Goh, K., Cusick, M., Barabasi, A., & Vidal, M. (2007). Drug-target network. *Nature Biotechnology, 25,* 1119–1126. doi:10.1038/nbt1338

Zhu, S., Okuno, Y., Tsujimoto, G., & Mamitsuka, H. (2005). A probabilistic model for mining implicit 'chemical compound-gene' relations from literature. *Bioinformatics (Oxford, England), 21*(Suppl 2), ii245–ii251. doi:10.1093/bioinformatics/bti1141

Chapter 17

Chemoinformatics on Metabolic Pathways:
Attaching Biochemical Information on Putative Enzymatic Reactions

Masahiro Hattori
Tokyo University of Technology, Japan

Masaaki Kotera
Kyoto University, Japan

ABSTRACT

Chemical genomics is one of the cutting-edge research areas in the post-genomic era, which requires a sophisticated integration of heterogeneous information, i.e., genomic and chemical information. Enzymes play key roles for dynamic behavior of living organisms, linking information in the chemical space and genomic space. In this chapter, the authors report our recent efforts in this area, including the development of a similarity measure between two chemical compounds, a prediction system of a plausible enzyme for a given substrate and product pair, and two different approaches to predict the fate of a given compound in a metabolic pathway. General problems and possible future directions are also discussed, in hope to attract more activities from many researchers in this research area.

INTRODUCTION

A cellular system is composed of a considerable number of elements including genes, proteins or enzymes, and metabolites. They are dynamically connected and interdependent to construct a network diagram, which is referred to as a metabolic pathway. The significance of a systems approach has been recognized for understanding the activities of living organisms (Eisenberg *et al.*, 2000; Kanehisa, 2001; Kanehisa & Bork, 2003).

Among the elements mentioned above, genetic information has been studied from a holistic viewpoint since the late 1990s. Whole genome sequencing projects have uncovered the genomic aspects of biological substances, *i.e.*, the repertoire of genes and their products encoded on each genome, generating an exhaustive encyclopedia containing macromolecules such as DNA, RNA and proteins. The information of molecular interactions like metabolic pathways and regulatory systems has also been compiled from text-based literature and stored in several computer data-

DOI: 10.4018/978-1-61520-911-8.ch017

bases including KEGG (Kanehisa & Goto, 2000; Kanehisa *et al.*, 2006 & 2008), EcoCyc (Keseler *et al.*, 2009), MetaCyc (Caspi *et al.*, 2008), and BioCarta (BioCarta LLC). Motivation of the early genome analyses was mainly to obtain the whole set of fundamental molecules in biology, hoping that the interaction network of them would be depicted by collecting the text information on pathways. Initially, pathway information was utilized to improve genome annotation in order to add in-depth information for macromolecules (Kanehisa, 1997).

As we gain more omic-scale information, *e.g.*, proteome (Wilkins *et al.*, 1996), transcriptome (Velculescu *et al.*, 1997), metabolome (Tweeddale *et al.*, 1998), and interactome (Sanchez *et al.*, 1999), studies based on those holistic viewpoints has become applicable to the other substances in living organisms. As a result, the emerging main stream has become what is referred to as chemical genomics, integrating heterogeneous information including genomic, chemical and interaction data (Yamanishi *et al.*, 2004). It has been shown that a sophisticated method combined with both genomic and chemical knowledge is also applicable to computationally predict missing enzymes in particular species in a given metabolic pathway (Yamanishi *et al.*, 2007).

Attaching biological information on a given genome sequence is referred to as genome annotation, which is mostly based on the assumption that if the sequences of genes are similar enough, then their functions would be identical. In the case of annotating enzyme genes, the function is generally inferred by a previously known enzyme reaction. The missing-enyzme finding problem mentioned above suggests putative genes on pre-defined enzyme reactions, which can also be described as one of the genome annotation approaches.

On the contrary, chemoinformatics has the potential to provide a reverse method of annotation, which we refer to as a "chemical annotation" approach. This is based on the assumption that when an enzyme reaction is found to be similar to a previously known reaction, then we can gain a clue what the enzyme or enzyme gene could be. The topic can be divided into two parts: how to find pairs of metabolites that are connected in metabolic pathways, and what kind of intermediate enzymes would catalyze those reactions. These are the main themes of this chapter.

Here we report our recent efforts concerning this research area. First, we have defined a similarity measure of chemical substances for metabolic pathway analysis, and elucidated the strong correlation among chemical and genomic structures *i.e.*, metabolic pathways and operons (Hattori *et al.*, 2003). As an application of this, we have developed a prediction system of enzymes for a given partial reaction equation (Kotera *et al.*, 2004). The next achievement was the prediction scheme of biodegradation pathways for a given set of xenobiotics of which molecular fate is unknown (Oh *et al.*, 2007). This type of research will be fundamental in predicting the bioremediation capability of environment chemicals by a bacterial community. The most recent work is the GREP system, where possible enzyme reaction equations are generated from the given set of compounds (Kotera *et al.*, 2008).

EC NUMBER CLASSIFICATION

The topics presented are currently performed through the Enzyme Committee (EC) number classification, with which reaction mechanisms and, more importantly, the corresponding enzyme genes in the genome could be deduced. The EC number classification has been developed by the Enzyme Committee in IUBMB to classify all characterized enzymes (Tipton & Boyce, 2005), and been utilized in many fields of analyses including chemoinformatics. Since the EC classification was established for enzyme nomenclature, every EC number is associated with a recommended name for the respective enzyme. Essentially, this Enzyme Nomenclature system has a strict policy that enzymes should be registered in the EC classification only when they have sufficient experimental evidence and are reported to scientific journals.

Enzymes are categorized in a hierarchical manner, based on the chemical reactions they catalyze. If different proteins, *i.e.*, taken from different organisms, catalyze the same reaction, then they are given the same EC number. Hence, EC numbers have aspects for identifiers of enzyme-catalyzed reactions, rather than those for enzyme proteins. An EC number consists of four digits where each number is concatenated with a period to form a string like "*A.B.C.D*". Here the first digit (*A*) represents one of the six classes of enzymatic reactions, corresponding to oxidoreductases, transferases, hydrolases, lyases, isomerases and ligases, respectively. The second (*B*) and the third digits (*C*) of EC numbers are subclasses and sub-subclasses, representing more detailed classification of enzymatic reactions. The fourth digit (*D*) is numbered serially in accordance with its substrate specificity or the enzyme protein family information. Therefore, when we analyze the chemical aspect of enzyme reactions without considering the variety of substrates or protein families, it is usually sufficient to take only the first three digits of EC numbers into account.

Despite that the EC number has the highest coverage on enzyme varieties and is in widespread use, it contains several drawbacks for computational approaches, which can be summarized as follows. Perhaps the most important thing is that the classification criteria for digits (*B*) and (*C*) are different depending on the value of digit (*A*). In the case of the first class (EC1, oxidoreductases), for instance, subclasses classify enzymes based on the functional groups of substrates directly involved in the reaction, and sub-subclasses are based on a variety of cofactors. On the other hand, in the case of the second class (EC2, transferases), subclasses and sub-subclasses discriminate transferred groups. It is natural that well-known enzyme families are classified in detail, while others not, but this also makes it difficult to analyze the EC classification in a systematic way. The next thing to be mentioned is that reactions can be characterized by various aspects, while enzyme main classes can only be classified in a one-way manner because of the hierarchical EC classification architecture. For example, if an enzyme catalyzes a chemical transformation containing both oxidoreductase (EC1) and lyase (EC4) reaction steps, there is a tendency that the enzyme is categorized only into EC1.

In addition, the strict policy that confirmed experimental evidence is required for an enzyme to be registered in the EC makes many enzymatic reactions remain to be assigned. This policy is absolutely necessary for creating a reliable library of enzymes present in nature; however, as long as enzyme gene annotation is based solely on the official EC numbers with confirmed evidences, it will be impossible to link putative genes to putative functions.

Still, the EC number has a long history and has been a *de-facto* standard to analyze metabolic pathways. The EC number is the most comprehensive way to know how each enzyme catalyzes its specific reaction; thus it plays a key role in linking genomic and chemical data in metabolic pathways. There are many reactions that we know to occur, but no corresponding enzyme has been identified in any organisms yet. In such cases, a computational way to reveal possible enzymes is through the chemical annotation approach: to find enzymes that catalyze similar reactions. Predicting the corresponding EC sub-subclass from a given partial reaction equation is currently the practical way to do this.

TRANSFORMATION PATTERN AND THE REACTION MECHANISM

All enzymatic reactions can be represented as graph objects, in which nodes are substrates and products, and edges are the binary relations between substrates and products. Here, the binary relation between a substrate and a product is defined as a reactant pair that shares the same atoms or atom groups before and after a reaction. The schematic representations of the typical enzymatic reactions and their corresponding pattern of atom sharing are shown in Figure 1, making it clear that every enzymatic reaction can be represented by the set of binary relations between substrates and products. For each binary relation, we can assume a transformation pattern of chemical structures, which describe the enzymatic reaction activities as the addition and the elimination of some specific atom groups. This observation was a strong motivation for us to develop the RPAIR database and the RDM pattern method, which are described below.

We have been checking the reaction information in the KEGG database, and manually assigning binary relation information in accordance with the presence of the shared atom groups. Manual assignment of binary relations should be required to keep high accuracy of computational results because of difficulties in automatic relation assignments. The details of manual assignment are given in the next section.

From the binary relation information, we can compute the atom alignment between the substrate and the product described in the record of each reaction entry. This calculation is successfully performed by the SIMCOMP algorithm using the KEGG Atom types. SIMCOMP is based on a graph comparison algorithm to search for the maximal common subgraph $MCS(G_1, G_2)$ from two given chemical graphs $G_1(V_1, E_1)$ and $G_2(V_2, E_2)$ (Hattori *et al.*, 2003; Hattori *et al.*, 2010), and the similarity measure for chemical compounds is based on the Jaccard coefficient (Jaccard, 1912; Watson, 1983). Here V and E are the sets of vertices and edges, that is, the sets of atoms and bonds found in each chemical compound, respectively. In our application the KEGG Atom types are used for the labeling function $L(V)$ of vertices V, which is one of the sophisticated representations of atoms defined by physicochemical properties around

Figure 1. Typical reactant pair connections for the six EC classes

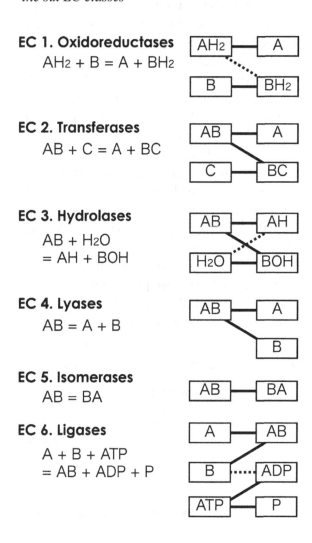

each atom. The maximal common subgraph is $MCS(G_1, G_2)$ induced from two given chemical graphs $G_1(\{L(V_1)\}, E_1)$ and $G_2(\{L(V_2)\}, E_2)$. The calculation of MCS is known to be NP-hard and very time consuming; however, the KEGG Atom types, that is, the detailed information of atom properties $L(V)$, can effectively reduce the possibilities of unlikely atom matching and work as a heuristic. This is how we have simultaneously done both reduction of time complexity and searching for biochemically meaningful substructures, which we refer to as atom alignments. This atom alignment information has been stored in the

RPAIR database (Oh *et al.*, 2007), which has been developed as one of the subcategories in the KEGG LIGAND database (Goto *et al.*, 1998 & 2002). Our algorithm described above is similar to others based on the MCS algorithm that have ever been introduced in the past (Blower & Dana, 1986; Chen *et al.*, 2002; Kuhl *et al.*, 1984; McGregor & Willett, 1981; Moock *et al.*, 1988; Raymond & Willett, 2002; Raymond *et al.*, 2002a & 2002b; Takahashi *et al.*, 1987), but our application has been optimized for various applications in chemical genomics research including themes such as EC number prediction and pathway prediction.

After obtaining an atom alignment, we can obtain the structural change within chemical compounds by comparing the matched subgraph and the unmatched subgraph, as shown in Figure 2. Here the matched subgraph is corresponding to the conserved atom group under the enzymatic

reaction, and the unmatched subgraph of each chemical compound is the eliminated or added atom groups. In the case of transferase reactions (EC2) in Figure 1, a substructure *B* of a compound *AB* is transferred to a compound *C*, producing compounds *A* and *BC*. When we focus on the pair *AB-A*, the substructure *A* is conserved and the substructure *B* is eliminated. When we focus on the pair *AB-BC*, where *B* corresponds to a transferred group, *B* is conserved while *A* and *C* are eliminated or added. These observations indicate that such conserved and non-conserved substructures represent the reaction type. More importantly, the boundary area between the conserved and the non-conserved substructures is often the reaction center on which the enzyme acts (Figure 2). In such a way, we regard enzymatic reaction mechanisms as a set of chemical graph changes. In the RPAIR database, we have also assigned what we call the

Figure 2. An example of RDM pattern for a reactant pair

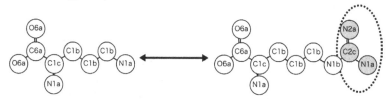

Reactant pair

KEGG atom types representation

RDM pattern

R:D:M = N1a-N1b:*-C2c:C1b-C1b

RDM chemical transformation patterns (Kotera *et al.*, 2004) around the boundary atoms between the conserved and non-conserved substructures as described in Figure 2. Labels of nodes (KEGG Atom Types) in the molecular graphs discriminate the atomic elements with functional group information. An RDM pattern describes the changes of the labels around the reaction centers during a reaction. An RDM pattern consists of three different types of atoms: R, D and M atoms. An R atom is a reaction center. D and M atoms are neighboring atoms of the R atom in the different (D) and matched (M) subgraphs, respectively. RDM patterns are described as a string formatted as "R:D:M=N1a-N1b:*-C2c:C1b-C1b", where "*" means a null node.

MANY HEURISTICS IN COMPUTATION

For characterizing a reaction mechanism, it is essential to determine which atom is the reaction center. This requires several procedures, such as to find reactant pairs from a reaction equation,

to calculate atom alignments for the pairs, and to extract reaction patterns from the alignments. These processes involved developments of some computational algorithms with many heuristics. In some cases, manual refinement of the computational result using expert knowledge was required.

The first heuristic is in finding reactant pairs for each enzymatic reaction, which needs expert knowledge in many cases. This is one of the differences compared with organic chemistry, where main substrate and product are mostly obvious. There are many cases where a reaction equation contains chemical compounds that are structurally similar to each other, which makes it difficult to automatically find reactant pairs. One of the most typical examples is the reaction R04355 (Acetyl-ACP (C03939) + Malonyl-ACP (C01209) <=> Acetoacetyl-ACP (C05744) + CO2 (C00011) + ACP (C00229)) in the KEGG LIGAND database (Figure 3). Here, ACP is an abbreviation of Acyl-Carrier Protein, and Rnnnnn and Cnnnnn are entry IDs in KEGG LIGAND. The reactant pairs for this reaction should be as follows: C01209-C00011, C01209-C05744, C03939-C05744 and C03939-C00229 (Figure 3a). The most confusing thing to

Figure 3. An example of difficult cases to find reactant pairs from a reaction

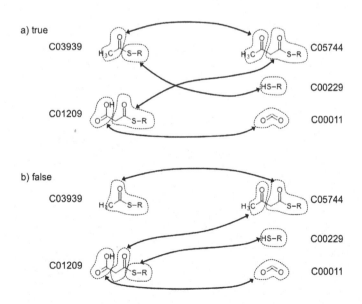

prevent their automatic finding is that the resulting structure of C01209 is identical to that of C03939 after the decarboxylation of C01209, which is described as the first pair. Thus, there is another mathematically possible answer: C01209-C00011, C01209-C00229, C01209-C05744 and C03939-C05744 (Figure 3b). This example illustrates the difficulty in making a correct answer from the viewpoint of its reaction mechanism.

Structurally related compounds in a reaction equation may additionally require expert knowledge, when making an atom alignment between a substrate and a product. As long as we adhere the strategy to obtain maximum common substructure between two compounds, the mathematically optimal solution does not always describe what occurs in nature, as typified by the reactant pair of Succinyl-CoA(C00091) and 3-Oxoadipyl-CoA (C02232) taken from the reaction R00829 ((Succinyl-CoA(C00091)+Acetyl-CoA(C00024) <=> CoA (C00010) + 3-Oxoadipyl-CoA (C02232)), whose graphical representation can be seen in Figure 4. In this reaction, the succinyl group of C00091 is transferred to another substrate Acetyl-CoA (C00024), and only the succinyl

group of C02232 should be aligned to that of C00091. However, the mathematically optimal solution without any supervised knowledge results in a larger atom alignment including the CoA group. This type of problem will be raised regardless of the alignment method itself.

Finally, although we put it simply in the previous section, extracting the reaction pattern from an atom alignment result is actually more difficult, as shown below. The simplest definition of a reaction center is the boundary atoms between the conserved and the non-conserved atom groups in a reactant pair, although there are many cases where reaction centers cannot be defined by such a definition. Such cases involve hydrogenation, tautmeric change, intramolecular transfer of a group, *etc*. This indicates we should check the bond valences (orders), aromaticities, and sometimes atom species around R atoms as well. The simplified pseudo-code algorithm we used in our study is described here to grasp the complicated heuristics involved. Eventually the problem is, given an aligned atom pair (a_1, a_2), to test if they could be R atoms or not with a set of IF-THEN rules, which enable us to obtain practically correct

Figure 4. An example of difficult cases in proper atom alignment

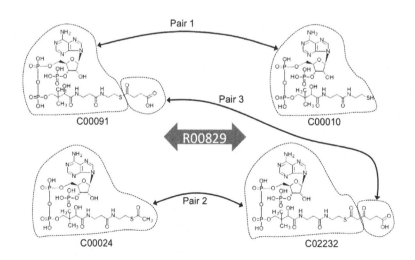

reaction centers. When at least one rule is accepted, both atoms a_1 and a_2 can be assigned as R atoms, and thus the RDM chemical transformation pattern will be obtained. In this scheme, the last rule for "Case 5" is used only to initially make

the RPAIR database, and it is never used in further applications such as E-zyme. The requirement of this last rule comes from a defectiveness or ambiguity in the database, that is, some reactions are incompletely described or some compounds are

Box 1. RDM assignment scheme we used

```
INPUT an atom alignment aa(q₁, q₂) between two structures q₁ and q₂

ALGORITHM
For each aligned atom pair (a₁ₖ, a₂ₖ) in aa(q₁, q₂)
        make a provisional RDM pattern rdm(a₁ₖ, a₂ₖ) around a₁ₖ and a₂ₖ;
        => "a₁ₖ-a₂ₖ:D₁₁+D₁₂+...-D₂₁+D₂₂+...:M₁₁+M₁₂+...-M₂₁+M₂₂+..."
        if (RatomOrNot(rdm(a₁ₖ, a₂ₖ))) then output rdm(a₁ₖ, a₂ₖ)
end
Ratom Or Not(rdm(a₁, a₂))
{
# After this, D₁ or D₂ means the set of D atoms on either side;
# D₁/₂ represents 'D₁ and D₂'. a₁/₂, M₁, M₂, and M₁/₂ are defined
# in a similar manner.
# Case 1: R₁-R₂:D₁₁+D₁₂-*+*:M₁₁+M₁₂+...-M₂₁+M₂₂+...
if (the numbers of D₁/₂ are different) then return true # (7824)
# Case 2: R₁-R₂:D₁+*-*+D₂:M₁₁+M₁₂+...-M₂₁+M₂₂+...
if (the number of D₁/₂ are same & they are not zero)
then
        if (neither D₁ᵢ nor D₂ⱼ are 'R'-represented) then return true # (4685)
endif
# Case 3: R₁-R₂:*-*:M₁₁+M₁₂+...-M₂₁+M₂₂+...
if (the number of D₁/₂ are zero & the number of M₁/₂ are not one)
then
        if (the valences of a₁/₂ are changed)
        then
                if (a₁ or a₂ has aromaticity & M₁/₂ have D atoms &
                the bond number of a-M are changed) then return true # (156)
                if (only one atom out of a₁/₂ has aromaticity &
                it has at least one aromatic M atom) then return true # (302)
                if (both a₁/₂ have no aromaticity) then return true # (1529)
        else
                if (the charge of a₁/₂ are changed) then return true # (7)
                if (only one atom out of a₁/₂ has aromaticity &
                its bond number to non-aromatic atom are changed)
                        then return true # (82)
        endif
endif
# Case 4: R₁-R₂:*-*:M₁-M₂
if (the number of D₁/₂ are zero & the number of M₁/₂ are one)
then
        if (a₁/₂ and M₁/₂ are all oxygens) then return true # (19)
        if (a₁/₂ are not oxygens & the bond number of a-M are changed)
        then
                if (M₁/₂ have D atoms) then return true # (55)
                if (M₁/₂ have no D atoms &
                the bond number of a-M is more than 2) then return true # (8)
        endif
endif
# Case 5: This is a case for manual assignments.
if (a₁/₂ have already been annotated as R by hand) then return true # (130)
# If none of the case above are applicable, then these are not R atoms.
return false
}
```

polymers. It means there may be other heuristics needed to properly define these cases. (see Box 1)

Here, the function RatomOrNot(*rdm*) returns *true* only when a provisional *rdm* can be accepted as an empirically correct one. The numbers in parentheses shown by boldface are the numbers of each case adapted when making the RPAIR database (as of 2009/6/23). The total number of R atoms is 14,797 from 11,353 RPAIR entries.

ACTUAL RESEARCH EXAMPLES OF REACTION PREDICTION

Chemical annotation involves two different topics as mentioned above: the first is to predict the enzymes for a given partial equation, *i.e.*, a given pair of a substrate and a product, and the second is to predict the fate of a given compound while reducing the computational complexity. In this section we describe an overview of the "E-zyme" system for the first topic. Many efforts have been made by many researchers for the second issue (Darvas, 1988; Ellis *et al.*, 2006; Hou *et al.*, 2004; Klopman & Tu, 1997; Langowski & Long, 2002; Talafous *et al.*, 1994). This is essentially the same as the first issue, but there occurs another problem: how to reduce the computational complexity. We summarize two different approaches for the second issue.

The E-zyme System

The E-zyme system is designed for the practical situation where users want to identify enzymes (enzyme genes, proteins or reaction mechanisms) from only a partial reaction equation (Kotera *et al.*, 2004). Users can input any compound pairs, and obtain the candidate EC classifications for the clue to identify the enzyme genes or proteins. Here is the pseudo-code representation of E-zyme, whose inner workings are depicted in Figure 5. (see Box 2)

This system needs the library of the RDM chemical transformation patterns calculated in advance, which is compared with the query transformation pattern, resulting in a list of possible EC classifications with specific scores. Recently, we have done a significant improvement in this E-zyme system, where a more complicated voting scheme and the EC-RDM profile based scoring system are applied to achieve higher coverage with a higher accuracy rate (Yamanishi *et al.*, 2009). The refined method is as described below. (see Box 3)

Here, the function match(rdm_1, rdm_2, *mc*) returns *true* only when two given patterns rdm_1 and rdm_2 are matched under a given matching condition *mc* like 'rd', which will work as the scope of each matching test. For instance, if *mc* is 'rd', then only R and D atoms from rdm patterns rdm_1

Figure 5. E-zyme scheme

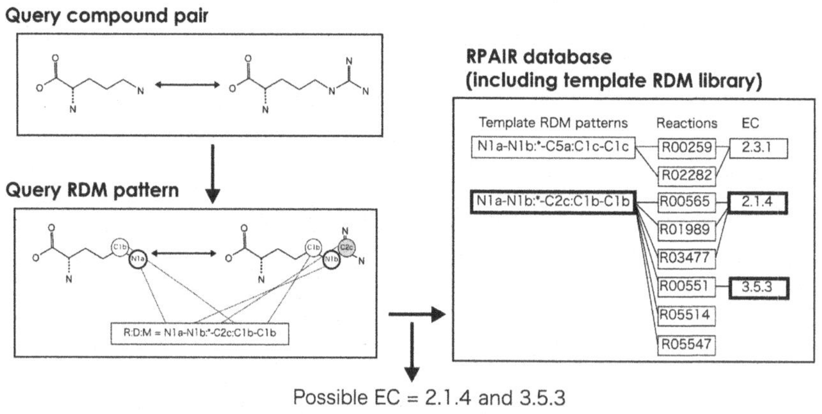

Possible EC = 2.1.4 and 3.5.3

Box 2. E-zyme scheme of Kotera-2004

```
INPUT two query structures q₁ and q₂
ALGORITHM
compute an atom alignment aa(q₁, q₂) from q₁ and q₂
extract an rdm pattern rdm(q₁, q₂) from aa(q₁, q₂)
foreach rdm pattern rdm_db in the RPAIR database
        if (rdm(q₁,q₂) = rdm_db) then store relevant ECs into {EC}.
end
if ({EC} is not empty)
then
        sort and output {EC_k} according to their frequencies
else
        output message that prediction was unsuccessful.
endif
```

Box 3. E-zyme scheme of Yamanishi-2009

```
INPUT two query structures q₁ and q₂
ALGORITHM
compute an atom alignment aa(q₁, q₂) from q₁ and q₂
extract an rdm pattern rdm(q₁, q₂) from aa(q₁, q₂)
foreach matching condition mc in 'rdm', 'rd', 'dm', 'rm', 'r', 'd', and 'm'
        foreach rdm pattern rdm_db in RPAIR database
                if (match(rdm(q₁,q₂), rdm_db, mc))
                    then store relevant ECs into {EC}.
        end
        if ({EC} is not empty) then break
end
if ({EC} is not empty)
then
        foreachEC_k in {EC}
                calculate its weighted score S(EC_k), which is based on the sum
                    of correlation coefficients between V_rdm(q1, q2) and any V_rdmdb.
        end
        sort and output {EC_k} according to their weighted scores S(EC_k)
else
        output message that prediction failed.
endif
```

and rdm_2 are used to test whether they are matched or not. V_{rdm} is a vector $(x_1, x_2,..., x_{|EC|})^T$, where x_k is the number of observed reactions having *rdm* pattern for k-th EC sub-subclass.

Prediction of Bacterial Biodegradation Pathways

The second issue, predicting the fate of a given compound, has been also tackled with the RDM patterns (Oh *et al.*, 2007; Moriya *et al.*, 2010), as an extension of the E-zyme approach. Because of the problem of computational time, we focused on bacterial biodegradation pathways of a query compound (such as new xenobiotics). We tested two different types of structure matching: a local substructure matching against the known RDM pattern library, and global structure matching of the query compound and the database compound. The former is for inferring a converted structure based on the matched RDM pattern, and the latter is to prioritize more similar compounds among many possible candidates.

The overall prediction scheme is illustrated in Figure 6, and the detailed pseudo-code representation is shown below. In the first step, the query compound structure is compared with those in the KEGG bacterial xenobiotics category. In the second step, possible RDM patterns on the query compound are selected from the RDM pattern library based on the structurally similar compounds containing the corresponding RDM patterns with the use of the SIMCOMP program. The third step is to obtain the plausible products according to the selected RDM patterns. The generated products become the next query compound and the prediction is iterated if possible. (see Box 4)

The GREP Approach

The GREP (Generator of Reaction Equations & Pathways) method has a different strategy to find plausible enzyme reactions and the putative EC classifications (Kotera *et al.*, 2008). Different from the above E-zyme approach, this method is designed for finding all possible metabolic pathways in a given set of compounds, without defining transformation patterns in advance, nor defining possible intermediate compounds. In this method, structurally related compounds are chosen and compared to estimate if one of these compounds can be a product (or precursor) of the other in an enzyme reaction. In this process, chemical bonds, functional groups, or substructures involved in the putative reaction are checked, and the fate of the remaining substructures is also suggested by construction of a putative reaction equation. The process of generating an equation is described in Figure 7, and the pseudo-code is shown below. If the equation is regarded as possible, the corresponding EC classification is also given simultaneously. Estimation of whether or not the putative reaction is likely to occur and the possible EC classes are performed with the decision tree and random forest methods. (see Box 5)

The above pseudo-code describes the first step of generating reaction equation shown in Figure 7. In the next step, an appropriate set of cofactors will be added if necessary. Here, $A \cap B$ denotes the size of the MCS between two compounds A and B without hydrogen atoms, and $A \backslash B$ represents the size of substructure found in A, which is not present in B. For example, suppose that A and B

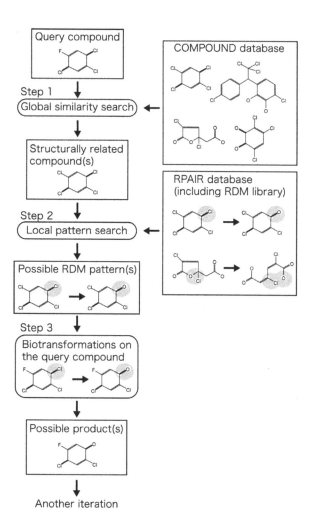

Figure 6. Prediction scheme for bacterial biodegradation

are D-glucose and D-glucose 6-phosphate, respectively, then $A \cap B = 12$ (corresponding to the C_6O_6 of the glucose residue), $B \backslash A = 4$ (corresponding to the phosphate residue) and $A \backslash B = 0$. The ternary operators $A \cap B \cap C$ and $A \backslash B \backslash C$ represent the size of MCS among three compounds A, B and C, and the size of substructure found in A, which is present in neither B nor C, respectively.

Toward the Integration of These Methods

Although both the E-zyme and GREP approaches result in proposing possible EC classifications,

Box 4. Pathway prediction scheme of Oh-2007

```
INPUT one query structure: q

ALGORITHM
call Search&Generate (q)
rank all outputs according to path length or average similarity score.
Search&Generate (q)
{
push q into {Q}
search compounds {C_i} similar to q against KEGG Compound DB
search rpairs {RP_j(C_i, C_i^counter)} related to each C_i against RPAIR DB
extract rdm patterns {rdm_k(C_i, C_i^counter)} from {RP_j(C_i, C_i^counter)}
foreach rdm pattern rdm_k in {rdm_k(C_i, C_i^counter)}
        if (rdm_k is applicable to q)
        then
                generate new compound q* from q with rdm_k
                if (q* is de novo structure) then call Search&Generate (q*)
        endif
end
if (no Search&Generate (q*) was called)
then
        output {Q}
endif
pop q from {Q}
}
```

Note: DB is an abbreviation for database in this algorithm

they differ in how to find the reaction transformation patterns and the scope of the compounds. The major advantages of the E-zyme approach are that user can input any chemical structure, regardless of the occurrence in the database, and also that the putative metabolic pathway can include compounds that are not registered in the database. This advantage is a two-edged blade, which inevitably causes an explosion in computational time. The E-zyme approach avoids this by limiting the scope of pathways to bacterial biodegradation. On the other hand, the GREP approach avoids this by considering only compounds in the database and pre-calculating candidates *a priori* to the user input. This GREP strategy has another advantage, *i.e.*, generating the whole reaction equation including not only main substrates and products but also cofactors. Predicting the cofactor selectivity of enzymes would contribute to understanding the metabolic balance in an organism.

Now we are integrating these two strategies to put their respective advantages together to obtain higher accuracy and coverage with improved

usability. For example, consider the case when a user wants to predict the reaction mechanism in terms of the EC number classification, and has all knowledge of a reaction equation including all of the substrates, products and the cofactor information. In the current E-zyme system, we offer only the multiple pair mode predicting the EC number from given multiple compound pairs. This process could be improved if a user can input the query reaction equation as it is, and it is natural that the prediction result will become more precise than the single pair prediction. The pre-calculated GREP result could be used to support the user input by guessing cofactors or products, and predict the fate of the query compounds at the same time.

FUTURE RESEARCH DIRECTIONS

Not limited to this particular problem, informatics approaches in general cannot guarantee that the predicted result actually occurs, so it is important to estimate the reaction's likelihood to occur in nature, and to understand the underlying assump-

Figure 7. GREP scheme

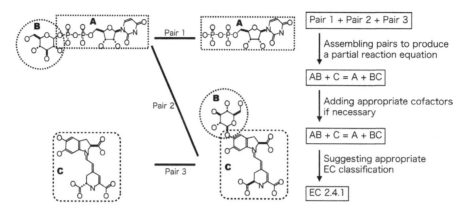

Box 5. GREP scheme of Kotera-2008

```
# Note: in the algorithms, set union, intersection and difference return
# the size of their respective operations
INPUT a set of compounds {S}
                 ALGORITHM
foreach compound pair (A, B) in {S}, where (A≠B&A∩B ≥ 5)
          if (PairTest1(A, B)) then store "A-B" into {P}
          foreachC in {S}, where (A≠B≠C&A∩C ≥ 5)
                    if (PairTest2(A, B, C)) then store "A-B & A-C" into {P}
                    if (B\A ≥ 2 &C\A ≤ 1)
                    then
                                  foreachD in {S}, where (A≠B≠C≠D&B∩D ≥ 5)
                                        if (PairTest3(A, B, C, D))
                                          then store "A-B & A-C & B-D" into {P}
                                  end
                    endif
          end
end
output a set of possible partial reactions {P} to the next step
# Generate a partial reaction "A = B" with a pair AB
PairTest1(A, B)
{
          if (A\B ≤ 1 &B\A ≤ 1) then return true
          if (A\B ≤ 4 &B\A = 0) then return true
          if (A\B = 0 &B\A ≤ 4) then return true
          return false
}
# Generate a partial reaction "A = B + C" with pairs AB and AC
PairTest2(A, B, C)
{
          if (A∩B∩C ≥ 2) then return false
          if (A\B\C ≥ 2) then return false
          if (B\A ≤ 1 &C\A ≤ 1) then return true
          return false
}
# Generate a partial reaction "A + D = B + C" with pairs AB, AC and BD
PairTest3(A, B, C, D)
{
          if (A∩B∩D ≥ 1) then return false
          if (B\A\D ≤ 1 &D\B) then return true
          return false
}
```

tion of the reaction system under investigation. All of these approaches mentioned above are based on the assumption that the query compound will mimic the biotransformation patterns of the compounds in the library. This assumption may have to be reconsidered for secondary metabolism including many reaction patterns that are not seen in primary metabolism, which provides the cases where no prediction is obtained if the query pattern is not in the library. This is a trade-off between the coverage and the precision. Future research directions involve how to solve this problem as well as the development of an efficient algorithm to reduce the computational amount.

All of these approaches explained in this chapter are based solely on the overall reaction and take nothing else into account, such as the detailed reaction steps, the sequence, or the structure of the putative enzymes. The freedom from such constrains will prove advantageous because there are many cases where only compound structures are known. This does not mean that we should not incorporate these types of information. Rather, incorporating protein sequence or structure information would be one of many possible intriguing developments for connecting the genomic, proteomic and metabolomic spaces. Integration of genomic and chemoinformatics analyses have been reported recently to elucidate the repertoire of possible structures of some specific compounds, such as N-glycans (Kawano *et al.*, 2005) and polyketides (Minowa *et al.*, 2007).

It should also be mentioned that confirmed negative data (*e.g.*, reactions that do not occur) is rare to find in published articles or databases, which makes it difficult to estimate if a negative prediction result is truly negative and *vice versa*. For example, intramolecular transfer of a group has to be estimated carefully because a compound usually has many positional (constitutional) isomers that cannot be converted into each other in a single reaction. As another example, a group of compounds sharing a core substructure with

diverse side-chain modifications usually results in a grid-shaped pathway in the pathway prediction. Some of these are known to be true, others false, and sometimes depend on organisms. These confusing issues may be resolved by taking additional information into account, such as the genome content in an organism, the context in the pathway, and the reactivity of the compounds, all of which will be input in future developments.

An effective algorithm for chemical structure comparison is also the remaining issue. The maximum-common-substructure strategy is generally successful in both of our approaches, but it fails in some cases. One of these includes compounds containing multiple occurrences of identical or similar substructures, which cause difficulties in determining which substructures come from which compound. It is also very difficult to trace multi-step reactions repeatedly incorporating small units, such as terpenoid or polyketide synthesis. Even in the case where the total prediction of the synthetic pathway may not be feasible, the development of the prediction algorithm to quickly divide a compound into small units with known origin would be valuable. This approach is analogous to the comprehensive analyses of approved drug structures, which are classified based on their core skeleton structures and/or peripheral fragments (Bemis & Murcko, 1996 & 1999; Shigemizu *et al.*, 2009). The most interesting and easiest manner from the viewpoint of chemoinformatics may be to apply the similar method to a set of metabolites.

Here we would like to introduce some tips for the researchers considering reaction prediction. First, there usually occur conflicting cases between the human-curated classification such as the EC and the computationally generated classification. There are cases where a reaction equation corresponds to more than one EC number or sub-subclass, the reason of which includes the enzyme protein architecture and the reaction mechanisms. It is hard to determine which may have to be corrected without expert knowledge. It should also be noted that the classification scheme or criteria may change according to the purpose. In this context, the E-zyme adopts a unified classification scheme over all enzymes, while the GREP strategy implements different criteria in different classes just as the EC does.

Secondly, we have to be careful with the intrinsic fidelity of the data. There are cases where the reported equations contain incomplete structures, or the equations are stoichiometrically unbalanced. There may be fewer intermediates being registered in the database than required for the prediction. We should also take care of the stereo- and ethene-based isomers. Although the MDL format allows 3D coordinates, most of the files in databases including KEGG are depicted as 2D graph objects on a plane to simplify the chemical substance. The 2D description seems to distinguish these isomers, since parts of chemical bonds can be described with special notation, namely bonds above and below the stereocenter, to exemplify the chirality of each structure when needed. However, the E-zyme currently ignores this and thus does not detect the chirality change occurring in epimerase or cis-trans isomerase reactions. The GREP implements a simple check of the bond direction, although this sometimes fails. This comes from not only our methodology but also from the database, in which there are compounds for which the stereoisomerism is not clear. We are trying to solve this problem from both sides: to implement a strict algorithm to check the chirality, and to refine the compound entries in the database.

CONCLUSION

In this chapter the authors overviewed the reaction prediction problems in chemoinformatics and some practical research efforts we have made. This research field connects genomics to metabolomics, holding enormous potential towards environmental problems, human health, *etc.*, although many issues remain unexplored. The authors would be grateful if this chapter served as a stepping-stone for those who hope to enter this intriguing research area.

AVAILABILITY

The EC number prediction system is named as "E-zyme" and is publicly available on the GenomeNet web site from the following URL: http://www.genome.jp/tools/e-zyme/. The GREP system is available at http://bisscat.org/GREP/.

REFERENCES

Bemis, G. W., & Murcko, M. A. (1996). The properties of known drugs. 1. Molecular frameworks. *Journal of Medicinal Chemistry*, *39*(15), 2887–2893. doi:10.1021/jm9602928

Bemis, G. W., & Murcko, M. A. (1999). Properties of known drugs. 2. Side chains. *Journal of Medicinal Chemistry*, *42*(25), 5095–5099. doi:10.1021/jm9903996

BioCarta. Retrieved from http://www.biocarta.com/. BioCarta LLC, San Diego, CA

Blower, P. E., & Dana, R. C. (1986). Creation of a chemical reaction database from the primary literature. In Willett, P. (Ed.), *Modern Approaches to Chemical Reaction Searching* (pp. 146–164). Aldershot, England: Gower Publishing Company.

Caspi, R., Foerster, H., Fulcher, C. A., Kaipa, P., Krummenacker, M., & Latendresse, M. (2008). The MetaCyc database of metabolic pathways and enzymes and the BioCyc collection of pathway/genome databases. *Nucleic Acids Research*, *36*(Database issue), D623–D631. doi:10.1093/nar/gkm900

Chen, L., Nourse, J. G., Christie, B. D., Leland, B. A., & Grier, D. L. (2002). Over 20 years of reaction access systems from MDL: a novel reaction substructure search algorithm. *Journal of Chemical Information and Computer Sciences*, *42*(6), 1296–1310. doi:10.1021/ci020023s

Darvas, F. (1988). Predicting metabolic pathways by logic programming. *Journal of Molecular Graphics & Modelling*, *6*(2), 80–86.

Eisenberg, D., Marcotte, E. M., Xenarios, I., & Yeates, T. O. (2000). Protein function in the post-genomic era. *Nature*, *405*(6788), 823–826. doi:10.1038/35015694

Ellis, L. B. M., Roe, D., & Wackett, L. P. (2006). The university of Minnesota biocatalysis/biodegradation database: the first decade. *Nucleic Acids Research*, *34*(Database issue), D517–D521. doi:10.1093/nar/gkj076

Goto, S., Nishioka, T., & Kanehisa, M. (1998). LIGAND: chemical database for enzyme reactions. *Bioinformatics (Oxford, England)*, *14*(7), 591–599. doi:10.1093/bioinformatics/14.7.591

Goto, S., Okuno, Y., Hattori, M., Nishioka, T., & Kanehisa, M. (2002). LIGAND: database of chemical compounds and reactions in biological pathways. *Nucleic Acids Research*, *30*(1), 402–404. doi:10.1093/nar/30.1.402

Hattori, M., Okuno, Y., Goto, S., & Kanehisa, M. (2003). Development of a chemical structure comparison method for integrated analysis of chemical and genomic information in the metabolic pathways. *Journal of the American Chemical Society*, *125*(39), 11853–11865. doi:10.1021/ja036030u

Hattori, M., Tanaka, N., Kanehisa, M., & Goto, S. (2010). SIMCOMP/SUBCOMP: chemical structure search servers for network analyses. *Nucleic Acids Research, Advance Access published on May 11, 2010.*

Hou, B. K., Ellis, L. B. M., & Wackett, L. P. (2004). Encoding microbial metabolic logic: predicting biodegradation. *Journal of Industrial Microbiology & Biotechnology*, *31*(6), 261–272. doi:10.1007/s10295-004-0144-7

Jaccard, P. (1912). The distribution of the flora of the alpine zone. *The New Phytologist*, *11*(2), 37–50. doi:10.1111/j.1469-8137.1912.tb05611.x

Kanehisa, M. (1997). A database for post-genome analysis. *Trends in Genetics*, *13*(9), 375–376. doi:10.1016/S0168-9525(97)01223-7

Kanehisa, M. (2001). Prediction of higher order functional networks from genomic data. *Pharmacogenomics*, *2*(4), 373–385. doi:10.1517/14622416.2.4.373

Kanehisa, M., Araki, M., Goto, S., Hattori, M., Hirakawa, M., & Itoh, M. (2008). KEGG for linking genomes to life and the environment. *Nucleic Acids Research*, *36*(Database issue), D480–D484. doi:10.1093/nar/gkm882

Kanehisa, M., & Bork, P. (2003). Bioinformatics in the post-sequence era. *Nature Genetics*, *33*(3s), 305–310. doi:10.1038/ng1109

Kanehisa, M., & Goto, S. (2000). KEGG: Kyoto encyclopedia of genes and genomes. *Nucleic Acids Research*, *28*(1), 27–30. doi:10.1093/nar/28.1.27

Kanehisa, M., Goto, S., Hattori, M., Aoki-Kinoshita, K. F., Itoh, M., & Kawashima, S. (2006). From genomics to chemical genomics: new developments in KEGG. *Nucleic Acids Research*, *34*(Database issue), D354–D357. doi:10.1093/nar/gkj102

Kawano, S., Hashimoto, K., Miyama, T., Goto, S., & Kanehisa, M. (2005). Prediction of glycan structures from gene expression data based on glycosyltransferase reactions. *Bioinformatics (Oxford, England)*, *21*(21), 3976–3982. doi:10.1093/bioinformatics/bti666

Keseler, I. M., Bonavides-Martinez, C., Collado-Vides, J., Gama-Castro, S., Gunsalus, R. P., & Johnson, D. A. (2009). EcoCyc: A comprehensive view of *Escherichia coli* biology. *Nucleic Acids Research, 37*(Database issue), D464–D470. doi:10.1093/nar/gkn751

Klopman, G., & Tu, M. (1997). Structure-biodegradability study and computer automated prediction of aerobic biodegradation of chemicals. *Environmental Toxicology and Chemistry, 16*(9), 1829–1835.

Kotera, M., McDonald, A. G., Boyce, S., & Tipton, K. F. (2008). Eliciting possible reaction equations and metabolic pathways involving orphan metabolites. *Journal of Chemical Information and Modeling, 48*(12), 2335–2349. doi:10.1021/ci800213g

Kotera, M., Okuno, Y., Hattori, M., Goto, S., & Kanehisa, M. (2004). Computational assignment of the EC numbers for genomic-scale analysis of enzymatic reactions. *Journal of the American Chemical Society, 126*(50), 16487–16498. doi:10.1021/ja0466457

Kuhl, F. S., Crippen, G. M., & Friesen, D. K. (1984). A combinatorial algorithm for calculating ligand binding. *Journal of Computational Chemistry, 5*(1), 24–34. doi:10.1002/jcc.540050105

Langowski, J., & Long, A. (2002). Computer systems for the prediction of xenobiotic metabolism. *Advanced Drug Delivery Reviews, 54*(3), 407–415. doi:10.1016/S0169-409X(02)00011-X

McGregor, J. J., & Willett, P. (1981). Use of a maximal common subgraph algorithm in the automatic identification of the ostensible bond changes occurring in chemical reactions. *Journal of Chemical Information and Computer Sciences, 21*(3), 137–140. doi:10.1021/ci00031a005

Minowa, Y., Araki, M., & Kanehisa, M. (2007). Comprehensive analysis of distinctive polyketide and nonribosomal peptide structural motifs encoded in microbial genomes. *Journal of Molecular Biology, 368*(5), 1500–1517. doi:10.1016/j.jmb.2007.02.099

Moock, T. E., Nourse, J. G., Grier, D., & Hounshell, W. D. (1988). The implementation of atom-atom mapping and related features in the reaction access system (REACCS) . In Warr, W. A. (Ed.), *Chemical Structures, The International Language of Chemistry* (pp. 303–313). Berlin, Germany: Springer-Verlag.

Moriya, Y., Shigemizu, D., Hattori, M., Tokimatsu, T., Kotera, M., Goto, S., & Kanehisa, M. (2010). PathPred: an enzyme-catalyzed metabolic pathway prediction server. *Nucleic Acids Research, Advance Access published on April 30, 2010.*

Oh, M., Yamada, T., Hattori, M., Goto, S., & Kanehisa, M. (2007). Systematic analysis of enzyme-catalyzed reaction patterns and prediction of microbial biodegradation pathways. *Journal of Chemical Information and Modeling, 47*(4), 1702–1712. doi:10.1021/ci700006f

Raymond, J. W., Gardiner, E. J., & Willett, P. (2002a). RASCAL: Calculation of graph similarity using maximum common edge subgraphs. *The Computer Journal, 45*(6), 631–644. doi:10.1093/comjnl/45.6.631

Raymond, J. W., Gardiner, E. J., & Willett, P. (2002b). Heuristics for similarity searching of chemical graphs using a maximum common edge subgraph algorithm. *Journal of Chemical Information and Computer Sciences, 42*(2), 305–316. doi:10.1021/ci010381f

Raymond, J. W., & Willett, P. (2002). Maximum common subgraph isomorphism algorithms for the matching of chemical structures. *Journal of Computer-Aided Molecular Design, 16*(7), 521–533. doi:10.1023/A:1021271615909

Sanchez, C., Lachaize, C., Janody, F., Bellon, B., Röder, L., & Euzenat, J. (1999). Grasping at molecular interactions and genetic networks in *Drosophila melanogaster* using FlyNets, an internet database. *Nucleic Acids Research, 27*(1), 89–94. doi:10.1093/nar/27.1.89

Shigemizu, D., Araki, M., Okuda, S., Goto, S., & Kanehisa, M. (2009). Extraction and analysis of chemical modification patterns in drug development. *Journal of Chemical Information and Modeling, 49*(4), 1122–1129. doi:10.1021/ci8003804

Takahashi, Y., Maeda, S., & Sasaki, S. (1987). Automated recognition of common geometrical patterns among a variety of three-dimensional molecular structures. *Analytica Chimica Acta, 200*(11), 363–377. doi:10.1016/S0003-2670(00)83783-6

Talafous, J., Sayre, L. M., Mieyal, J. J., & Klopman, G. (1994). META.2. A dictionary model of mammalian xenobiotic metabolism. *Journal of Chemical Information and Computer Sciences, 34*(6), 1326–1333. doi:10.1021/ci00022a015

Tipton, K. F., & Boyce, S. (2005). Enzyme classification and nomenclature . In *John Wiley & Sons, Inc., Encyclopedia of Life Sciences*. Chichester, England: Wiley.

Tweeddale, H., Notley-McRobb, L., & Ferenci, T. (1998). Effect of slow growth on metabolism of *Escherichia coli*, as revealed by global metabolite pool ("metabolome") analysis. *Journal of Bacteriology, 180*(19), 5109–5116.

Velculescu, V. E., Zhang, L., Zhou, W., Vogelstein, J., Basrai, M. A., & Bassett, D. E. Jr (1997). Characterization of the yeast transcriptome. *Cell, 88*(2), 243–251. doi:10.1016/S0092-8674(00)81845-0

Watson, G. A. (1983). An algorithm for the single facility location problem using the Jaccard metric. *SIAM Journal on Scientific and Statistical Computing, 4*(4), 748–756. doi:10.1137/0904052

Wilkins, M. R., Sanchez, J. C., Gooley, A. A., Appel, R. D., Humphery-Smith, I., Hochstrasser, D. F., & Williams, K. L. (1996). Progress with proteome projects: why all proteins expressed by a genome should be identified and how to do it. *Biotechnology & Genetic Engineering Reviews, 13*, 19–50.

Yamanishi, Y., Hattori, M., Kotera, M., Goto, S., & Kanehisa, M. (2009). E-zyme: predicting potential EC numbers from the chemical transformation pattern of substrate-product pairs. *Bioinformatics (Oxford, England), 25*(12), i179–i186. doi:10.1093/bioinformatics/btp223

Yamanishi, Y., Mihara, H., Osaki, M., Muramatsu, H., Esaki, N., & Sato, T. (2007). Prediction of missing enzyme genes in bacterial metabolic network: a reconstruction of lysine degradation pathway of Pseudomonas aeruginosa. *The FEBS Journal, 274*(9), 2262–2273. doi:10.1111/j.1742-4658.2007.05763.x

Yamanishi, Y., Vert, J.-P., & Kanehisa, M. (2004). Heterogeneous data comparison and gene selection with kernel canonical correlation analysis . In Scholkopf, B., Tsuda, K., & Vert, J.-P. (Eds.), *Kernel Methods in Computational Biology* (pp. 209–230). Cambridge, MA: MIT Press.

ADDITIONAL READING

Arakawa, M., Hasegawa, K., & Funatsu, K. (2003). Novel alignment method of small molecules using the Hopfield neural network. *Journal of Chemical Information and Computer Sciences, 43*(5), 1390–1395. doi:10.1021/ci0300011

Arakawa, M., Hasegawa, K., & Funatsu, K. (2003). Application of the novel molecular alignment method using the Hopfield neural network to 3D-QSAR. *Journal of Chemical Information and Computer Sciences, 43*(5), 1396–1402. doi:10.1021/ci030005q

Barrett, A. J., Canter, C. R., Liebecq, C., Moss, G. P., Saenger, W., & Sharon, N. (1992). *Enzyme Nomenclature*. San Diego, CA: Academic Press.

Bender, A., Mussa, H. Y., Glen, R. C., & Reiling, S. (2004). Molecular similarity searching using atom environments, information-based feature selection, and a naive Bayesian classifier. *Journal of Chemical Information and Computer Sciences, 44*(1), 170–178. doi:10.1021/ci034207y

Bender, A., Mussa, H. Y., Glen, R. C., & Reiling, S. (2004). Similarity searching of chemical databases using atom environment descriptors (MOLPRINT 2D): evaluation of performance. *Journal of Chemical Information and Computer Sciences, 44*(5), 1708–1718. doi:10.1021/ci0498719

Bocker, A., Derksen, S., Schmidt, E., Teckentrup, A., & Schneider, G. (2005). A hierarchical clustering approach for large compound libraries. *Journal of Chemical Information and Modeling, 45*(4), 807–815. doi:10.1021/ci0500029

Bron, C., & Kerbosch, J. (1973). Algorithm 457: Finding all cliques of an undirected graph. *Communications of the ACM, 16*(9), 575–577. doi:10.1145/362342.362367

Brooksbank, C., Cameron, G., & Thornton, J. (2005). The European bioinformatics institute's data resources: towards systems biology. *Nucleic Acids Research, 33*(Database issue), D46–D53. doi:10.1093/nar/gki026

Brown, R. D., & Martin, Y. C. (1996). Use of structure – activity data to compare structure-based clustering methods and descriptors for use in compound selection. *Journal of Chemical Information and Computer Sciences, 36*(3), 572–584. doi:10.1021/ci9501047

Brown, R. D., & Martin, Y. C. (1997). The information content of 2D and 3D structural descriptors relevant to ligand-receptor binding. *Journal of Chemical Information and Computer Sciences, 37*(1), 1–9. doi:10.1021/ci960373c

Cosgrove, D. A., & Willett, P. (1998). SLASH: a program for analysing the functional groups in molecules. *Journal of Molecular Graphics & Modelling, 16*(1), 19–32. doi:10.1016/S1093-3263(98)00014-X

Darvas, F. (1988). Predicting metabolic pathways by logic programming. *Journal of Molecular Graphics, 6*(2), 80–86. doi:10.1016/0263-7855(88)85004-5

Dobson, C. M. (2004). Chemical space and biology. *Nature, 432*(7019), 824–828. doi:10.1038/nature03192

Douguet, D., Munier-Lehmann, H., Labesse, G., & Pochet, S. (2005). LEA3D: a computer-aided ligand design for structure-based drug design. *Journal of Medicinal Chemistry, 48*(7), 2457–2468. doi:10.1021/jm0492296

Erlanson, D. A., McDowell, R. S., & O'Brien, T. (2004). Fragment-based drug discovery. *Journal of Medicinal Chemistry, 47*(14), 3463–3482. doi:10.1021/jm040031v

Feldman, H. J., Dumontier, M., Ling, S., Haider, N., & Hogue, C. W. V. (2005). CO: A chemical ontology for identification of functional groups and semantic comparison of small molecules. *FEBS Letters, 579*(21), 4685–4691. doi:10.1016/j.febslet.2005.07.039

Flower, D. R. (1998). On the properties of bit string-based measures of chemical similarity. *Journal of Chemical Information and Computer Sciences, 38*(3), 379–386. doi:10.1021/ci970437z

Goto, S., Bono, H., Ogata, H., Fujibuchi, W., Nishioka, T., Sato, K., & Kanehisa, M. (1996). Organizing and computing metabolic pathway data in terms of binary relations. *Pacific Symposium on Biocomputing 2*, 175-186.

Guimera, R., & Amaral, L. A. N. (2005). Functional cartography of complex metabolic networks. *Nature, 433*(7028), 895–900. doi:10.1038/nature03288

Hashimoto, K., Goto, S., Kawano, S., Aoki-Kinoshita, K. F., Ueda, N., & Hamajima, M. (2006). KEGG as a glycome informatics resource. *Glycobiology, 16*(5), 63R–70R. doi:10.1093/glycob/cwj010

Ihmels, J., Friedlander, G., Bergmann, S., Sarig, O., Ziv, Y., & Barkai, N. (2002). Revealing modular or-ganization in the yeast transcriptional network. *Nature Genetics, 31*(4), 370–377.

Janssen, D. B., Dinkla, I. J. T., Poelarends, G. J., & Terpstra, P. (2005). Bacterial degradation of xenobiotic compounds: evolution and distribution of novel enzyme activities. *Environmental Microbiology, 7*(12), 1868–1882. doi:10.1111/j.1462-2920.2005.00966.x

Karp, P. D. (2004). Call for an enzyme genomics initiative. *Genome Biology, 5*, 401. doi:10.1186/gb-2004-5-8-401

Kelley, B. P., Sharan, R., Karp, R. M., Sittler, T., Root, D. E., Stockwell, B. R., & Ideker, T. (2003). Conserved pathways within bacteria and yeast as revealed by global protein network alignment. *Proceedings of the National Academy of Sciences of the United States of America, 100*(20), 11394–11399. doi:10.1073/pnas.1534710100

Kharchenko, P., Vitkup, D., & Church, G. M. (2004). Filling gaps in a metabolic network using expression information. *Bioinformatics (Oxford, England), 20*(Suppl 1), i178–i185. doi:10.1093/bioinformatics/bth930

Koch, M. A., Schuffenhauer, A., Scheck, M., Wetzel, S., Casaulta, M., & Odermatt, A. (2005). Charting biologically relevant chemical space: a structural classification of natural products (SCONP). *Proceedings of the National Academy of Sciences of the United States of America, 102*(48), 17272–17277. doi:10.1073/pnas.0503647102

Kogej, T., Engkvist, O., Blomberg, N., & Muresan, S. (2006). Multifingerprint based similarity searches for targeted class compound selection. *Journal of Chemical Information and Modeling, 46*(3), 1201–1213. doi:10.1021/ci0504723

Koyuturk, M., Grama, A., & Szpankowski, W. (2004). An efficient algorithm for detecting frequent sub-graphs in biological networks. *Bioinformatics (Oxford, England), 20*(Suppl 1), i200–i207. doi:10.1093/bioinformatics/bth919

Lameijer, E. W., Kok, J. N., Back, T., & Ijzerman, A. P. (2006). Mining a chemical database for fragment co-occurrence: discovery of "chemical cliches". *Journal of Chemical Information and Modeling, 46*(2), 553–562. doi:10.1021/ci050370c

Latino, D. A., Zhang, Q. Y., & Aires-de-Sousa, J. (2008). Genome-scale classification of metabolic reactions and assignment of EC numbers with self-organizing maps. *Bioinformatics (Oxford, England), 24*(19), 2236–2244. doi:10.1093/bioinformatics/btn405

Lewell, X. Q., Judd, D. B., Watson, S. P., & Hann, M. M. (1998). RECAP--retrosynthetic combinatorial analysis procedure: a powerful new technique for identifying privileged molecular fragments with useful applications in combinatorial chemistry. *Journal of Chemical Information and Computer Sciences, 38*(3), 511–522. doi:10.1021/ci970429i

Mavrovouniotis, M., & Stephanopoulos, G. (1992). Synthesis of biochemical production routes. *Computers & Chemical Engineering, 16*(6), 605–619. doi:10.1016/0098-1354(92)80071-G

Mayeno, A. N., Yang, R. S. H., & Reisfeld, B. (2005). Biochemical reactions network modeling: predicting metabolism of organic chemical mixtures. *Environmental Science & Technology, 39*(14), 5363–5371. doi:10.1021/es0479991

McShan, D. C., Updadhayaya, M., & Shah, I. (2004). Symbolic inference of xenobiotics metabolism. *Pacific Symposium on Biocomputing, 9*, 545-556.

Miller, M. A. (2002). Chemical database techniques in drug discovery. *Nature Reviews. Drug Discovery, 1*(3), 220–227. doi:10.1038/nrd745

Morphy, R., & Rankovic, Z. (2005). Designed multiple ligands. An emerging drug discovery paradigm. *Journal of Medicinal Chemistry, 48*(21), 6523–6543. doi:10.1021/jm058225d

Muller, G. (2003). Medicinal chemistry of target family-directed masterkeys. *Drug Discovery Today, 8*(15), 681–691. doi:10.1016/S1359-6446(03)02781-8

Muresan, S., & Sadowski, J. (2005). "In-house likeness": comparison of large compound collections using artificial neural networks. *Journal of Chemical Information and Modeling, 45*(4), 888–893. doi:10.1021/ci049702o

Nobeli, I., Ponstingl, H., Krissinel, E. B., & Thornton, J. M. (2003). A structure-based anatomy of the *E. coli* metabolome. *Journal of Molecular Biology, 334*(4), 697–719. doi:10.1016/j.jmb.2003.10.008

Nobeli, I., & Thornton, J. M. (2006). A bioinformatician's view of the metabolome. *BioEssays, 28*(5), 534–545. doi:10.1002/bies.20414

Ogata, H., Fujibuchi, W., Goto, S., & Kanehisa, M. (2000). A heuristic graph comparison algorithm and its application to detect functionally related enzyme clusters. *Nucleic Acids Research, 28*(20), 4021–4028. doi:10.1093/nar/28.20.4021

Pinter, R. Y., Rokhlenko, O., Yeger-Lotem, E., & Ziv-Ukelson, M. (2005). Alignment of metabolic pathways. *Bioinformatics (Oxford, England), 21*(16), 3401–3408. doi:10.1093/bioinformatics/bti554

Rees, D. C., Congreve, M., Murray, C. W., & Carr, R. (2004). Fragment-based lead discovery. *Nature Reviews. Drug Discovery, 3*(8), 660–672. doi:10.1038/nrd1467

Roberts, G., Myatt, G. J., Johnson, W. P., Cross, K. P., & Blower, P. E. Jr. (2000). LeadScope: software for exploring large sets of screening data. *Journal of Chemical Information and Computer Sciences, 40*(6), 1302–1314. doi:10.1021/ci0000631

Schneider, G., Lee, M. L., Stahl, M., & Schneider, P. (2000). De novo design of molecular architectures by evolutionary assembly of drug-derived building blocks. *Journal of Computer-Aided Molecular Design, 14*(5), 487–494. doi:10.1023/A:1008184403558

Schneider, G., Neidhart, W., Giller, T., & Schmid, G. (1999). "Scaffold-hopping" by topological pharmacophore search: A contribution to virtual screening. *Angewandte Chemie International Edition, 38*(19), 2894–2896. doi:10.1002/(SICI)1521-3773(19991004)38:19<2894::AID-ANIE2894>3.0.CO;2-F

Schnur, D. M., Hermsmeier, M. A., & Tebben, A. J. (2006). Are target-family-privileged substructures truly privileged? *Journal of Medicinal Chemistry, 49*(6), 2000–2009. doi:10.1021/jm0502900

Sharan, R., Suthram, S., Kelley, R. M., Kuhn, T., McCuine, S., & Uetz, P. (2005). Conserved patterns of protein interaction in multiple species. *Proceedings of the National Academy of Sciences of the United States of America, 102*(6), 1974–1979. doi:10.1073/pnas.0409522102

Sheridan, R. P., & Miller, M. D. (1998). A method for visualizing recurrent topological substructures in sets of active molecules. *Journal of Chemical Information and Computer Sciences, 38*(5), 915–924. doi:10.1021/ci980044f

Stockwell, B. R. (2000). Chemical genetics: ligand-based discovery of gene function. *Nature Reviews. Genetics, 1*(2), 116–125. doi:10.1038/35038557

Tipton, K. F., & Boyce, S. (2000). History of the enzyme nomenclature system. *Bioinformatics (Oxford, England), 16*(1), 34–40. doi:10.1093/bioinformatics/16.1.34

Trepalin, S., & Osadchiy, N. (2005). The centroidal algorithm in molecular similarity and diversity calculations on confidential datasets. *Journal of Computer-Aided Molecular Design, 19*(9-10), 715–729. doi:10.1007/s10822-005-9023-1

Willett, P. (1995). Searching for pharmacophoric patterns in databases of three-dimensional chemical structures. *Journal of Molecular Recognition, 8*(5), 290–303. doi:10.1002/jmr.300080503

Willett, P., Barnard, J., & Downs, G. M. (1998). Chemical similarity searching. *Journal of Chemical Information and Computer Sciences, 38*(6), 983–996. doi:10.1021/ci9800211

Willett, P., Winterman, V., & Bawden, D. (1986). Implementation of nearest-neighbor searching in an online chemical structure search system. *Journal of Chemical Information and Computer Sciences, 26*(1), 36–41. doi:10.1021/ci00049a008

Xue, L., Godden, J. W., Stahura, F. L., & Bajorath, J. (2003). Design and evaluation of a molecular fingerprint involving the transformation of property descriptor values into a binary classification scheme. *Journal of Chemical Information and Computer Sciences, 43*(4), 1151–1157. doi:10.1021/ci030285+

Yamada, T., Kanehisa, M., & Goto, S. (2006). Extraction of phylogenetic network modules from the metabolic network. *BMC Bioinformatics, 7*, 130. doi:10.1186/1471-2105-7-130

Yamanishi, Y., Vert, J.-P., & Kanehisa, M. (2004). Protein network inference from multiple genomic data: a supervised approach. *Bioinformatics (Oxford, England), 20*(Suppl 1), i363–i370. doi:10.1093/bioinformatics/bth910

Yamanishi, Y., Vert, J.-P., & Kanehisa, M. (2005). Supervised enzyme network inference from the integration of genomic data and chemical information. *Bioinformatics (Oxford, England), 21*(Suppl 1), i468–i477. doi:10.1093/bioinformatics/bti1012

Zhang, Q. Y., & Aires-de-Sousa, J. (2005). Structure-based classification of chemical reactions without assignment of reaction. *Journal of Chemical Information and Modeling, 45*(6), 1775–1783. doi:10.1021/ci0502707

Compilation of References

Adamson, G. W., Lynch, M. F., & Town, W. G. (1971). Analysis of Structural Characteristics of Chemical Compounds in a Large Computer-based File. Part II. Atom-Centred Fragments. Journal of the Chemical Society, (C), 3702–3706.

Aggarwal, C. C., Hinneburg, A., & Keim, D. A. (2001). On the Surprising Behavior of Distance Metrics in High Dimensional Space. International Conference on Database Theory (pp. 420-434). New York: Springer.

Agrafiotis, D., & Xu, H. (2003). A Geodisc Framework for Analyzing Molecular Similarities. Journal of Chemical Information and Computer Sciences, 43, 475–484. doi:10.1021/ci025631m

Aha, D. W. (1997). Lazy learning. Artificial Intelligence Review, 11, 7–10. doi:10.1023/A:1006538427943

Ajay & Murcko, M. A. (1995). Computational methods to predict binding free energy in ligand-receptor complexes. Journal of Medicinal Chemistry, 38, 4953–4967. doi:10.1021/jm00026a001

Akaike, H. (1974). A New Look at the Statistical Model Identification. IEEE Transactions on Automatic Control, AC-19, 716–723. doi:10.1109/TAC.1974.1100705

Akula, N., Lecanu, L., Greeson, J., & Papadopoulos, V. (2006). 3D QSAR studies of AChE inhibitors based on molecular docking scores and CoMFA. Bioorganic & Medicinal Chemistry Letters, 16(24), 6277–6280.

Akutsu, T., & Fukagawa, D. (2005). Inferring a graph from path frequency. In Proceedings of the sixteenth annual symposium on combinatorial pattern matching (p. 371-382). New York: Springer.

Albert, R., & Barabási, A.-L. (2002). Statistical mechanics of complex networks. Reviews of Modern Physics, 74, 47.

Altman, D. G., & Bland, J. M. (1994). Diagnostic tests 1: sensitivity and specificity. British Medical Journal, 308, 1552.

Alvarez, J., & Shoichet, B. (Eds.). (2005). Virtual Screening in Drug Discovery. Boca Raton, FL: CRC Press.

Ames, B. N., Lee, F. D., & Durston, W. E. (1973). An improved bacterial test system for the detection and classification of mutagens and carcinogens. Proceedings of the National Academy of Sciences of the United States of America, 70(3), 782–786. doi:10.1073/pnas.70.3.782

Andersson, C. A., & Bro, R. (1998). Improving the speed of multi-way algorithms: Part I. Tucker3. Chemometrics and Intelligent Laboratory Systems, 42, 93–103.

Angelopoulos, N., & Cussens, J. (2005a, July-August). Exploiting informative priors for Bayesian classification and regression trees. In Artificial Intelligence; Proceedings of the 19th International Joint Conference (p. 641-646). Edinburgh, UK: IJCAI.

Angelopoulos, N., & Cussens, J. (2005b, August). Tempering for Bayesian C&RT. In Machine Learning; Proceeding of the 22nd International Conference (p. 17-24). Bonn, Germany: ACM.

Arakawa, M., Hasegawa, K., & Funatsu, K. (2006). QSAR study of anti-HIV HEPT analogues based on multi-objective genetic programming and counter-propagation neural network. Chemometrics and Intelligent Laboratory Systems, 83(2), 91–98.

Arakawa, M., Hasegawa, K., & Funatsu, K. (2007). The recent trend in QSAR modeling -variable selection and 3D-QSAR methods. Current Computer-aided Drug Design, 3(4), 254–262.

Araneda, R. C., Kini, A. D., & Firestein, S. (2000, Dec). The molecular receptive range of an odorant receptor. Nature Neuroscience, 3(12), 1248–1255. Available from http://dx.doi.org/10.1038/81774. doi:10.1038/81774

Arevian, A. C., Kapoor, V., & Urban, N. N. (2008, Jan). Activity-dependent gating of lateral inhibition in the mouse olfactory bulb. Nature Neuroscience, 11(1), 80–87. Available from http://dx.doi.org/10.1038/nn2030. doi:10.1038/nn2030

Arita, M. (2003). In silico atomic tracing by substrate-product relationships in escherichia coli intermediary metabolism. Genome Research, 13, 2455–2466. doi:10.1101/gr.1212003

Armengol, E., & Plaza, E. (2003). Discovery of toxicological patterns with lazy learning. Knowledge-Based Intellignet Information and Engineering Systems, Pt 2 . Proceedings, 2774, 919–926.

Aronszajn, N. (1950). Theory of reproducing kernels. Transactions of the American Mathematical Society, 68, 337–404.

Atilgan, A. R., Akan, P., & Baysal, C. (2004). Small-world communication of residues and significance for protein dynamics. Biophysical Journal, 86, 85–91.

Atkeson, C. G. (1992). In Casdagli, M., & Eubank, S. G. (Eds.), Memory-based approaches to approximating continous functions (pp. 503–521). Redwood City, CA: Addison-Wesley.

Atkeson, C. G., Moore, A. W., & Schaal, S. (1997a). Locally weighted learning. Artificial Intelligence Review, 11, 11–73. doi:10.1023/A:1006559212014

Atkeson, C. G., Moore, A. W., & Schaal, S. (1997b). Locally weighted learning for control. Artificial Intelligence Review, 11, 75–113. doi:10.1023/A:1006511328852

Baffi, G., Martin, E. B., & Morris, A. J. (1999). Non-linear projection to latent structures revisited: the quadratic PLS algorithm. Computers & Chemical Engineering, 23, 395–411.

Baffi, G., Martin, E., & Morris, A. (2000). Non-linear dynamic projection to latent structures modelling. Chemometrics and Intelligent Laboratory Systems, 52, 5–22.

Bagler, G. & Sinha, S. (2005). Network properties of protein structures. Physica A: Statistical Mechanics and its Applications, 346, 27-33.

Bajorath, J. (2001). Selected concepts and investigations in compound classification, molecular descriptor analysis, and virtual screening. Journal of Chemical Information and Computer Sciences, 41(2), 233–245. doi:10.1021/ci0001482

Bajorath, J. (2002). Integration of virtual and high-throughput screening. Nature Reviews. Drug Discovery, 1, 882–894. doi:10.1038/nrd941

Bakir, G., Alexander, Z., & Tsuda, K. (2004). Learning to find graph pre-images. In Dagm-symposium (pp. 253–261).

Balaban, A. T., Motoc, I., Bonchev, D., & Mekenyan, O. (1983). Topological Indices for Structure-Activity Correlations. In M.Charton & I. Motoc (Eds.), Steric Effects in Drug Design (Topics in Current Chemistry, Vol. 114) (pp. 21-55). Berlin, Germany: Springer-Verlag.

Balon, K., Riebesehl, B., & Müller, B. (1999). Drug liposome partitioning as a tool for the prediction of human passive intestinal absorption. Pharmaceutical Research, 16, 882–888. doi:10.1023/A:1018882221008

Barber, C. B., David, P. D., & Hannu, H. (1996). The quickhull algorithm for convex hulls . ACM Transactions on Mathematical Software, 22, 469–483.

Barker, J. A., & Thornton, J. M. (2003). An algorithm for constraint-based structural template matching: application to 3D templates with statistical analysis. (pp.1644-1649).

Barker, M., & Rayens, W. (2003). Partial least squares for discrimination. Journal of Chemometrics, 17, 166–173.

Basak, S. C., Gute, B. D., & Grunwald, G. D. (1997). Use of Topostructural, Topochemical, and Geometric Parameters in the Prediction of Vapor Pressure: A Hierarchical QSAR Approach. Journal of Chemical Information and Computer Sciences, 37, 651–655. doi:10.1021/ci960176d

Basak, S. C., Magnuson, V. R., & Veith, G. D. (1987). Topological Indices: Their Nature, Mutual Relatedness, and Applications. In X.J.R.Avula, G. Leitmann, C. D. Jr. Mote, & E. Y. Rodin (Eds.), Mathematical Modelling in Science and Technology (pp. 300-305). Oxford, UK: Pergamon Press.

Baskin, I. I., Palyulin, V. A., & Zefirov, N. S. (2008). Neural networks in building QSAR models. Methods in Molecular Biology (Clifton, N.J.), 458, 137–158.

Baumann, K. (2002). An Alignment-Independent Versatile Structure Descriptor for QSAR and QSPR Based on the Distribution of Molecular Features. Journal of Chemical Information and Computer Sciences, 42, 26–35. doi:10.1021/ci990070t

Baumann, K., Albert, H., & von Korff, M. (2002). A systematic evaluation of the benefits and hazards of variable selection in latent variable regression. Part I. Search algorithm, theory and simulations. Journal of Chemometrics, 16, 339–350. doi:10.1002/cem.730

Baumgarte, F. (2002, Oct.). Improved audio coding using a psychoacoustic model based on a cochlear filter bank. IEEE Transactions on Speech and Audio Processing, 10(7), 495–503. doi:10.1109/TSA.2002.804536

Belkin, N. J., Kantor, P., Fox, E. A., & Shaw, J. B. (1995). Combining the evidence of multiple query representations for information retrieval. Information Processing & Management, 31, 431–448. doi:10.1016/0306-4573(94)00057-A

Bemis, G. W., & Murcko, M. A. (1996). The properties of known drugs. 1. Molecular frameworks. Journal of Medicinal Chemistry, 39(15), 2887–2893. doi:10.1021/jm9602928

Bemis, G. W., & Murcko, M. A. (1999). Properties of known drugs. 2. Side chains. Journal of Medicinal Chemistry, 42(25), 5095–5099. doi:10.1021/jm9903996

Bender, A., & Glen, R. C. (2004). Molecular similarity: a key technique in molecular informatics. Organic & Biomolecular Chemistry, 2, 3204–3218. doi:10.1039/b409813g

Bender, A., Mussa, H. Y., Glen, R. C., & Reiling, S. (2004). Molecular similarity searching using atom environments, information-based feature selection, and a naïve Bayesian classifier. Journal of Chemical Information and Computer Sciences, 44, 170–178. doi:10.1021/ci034207y

Ben-Hur, A., Ong, C. S., Sonnenburg, S., Scholkopf, B., & Ratsch, G. (2008). Support vector machines and kernels for computational biology. PLoS Computational Biology, 4(10). doi:10.1371/journal.pcbi.1000173

Bennett, R. N., & Wallsgrove, R. M. (1994). Secondary metabolites in plant defence mechanisms. The New Phytologist, 127(4), 617–633. doi:10.1111/j.1469-8137.1994.tb02968.x

Berglund, A., & Wold, S. (1997). INLR, Implicit Nonlinear Latent Variable Regression. Journal of Chemometrics, 11(2), 141–156.

Berglund, A., & Wold, S. (1999). A Serial Extension of Multiblock PLS. Journal of Chemometrics, 13, 461–471.

Berglund, A., Kettaneh, N., Uppgard, L., Wold, S., Bendwell, N., & Cameron, D. R. (2001). The GIFI approach to non-linear PLS modeling. Journal of Chemometrics, 15(4), 321–336.

Berthold, M., & Borgelt, C. (2002). Mining Molecular Fragments: Finding Relevant Substructures of Molecules. International Conference on Data Mining (pp. 51-58). Washington, DC: IEEE Press.

Beyer, K. S., Goldstein, J., Ramakrishnan, R., & Shaft, U. (1999). When is nearest neighbor meaningful. International Conference on Database Theory, (pp. 217-235). Washington, DC: IEEE Press.

Bhonsle, J. B., & Huddler, D. (2008). Novel method for mining QSPR-relevant conformations. Chemical Engineering Communications, 195(11), 1396–1423.

Bhonsle, J. B., Wang, Z., Tamamura, H., Fujii, N., Peiper, S. C., & Trent, J. O. (2005). A simple, automated quasi-4D-QSAR, quasi-multi way PLS approach to develop highly predictive QSAR models for highly flexible CXCR4 inhibitor cyclic pentapeptide ligands using scripted common molecular modeling tools. QSAR & Combinatorial Science, 24(5), 620–630.

Binkowski, A. T., Adamian, L., & Liang, J. (2003). Inferring functional relationships of proteins from local sequence and spatial surface patterns. Journal of Molecular Biology, 332, 505–526. doi:10.1016/S0022-2836(03)00882-9

BioCarta. Retrieved from http://www.biocarta.com/. BioCarta LLC, San Diego, CA

Bishop, C. (1995). Neural networks for pattern recognition. Oxford: Clarendon Press.

Bishop, C. (2006). Pattern recognition and machine learning. New York: Springer.

Bleakley, K., Biau, G., & Vert, J.-P. (2007). Supervised reconstruction of biological networks with local models. Bioinformatics (Oxford, England), 23, i57–i65. doi:10.1093/bioinformatics/btm204

Blower, P. E., & Dana, R. C. (1986). Creation of a chemical reaction database from the primary literature . In Willett, P. (Ed.), Modern Approaches to Chemical Reaction Searching (pp. 146–164). Aldershot, England: Gower Publishing Company.

Blundell, T. L., Jhoti, H., & Abell, C. (2002). High-throughput crystallography for lead discovery in drug design. Nature Reviews. Drug Discovery, 1, 45–54. doi:10.1038/nrd706

Bock, J. R., & Gough, D. A. (2005). Virtual screen for ligands of orphan g protein-coupled receptors. Journal of Chemical Information and Modeling, 45(5), 1402–1414. doi:10.1021/ci050006d

Bock, M. E., Garutti, C., & Guerra, C. (2007). Discovery of Similar Regions on Protein Surfaces. Journal of Computational Biology, 14(3), 285–299. doi:10.1089/cmb.2006.0145

Bohac, M., Loeprecht, B., Damborsky, J., & Schuurmann, G. (2002). Impact of orthogonal signal correction (OSC) on the predictive ability of CoMFA models for the ciliate toxicity of nitrobenzenes. Quantitative Structure-Activity Relationships, 21(1), 3–11.

Böhm, H. J., & Schneider, G. (2000). Virtual screening for bioactive molecules. Weinheim: Wiley-VCH.

Böhm, M., & Klebe, G. (2002). Development of a new hydrogen-bond descriptor and their application to comparative mean field analysis. Journal of Medicinal Chemistry, 45, 1585–1597. doi:10.1021/jm011039x

Bolis, G., Dipace, L., & Fabrocini, F. (1991). A Machine Learning Approach to Computer-Aided Molecular Design. Journal of Computer-Aided Molecular Design, 5, 617–628. doi:10.1007/BF00135318

Bonchev, D., & Rouvray, D. H. (Eds.). (1990). Chemical Graph Theory: Introduction and Fundamentals (Vol. 1). London, UK: Gordon and Breach Science Publishers.

Bonchev, D., & Trinajstic, N. (1977). Information Theory, Distance Matrix, and Molecular Branching. The Journal of Chemical Physics, 67, 4517–4533. doi:10.1063/1.434593

Borgwardt, K. M., & Kriegel, H.-P. (2005). Shortest-path kernels on graphs. In Proceedings of the international conference on data mining (Vol. 0, pp. 74-81).

Borgwardt, K. M., Ong, C. S., Schönauer, S., Vishwanathan, S. V. N., Smola, A. J., & Kriegel, H. P. (2005). Protein function prediction via graph kernels. Bioinformatics (Oxford, England), 21(suppl 1), i47–i56. doi:10.1093/bioinformatics/bti1007

Boser, B., Guyon, M., & Vapnik, V. (1992). A training algorithm for optimal margin classifiers. In D. Haussler (Ed.), Proc. 5th ann. acm workshop on comp. learning theory. Pittsburgh, PA: ACM Press.

Boulesteix, A. L., & Strimmer, K. (2007). Partial Least Squares: A Versatile Tool for the Analysis of High-Dimensional Genomic Data. Briefings in Bioinformatics, 8(1), 32–44.

Brüderle, D., Meier, K., Mueller, E., & Schemmel, J. (2007). Verifying the biological relevance of a neuromorphic hardware device. BMC Neuroscience, 8(Suppl 2), P10.

Brandes, U. (2001). A faster algorithm for betweenness centrality. The Journal of Mathematical Sociology, 25, 163–177.

Breiman, L. (1995). Better subset regression using the nonnegative garotte. Technometrics, 37, 373–384. doi:10.2307/1269730

Breiman, L. (1998). Arcing classifiers. Annals of Statistics, 3(26), 801–849.

Breiman, L. (2001). Random forests. Machine Learning, 45(1), 5–32. doi:10.1023/A:1010933404324

Breiman, L., Friedman, J. H., Olshen, R. A., & Stone, C. J. (1984). Classification and regression trees. Belmond, CA, USA: Wadsworth International. 22

Breneman, C. M., Thompson, T. R., Rhem, M., & Dung, M. (1995). Electron Density Modeling of Large Systems using the Transferable Atom Equivalent Method. Computers & Chemistry, 19, 161–169. doi:10.1016/0097-8485(94)00052-G

Bringmann, B., Zimmermann, A., Raedt, L. D., & Nijssen, S. (2006). Don't be afraid of simpler patterns. In 10th european conference on principles and practice of knowledge discovery in databases (pkdd) (pp. 55–66).

Bro, R. (1996). Multiway calibration. Multilinear PLS. Journal of Chemometrics, 10(1), 47–61.

Bro, R. (1997). Tutorial PARAFAC. Tutorial and applications. Chemometrics and Intelligent Laboratory Systems, 38, 149–171.

Bro, R. (2006). Review on multiway analysis in chemistry-2000-2005. Critical Reviews in Analytical Chemistry, 36(3-4), 279–293.

Brooijmans, N., & Kuntz, I. D. (2003). Molecular recognition and docking algorithms. Annual Review of Biophysics and Biomolecular Structure, 32, 335–373. doi:10.1146/annurev.biophys.32.110601.142532

Broto, P., & Devillers, J. (1990). Autocorrelation of Properties Distributed on Molecular Graphs . In Karcher, W., & Devillers, J. (Eds.), Practical Applications of Quantitative Structure-Activity Relationships (QSAR) in Environmental Chemistry and Toxicology (pp. 105–127). Dordrecht, The Netherlands: Kluwer.

Broto, P., Moreau, G., & Vandycke, C. (1984a). Molecular Structures: Perception, Autocorrelation Descriptor and SAR Studies. Autocorrelation Descriptor. European Journal of Medicinal Chemistry, 19, 66–70.

Broto, P., Moreau, G., & Vandycke, C. (1984b). Molecular Structures: Perception, Autocorrelation Descriptor and SAR Studies. Use of the Autocorrelation Descriptors in the QSAR Study of Two Non-Narcotic Analgesic Series. European Journal of Medicinal Chemistry, 19, 79–84.

Brown, F. K. (1998). Chemoinformatics: what is it and how does it impact drug discovery. Annual Reports in Medicinal Chemistry, 33, 375–384. doi:10.1016/S0065-7743(08)61100-8

Brown, R. D., & Martin, Y. C. (1996). Use of structure-activity data to compare structure-based clustering methods and descriptors for use in compound selection. Journal of Chemical Information and Computer Sciences, 36, 572–584. doi:10.1021/ci9501047

Brown, R. D., & Martin, Y. C. (1997). The information content of 2D and 3D structural descriptors relevant to ligand-receptor binding. Journal of Chemical Information and Computer Sciences, 37, 1–9. doi:10.1021/ci960373c

Buntine, W. L. (1992, June). Learning classification trees. Statistics and Computing, 2(2), 63–73. doi:10.1007/BF01889584

Buntine, W., & Caruana, R. (1992). Introduction to IND version 2.1 and recursive partitioning. CA: Moffet Field.

Burges, C. J. C. (1998). A tutorial on support vector machines for pattern recognition. Data Mining and Knowledge Discovery, 2, 121–167. doi:10.1023/A:1009715923555

Butler, N., & Denham, M. (2000). The peculiar shrinkage properties of partial least squares regression. Journal of the Royal Statistical Society. Series B. Methodological, 62, 585–593.

Bylesjoe, M., Eriksson, D., Kusano, M., Moritz, T., & Trygg, J. (2007). Data integration in plant biology: the O2PLS method for combined modeling of transcript and metabolite data. The Plant Journal, 52(6), 1181–1191.

Bylesjoe, M., Rantalainen, M., Cloarec, O., Nicholson, J. K., Holmes, E., & Trygg, J. (2007). OPLS discriminant analysis: combining the strengths of PLS-DA and SIMCA classification. Journal of Chemometrics, 20(8-10), 341–351.

Byvatov, E., & Schneider, G. (2004). Svm-based feature selection for characterization of focused compound collections. Journal of Chemical Information and Computer Sciences, 44(3), 993–999. doi:10.1021/ci0342876

Cai, L., & Hofmann, T. (2004). Hierarchical document categorization with support vector machines. In Acm 13th conference on information and knowledge management (p. 78-87). New York: ACM Press.

Carhart, R. E., Smith, D. H., & Venkataraghavan, R. (1985). Atom Pairs as Molecular Features in Structure-Activity Studies: Definition and Applications. Journal of Chemical Information and Computer Sciences, 25, 64–73. doi:10.1021/ci00046a002

Carter, C. W. Jr, LeFebvre, B. C., Cammer, S. A., Tropsha, A., & Edgell, M. H. (2001). Four-body potentials reveal protein-specific correlations to stability changes caused by hydrophobic core mutations. Journal of Molecular Biology, 311, 625–638. doi:10.1006/jmbi.2001.4906

Caspi, R., Foerster, H., Fulcher, C. A., Kaipa, P., Krummenacker, M., & Latendresse, M. (2008). The MetaCyc database of metabolic pathways and enzymes and the BioCyc collection of pathway/genome databases. Nucleic Acids Research, 36(Database issue), D623–D631. doi:10.1093/nar/gkm900

Catana, C., & Stouten, P. F. W. (2007). Novel, Customizable Scoring Functions, Parameterized Using N-PLS, for Structure-Based Drug Discovery. Journal of Chemical Information and Modeling, 47(1), 85–91.

Cavalli, A., Poluzzi, E., De, P. F., & Recanatini, M. (2002). Toward a pharmacophore for drugs inducing the long QT syndrome: insights from a CoMFA study of HERG K(+) channel blockers. Journal of Medicinal Chemistry, 45, 3844–3853. doi:10.1021/jm0208875

Centner, V., Massart, D. L., de Noord, O. E., De Jong, S., Vandeginste, B. G. M., & Sterna, C. (1996). Elimination of Uniformative Variables for Multivariate Calibration. Analytical Chemistry, 68, 3851–3858. doi:10.1021/ac960321m

Chan, M., Tan, D. S., & Sim, T. S. (2007). Plasmodium falciparum pyruvate kinase as a novel target for antimalarial drug-screening. Travel Medicine and Infectious Disease, 5(2), 125–131. doi:10.1016/j.tmaid.2006.01.015

Charton, M., & Charton, B. I. (2002). Advances in Quantitative Structure-Property Relationships. Amsterdam, The Netherlands: JAI Press.

ChemDiv. (2005). Retrieved from http://www.chemdiv.com

Chen, B., Harrison, R. F., Papadatos, G., Willett, P., Wood, D. J., & Lewell, X. Q. (2007). Evaluation of machine-learning methods for ligand-based virtual screening. Journal of Computer-Aided Molecular Design, 21, 53–62. doi:10.1007/s10822-006-9096-5

Chen, B., Harrison, R. F., Pasupa, K., Willett, P., Wilton, D. J., & Wood, D. J. (2006). Virtual screening using binary kernel discrimination: Effect of noisy training data and the optimization of performance. *Journal of Chemical Information and Modeling, 46*, 478–486. doi:10.1021/ci0505426

Chen, L., Nourse, J. G., Christie, B. D., Leland, B. A., & Grier, D. L. (2002). Over 20 years of reaction access systems from MDL: a novel reaction substructure search algorithm. *Journal of Chemical Information and Computer Sciences, 42*(6), 1296–1310. doi:10.1021/ci020023s

Chen, X., Rusinko, A., Tropsha, A., & Young, S. S. (1999). Automated Pharmacophore Identification for Large Chemical Data Sets. *Journal of Chemical Information and Computer Sciences, 39*, 887–896. doi:10.1021/ci990327n

Cheng, A., Coleman, R., Smyth, K., Cao, Q., Soulard, P., & Caffrey, D. (2007). Structure-based maximal affinity model predicts small-molecule druggability. *Nature Biotechnology, 25*, 71–75. doi:10.1038/nbt1273

Cherkassky, V., & Mulier, F. (2007). *Learning From Data*. New York, NY: Wiley Inter-Science. doi:10.1002/9780470140529

Chiang, L. H., Leardi, R., Pell, R. J., & Seasholtz, M. B. (2006). Industrial experiences with multivariate statistical analysis of batch process data. *Chemometrics and Intelligent Laboratory Systems, 81*(2), 109–119.

Chipman, H., George, E., & McCulloch, R. (1998). Bayesian CART model search (with discussion). *Journal of the American Statistical Association, 93*, 935–960. doi:10.2307/2669832

Christofk, H. R., Vander Heiden, M. G., Harris, M. H., Ramanathan, A., Gerszten, R. E., & Wei, R. (2008). The M2 splice isoform of pyruvate kinase is important for cancer metabolism and tumour growth. *Nature, 452*, 230–233. doi:10.1038/nature06734

Clark, D. E., & Grootenhuis, P. D. (2002). Progress in computational methods for the prediction of ADMET properties. *Current Opinion in Drug Discovery & Development, 5*, 382–390.

Cleland, T. A., & Linster, C. (2005, Nov). Computation in the olfactory system. *Chemical Senses, 30*(9), 801–813. Available from http://dx.doi.org/10.1093/chemse/bji072. doi:10.1093/chemse/bji072

Clementi, S., Cruciani, G., Riganelli, D., Valigi, R., Costantino, G., & Baroni, M. (1993). Autocorrelation as a Tool for a Congruent Description of Molecules in 3D QSAR Studies. *Pharmaceutical and Pharmacological Letters, 3*, 5–8.

Clerc, J. T., & Terkovics, A. L. (1990). Versatile Topological Structure Descriptor for Quantitative Structure/Property Studies. *Analytica Chimica Acta, 235*, 93–102. doi:10.1016/S0003-2670(00)82065-6

Cleveland, W. (1979). Robust Locally Weighted Regression and Smoothing Scatterplots. *Journal of the American Statistical Association, 74*(368), 829–836.

Cleveland, W. S. (1981). Lowess - A Program for Smoothing Scatterplots by Robust Locally Weighted Regression. *The American Statistician, 35*, 54. doi:10.2307/2683591

Cohen, C., Samuel, G., & Fischel, O. (2004). Computer-assisted drug design, where do we stand? *Chimica Oggi-Chemistry Today, Mar.*, 22-26.

Cohen, W., & Singer, Y. (1999). A simple, fast and effective rule learner. In *Proceedings of the 16th national conference on artificial intelligence*.

Collins, M., & Duffy, N. (2002). Convolution kernels for natural language. [Cambridge, MA: MIT Press.]. *Advances in Neural Information Processing Systems, 14*, 625–632.

Conesa, A., Gotz, S., Garcia-Gomez, J. M., Terol, J., Talon, M., & Robles, M. (2005). Blast2GO: a universal tool for annotation, visualization and analysis in functional genomics research. *Bioinformatics (Oxford, England), 21*, 3674–3676.

Consonni, V., Todeschini, R., & Pavan, M. (2002). Structure/Response Correlations and Similarity/Diversity Analysis by GETAWAY Descriptors. Part 1. Theory of the Novel 3D Molecular Descriptors. Journal of Chemical Information and Computer Sciences, 42, 682–692. doi:10.1021/ci015504a

Cordella, L. P., Foggia, P., Sansone, C., & Tortorella, F. (1998). Graph Matching: a Fast Algorithm and its Evaluation. In Proc. 14th Int. Conf. On Pattern Recognition.

Cordella, L. P., Foggia, P., Sansone, C., & Vento, M. (1999). Performance evaluation of the VF graph matching algorithm. In Proceedings of the 10th International Conference on Image Analysis and Processing. IEEE Computer Society.

Cordella, L. P., Foggia, P., Sansone, C., & Vento, M. (2001). An improved algorithm for matching large graphs. In Proc. of the 3rd IAPR-TC-15 International Workshop on Graph-based Representation. Italy.

Cortes, C., & Vapnik, V. (1995, September). Support-vector networks. Machine Learning, 20(3), 273–297. doi:10.1007/BF00994018

Costa, V. S., Srinivasan, A., Camacho, R., Blockeel, H., Demoae, B., & Janssens, G. (2003). Query transformations for improving the efficiency of ILP systems. Journal of Machine Learning Research, 4, 465–491. doi:10.1162/153244304773936027

Couto, A., Alenius, M., & Dickson, B. J. (2005, Sep). Molecular, anatomical, and functional organization of the Drosophila olfactory system. Current Biology, 15(17), 1535–1547. Available from http://dx.doi.org/10.1016/j.cub.2005.07.034. doi:10.1016/j.cub.2005.07.034

Cramer, R. D. III, Bunce, J. D., Patterson, D. E., & Frank, I. E. (1988). Crossvalidation, Bootstrapping and Partial Least Squares Compared with Multiple Regression in Conventional QSAR Studies. Quantitative Structure-Activity Relationships, 7, 18–25. doi:10.1002/qsar.19880070105

Cramer, R. D., Patterson, D. E., & Bunce, J. D. (1988). Comparative molecular field analysis (CoMFA). 1. Effect of shape on binding of steroids to carrier proteins. Journal of the American Chemical Society, 110(18), 5959–5967.

Cramer, R., Redl, G., & Berkoff, C. (1974). Substructural analysis. A novel approach to the problem of drug design. Journal of Medicinal Chemistry, 17, 533–535. doi:10.1021/jm00251a014

Cristianini, N., & Shawe-Taylor, J. (2000). An Introduction to Support Vector Machines and Other Kernel-based Learning Methodsq. New York: Cambridge University Press.

Crivori, P., Zamora, I., Speed, B., Orrenius, C., & Poggesi, I. (2004). Model based on GRID-derived descriptors for estimating CYP3A4 enzyme stability of potential drug candidates. Journal of Computer-Aided Molecular Design, 18(3), 155–166.

Cronin, M. T. D., Aptula, A. O., Duffy, J. C., Netzeva, T. I., Rowe, P. H., & Valkova, I. V. (2002). Comparative assessment of methods to develop QSARs for the prediction of the toxicity of phenols to Tetrahymena pyriformis. Chemosphere, 49, 1201–1221. doi:10.1016/S0045-6535(02)00508-8

Crowe, J. E., Lynch, M. F., & Town, W. G. (1970). Analysis of Structural Characteristics of Chemical Compounds in a Large Computer-based File. Part 1. Non-cyclic Fragments. Journal of the Chemical Society, (C), 990–997.

Crum-Brown, A. (1867). On an application of mathematics to chemistry. Proceedings of the Royal Society of Edinburgh, VI(73), 89–90.

Crum-Brown, A., & Fraser, T. R. (1868). On the connection between chemical constitution and physiological action. Part 1. On the physiological action of salts of the ammonium bases, derived from strychnia, brucia, thebia, codeia, morphia and nicotia. Transactions of the Royal Society of Edinburgh, 25, 151–203.

Cybenko, G. (1989). Approximation by superpositions of a sigmoidal function. Mathematics of Control, Signals, and Systems, 2(4), 303–314. doi:10.1007/BF02551274

Czarnik, A. W. (1997). Encoding methods for combinatorial chemistry. Current Opinion in Chemical Biology, 1, 60–66. doi:10.1016/S1367-5931(97)80109-3

Daren, Z. (2001). QSPR studies of PCBs by the combination of genetic algorithms and PLS analysis. Computers & Chemistry, 25(2), 197–204.

Darvas, F. (1988). Predicting metabolic pathways by logic programming. Journal of Molecular Graphics & Modelling, 6(2), 80–86.

Davis, A. M., & Riley, R. J. (2004). Predictive ADMET studies, the challenges and the opportunities. Current Opinion in Chemical Biology, 8, 378–386. doi:10.1016/j.cbpa.2004.06.005

De Bie, T., Cristianini, N., & Rosipal, R. (2005). Eigenproblems in Pattern Recognition. In Bayro-Corrochano, E. (Ed.), Handbook of Geometric Computing: Applications in Pattern Recognition, Computer Vision, Neuralcomputing, and Robotics (pp. 129–170). New York: Springer.

de Bruyne, M., Foster, K., & Carlson, J. R. (2001, May). Odor coding in the Drosophila antenna. Neuron, 30(2), 537–552. doi:10.1016/S0896-6273(01)00289-6

de Cerqueira, L. P., Golbraikh, A., Oloff, S., Xiao, Y., & Tropsha, A. (2006). Combinatorial QSAR modeling of P-glycoprotein substrates. Journal of Chemical Information and Modeling, 46, 1245–1254. doi:10.1021/ci0504317

De Jong, S. (1993). SIMPLS: an alternative approach to partial least squares regression. Chemometrics and Intelligent Laboratory Systems, 18, 251–263.

Dearden, J. C. (2007). In silico prediction of ADMET properties: how far have we come? Expert Opinion on Drug Metabolism & Toxicology, 3, 635–639. doi:10.1517/17425255.3.5.635

Debnath, A. K., de Compadre, R. L. L., Debnath, G., Schusterman, A. J., & Hansch, C. (1991). Structure-activity relationship of mutagenic aromatic and heteroaromatics nitro compounds. correlation with molecular orbital energies and hydrophobicity. Journal of Medicinal Chemistry, 34(2), 786–797. doi:10.1021/jm00106a046

Deeb, O., Hemmateenejad, B., Jaber, A., Garduno-Juarez, R., & Miri, R. (2007). Effect of the electronic and physicochemical parameters on the carcinogenesis activity of some sulfa drugs using QSAR analysis based on genetic-MLR and genetic-PLS. Chemosphere, 67(11), 2122–2130.

Demiriz, A., Bennet, K., & Shawe-Taylor, J. (2002). Linear programming boosting via column generation. Machine Learning, 46(1-3), 225–254. doi:10.1023/A:1012470815092

Deng, W., Breneman, C., & Embrechts, M. J. (2004). Predicting protein-ligand binding affinities using novel geometrical descriptors and machine-learning methods. Journal of Chemical Information and Computer Sciences, 44(2), 699–703.

Denison, D. G. T., Holmes, C. C., Mallick, B. K., & Smith, A. F. M. (2002). Bayesian methods for nonlinear classification and regression. Chichester, UK: Wiley.

Deshpande, M. Kuramochi, M. & Karypis, G. (2003) Frequent Sub-Structure-Based Approaches for Classifying Chemical Compounds. International Conference on Data Mining (pp. 35-42). Washington, DC: IEEE Press.

Deshpande, M., Kuramochi, M., Wale, N., & Karypis, G. (2005). Frequent sub-structure-based approaches for classifying chemical compounds. IEEE Transactions on Knowledge and Data Engineering, 17(8), 1036–1050. doi:10.1109/TKDE.2005.127

Devillers, J. (Ed.). (1996). Genetic Algorithms in Molecular Modeling. New York: Academic Press.

Devillers, J. (Ed.). (1998). Comparative QSAR. Washington, DC: Taylor & Francis.

Devos, D., & Valencia, A. (2000). Practical limits of function prediction. Proteins: Structure, Function, and Bioinformatics, 41, 98–107.

Dimmock, J. R., Puthucode, R. N., Smith, J. M., Hetherington, M., Quail, J. W., & Pugazhenthi, U. (1996). (Aryloxy)aryl semicarbazones and related compounds: A novel class of anticonvulsant agents possessing high activity in the maximal electroshock screen. Journal of Medicinal Chemistry, 39, 3984–3997. doi:10.1021/jm9603025

Dimmock, J. R., Vashishtha, S. C., & Stables, J. P. (2000). Anticonvulsant properties of various acetylhydrazones, oxamoylhydrazones and semicarbazones derived from aromatic and unsaturated carbonyl compounds. European Journal of Medicinal Chemistry, 35, 241–248. doi:10.1016/S0223-5234(00)00123-9

Diudea, M. V., & Gutman, I. (1998). Wiener-Type Topological Indices. Croatica Chemica Acta, 71, 21–51.

Dobson, C. (2004). Chemical space and biology. Nature, 432, 824–828. doi:10.1038/nature03192

Domingos, P., & Pazzani, M. (1996). Beyond independence: Conditions for the optimality of the simple Bayesian classifier . In Machine learning (pp. 105–112). San Francisco: Morgan Kaufmann.

Donoho, D., Johnsotne, I., Kerkyacharian, G., & Picard, D. (1995). Wavelet shrinkage: Asymptopia? (with discussion). Journal of the Royal Statistical Society. Series B. Methodological, 57, 301–369.

Doucet, J., Barbault, F., Xia, H., Panaye, A., & Fan, B. (2007). Nonlinear SVM approaches to QSPR/QSAR studies and drug design. Current Computer-aided Drug Design, 3(4), 263–289.

du Merle, O., Villeneuve, D., Desrosiers, J., & Hansen, P. (1999). Stabilized column generation. Discrete Mathematics, 194, 229–237. doi:10.1016/S0012-365X(98)00213-1

Duch, W., Swaminathan, K., & Meller, J. (2007). Artificial intelligence approaches for rational drug design and discovery. Current Pharmaceutical Design, 13, 1497–1508. doi:10.2174/138161207780765954

Ducrot, P., Andrianjara, C. R., & Wrigglesworth, R. (2001). CoMFA and CoMSIA 3D-quantitative structure-activity relationship model on benzodiazepine derivatives, inhibitors of phosphodiesterase IV. Journal of Computer-Aided Molecular Design, 15, 767–785. doi:10.1023/A:1013104713913

Duda, R. O., Hart, P. E., & Stork, D. G. (2000). Pattern classification (2nd ed.). New York: Wiley-Interscience.

Dutra, I. C., Page, D., & Shavilk, J. (2002). An emperical evaluation of bagging in inductive logic programming. In Proceedings of the international conference on inductive logic programming.

Dutta, D., Guha, R., Jurs, P. C., & Chen, T. (2006). Scalable Partitioning and Exploration of Chemical Spaces Using Geometric Hashing. Journal of Chemical Information and Modeling, 46, 321–333. doi:10.1021/ci050403o

Eckert, H., & Bajorath, J. (2007). Molecular similarity analysis in virtual screening: foundations, limitation and novel approaches. Drug Discovery Today, 12, 225–233. doi:10.1016/j.drudis.2007.01.011

Eckert, H., Vogt, I., & Bajorath, J. (2006). Mapping Algorithms for Molecular Similarity Analysis and Ligand-Based Virtual Screening: Design of DynaMAD and Comparison with MAD and DMC. Journal of Chemical Information and Modeling, 46, 1623–1634. doi:10.1021/ci060083o

Edgar, S. J., Holliday, J. D., & Willett, P. (2000). Effectiveness of retrieval in similarity searches of chemical databases: A review of performance measures. Journal of Molecular Graphics & Modelling, 18(4-5), 343–357. doi:10.1016/S1093-3263(00)00061-9

Edraki, N., Hemmateenejad, B., Miri, R., & Khoshneviszade, M. (2007). QSAR study of phenoxypyrimidine Derivatives as Potent Inhibitors of p38 Kinase using different chemometric tools. Chemical Biology & Drug Design, 70(6), 530–539.

Efron, B. (1982). The Jackknife, the Bootstrap and Other Resampling Planes. Philadelphia, PA: Society for Industrial and Applied Mathematics.

Efron, B. (1987). Better Bootstrap Confidence Intervals. Journal of the American Statistical Association, 82, 171–200. doi:10.2307/2289144

Efron, B., & Tibshirani, R. (1993). An introduction to bootstrap. New York: Chapman and Hall.

Efron, B., Hastie, T., Johnstone, I., & Tibshirani, R. (2004). Least angle regression. Annals of Statistics, 32(2), 407–499. doi:10.1214/009053604000000067

Efroymson, M. A. (1960). Multiple Regression Analysis. In Ralston, A., & Wilf, H. S. (Eds.), Mathematical Methods for Digital Computers. New York: Wiley.

Eidhammer, I., Jonasses, I., & Taylor, W. R. (2000). Structure Comparison and Structure Patterns. Journal of Computational Biology, 7(5), 685–716. doi:10.1089/106652701446152

Eisenberg, D., Marcotte, E. M., Xenarios, I., & Yeates, T. O. (2000). Protein function in the post-genomic era. Nature, 405(6788), 823–826. doi:10.1038/35015694

Eldén, L. (2004). Partial least squares vs. lanczos bidiagonalization i: Analysis of a projection method for multiple regression. Computational Statistics & Data Analysis, 46(1), 11–31. doi:10.1016/S0167-9473(03)00138-5

Ellis, L. B. M., Roe, D., & Wackett, L. P. (2006). The university of Minnesota biocatalysis/biodegradation database: the first decade. Nucleic Acids Research, 34(Database issue), D517–D521. doi:10.1093/nar/gkj076

Engelhardt, B. E., Jordan, M. I., Muratore, K. E., & Brenner, S. E. (2005). Protein molecular function prediction by Bayesian phylogenomics. PLoS Computational Biology, 1, e45.

Enslein, K. (1988). An overview of structure-activity relationships as an alternative to testing in animals for carcinogenicity, mutagenicity, dermal and eye irritation, and acute oral toxicity. Toxicology and Industrial Health, 4(4), 479–497.

Equbal, T., Silakari, O., & Ravikumar, M. (2008). Exploring three-dimensional quantitative structural activity relationship (3D-QSAR) analysis of SCH 66336 (Sarasar) analogues of farnesyltransferase inhibitors. European Journal of Medicinal Chemistry, 43(1), 204–209.

Erhan, D., LÕheureux, P.-J., Yue, S. Y., & Bengio, Y. (2006). Collaborative Þltering on a family of biological targets. Journal of Chemical Information and Modeling, 46(2), 626–635. doi:10.1021/ci050367t

Eriksson, L., Andersson, P. L., Johansson, E., & Tysklind, M. (2006). Megavariate analysis of environmental QSAR data. Part II - Investigating very complex problem formulations using hierarchical, non-linear and batch-wise extensions of PCA and PLS. Molecular Diversity, 10(2), 187–205.

Eriksson, L., Gottfries, J., Johansson, E., & Wold, S. (2004). Time-resolved QSAR: an approach to PLS modeling of three-way biological data. Chemometrics and Intelligent Laboratory Systems, 73(1), 73–84.

Eriksson, L., Johansson, E., Lindgren, F., & Wold, S. (2000). GIFI-PLS: modeling of non-linearities and discontinuities in QSAR. Quantitative Structure-Activity Relationships, 19(4), 345–355.

Eroes, D., Keri, G., Koevesdi, I., Szantai-Kis, C., Meszaros, G., & Oerfi, L. (2004). Comparison of predictive ability of water solubility QSPR models generated by MLR, PLS and ANN methods. Mini Reviews in Medicinal Chemistry, 4(2), 167–177.

Estrada, E. (2001). Generalization of topological indices. Chemical Physics Letters, 336, 248–252. doi:10.1016/S0009-2614(01)00127-0

Estrada, E. (2008). How the parts organize in the whole? A top-down view of molecular descriptors and properties for QSAR and drug design. Mini Reviews in Medicinal Chemistry, 8(3), 213–221.

Estrada, E., & Matamala, A. R. (2007). Generalized Topological Indices. Modeling Gas-Phase Rate Coefficients of Atmospheric Relevance. Journal of Chemical Information and Modeling, 47, 794–804. doi:10.1021/ci600448b

Fariselli, P., & Casadio, R. (1999). A neural network based predictor of residue contacts in proteins. Protein Engineering, 12, 15–21.

Fassihi, A., & Sabet, R. (2008). QSAR study of p56lck protein tyrosine kinase inhibitory activity of flavonoid derivatives using MLR and GA-PLS. International Journal of Molecular Sciences, 9(9), 1876–1892.

Fatemi, M. H., & Baher, E. (2009). A novel quantitative structure-activity relationship model for prediction of biomagnification factor of some organochlorine pollutants. Mol.Divers.

Fatemi, M. H., & Gharaghani, S. (2007). A novel QSAR model for prediction of apoptosis-inducing activity of 4-aryl-4-H-chromenes based on support vector machine. Bioorganic & Medicinal Chemistry, 15, 7746–7754. doi:10.1016/j.bmc.2007.08.057

Faulon, J., Misra, M., Martin, S., Sale, K., & Sapra, R. (2008). Genome scale enzymeÐmetabolite and drugÐtarget interaction predictions using the signature molecular descriptor. Bioinformatics (Oxford, England), 24, 225–233. doi:10.1093/bioinformatics/btm580

Fechner, N., Jahn, A., Hinselmann, G., & Zell, A. (2009, Feb). Atomic local neighborhood flexibility incorporation into a structured similarity measure for qsar. J Chem Inf Model. Retrieved from http://dx.doi.org/10.1021/ci800329r

Fechner, U., Franke, L., Renner, S., Schneider, P., & Schneider, G. (2003, Oct). Comparison of correlation vector methods for ligand-based similarity searching. Journal of Computer-Aided Molecular Design, 17(10), 687–698. doi:10.1023/B:JCAM.0000017375.61558.ad

Feher, M. (2006). Consensus scoring for protein-ligand interactions. Drug Discovery Today, 11, 421–428. doi:10.1016/j.drudis.2006.03.009

Feng, B. Y., & Shoichet, B. K. (2006). A detergent-based assay for the detection of promiscuous inhibitors. Nature Protocols, 1, 550–553. doi:10.1038/nprot.2006.77

Ferreira da Cunha, E. F., Martins, R. C. A., Albuquerque, M. G., & Bicca de Alencastro, R. (2004). LIV-3D-QSAR model for estrogen receptor ligands. Journal of Molecular Modeling, 10(4), 297–304.

Figueras, J. (1996). Ring Perception Using Breadth–First Search. Journal of Chemical Information and Computer Sciences, 36, 986–991. doi:10.1021/ci960013p

Finn, P., Muggleton, S., Page, D., & Srinivasan, A. (1998). Pharmacophore discovery using the inductive logic programming language Progol. Machine Learning, 30, 241–270. doi:10.1023/A:1007460424845

Firestein, S. (2001, Sep). How the olfactory system makes sense of scents. Nature, 413(6852), 211–218. Available from http://dx.doi.org/10.1038/35093026. doi:10.1038/35093026

Flach, P., & Lachiche, N. (2004). Naive bayesian classifcation of structured data. Machine Learning, 57(3), 233–269. doi:10.1023/B:MACH.0000039778.69032.ab

Flickinger, B. (2001) Using metabolism data in early development. Drug Discovery and Development, September 2001.

Fort, G., & Lambert-Lacroix, S. (2005). Classification using partial least squares with penalized logistic regression. Bioinformatics (Oxford, England), 21(7), 1104–1111.

Fox, S., Farr-Jones, S., & Yund, M. A. (1999). High-throughput screening for drug discovery: Continually transitioning into new technologies. Journal of Biomolecular Screening, 4, 183–186. doi:10.1177/108705719900400405

Fradkin, D., & Muchnik, I. (2006). Support vector machines for classification. Discrete Methods in Epidemiology, 70, 13–20.

Frank, E., & Friedman, J. H. (1993). A statistical view of some chemometrics regression tools. Technometrics, 35(2), 109–135. doi:10.2307/1269656

Frank, E., & Witten, I. (1998). Generating accurate rule sets without global optimization. In Proceedings of the fifteenth international conference on machine learning (icml-98) (pp. 144–151).

Frank, I. (1995). Modern nonlinear regression methods. Chemometrics and Intelligent Laboratory Systems, 27, 1–9.

Frank, I., & Friedman, J. (1993). A Statistical View of Some Chemometrics Regression Tools. Technometrics, 35, 109–147.

Free, S. M., & Wilson, J. W. (1964). A Mathematical Contribution to Structure-Activity Studies. Journal of Medicinal Chemistry, 7, 395–399. doi:10.1021/jm00334a001

Freeman, L. (1977). A set of masures of centrality based on betweenness. Sociometry, 40, 35–41.

Freund, Y., & Schapire, R. E. (1995). A decision-theoretic generalization of on-line learning and an application to boosting. In Computational learning theory: Eurocolt '95 (pp. 23–37). New York: Springer-Verlag.

Freund, Y., & Schapire, R. E. (1999b). A short introduction to boosting. Journal of Japanese Society for Artificial Intelligence, 5(14), 771–780.

Freund, Y., & Shapire, R. (1999). Large margin classification using the perceptron algorithm. Machine Learning, 37(3), 277–296. doi:10.1023/A:1007662407062

Friedberg, I. (2006). Automated protein function prediction--the genomic challenge. Briefings in Bioinformatics, 7, 225–242.

Friedman, J. (1991). Multivariate Adaptive Regression Splines (with discussion). Annals of Statistics, 19, 1–141.

Friedman, J. H. (1988). Multivariate Adaptive Regression Splines (Tech. Rep. No. 102). Stanford, CA: Laboratory of Computational Statistics - Dept. of Statistics.

Friedman, J., & Stuetzle, W. (1981). Projection pursuit regression. Journal of the American Statistical Association, 76(376), 817–823.

Friedrich, R. W., & Korsching, S. I. (1997, May). Combinatorial and chemotopic odorant coding in the zebrafish olfactory bulb visualized by optical imaging. Neuron, 18(5), 737–752. doi:10.1016/S0896-6273(00)80314-1

Fröhlich, H. (2006). Kernel methods in chemo- and bioinformatics. Berlin: Logos-Verlag. (PhD-Thesis)

Fröhlich, H., & Zell, A. (2005). Efficient parameter selection for support vector machines in classification and regression via model-based global optimization. In Proc. int. joint conf. neural networks (pp. 1431 - 1438).

Fröhlich, H., Wegner, J., & Zell, A. (2005). Assignment kernels for chemical compounds. In Proc. int. joint conf. neural networks (pp. 913 - 918).

Fröhlich, H., Wegner, J., Sieker, F., & Zell, A. (2005). Optimal assignment kernels for attributed molecular graphs . In Raedt, L. D., & Wrobel, S. (Eds.), Proc. int. conf. machine learning (pp. 225–232). ACM Press.

Fröhlich, H., Wegner, J., Sieker, F., & Zell, A. (2006). Kernel functions for attributed molecular graphs - a new similarity based approach to adme prediction in classification and regression. QSAR & Combinatorial Science, 25(4), 317–326. doi:10.1002/qsar.200510135

Fujita, T., Iwasa, J., & Hansch, C. (1964). A New Substituent Constant, p, Derived from Partition Coefficients. Journal of the American Chemical Society, 86, 5175–5180. doi:10.1021/ja01077a028

Galvez, J., Julian-Ortiz, J. V., & Garcia-Domenech, R. (2001). General Topological Patterns of Known Drugs. Journal of Molecular Graphics & Modelling, 20(1), 84–94. doi:10.1016/S1093-3263(01)00103-6

Gao, H. (2001). Application of BCUT metrics and genetic algorithm in binary QSAR analysis. Journal of Chemical Information and Computer Sciences, 41, 402–407. doi:10.1021/ci000306p

Gärtner, T. (2002). Exponential and Geometric Kernels for Graphs. In NIPS Workshop on Unreal Data: Principles of Modeling Nonvectorial Data.

Gärtner, T., Flach, P., & Wrobel, S. (2003). On graph kernels: Hardness results and efficient alternatives. In Proceedings of the sixteenth annual conference on computational learning theory (pp. 129–143).

Gärtner, T., Flach, P., Kowalczyk, A., & Smola, A. (2002). Multi-Instance Kernels. In C. Sammut & A. Hoffmann (Eds.), Proceedings of the Nineteenth International Conference on Machine Learning (pp. 179-186). San Francisco: Morgan Kaufmann.

Gasteiger, J. (Ed.). (2003). Handbook of Chemoinformatics. From Data to Knowledge in 4 Volumes. Weinheim, Germany: Wiley-VCH.

Gasteiger, J., & Engel, T. (2003). Chemoinformatics: a textbook. New York: Wiley-VCH. doi:10.1002/9783527618279

Gasteiger, J., & Marsili, M. (1978). A New Model for Calculating Atomic Charges in Molecules. Tetrahedron Letters, 34, 3181–3184. doi:10.1016/S0040-4039(01)94977-9

Gasteiger, J., & Marsili, M. (1980). Iterative Partial Equalization of Orbital Electronegativity: A Rapid Access to Atomic Charges. Tetrahedron, 36, 3219–3228. doi:10.1016/0040-4020(80)80168-2

Geary, R. C. (1954). The Contiguity Ratio and Statistical Mapping. Incorp. Statist., 5, 115–145. doi:10.2307/2986645

Gedeck, P., & Lewis, R. A. (2008). Exploiting QSAR models in lead optimization. Current Opinion in Drug Discovery & Development, 11(4), 569–575.

Geladi, P., & Kowalski, B. R. (1986). Partial least-squares regression: a tutorial. Analytica Chimica Acta, 185, 1–17.

Gellert, E., & Rudzats, R. (1964). The antileukemia activity of tylocrebrine. Journal of Medicinal Chemistry, 15, 361–362. doi:10.1021/jm00333a029

Ghose, A. K., Viswanadhan, V. N., & Wendoloski, J. J. (1998). Prediction of Hydrophobic (Lipophilic) Properties of Small Organic Molecules Using Fragmental Methods: An Analysis of ALOGP and CLOGP Methods. The Journal of Physical Chemistry A, 102, 3762–3772. doi:10.1021/jp980230o

Ghuloum, A., Sage, C., & Jain, A. (1999). Molecular hashkeys: a novel method for molecular characterization and its application for predicting important pharmaceutical properties of molecules. Journal of Medicinal Chemistry, 42(10), 1739–1748. doi:10.1021/jm980527a

Gillet, V., Willett, P., & Bradshaw, J. (2003). Similarity searching using reduced graphs. Journal of Chemical Information and Computer Sciences, 43, 338–345. doi:10.1021/ci025592e

Ginn, C. M. R., Willett, P., & Bradshaw, J. (2000). Combination of molecular similarity measures using data fusion. Perspectives in Drug Discovery and Design, 20, 1–16. doi:10.1023/A:1008752200506

Girosi, F. (1998). An Equivalence Between Sparse Approximation and Support Vector Machines. Neural Computation, 10(6), 1455–1480.

Girosi, F., Jones, M., & Poggio, T. (1995). Regularization theory and neural networks architectures. Neural Computation, 7, 219–269. doi:10.1162/neco.1995.7.2.219

Givehchi, A., & Schneider, G. (2005). Multi-space classification for predicting GPCR ligands. Molecular Diversity, 9, 371–383. doi:10.1007/s11030-005-6293-4

Glen, R. C., & Adams, S. E. (2006). Similarity metrics and descriptor spaces - which combinations to choose? QSAR & Combinatorial Science, 25, 1133–1142. doi:10.1002/qsar.200610097

Glick, M., Jenkins, J. L., Nettles, J. H., Hitchings, H., & Davies, J. W. (2006). Enrichment of high-throughput screening data with increasing levels of noise using support vector machines, recursive partitioning, and Laplacian-modified naive Bayesian classifiers. Journal of Chemical Information and Modeling, 46, 193–200. doi:10.1021/ci050374h

Glusman, G., Yanai, I., Rubin, I., & Lancet, D. (2001, May). The complete human olfactory subgenome. Genome Research, 11(5), 685–702. Available from http://dx.doi.org/10.1101/gr.171001. doi:10.1101/gr.171001

Gnanadesikan, R. (1977). Methods for Statistical Data Analysis of Multivariate Observations. New York: Wiley.

Godden, J., & Bajorath, J. (2006). A distance function for retrieval of active molecules from complex chemical space representations. Journal of Chemical Information and Modeling, 46, 1094–1097. doi:10.1021/ci050510i

Godden, J., Furr, J., Xue, L., Stahura, F., & Bajorath, J. (2003). Recursive median partitioning for virtual screening of large databases. Journal of Chemical Information and Computer Sciences, 43, 182–188. doi:10.1021/ci0203848

Gohlke, H., Dullweber, F., Kamm, W., März, J., Kissel, T., & Klebe, G. (2001). Prediction of human intestinal absorption using a combined simmulated annealing/backpropagation neural network approach . In Hültje, H. D., & Sippl, W. (Eds.), Rational approaches drug des (pp. 261–270). Barcelona: Prous Science Press.

Golbraikh, A., & Tropsha, A. (2002). Beware of q(2)! Journal of Molecular Graphics & Modelling, 20, 269–276. doi:10.1016/S1093-3263(01)00123-1

Golbraikh, A., Bonchev, D., & Tropsha, A. (2002). Novel ZE-isomerism descriptors derived from molecular topology and their application to QSAR analysis. Journal of Chemical Information and Computer Sciences, 42, 769–787. doi:10.1021/ci0103469

Golbraikh, A., Shen, M., Xiao, Z. Y., Xiao, Y. D., Lee, K. H., & Tropsha, A. (2003). Rational selection of training and test sets for the development of validated QSAR models. Journal of Computer-Aided Molecular Design, 17, 241–253. doi:10.1023/A:1025386326946

Goldberg, D. E. (1989). Genetic Algorithms in Search, Optimization and Machine Learning. Massachusetts, MA: Addison-Wesley.

Gomeni, R., Bani, M., D'Angeli, C., Corsi, M., & Bye, A. (2001). Computer assisted drug development (CADD): An emerging technology for accelerating drug development. Clinical Pharmacology and Therapeutics, 69, 3.

Gonzalez, M. P., Teran, C., Saiz-Urra, L., & Teijeira, M. (2008). Variable selection methods in QSAR: an overview. Current Topics in Medicinal Chemistry, 8(18), 1606–1627.

Goodford, P. J. (1985). A Computational Procedure for Determining Energetically Favorable Binding Sites on Biologically Important Macromolecules. Journal of Medicinal Chemistry, 28, 849–857. doi:10.1021/jm00145a002

Goto, S., Nishioka, T., & Kanehisa, M. (1998). LIGAND: chemical database for enzyme reactions. Bioinformatics (Oxford, England), 14(7), 591–599. doi:10.1093/bioinformatics/14.7.591

Goto, S., Okuno, Y., Hattori, M., Nishioka, T., & Kanehisa, M. (2002). LIGAND: database of chemical compounds and reactions in biological pathways. Nucleic Acids Research, 30(1), 402–404. doi:10.1093/nar/30.1.402

Goutis, C. (1996). Partial least squares yields shrinkage estimators. Annals of Statistics, 24, 816–824.

Goyal, K., Mohanty, D. & Mande, S.C. (2007). PAR-3D: a server to predict protein active site residues. gkm252.

Grover, M., Singh, B., Bakshi, M., & Singh, S. (2000). Quantitative structure-property relationships in pharmaceutical research – part 1. Pharmaceutical Science & Technology Today, 3(1), 28–35. doi:10.1016/S1461-5347(99)00214-X

Guba, W., & Cruciani, G. (2000). Molecular field-derived descriptors for the multivariate modeling of pharmacokinetic data . In Gundertofte, K., & Jorgensen, F. (Eds.), Molecular modeling and prediction of bioactivity (pp. 89–94). New York: Kluwer Academic/Plenum Publishers.

Gunther, S., Guenther, S., Kuhn, M., Dunkel, M., & al. et. (2007). Supertarget and matador: resources for exploring drug-target relationships. Nucleic Acids Res.

Guo, Y., Xiao, J., Guo, Z., Chu, F., Cheng, Y., & Wu, S. (2005). Exploration of a binding mode of indole amide analogues as potent histone deacetylase inhibitors and 3D-QSAR analyses. Bioorganic & Medicinal Chemistry, 13(18), 5424–5434.

Gupta, N., Mangal, N., & Biswas, S. (2005). Evolution and similarity evaluation of protein structures in contact map space. Proteins, 59, 196–204.

Guyon, I., Weston, J., Barnhill, S., & Vapnik, V. (2002). Gene Selection for Cancer Classification using Support Vector Machines. Machine Learning, 46, 389–422. doi:10.1023/A:1012487302797

Hadjiprocopis, A., & Smith, P. (1997). Feed forward neural network entities. In J. Mira, R. Moreno-Diaz, & J. Cabestany (Eds.), Biological and artificial computation: From neuroscience to technology (pp. 349{359). Spain: Springer-Verlag.

Hage, P., & Harary, F. (1995). Eccentricity and centrality in networks. Social Networks, 17, 57–63.

Halgren, T. A. (1998). Merck molecular force field. I–V. MMFF94 Basics and Parameters. Journal of Computational Chemistry, 17, 490–641. doi:10.1002/(SICI)1096-987X(199604)17:5/6<490::AID-JCC1>3.0.CO;2-P

Hallem, E. A., & Carlson, J. R. (2006, Apr). Coding of odors by a receptor repertoire. Cell, 125(1), 143–160. Available from http://dx.doi.org/10.1016/j.cell.2006.01.050. doi:10.1016/j.cell.2006.01.050

Halperin, D., & Overmars, M. (1994). Spheres, Molecules, and Hidden Surface Removal. 10th Annual Symposium on Computational Geometry (pp. 113-122). New York: ACM Press

Hammett, L. P. (1935). Reaction Rates and Indicator Acidities. Chemical Reviews, 17, 67–79. doi:10.1021/cr60053a006

Hammett, L. P. (1937). The Effect of Structure upon the Reactions of Organic Compounds. Benzene Derivatives. Journal of the American Chemical Society, 59, 96–103. doi:10.1021/ja01280a022

Hammett, L. P. (1938). Linear Free Energy Relationships in Rate and Equilibrium Phenomena. Transactions of the Faraday Society, 34, 156–165. doi:10.1039/tf9383400156

Han, L. Y., Zheng, C. J., Xie, B., Jia, J., Ma, X. H., & Zhu, F. (2007). Support vector machines approach for predicting druggable proteins: recent progress in its exploration and investigation of its usefulness. Drug Discovery Today, 12(7 & 8), 304–313.

Hansch, C. (1969). Quantitative approach to biochemical structure-activity relationships. Accounts of Chemical Research, 2(8), 232–239.

Hansch, C., & Fujita, T. (1964). ρ-σ-π Analysis; method for the correlation of biological activity and chemical structure. Journal of the American Chemical Society, 86(8), 1616–1626.

Hansch, C., & Fujita, T. (1995). Status of QSAR at the end of the twentieth century. Classical and Three-Dimensional Qsar in Agrochemistry, 606, 1–12. doi:10.1021/bk-1995-0606.ch001

Hansch, C., & Leo, A. (1995). Exploring QSAR. Fundamentals and Applications in Chemistry and Biology. Washington, DC: American Chemical Society.

Hansch, C., Hoekman, D., & Gao, H. (1996). Comparative QSAR: Toward a deeper understanding of chemicobiological interactions. Chemical Reviews, 96, 1045–1076. doi:10.1021/cr9400976

Hansch, C., Maloney, P. P., Fujita, T., & Muir, R. M. (1962). Correlation of Biological Activity of Phenoxyacetic Acids with Hammett Substituent Constants and Partition Coefficients. Nature, 194, 178–180. doi:10.1038/194178b0

Hansch, C., Muir, R. M., Fujita, T., Maloney, P. P., Geiger, E., & Streich, M. (1963). The Correlation of Biological Activity of Plant Growth Regulators and Chloromycetin Derivatives with Hammett Constants and Partition Coefficients. Journal of the American Chemical Society, 85, 2817–2824. doi:10.1021/ja00901a033

Hardoon, D., Ajanki, A., Puolamaki, K., Shawe-Taylor, J., & Kaski, S. (2007). Information Retrieval by Inferring Implicit Queries from Eye Movements. In Proceedings of the 11th International Conference on Artificial Intelligence and Statistics (AISTATS).

Harper, G., Bradshaw, J., Gittins, J. C., Green, D. V. S., & Leach, A. R. (2001). Prediction of biological activity for high-throughput screening using binary kernel discrimination. Journal of Chemical Information and Computer Sciences, 41, 1295–1300. doi:10.1021/ci000397q

Hasegawa, K., & Funatsu, K. (2000). Partial least squares modeling and genetic algorithm optimization in quantitative structure-activity relationships. SAR and QSAR in Environmental Research, 11(3-4), 189–209.

Hasegawa, K., & Funatsu, K. (2009in press). Data Modeling and Chemical Interpretation of ADME Properties Using Regression and Rule Mining Techniques . In Gary, W. C. (Ed.), Frontier in Drug Design & Discovery 4. Bentham Science Publisher.

Hasegawa, K., Arakawa, M., & Funatsu, K. (2000). Rational choice of bioactive conformations through use of conformation analysis and 3-way partial least squares modeling. Chemometrics and Intelligent Laboratory Systems, 50(2), 253–261.

Hasegawa, K., Arakawa, M., & Funatsu, K. (2003). Simultaneous determination of bioactive conformations and alignment rules by multi-way PLS modeling. Computational Biology and Chemistry, 27(3), 211–216.

Hasegawa, K., Arakawa, M., & Funatsu, K. (2005). Novel computational approaches in QSAR and molecular design based on GA, multi-way PLS and NN. Current Computer-aided Drug Design, 1(2), 129–145.

Hasegawa, K., Deushi, T., Yoshida, H., Miyashita, Y., & Sasaki, S. (1994). Chemometric QSAR studies of antifungal azoxy compounds. Journal of Computer-Aided Molecular Design, 8(4), 449–456.

Hasegawa, K., Kimura, T., & Funatsu, K. (1997). Nonlinear CoMFA using QPLS as a novel 3D-QSAR approach. Quantitative Structure-Activity Relationships, 16(3), 219–223.

Hasegawa, K., Kimura, T., Miyashita, Y., & Funatsu, K. (1996). Nonlinear Partial Least Squares Modeling of Phenyl Alkylamines with the Monoamine Oxidase Inhibitory Activities. Journal of Chemical Information and Computer Sciences, 36(5), 1025–1029.

Hasegawa, K., Matsuoka, S., Arakawa, M., & Funatsu, K. (2002). New molecular surface-based 3D-QSAR method using kohonen neural network and 3-way PLS. Computers & Chemistry, 26(6), 583–589.

Hasegawa, K., Matsuoka, S., Arakawa, M., & Funatsu, K. (2003). Multi-way PLS modeling of structure-activity data by incorporating electrostatic and lipophilic potentials on molecular surface. Computational Biology and Chemistry, 27(3), 381–386.

Hasegawa, K., Miyashita, Y., & Funatsu, K. (1997). GA Strategy for Variable Selection in QSAR Studies: GA Based PLS Analysis of Calcium Channel Antagonists. Journal of Chemical Information and Computer Sciences, 37(2), 306–310.

Hastie, T., & Tibshirani, R. (1990). Generalized Additive Models. New York: Chapman & Hall/CRC.

Hattori, M., Okuno, Y., Goto, S., & Kanehisa, M. (2003). Development of a chemical structure comparison method for integrated analysis of chemical and genomic information in the metabolic pathways. Journal of the American Chemical Society, 125(39), 11853–11865. doi:10.1021/ja036030u

Hattori, M., Tanaka, N., Kanehisa, M., & Goto, S. (2010). SIMCOMP/SUBCOMP: chemical structure search servers for network analyses. Nucleic Acids Research, Advance Access published on May 11, 2010.

Haussler, D. (1999). Convolution Kernels on Discrete Structures (Tech. Rep. No. UCSC-CRL-99-10). UC Santa Cruz.

Hawkins, T., Luban, S., & Kihara, D. (2006). Enhanced automated function prediction using distantly related sequences and contextual association by PFP. Protein Science, 15, 1550–1556.

Haykin, S. (1999). Neural Networks: A Comprehensive Foundation (2nd ed.). Upper Saddle River, NJ: Prentice-Hall.

Hebert, M., Ikeuchi, K., & Delingette, H. (1995). A spherical representation for recognition of free-form surfaces. IEEE Transactions on Pattern Analysis and Machine Intelligence, 17(7), 681–689. doi:10.1109/34.391410

Heijden, F.v.d., Duin, R.P.W., Ridder, D.d. & Tax, D.M.J. (2004). Classification. parameter estimation and state estimation - an engineering approach using Matlab.

Helma, C. (2006). Lazy structure-activity relationships (lazar) for the prediction of rodent carcinogenicity and Salmonella mutagenicity. Mol.Divers Online First, 1-12.

Helma, C., Cramer, T., Kramer, S., & Raedt, L. (2004). Data mining and machine learning techniques for the identification of mutagenicity inducing substructures and structure activity relationships of noncongeneric compounds. Journal of Chemical Information and Computer Sciences, 44, 1402–1411. doi:10.1021/ci034254q

Helma, C., King, R., Kramer, S., & Srinivasan, A. (2001). The Predictive Toxicology Challenge 2000-2001. Bioinformatics (Oxford, England), 17(1), 107–108. doi:10.1093/bioinformatics/17.1.107

Hiden, H., McKay, B., Willis, M., & Montague, G. (1998). Non-linear partial least squares using genetic programming. In J. Koza (Ed.), Genetic Programming: Proceedings of the Third Annual Conference. San Francisco: Morgan Kaufmann.

Hildebrand, J. G., & Shepherd, G. M. (1997). Mechanisms of olfactory discrimination: converging evidence for common principles across phyla. Annual Review of Neuroscience, 20, 595–631. Available from http://dx.doi.org/10.1146/annurev.neuro.20.1.595. doi:10.1146/annurev.neuro.20.1.595

Ho, R. L., & Lieu, C. A. (2008). Systems biology: an evolving approach in drug discovery and development. Drugs in R&D., 9(4), 203–216.

Hoche, S., & Wrobel, S. (2001). Relational learning using constrained confidence-rated boosting. In C. Rouveirol & M. Sebag (Eds.), Proceedings of the eleventh international conference on inductive logic programming ILP (pp. 51–64). New York: Springer-Verlag.

Hoche, S., & Wrobel, S. (2002). Scaling boosting by margin-based inclusion of features and relations. In Proceedings of the 13th european conference on machine learning (ECML) (pp. 148–160).

Hoffman, B. T., Kopajtic, T., Katz, J. L., & Newman, A. H. (2000). 2D QSAR Modeling and Preliminary Database Searching for Dopamine Transporter Inhibitors Using Genetic Algorithm Variable Selection of Molconn Z Descriptors. Journal of Medicinal Chemistry, 43(22), 4151–4159.

Hoffman, B., Cho, S. J., Zheng, W. F., Wyrick, S., Nichols, D. E., & Mailman, R. B. (1999). Quantitative structure-activity relationship modeling of dopamine D-1 antagonists using comparative molecular field analysis, genetic algorithms-partial least-squares, and K nearest neighbor methods. Journal of Medicinal Chemistry, 42, 3217–3226. doi:10.1021/jm980415j

Hollas, B. (2002). Correlation Properties of the Autocorrelation Descriptor for Molecules. MATCH Communications in Mathematical and in Computer Chemistry, 45, 27–33.

Holliday, J. D., & Willett, P. (1997). Using a genetic algorithm to identify common structural features in sets of ligands. Journal of Molecular Graphics & Modelling, 15(4), 221–232. doi:10.1016/S1093-3263(97)00080-6

Holm, L., Kaariainen, S., Rosenstrom, P. & Schenkel, A. (2008). Searching protein structure databases with DaliLite (vol.3,pp. 2780-2781).

Hong, H., Tong, W., Xie, Q., Fang, H., & Perkins, R. (2005). An in silico ensemble method for lead discovery: decision forest. SAR and QSAR in Environmental Research, 16, 339–347. doi:10.1080/10659360500203022

Hong, X., & Hopfinger, A. J. (2003). 3D-Pharmacophores of Flavonoid Binding at the Benzodiazepine GABAA Receptor Site Using 4D-QSAR Analysis. Journal of Chemical Information and Computer Sciences, 43, 324–336. doi:10.1021/ci0200321

Hopfinger, A. J. (1980). A QSAR investigation of dihydrofolate reductase inhibition by Baker triazines based upon molecular shape analysis. Journal of the American Chemical Society, 102, 7196–7206. doi:10.1021/ja00544a005

Hopfinger, A. J., Wang, S., Tokarski, J. S., Jin, B., Albuquerque, M., & Madhav, P. J. (1997). Construction of 3D-QSAR models using the 4D-QSAR analysis formalism. Journal of the American Chemical Society, 119, 10509–10524. doi:10.1021/ja9718937

Horváth, T., Gärtner, T., & Wrobel, S. (2004). Cyclic pattern kernels for predictive graph mining. In Proceedings of the tenth acm sigkdd international conference on knowledge discovery and data minig (p. 158-167).New York: ACM Press.

Höskuldsson, A. (1988). PLS Regression Methods. Journal of Chemometrics, 2, 211–228. doi:10.1002/cem.1180020306

Höskuldsson, A. (1992). Quadratic PLS regression. Journal of Chemometrics, 6(6), 307–334.

Hotelling, H. (1933). Analysis of a complex of statistical variables into principal components. J. Educat. Psychol., 24, 417 - 441 &498 - 520.

Hou, B. K., Ellis, L. B. M., & Wackett, L. P. (2004). Encoding microbial metabolic logic: predicting biodegradation. Journal of Industrial Microbiology & Biotechnology, 31(6), 261–272. doi:10.1007/s10295-004-0144-7

Hou, T., & Xu, X. (2003). Adme evaluation in drug discovery. 3. modelling blood-brain barrier partitioning using simple molecular descriptors. Journal of Chemical Information and Computer Sciences, 43(6), 2137–2152. doi:10.1021/ci034134i

Hou, T.-J., Xia, K., Zhang, W., & Xu, X. (2004). ADME Evaluation in Drug Discovery. 4. Prediction of Aqueous Solubility Based on Atom Contribution Approach. Journal of Chemical Information and Computer Sciences, 44, 266–275. doi:10.1021/ci034184n

Hsin, K.-Y. (2009). Eduliss 2, Edinburgh University Ligand Selection System. Retrieved from http://eduliss.bch.ed.ac.uk/

Hsu, D. F., & Taksa, I. (2005). Comparing rank and score combination methods for data fusion in information retrieval. Information Retrieval, 8, 449–480. doi:10.1007/s10791-005-6994-4

Hu, L., Wu, H., Lin, W., Jiang, J., & Yu, R. (2007). Quantitative structure-activity relationship studies for the binding affinities of imidazobenzodiazepines for the α6 benzodiazepine receptor isoform utilizing optimized blockwise variable combination by particle swarm optimization for partial least squares modeling. QSAR & Combinatorial Science, 26(1), 92–101.

Huan, J., Wang, W., Bandyopadhyay, D., Snoeyink, J., Prins, J., & Tropsha, A. (2004). Mining protein family specific residue packing patterns from protein structure graphs. In Proceedings of the eighth annual international conference on Resaerch in computational molecular biology. ACM, San Diego, CA, USA.

Huang, B., & Schroeder, M., M. (2006). Ligsite: predicting ligand binding sites using the connolly surface and degree of conservation. BMC Structural Biology, 6(19).

Huang, J. Y., & Brutlag, D. L. (2001). The EMOTIF database. Nucleic Acids Research, 29, 202–204.

Huang, Q.-G., Song, W.-L., & Wang, L.-S. (1997). Quantitative Relationship Between the Physiochemical Characteristics as well as Genotoxicity of Organic Pollutants and Molecular Autocorrelation Topological Descriptors. Chemosphere, 35, 2849–2855. doi:10.1016/S0045-6535(97)00345-7

Huang, X., Liu, T., Gu, J., Luo, X., Ji, R., & Cao, Y. (2001). 3D-QSAR models of flavonoids binding at benzodiazepine site in GABAA receptors. Journal of Medicinal Chemistry, 44, 1883–1891. doi:10.1021/jm000557p

Hulland, J. (1999). Use of partial least squares (PLS) in strategic management research: A review of four recent studies. Strategic Management Journal, 20, 195–204.

Hulo, N., Bairoch, A., Bulliard, V., Cerutti, L., De Castro, E., & Langendijk-Genevaux, P. S. (2006). The PROSITE database. Nucleic Acids Research, 34, D227–D230.

Huuskonen, J. (2000). Estimation of Aqueous Solubility for Diverse Set of Organic Compounds Based on Molecular Topology. Journal of Chemical Information and Computer Sciences, 40, 773–777. doi:10.1021/ci9901338

Inglese, J., Auld, D. S., Jadhav, A., Johnson, R. L., Simeonov, A., & Yasgar, A. (2006). Quantitative high-throughput screening: a titration-based approach that e±ciently identifies biological activities in large chemical libraries. Proceedings of the National Academy of Sciences of the United States of America, 103, 11473–11478. doi:10.1073/pnas.0604348103

Inokuchi, A. (2005). Mining generalized substructures from a set of labeled graphs. In Proceedings of the 4th ieee internatinal conference on data mining (p. 415-418). New York: IEEE Computer Society.

Inokuchi, A., Washio, T., & Motoda, H. (2000). An Apriori-based algorithm for mining frequent substructures from graph data. In Proceedings of the fourth european conference on machine learning and principles and practice of knowledge discovery in databases (pp. 13-23).

Irwin, J., & Shoichet, B. (2005). ZINC - a free database of commercially available compounds for virtual screening. Journal of Chemical Information and Modeling, 45, 177–182. doi:10.1021/ci049714+

Ishibashi, K., Hatano, K., & Takeda, M. (2008). Online learning of approximate maximum p-norm margin classiÞers with biases. In Proceedings of the 21st annual conference on learning theory (colt2008).

Ivanciuc, O. (2001). 3D QSAR Models . In Diudea, M. V. (Ed.), QSPR / QSAR Studies by Molecular Descriptors (pp. 233–280). Huntington, NY: Nova Science.

Ivanciuc, O. (2007). Applications of support vector machines in chemistry. In K. B. Lipkowitz & T. R. Cundari (Eds.), Reviews in computational chemistry (Vol. 23, pp. 291–400). Weiheim: Wiley-VCH.

Ivanciuc, O., & Balaban, A. T. (1999). The Graph Description of Chemical Structures . In Devillers, J., & Balaban, A. T. (Eds.), Topological Indices and Related Descriptors in QSAR and QSPR (pp. 59–167). Amsterdam, The Netherlands: Gordon & Breach Science Publishers.

Ivanisenko, V.A., Pintus, S.S., Grigorovich, D.A.& Kolchanov, N.A. (2004). PDBSiteScan: a program for searching for active, binding and posttranslational modification sites in the 3D structures of proteins. W549-554.

Ivanov, V. (1976). The Theory of Approximate Methods and Their Application to the Numerical Solution of Singular Integral Equations. Nordhoff International.

Jaccard, P. (1912). The distribution of the flora of the alpine zone. The New Phytologist, 11(2), 37–50. doi:10.1111/j.1469-8137.1912.tb05611.x

Jacob, L., & Vert, J.-P. (2008). Protein-ligand interaction prediction: an improved chemogenomics approach. Bioinformatics (Oxford, England), 24, 2149–2156. doi:10.1093/bioinformatics/btn409

Jain, A. N., & Nicholls, A. (2008). Recommendations for evaluation of computational methods. Journal of Computer-Aided Molecular Design, 22, 133–139. doi:10.1007/s10822-008-9196-5

Jain, A., & Dubes, R. (1988). Algorithms for Clustering Data. Upper Saddle River, NJ: Prentice Hall.

Jain, A., Koile, K., & Chapman, D. (1994). Compass: predicting biological activities from molecular surface properties. performance comparisons on a steroid benchmark. Journal of Medicinal Chemistry, 37(15), 2315–2327. doi:10.1021/jm00041a010

Jalali-Heravi, M., & Kyani, A. (2007). Application of genetic algorithm-kernel partial least square as a novel nonlinear feature selection method: Activity of carbonic anhydrase II inhibitors. European Journal of Medicinal Chemistry, 42(5), 649–659.

Jenkins, J. L., Bender, A., & Davies, J. W. (2006). In silico target fishing: Predicting biological targets from chemical structure. Drug Discovery Today. Technologies, 3, 413–421. doi:10.1016/j.ddtec.2006.12.008

Joachims, T. (1998). Making large-scale support vector machine learning practical. In B. Schaolkopf, C. J. Burges, & A. J. Smola (Eds.), Advances in kernel methods: Support vector machines (p. 169-184). Cambridge, MA: MIT Press. Retrieved from http://citeseer.ist.psu.edu/joachims98making.html

Joachims, T. (2003). Learning to classify text using support vector machines: Methods, theory and algorithms. New York: Kluwer Academic Publishers.

Johnson, B. A., Farahbod, H., Saber, S., & Leon, M. (2005, Mar). Effects of functional group position on spatial representations of aliphatic odorants in the rat olfactory bulb. The Journal of Comparative Neurology, 483(2), 192–204. Available from http://dx.doi.org/10.1002/cne.20415. doi:10.1002/cne.20415

Johnson, M. A., & Maggiora, G. M. (Eds.). (1990). Concepts and Applications of Molecular Similarity. New York: John Wiley.

Johnson, M., & Maggiora, G. (Eds.). (1990). Concepts and applications of molecular similarity. New York: John Wiley & Sons.

Joliffe, I. T. (2002). Principle Component Analysis (2nd ed., Vol. 29). New York: Springer.

Jolli®e, I. T. (2002). Principal component analysis (2nd ed.). New York: Springer.

Jolliffe, I. T. (1986). Principal Component Analysis. New York: Springer-Verlag.

Jong, S. (1993). SIMPLS: an alternative approach to partial least squares regression. Chemometrics and Intelligent Laboratory Systems, 18, 251–263.

Jorissen, R. N., & Gilson, M. K. (2005). Virtual screening of molecular databases using a support vector machine. Journal of Chemical Information and Modeling, 45, 549–561. doi:10.1021/ci049641u

Kanehisa, M. (1997). A database for post-genome analysis. Trends in Genetics, 13(9), 375–376. doi:10.1016/S0168-9525(97)01223-7

Kanehisa, M. (2001). Prediction of higher order functional networks from genomic data. Pharmacogenomics, 2(4), 373–385. doi:10.1517/14622416.2.4.373

Kanehisa, M., & Bork, P. (2003). Bioinformatics in the post-sequence era. Nature Genetics, 33(3s), 305–310. doi:10.1038/ng1109

Kanehisa, M., & Goto, S. (2000). KEGG: Kyoto encyclopedia of genes and genomes. Nucleic Acids Research, 28(1), 27–30. doi:10.1093/nar/28.1.27

Kanehisa, M., Araki, M., Goto, S., Hattori, M., Hirakawa, M., & Itoh, M. (2008). Kegg for linking genomes to life and the environment. Nucleic Acids Research, 36(Database issue), D480–D485. doi:10.1093/nar/gkm882

Kanehisa, M., Goto, S., Hattori, M., Aoki-Kinoshita, K., Itoh, M., & Kawashima, S. (2006). From genomics to chemical genomics: new developments in KEGG. Nucleic Acids Research, 34, D354–D357. doi:10.1093/nar/gkj102

Kansy, M., Senner, F., & Gubernator, K. (1998). Physicochemical high throughput screening: Parallel artificial membrane permeation assay in the description of passive absorption processes. Journal of Medicinal Chemistry, 41, 1007–1010. doi:10.1021/jm970530e

Kashima, H., & Koyanagi, T. (2002). Kernels for semistructured date. In Proceedings of the nineteenth international conference on machine learning (pp. 291–298). San Francisco, CA: Morgan Kaufmann.

Kashima, H., Kato, T., Yamanishi, Y., Sugiyama, M., & Tsuda, K. (2009). Link propagation: A fast semi-supervised learning algorithm for link prediction. In Proceedings of the 2009 siam conference on data mining (p. 1099-1110).

Kashima, H., Tsuda, K., & Inokuchi, A. (2003). Marginalized Kernels between Labeled Graphs. In T. Faucett & N. Mishra (Eds.), Proceedings of the Twentieth International Conference on Machine Learning (pp. 321-328). New York: AAAI Press.

Kashima, H., Tsuda, K., & Inokuchi, A. (2004). Kernels for graphs . In Schölkopf, B., Tsuda, K., & Vert, J. P. (Eds.), Kernel methods in computational biology (pp. 155–170). Cambridge, MA: MIT Press.

Kass, M., Witkin, A., & Terzopoulos, D. (1987). Snakes - active contour models. International Journal of Computer Vision, 1(4), 321–331. doi:10.1007/BF00133570

Kawano, S., Hashimoto, K., Miyama, T., Goto, S., & Kanehisa, M. (2005). Prediction of glycan structures from gene expression data based on glycosyltransferase reactions. Bioinformatics (Oxford, England), 21(21), 3976–3982. doi:10.1093/bioinformatics/bti666

Kazius, J., Nijssen, S., Kok, J., Bäck, T., & Ijzerman, A. (2006). Substructure mining using elaborate chemical representation. Journal of Chemical Information and Modeling, 46, 597–605. doi:10.1021/ci0503715

Keiser, M., Roth, B., Armbruster, B., Ernsberger, P., Irwin, J., & Shoichet, B. (2007). Relating protein pharmacology by ligand chemistry. Nature Biotechnology, 25, 197–206. doi:10.1038/nbt1284

Kellogg, G. E. (2002). MolConnZ (Version 4.05). Quincy, MA: Hall Associates Consulting.

Keseler, I. M., Bonavides-Martinez, C., Collado-Vides, J., Gama-Castro, S., Gunsalus, R. P., & Johnson, D. A. (2009). EcoCyc: A comprehensive view of Escherichia coli biology. Nucleic Acids Research, 37(Database issue), D464–D470. doi:10.1093/nar/gkn751

Keseler, I., Collado-Vides, J., Gama-Castro, S., Ingraham, J., Paley, S., & Paulsen, I. (2005). EcoCyc: a comprehensive database resource for escherichia coli. Nucleic Acids Research, 33, D334–D337. doi:10.1093/nar/gki108

Kier, L. B. (1971). Molecular Orbital Theory in Drug Research. New York: Academic Press.

Kier, L. B., & Hall, L. H. (1976). Molecular Connectivity in Chemistry and Drug Research. New York: Academic Press.

Kier, L. B., & Hall, L. H. (1990). An Electrotopological-State Index for Atoms in Molecules. Pharmaceutical Research, 7, 801–807. doi:10.1023/A:1015952613760

Kim, S. J., Koh, K., Lustig, M., Boyd, S., & Gorinevsky, D. (2007). An interior-point method for large-scale \ ell_1-regularized least squares. IEEE Jounal of Selected Topics in Signal Processing, 1(4), 606–617. doi:10.1109/JSTSP.2007.910971

Kimeldorf, G., & Wahba, G. (1971). Some results on Tchebycheffian spline functions. Journal of Mathematical Analysis and Applications, 33, 82–95.

Kimura, T., Miyashita, Y., Funatsu, K., & Sasaki, S. (1996). Quantitative Structure-Activity Relationships of the Synthetic Substrates for Elastase Enzyme Using Nonlinear Partial Least Squares Regression. Journal of Chemical Information and Computer Sciences, 36(2), 185–189.

King, R. D., Muggleton, S., Lewis, R. A., & Sternberg, M. J. E. (1992). Drug Design by Machine Learning - the Use of Inductive Logic Programming to Model the Structure-Activity-Relationships of Trimethoprim Analogs Binding to Dihydrofolate-Reductase. Proceedings of the National Academy of Sciences of the United States of America, 89, 11322–11326. doi:10.1073/pnas.89.23.11322

King, R. D., Muggleton, S., Srinivasan, A., & Sternberg, M. (1996). Structure-activity relationships derived by machine learning: The use of atoms and their bond connectivities to predict mutagenicity by inductive logic programming. Proceedings of the National Academy of Sciences of the United States of America, 93(1), 438–442. doi:10.1073/pnas.93.1.438

Klein, Ch. Th., Kaiser, D., & Ecker, G. (2004). Topological Distance Based 3D Descriptors for Use in QSAR and Diversity Analysis. Journal of Chemical Information and Computer Sciences, 44, 200–209. doi:10.1021/ci0256236

Klein, T. E., Huang, C., Ferrin, T. E., Langridge, R., & Hansch, C. (1986). Computer-Assisted Drug Receptor Mapping Analysis. ACS Symposium Series. American Chemical Society, 306, 147–158. doi:10.1021/bk-1986-0306.ch013

Klopman, G., & Tu, M. (1997). Structure-biodegradability study and computer automated prediction of aerobic biodegradation of chemicals. Environmental Toxicology and Chemistry, 16(9), 1829–1835.

Kohavi, R., & John, G. H. (1997). Wrappers for feature subset selection. Artificial Intelligence, 1-2, 273–324. doi:10.1016/S0004-3702(97)00043-X

Kohonen, T. (1982, January). Self-organized formation of topologically correct feature maps. Biological Cybernetics, V43(1), 59–69. Available from http://dx.doi.org/10.1007/BF00337288. doi:10.1007/BF00337288

Kohonen, T. (1989). Self-organization and associative memory. Berlin: Springer.

Kondor, R., & Lafferty, J. (2002). Diffusion kernels on graphs and other discrete input spaces. In T. Faucett & N. Mishra (Eds.), Proceedings of the twentieth international conference on machine learning (p. 321-328). Boston: AAAI Press.

Kotera, M., McDonald, A. G., Boyce, S., & Tipton, K. F. (2008). Eliciting possible reaction equations and metabolic pathways involving orphan metabolites. Journal of Chemical Information and Modeling, 48(12), 2335–2349. doi:10.1021/ci800213g

Kotera, M., Okuno, Y., Hattori, M., Goto, S., & Kanehisa, M. (2004). Computational assignment of the EC numbers for genomic-scale analysis of enzymatic reactions. Journal of the American Chemical Society, 126(50), 16487–16498. doi:10.1021/ja0466457

Kovatcheva, A., Golbraikh, A., Oloff, S., Feng, J., Zheng, W., & Tropsha, A. (2005). QSAR modeling of datasets with enantioselective compounds using chirality sensitive molecular descriptors. SAR and QSAR in Environmental Research, 16, 93–102. doi:10.1080/10629360412331319844

Kovatcheva, A., Golbraikh, A., Oloff, S., Xiao, Y. D., Zheng, W. F., & Wolschann, P. (2004). Combinatorial QSAR of ambergris fragrance compounds. Journal of Chemical Information and Computer Sciences, 44, 582–595. doi:10.1021/ci034203t

Kramer, B., Rarey, M., & Lengauer, T. (1999). Evaluation of the FLEXX incremental construction algorithm for protein ligand docking. Proteins, 37(2), 228–241. doi:10.1002/(SICI)1097-0134(19991101)37:2<228::AID-PROT8>3.0.CO;2-8

Krämer, N. (2007). An overview on the shrinkage properties of partial least squares regression. Computational Statistics, 22(2), 249–273.

Krämer, N., & Braun, M. (2007). Kernelizing PLS, Degrees of Freedom, and Efficient Model Selection. In Z. Ghahramani (Ed.), Proceedings of the 24th International Conference on Machine Learning (pp. 441–448).

Kramer, S. (2000). Relational learning vs. propositionalization. AI Communications, 13(4), 275–276.

Kramer, S., & De Raedt, L. (2001). Feature construction with version spaces for biochemical application. In Proceedings of the eighteenth international conference on machine learning (pp. 258–265).

Kramer, S., Lavrac, N., & Flach, P. (2001). Propositionalisation approaches to Relational Data Mining . In Dzeroski, S., & Larac, N. (Eds.), Relational data mining (pp. 262–291). Berlin: Springer.

Kramer, S., Raedt, L., & Helma, C. (2001). Molecular feature mining in HIV data. In Proceedings of the 7th acm sigkdd international conference on knowledge discovery and data mining. New York: ACM Press.

Kratochwil, N. A., Huber, W., Muller, F., Kansy, M., & Gerber, P. R. (2002). Predicting plasma protein binding of drugs: a new approach. Biochemical Pharmacology, 64(9), 1355–1374.

Kratochwil, N., Malherbe, P., Lindemann, L., Ebeling, M., Hoener, M., & Muhlemann, A. (2005). An automated system for the analysis of g protein-coupled receptor transmembrane binding pockets: Alignment, receptor-based pharmacophores, and their application. Journal of Chemical Information and Modeling, 45(5), 1324–1336. doi:10.1021/ci050221u

Kreher, D. L., & Stinson, D. R. (1998). Combinatorial Algorithms: Generation, Enumeration and Search. CRC Press.

Kriegl, J. M., Eriksson, L., Arnhold, T., Beck, B., Johansson, E., & Fox, T. (2005). Multivariate modeling of cytochrome P450 3A4 inhibition. European Journal of Pharmaceutical Sciences, 24(5), 451–463.

Krogel, M. A., & Wrobel, S. (200). Transformation-based learning using multirelational aggregation. In S. Rouveirol & M. Sebag (Eds.), Proceedings of the eleventh international conference on inductive logic programming (ILP). New York: Springer.

Krogel, M. A., Rawles, S., Zelezny, F., Flach, P. A., Lavrac, N., & Wrobel, S. (2003). Comparative evaluation of approaches to propositionalization. In Proceedings of the 13th international conference on inductive logic programming. New York: Springer-Verlag.

Krzanowski, W. J. (1988). Principles of Multivariate Analysis. New York: Oxford Univ. Press.

Kubinyi, H. (1986). Quantitative Relationships Between Chemical-Structure and Biological-Activity. Chemie in Unserer Zeit, 20, 191–202. doi:10.1002/ciuz.19860200605

Kubinyi, H. (1990). Quantitative Structure-Activity-Relationships (Qsar) and Molecular Modeling in Cancer-Research. Journal of Cancer Research and Clinical Oncology, 116, 529–537. doi:10.1007/BF01637071

Kubinyi, H. (1994). Variable Selection in QSAR Studies. I. An Evolutionary Algorithm. Quantitative Structure-Activity Relationships, 13, 285–294.

Kubinyi, H. (1996). Evolutionary Variable Selection in Regression and PLS Analyses. Journal of Chemometrics, 10, 119–133. doi:10.1002/(SICI)1099-128X(199603)10:2<119::AID-CEM409>3.0.CO;2-4

Kubinyi, H. (2002). From Narcosis to Hyperspace: The History of QSAR. Quant. Struct. Act. Relat., 21, 348–356. doi:10.1002/1521-3838(200210)21:4<348::AID-QSAR348>3.0.CO;2-D

Kubinyi, H. (2003). Drug research: myths, hype and reality. Nature Reviews. Drug Discovery, 2, 665–668. Retrieved from http://home.t-online.de/home/kubinyi/nrdd-pub-08-03.pdf. doi:10.1038/nrd1156

Kubinyi, H. (2004). Changing paradigms in drug discovery. In M. H. et al. (Ed.), Proc. int. beilstein workshop (pp. 51 - 72). Berlin: Logos-Verlag.

Kubinyi, H. (Ed.). (1993). 3D QSAR in Drug Design. Theory, Methods, and Applications. Leiden, The Netherlands: ESCOM.

Kubinyi, H., Folkers, G., & Martin, Y. (1998). 3D QSAR in Drug Design. New York: Kluwer.

Kubinyi, H., Folkers, G., & Martin, Y. C. (Eds.). (1998). 3D QSAR in Drug Design (Vol. 3). Dordrecht, The Netherlands: Kluwer/ESCOM.

Küçükural, A. & Sezerman, O.U. (2009). Protein Strcuture Characterization Using Attributed Sub-Graph Matching Algorithms with Parallel Computing, (In preperation).

Küçükural, A. & Sezerman, U. (2009). Structural Alignment of Proteins Using Network Properties with Dynamic Programming, (In preperation).

Küçükural, A., Sezerman, O. U., & Ercil, A. (2008). Discrimination of Native Folds Using Network Properties of Protein Structures . In Brazma, A., Miyano, S., & Akutsu, T. (Eds.), APBC (pp. 59–68). Kyoto, Japan: Imperial College Press.

Kudo, T., Maeda, E., & Matsumoto, Y. (2005). An application of boosting to graph classification. [Cambridge, MA: MIT Press.]. Advances in Neural Information Processing Systems, 17, 729–736.

Kuhl, F. S., Crippen, G. M., & Friesen, D. K. (1984). A combinatorial algorithm for calculating ligand binding. Journal of Computational Chemistry, 5(1), 24–34. doi:10.1002/jcc.540050105

Kuhn, H. (1955). The hungarian method for the assignment problem. Naval Res. Logist. Quart., 2, 83–97. doi:10.1002/nav.3800020109

Kullback, S. (1997). Information theory and statistics. Mineola, MN: Dover Publications.

Kumar, R., Kulkarni, A., Jayaraman, V. K., & Kulkarni, B. D. (2004). Structure-Activity Relationships using Locally Linear Embedding Assisted by Support Vector and Lazy Learning Regressors. Internet Electron.J.Mol. Des., 3, 118–133.

Kuntz, I. D. (1992). Structure-based strategies for drug design and discovery. Science, 257, 1078–1082. doi:10.1126/science.257.5073.1078

Kuntz, I. D., Blaney, J. M., Oatley, S. J., Langridge, R., & Ferrin, T. E. (1982). A geometric approach to macromolecule-ligand interactions. Journal of Molecular Biology, 161, 269–288. doi:10.1016/0022-2836(82)90153-X

Kuzelka, O., & Zelenzy, F. (2009). Block-wise construction of acyclic relational features with monotone irreducibility and relevanve properties. In Proceedings of the 26th international conference on machine learning (pp. 569–575).

Labute, P. (1999). A new method for the determination of quantitative structure activity relationships . In Pacific symposium on biocomputing (Vol. 4, pp. 444–455). Binary QSAR.

Labute, P., & Williams, C. (2001). Flexible alignment of small molecules. Journal of Medicinal Chemistry, 44(10), 1483–1490. doi:10.1021/jm0002634

Lanczos, C. (1950). An iteration method for the solution of the eigenvalue problem of linear differential and integral operators. Journal of Research of the National Bureau of Standards, 45, 255–282.

Lander, E. (2001). Initial Sequencing and Analysis of the Human Genome. Nature, 409(6822), 860–921. doi:10.1038/35057062

Landon, M. R., & Schaus, S. E. (2006). JEDA: Joint entropy diversity analysis. An information-theoretic method for choosing diverse and representative subsets from combinatorial libraries. Molecular Diversity, 10, 333–339. doi:10.1007/s11030-006-9042-4

Landwehr, N., Kersting, K., & Raedt, L. D. (2005). nFOIL:integrating naive bayes and foil. In Proceedings of the twentieth national conference on artificial intelligence (AAAI-05).

Landwehr, N., Passerini, A., Raedt, L., & Frasconi, P. (2006). kFOIL: Learning simple relational kernels. In Proceedings of the national conference on artificial intelligence (AAAI) (Vol. 21, pp. 389–394).

Langowski, J., & Long, A. (2002). Computer systems for the prediction of xenobiotic metabolism. Advanced Drug Delivery Reviews, 54(3), 407–415. doi:10.1016/S0169-409X(02)00011-X

Larsen, S. B., Jorgensen, F. S., & Olsen, L. (2008). QSAR Models for the Human H+/Peptide Symporter, hPEPT1: Affinity Prediction Using Alignment-Independent Descriptors. Journal of Chemical Information and Modeling, 48(1), 233–241.

Laskowski, R. A., Watson, J. D., & Thornton, J. M. (2005). Protein Function Prediction Using Local 3D Templates. Journal of Molecular Biology, 351, 614–626.

Lavrac, N., & Dzeroski, S. (1994). Inductive logic programming. Ellis Horwood. Lavrac, N., Zelezny, F., & Flach, P. A. (200). RSD: Relational subgroup discovery through first-order feature construction. In S. Matwin & C. Sammut (Eds.), Proceedings of the twelfth international conference on inductive logic programming (ILP). New York: Springer.

Leach, A., & Gillet, V. J. (2003). An Introduction to Chemoinformatics. New York: Springer.

Leardi, R. (1994). Application of Genetic Algorithms to Feature Selection Under Full Validation Conditions and to Outlier Detection. Journal of Chemometrics, 8, 65–79. doi:10.1002/cem.1180080107

Leardi, R., Boggia, R., & Terrile, M. (1992). Genetic Algorithms as a Strategy for Feature Selection. Journal of Chemometrics, 6, 267–281. doi:10.1002/cem.1180060506

Lee, J. A., & Verleysen, M. (2007). Nonlinear Dimensionality Reduction. New York: Springer. doi:10.1007/978-0-387-39351-3

Lee, S., Wu, Q., Huntbatch, A., & Yang, G. (2007). Predictive K-PLSR Myocardial Contractility Modeling with Phase Contrast MR Velocity Mapping . In Ayache, N., Ourselin, S., & Maeder, A. (Eds.), Medical Image Computing and Computer-Assisted Intervention – MICCAI 2007 (pp. 866–873). New York: Springer.

Lehman Brothers and McKinsey&Company. (2001). The Fruits of Genomics.

Lemmen, C., & Lengauer, T. (2000). Computational methods for the structural alignment of molecules. Journal of Computer-Aided Molecular Design, 14(3), 215–232. doi:10.1023/A:1008194019144

Lengauer, T., Lemmen, C., Rarey, M., & Zimmermann, M. (2004). Novel technologies for virtual screening. Drug Discovery Today, 9, 27–34. doi:10.1016/S1359-6446(04)02939-3

Leslie, C., Eskin, E., & Noble, W. (2002). The spectrum kernel: A string kernel for SVM protein classification. In R. B. Altman, A. K. Dunker, L. Hunter, K. Lauerdale, & T. E. Klein (Eds.), Proceedings of the pacific symposium on biocomputing (pp. 566–575). Hackensack, NJ: World Scientific.

Leslie, C., Eskin, E., Weston, J., & Noble, W. (2003). Mismatch string kernels for svm protein classification . In Becker, S., Thrun, S., & Obermayer, K. (Eds.), Advances in neural information processing systems 15 (pp. 467–476). Cambridge, MA: MIT Press.

Leslie, C., Kuang, R., & Eskin, E. (2004). Inexact matching string kernels for protein classification . In Schölkopf, B., Tsuda, K., & Vert, J. P. (Eds.), Kernel methods in computational biology (pp. 95–112). Cambridge, MA: MIT Press.

Lewis, D. D. (1998). Naive (Bayes) at forty: The independence assumption in information retrieval. In Lecture notes in computer science [Berlin: Springer.]. Machine Learning, ECML-98, 4–15.

Li, H., Yap, C. W., Ung, C. Y., Xue, Y., Cao, Z. W., & Chen, Y. Z. (2005). Effect of selection of molecular descriptors on the prediction of blood-brain barrier penetrating and nonpenetrating agents by statistical learning methods. Journal of Chemical Information and Modeling, 45, 1376–1384. doi:10.1021/ci050135u

Li, H., Yap, C. W., Ung, C. Y., Xue, Y., Li, Z. R., & Han, L. Y. (2007). Machine learning approaches for predicting compounds that interact with therapeutic and ADMET related proteins. Journal of Pharmaceutical Sciences, 96, 2838–2860. doi:10.1002/jps.20985

Li, T., Mei, H., & Cong, P. (1999). Combining nonlinear PLS with the numeric genetic algorithm for QSAR. Chemometrics and Intelligent Laboratory Systems, 45, 177–184.

Li, Z., Fu, B., Wang, Y., & Liu, S. (2001). On Structural Parametrization and Molecular Modeling of Peptide Analogues by Molecular Electronegativity Edge Vector (VMEE): Estimation and Prediction for Biological Activity of Dipeptides. Journal of Chinese Chemical Society, 48, 937–944.

Liang, J., & Dill, K. A. (2001). Are Proteins Well-Packed? Biophysical Journal, 81, 751–766.

Liaw, A., & Wiener, M. (2002). Classification and regression by randomForest. R News, 2(3), 18.

Lindgren, F., Geladi, P., & Wold, S. (1993). The kernel algorithm for PLS. Journal of Chemometrics, 7(1), 45–59.

Lindgren, F., Geladi, P., & Wold, S. (1994). Kernel-based pls regression; cross-validation and applications to spectral data. Journal of Chemometrics, 8(6), 377–389.

Lindgren, F., Geladi, P., Rännar, S., & Wold, S. (1994). Interactive Variable Selection (IVS) for PLS. Part I: Theory and Algorithms. Journal of Chemometrics, 8, 349–363. doi:10.1002/cem.1180080505

Lindgren, F., Hansen, B., Karcher, W., Sjöström, M., & Eriksson, L. (1996). Model Validation by Permutation Tests: Applications to Variable Selection. Journal of Chemometrics, 10, 521–532. doi:10.1002/(SICI)1099-128X(199609)10:5/6<521::AID-CEM448>3.0.CO;2-J

Lindstrom, A., Pettersson, F., Almqvist, F., Berglund, A., Kihlberg, J., & Linusson, A. (2006). Hierarchical PLS Modeling for Predicting the Binding of a Comprehensive Set of Structurally Diverse Protein-Ligand Complexes. *Journal of Chemical Information and Modeling, 46*(3), 1154–1167.

Lingjærde, O., & Christophersen, N. (2000). Shrinkage Structure of Partial Least Squares. *Scandinavian Journal of Statistics, 27*, 459–473.

Linster, C., Sachse, S., & Galizia, C. G. (2005, Jun). Computational modeling suggests that response properties rather than spatial position determine connectivity between olfactory glomeruli. *Journal of Neurophysiology, 93*(6), 3410–3417. Available from http://dx.doi.org/10.1152/jn.01285.2004. doi:10.1152/jn.01285.2004

Lipinski, C. A., Lombardo, F., Dominy, B. W., & Feeney, P. J. (2001). Experimental and computational approaches to estimate solubility and permeability in drug discovery and development settings. *Advanced Drug Delivery Reviews, 46*(1-3), 3–26. doi:10.1016/S0169-409X(00)00129-0

Liu, H. X., Zhang, R. S., Yao, X. J., Liu, M. C., Hu, Z. D., & Fan, B. T. (2004). Prediction of the Isoelectric Point of an Amino Acid Based on GA-PLS and SVMs. *Journal of Chemical Information and Computer Sciences, 44*(1), 161–167.

Liu, J., Kern, P. S., Gerberick, G. F., Santos-Filho, O. A., Esposito, E. X., Hopfinger, A. J., & Tseng, Y. J. (2008). Categorical QSAR models for skin sensitization based on local lymph node assay measures and both ground and excited state 4D-fingerprint descriptors. *Journal of Computer-Aided Molecular Design, 22*(6-7), 345–366.

Livingstone, D. J. (2000). The characterization of chemical structures using molecular properties: A survey. *Journal of Chemical Information and Computer Sciences, 40*(2), 195–209. doi:10.1021/ci990162i

Livingstone, D. J., Manallack, D. T., & Tetko, I. V. (1997). Data modelling with neural networks: advantages and limitations. *Journal of Computer-Aided Molecular Design, 11*, 135–142. doi:10.1023/A:1008074223811

Lo, W.-C., Huang, P.-J., Chang, C.-H., & Lyu, P.-C. (2007). Protein structural similarity search by Ramachandran codes. *BMC Bioinformatics, 8*, 307.

Lobaugh, N., West, R., & McIntosh, A. (2001). Spatiotemporal analysis of experimental differences in event-related potential data with partial least squares. *Psychophysiology, 38*, 517–530.

Lobell, M., Hendrix, M., Hinzen, B., Keldenich, J., Meier, H., & Schmeck, C. (2006). In silico ADMET traffic lights as a tool for the prioritization of HTS hits. *ChemMedChem, 1*, 1229–1236. doi:10.1002/cmdc.200600168

Lodhi, H., & Muggleton, S. (2005). Is mutagenesis still challenging. In *International conference on inductive logic programming, (ILP - late-breaking papers)* (pp. 35–40).

Lodhi, H., Karakoulas, G., & Shawe-Taylor, J. (2002). Boosting strategy for classification. *Intelligent Data Analysis, 6*(2), 149–174.

Lodhi, H., Muggleton, S., & Sternberg, M. J. E. (2009). Learning large margin first order decision lists for multi-class classification. In *Proceedings of the twelfth international conference on discovery science (DS-09)* (Vol. LNAI).

Lodhi, H., Saunders, C., Shawe-Taylor, J., Cristianini, N., & Watkins, C. (2002). Text classification using string kernels. *Journal of Machine Learning Research, 2*, 419–444. doi:10.1162/153244302760200687

Luenberger, D. G. (1969). *Optimization by vector space methods.* New York: Wiley.

Luke, B. T. (1994). Evolutionary Programming Applied to the Development of Quantitative Structure-Activity Relationships and Quantitative Structure-Property Relationships. *Journal of Chemical Information and Computer Sciences, 34*, 1279–1287. doi:10.1021/ci00022a009

Mahé, P., & Vert, J. P. (2009). Graph kernels based on tree patterns for molecules. *Machine Learning, 75*(1), 3–35. doi:10.1007/s10994-008-5086-2

Mahé, P., & Vert, J.-P. (2008). Graph kernels based on tree patterns for molecules. *Machine Learning.*

Mahé, P., Ralaivola, L., Stoven, V., & Vert, J.-P. (2006). The pharmacophore kernel for virtual screening with support vector machines. Journal of Chemical Information and Modeling, 46(5), 2003–2014. doi:10.1021/ci060138m

Mahe, P., Ueda, N., Akutsu, T., Perret, J. L., & Vert, J. P. (2005). Graph kernels for molecular structure-activity relationship analysis with support vector machines. Journal of Chemical Information and Modeling, 45(4), 939–951. doi:10.1021/ci050039t

Mahe, P., Ueda, N., Akutsu, T., Perret, J. L., & Vert, J. P. (2004). Extensions of marginalized graph kernels. In R. Greiner & D. Schuurmans (Eds.), Proceedings of the twenty-first international conference on machine learning (ICML-2004) (pp. 552–559). New York: ACM Press.

Maldonado, A. G., Doucet, J. P., Petitjean, M., & Fan, B.-T. (2006). Molecular similarity and diversity in chemoinformatics: from theory to applications. Molecular Diversity, 10, 39–79. doi:10.1007/s11030-006-8697-1

Mallat, S. G. (1989). Multifrequency channel decompositions of images and wavelet models. IEEE Transactions on Acoustics, Speech, and Signal Processing, 37, 2091–2110. doi:10.1109/29.45554

Mandel, J. (1985). The Regression Analysis of Collinear Data. Journal of Research of the National Bureau of Standards, 90, 465–476.

Manne, R. (1987). Analysis of Two Partial-Least-Squares Algorithms for Multivariate Calibration. Chemometrics and Intelligent Laboratory Systems, 2, 187–197.

Marek, K., & Wojciech, R. (1998). Fast parallel algorithms for graph matching problems. New York: Oxford University Press, Inc.

Martens, H., & Naes, T. (Eds.). (1989). Multivnriate Calibration. New York: Wiley.

Martens, M., & Martens, H. (1986). Partial Least Squares Regression . In Piggott, J. (Ed.), Statistical Procedures in Food Research (pp. 293–359). London: Elsevier Applied Science.

Martin, A. C., Orengo, C. A., Hutchinson, E. G., Jones, S., Karmirantzou, M., & Laskowski, R. A. (1998). Protein folds and functions . Structure (London, England), 6, 875–884.

Martin, D., Berriman, M., & Barton, G. (2004). GOtcha: a new method for prediction of protein function assessed by the annotation of seven genomes. BMC Bioinformatics, 5, 178.

Martin, Y. (1998). Pharmacophore mapping . In Martin, Y., & Willett, P. (Eds.), Designing bioactive molecules (pp. 121–148). Oxford University Press.

Martin, Y. C. (1979). Advances in the Methodology of Quantitative Drug Design . In Ariëns, E. J. (Ed.), Drug Design (Vol. VIII, pp. 1–72). New York, NY: Academic Press.

Martin, Y. C. (1991). Computer-Assisted Rational Drug Design. Methods in Enzymology, 203, 587–613. doi:10.1016/0076-6879(91)03031-B

Martin, Y. C. (1992). 3D database searching in drug design. Journal of Medicinal Chemistry, 35, 2145–2154. doi:10.1021/jm00090a001

Martin, Y. C. (1998). 3D QSAR: Current State Scope, and Limitations. In H.Kubinyi, G. Folkers, & Y. C. Martin (Eds.), 3D QSAR in Drug Design (pp. 3-23). Dordrecht, The Netherlands: Kluwer/ESCOM.

Martin, Y. C. (2001). Diverse viewpoints on computational aspects of molecular diversity. Journal of Combinatorial Chemistry, 3, 231–250. doi:10.1021/cc000073e

Martin, Y. C., Kofron, J. L., & Traphagen, L. M. (2002). Do structurally similar molecules have similar biological activity? Journal of Medicinal Chemistry, 45, 4350–4358. doi:10.1021/jm020155c

Mason, J., Morize, I., Menard, P., Cheney, D., Hulme, C., & Labaudiniere, R. (1999). New 4-point pharmacophore method for molecular similarity and diversity applications: Overview of the method and applications, including a novel approach to the design of combinatorial libraries containing privileged substructures. Journal of Medicinal Chemistry, 42, 3251–3264. doi:10.1021/jm9806998

Matsuda, H., Taniguchi, F., & Hashimoto, A. (1997). An approach to detection of protein structural motifs using an encoding scheme of backbone conformations. Proc. of 2nd Pacific Symposium on Biocomputing (pp280-291).

Matter, H., & Pötter, T. (1999). Comparing 3D pharmacophore triplets and 2D fingerprints for selecting diverse compound subsets. Journal of Chemical Information and Computer Sciences, 39(6), 1211–1225. doi:10.1021/ci980185h

McGregor, J. J., & Willett, P. (1981). Use of a maximal common subgraph algorithm in the automatic identification of the ostensible bond changes occurring in chemical reactions. Journal of Chemical Information and Computer Sciences, 21(3), 137–140. doi:10.1021/ci00031a005

McInerney, T., & Terzopolous, D. (1999). Topology adaptive deformable surfaces for medical image volume segmentation. IEEE Transactions on Medical Imaging, 18(10), 840–850. doi:10.1109/42.811261

McLachlan, G. J. (2004). Discriminant Analysis and Statistical Pattern Recognition (1st ed.). New York: Wiley Interscience.

Mehlhorn, K., & Näher, S. (1999). The LEDA Platform of Combinatorial and Geometric Computing. Cambridge University Press.

Meister, M., & Bonhoeffer, T. (2001, Feb). Tuning and topography in an odor map on the rat olfactory bulb. The Journal of Neuroscience, 21(4), 1351–1360.

Melville, J. L., & Hirts, J. D. (2007). TMACC: Interpretable Correlation Descriptors for Quantitative Structure-Activity Relationships. Journal of Chemical Information and Modeling, 47, 626–634. doi:10.1021/ci6004178

Mercer, J. (1909). Functions of positive and negative type, and their connection with the theory of integral equations. Proceedings of the Royal Society of London. Series A, Containing Papers of a Mathematical and Physical Character, 83(559), 69–70. doi:10.1098/rspa.1909.0075

Meyer, D., Leisch, F., & Hornik, K. (2003, September). The support vector machine under test. Neurocomputing, 55, 169–186. Retrieved from http://dx.doi.org/10.1016/S0925-2312(03)00431-4. doi:10.1016/S0925-2312(03)00431-4

Michailidis, G., & De Leeuw, J. (1998). The Gifi System of Descriptive Multivariate Analysis. Statistical Science, 13(4), 307–336.

Micthell, T. (1982). Generalization as search. Artificial Intelligence, 18, 203–226. doi:10.1016/0004-3702(82)90040-6

Minowa, Y., Araki, M., & Kanehisa, M. (2007). Comprehensive analysis of distinctive polyketide and nonribosomal peptide structural motifs encoded in microbial genomes. Journal of Molecular Biology, 368(5), 1500–1517. doi:10.1016/j.jmb.2007.02.099

Mitchell, T., Buchanan, B., Dejong, G., Dietterich, T., Rosenbloom, P., & Waibel, A. (1989). Machine Learning. Annual Review of Computer Science, 4, 417–433. doi:10.1146/annurev.cs.04.060190.002221

Mitra, S., Fusi, S., & Indiveri, G. (2009, Feb.). Real-time classification of complex patterns using spike-based learning in neuromorphic VLSI. IEEE Transactions on Biomedical Circuits and Systems, 3(1), 32–42. doi:10.1109/TBCAS.2008.2005781

Miyashita, Y., & Sasaki, S. (Eds.). (1994). Chemical Pattern Recognition and Multivariate Aanalysis. Tokyo: Kyoritsu Publisher.

Miyashita, Y., Li, Z., & Sasaki, S. (1993). Chemical pattern recognition and multivariate analysis for QSAR studies. Trends in Analytical Chemistry, 12(2), 50–60.

Miyashita, Y., Ohsako, H., Takayama, C., & Sasaki, S. (1992). Multivariate structure-activity relationships analysis of fungicidal and herbicidal thiolcarbamates using partial least squares method. Quantitative Structure-Activity Relationships, 11(1), 17–22.

Miyazawa, S., & Jernigan, R. L. (1996). Residue-residue potentials with a favorable contact pair term and an unfavorable high packing density term. for simulation and threading . Journal of Molecular Biology, 256.

Mjolsness, E., & DeCoste, D. (2001). Machine learning for science: State of the art and future prospects. Science, 293, 2051–2055. doi:10.1126/science.293.5537.2051

Mohajeri, A., Hemmateenejad, B., Mehdipour, A., & Miri, R. (2008). Modeling calcium channel antagonistic activity of dihydropyridine derivatives using QTMS indices analyzed by GA-PLS and PC-GA-PLS. Journal of Molecular Graphics & Modelling, 26(7), 1057–1065.

Molecular operating environment (MOE 2007.09). (2007). Montreal, Canada: Chemical Computing Group.

Momma, M. (2005). Efficient Computations via Scalable Sparse Kernel Partial Least Squares and Boosted Latent Features. In Proceedings of SIGKDD International Conference on Knowledge and Data Mining (pp. 654–659). Chicago, IL.

Moock, T. E., Nourse, J. G., Grier, D., & Hounshell, W. D. (1988). The implementation of atom-atom mapping and related features in the reaction access system (REACCS) . In Warr, W. A. (Ed.), Chemical Structures, The International Language of Chemistry (pp. 303–313). Berlin, Germany: Springer-Verlag.

Moran, P. A. P. (1950). Notes on Continuous Stochastic Phenomena. Biometrika, 37, 17–23.

Moreau, G., & Broto, P. (1980). The Autocorrelation of a Topological Structure: A New Molecular Descriptor. New Journal of Chemistry, 4, 359–360.

Moreau, G., & Turpin, C. (1996). Use of similarity analysis to reduce large molecular libraries to smaller sets of representative molecules. Analusis, 24, M17–M21.

Mori, K., Takahashi, Y. K., Igarashi, K. M., & Yamaguchi, M. (2006, Apr). Maps of odorant molecular features in the mammalian olfactory bulb. Physiological Reviews, 86(2), 409–433. Available from http://dx.doi.org/10.1152/physrev.00021.2005. doi:10.1152/physrev.00021.2005

Morishita, S. (2001). Computing optimal hypotheses efficiently for boosting. In Discovery science (p. 471-481).

Moriya, Y., Shigemizu, D., Hattori, M., Tokimatsu, T., Kotera, M., Goto, S., & Kanehisa, M. (2010). PathPred: an enzyme-catalyzed metabolic pathway prediction server. Nucleic Acids Research, Advance Access published on April 30, 2010.

Moro, S., Bacilieri, M., Cacciari, B., & Spalluto, G. (2005). Autocorrelation of Molecular Electrostatic Potential Surface Properties Combined with Partial Least Squares Analysis as New Strategy for the Prediction of the Activity of Human A3 Adenosine Receptor Antagonists. Journal of Medicinal Chemistry, 48(18), 5698–5704.

Moss, G.P. (2006). Recommendations of the Nomenclature Committee. International Union of Biochemistry and Molecular Biology on the Nomenclature and Classification of Enzymes by the Reactions they Catalyse.

Mu, T., Nandi, A., & Rangayyan, R. (2007). Classification of breast masses via nonlinear transformation of features based on a kernel matrix. Medical & Biological Engineering & Computing, 45(8), 769–780.

Muggleton, S. (1995). Inverse entailment and progol. New Generation Computing, 13, 245–286. doi:10.1007/BF03037227

Muggleton, S., Lodhi, H., Amini, A., & Sternberg, M. J. E. (2005). Support Vector Inductive Logic Programming. In Proceedings of the eighth international conference on discovery science (Vol. LNCS, pp. 163–175). New York: Springer Verlag.

Næs, T., & Isaksson, T. (1992). Locally Weighted Regression in Diffuse Near-Infrared Transmittance Spectroscopy. Applied Spectroscopy, 46(1), 34–43.

Nagamine, N., & Sakakibara, Y. (2007). Statistical prediction of proteinÐchemical interactions based on chemical structure and mass spectrometry data. Bioinformatics (Oxford, England), 23, 2004–2012. doi:10.1093/bioinformatics/btm266

Nasney, J., & Livny, M. (2000). Managing network resources in Condor. In Proceedings of the ninth ieee symposium on high performance distributed computing (hpdc9) (pp. 298–299).

Newman, D. J., Cragg, G. M., & Snader, K. M. (2003). Natural products as sources of new drugs over the period 1981-2002. Journal of Natural Products, 66, 1022–1037. doi:10.1021/np030096l

Newman, M. E. J. (2003). A measure of betweenness centrality based on random walks arXiv.org:cond-mat/0309045.

Nidhi, Glick, M., Davies, J. W., & Jenkins, J. L. (2006). Prediction of biological targets for compounds using multiple-category Bayesian models trained on chemogenomics databases. Journal of Chemical Information and Modeling, 46, 1124–1133. doi:10.1021/ci060003g

Nigsch, F., Bender, A., Jenkins, J. L., & Mitchell, J. B. O. (2008). Ligand-target prediction using winnow and naive Bayesian algorithms and the implications of overall performance statistics. Journal of Chemical Information and Modeling, 48, 2313–2325. doi:10.1021/ci800079x

Nijssen, S., & Kok, J. (2004). A quickstart in frequent structure mining can make a difference. In Proceedings of the 10th acm sigkdd international conference on knowledge discovery and data mining (pp. 647–652). New York: ACM Press.

Nilsson, J., de Jong, S., & Smilde, A. (1997). Multiway Calibration in 3D QSAR. Journal of Chemometrics, 11, 511–524.

Nisius, B., Goller, A. H., & Bajorath, J. (2009). Combining cluster analysis, feature selection and multiple support vector machine models for the identification of human ether-a-go-go related gene channel blocking compounds. Chemical Biology & Drug Design, 73, 17–25. doi:10.1111/j.1747-0285.2008.00747.x

Nowicki, M. W., Tulloch, L. B., Worralll, L., McNae, I. W., Hannaert, V., & Michels, P. A. (2008). Design, synthesis and trypanocidal activity of lead compounds based on inhibitors of parasite glycolysis. Bioorganic & Medicinal Chemistry, 16(9), 5050–5061. doi:10.1016/j.bmc.2008.03.045

Nowozin, S., Bakir, G., & Tsuda, K. (2007). Discriminative subsequence mining for action classification. In Proceedings of the 11th ieee international conference on computer vision (iccv 2007) (pp. 1919–1923). IEEE Computer Society.

Nowozin, S., Tsuda, K., Uno, T., Kudo, T., & Bakir, G. (2007). Weighted substructure mining for image analysis . In Ieee computer society conference on computer vision and pattern recognition (cvpr). IEEE Computer Society.

Obrezanova, O., Gola, J. M., Champness, E. J., & Segall, M. D. (2008). Automatic QSAR modeling of ADME properties: blood-brain barrier penetration and aqueous solubility. Journal of Computer-Aided Molecular Design, 22, 431–440. doi:10.1007/s10822-008-9193-8

Oh, M., Yamada, T., Hattori, M., Goto, S., & Kanehisa, M. (2007). Systematic analysis of enzyme-catalyzed reaction patterns and prediction of microbial biodegradation pathways. Journal of Chemical Information and Modeling, 47(4), 1702–1712. doi:10.1021/ci700006f

Ohgaru, T., Shimizu, R., Okamoto, K., Kawashita, N., Kawase, M., & Shirakuni, Y. (2008). Enhancement of ordinal CoMFA by ridge logistic partial least squares. Journal of Chemical Information and Modeling, 48(4), 910–917.

Olender, T., Fuchs, T., Linhart, C., Shamir, R., Adams, M., & Kalush, F. (2004, Mar). The canine olfactory subgenome. Genomics, 83(3), 361–372. Available from http://dx.doi.org/10.1016/j.ygeno.2003.08.009. doi:10.1016/j.ygeno.2003.08.009

Oloff, S., Mailman, R. B., & Tropsha, A. (2005). Application of validated QSAR models of D1 dopaminergic antagonists for database mining. Journal of Medicinal Chemistry, 48, 7322–7332. doi:10.1021/jm049116m

Oloff, S., Zhang, S., Sukumar, N., Breneman, C., & Tropsha, A. (2006). Chemometric analysis of ligand receptor complementarity: identifying Complementary Ligands Based on Receptor Information (CoLiBRI). Journal of Chemical Information and Modeling, 46, 844–851. doi:10.1021/ci050065r

Oprea, T. I. (2003). Chemoinformatics and the Quest for Leads in Drug Discovery. In Gasteiger, J. (Ed.), Handbook of Chemoinformatics (pp. 1509–1531). Weinheim, Germany: Wiley-VCH. doi:10.1002/9783527618279.ch44b

Oprea, T. I. (2004). 3D QSAR modeling in drug design. In Bultinck, P., De Winter, H., Langenaeker, W., & Tollenaere, J. P. (Eds.), Computational Medicinal Chemistry for Drug Discovery (pp. 571–616). New York: Marcel Dekker.

Oprea, T. I., & Matter, H. (2004). Integrating virtual screening in lead discovery. Current Opinion in Chemical Biology, 8, 349–358. doi:10.1016/j.cbpa.2004.06.008

Oprea, T. I., Zamora, I., & Ungell, A. L. (2002). Pharmacokinetically based mapping device for chemical space navigation. Journal of Combinatorial Chemistry, 4, 258–266. doi:10.1021/cc010093w

Ormerod, A., Willett, P., & Bawden, D. (1989). Comparison of fragment weighting schemes for substructural analysis. Quantitative Structure-Activity Relationships, 8, 115–129. doi:10.1002/qsar.19890080207

Osborne, M., Presnell, B., & Turlach, B. (2000). On the lasso and its dual. IMA Journal of Numerical Analysis, 20, 389–404. doi:10.1093/imanum/20.3.389

Palm, K., Stenburg, P., Luthman, K., & Artursson, P. (1997). Polar molecular surface properties predict the intestinal absorption of drugs in humans. Pharmaceutical Research, 14, 586–571. doi:10.1023/A:1012188625088

Palsson, B. (2000). The challenges of in-silico biology. Nature Biotechnology, 18(11), 1147–1150. doi:10.1038/81125

Paolini, G. V., Shapland, R. H. B., van Hoorn, W. P., & Mason, J. S. (2006). Global mapping of pharmocological space. Nature Biotechnology, 24, 805–815. doi:10.1038/nbt1228

Pawlowski, K., & Godzik, A. (2001). Surface map comparison: studying function diversity of homologous proteins. Journal of Molecular Biology, 309, 793–806. doi:10.1006/jmbi.2001.4630

Pearlman, R. S., & Smith, K. (1998). Novel software tools for chemical diversity. Perspectives in Drug Discovery and Design, 9, 339–353. doi:10.1023/A:1027232610247

Pei, J., Han, J., Mortazavi-asl, B., Wang, J., Pinto, H., & Chen, Q. (2004). Mining sequential patterns by pattern-growth: The prefixspan approach. IEEE Transactions on Knowledge and Data Engineering, 16(11), 1424–1440. doi:10.1109/TKDE.2004.77

Peng, Y., Keenan, S. M., Zhang, Q., Kholodovych, V., & Welsh, W. J. (2005). 3D- QSAR Comparative Molecular Field Analysis on Opioid Receptor Antagonists: Pooling Data from Different Studies. Journal of Medicinal Chemistry, 48(5), 1620–1629.

Penzotti, J. E., Landrum, G. A., & Putta, S. (2004). Building predictive ADMET models for early decisions in drug discovery. Current Opinion in Drug Discovery & Development, 7, 49–61.

Perez, C., Pastor, M., Ortiz, A. R., & Gago, F. (1998). Comparative binding energy analysis of HIV-1 protease inhibitors: incorporation of solvent effects and validation as a powerful tool in receptor-based drug design. Journal of Medicinal Chemistry, 41, 836–852. doi:10.1021/jm970535b

Pettit, G. R., Goswami, A., Cragg, G. M., Schmidt, J. M., & Zou, J. C. (1984). Antineoplastic agents, 103. The isolation and structure of hypoestestatins 1 and 2 from the East African Hypoestes verticillaris. Journal of Natural Products, 47, 913–919. doi:10.1021/np50036a001

Pickett, S. D., Mason, J. S., & McLay, I. M. (1996). Diversity profiling and design using 3D pharmacophores: Pharmacophores-Derived Queries (PQD). Journal of Chemical Information and Computer Sciences, 36(6), 1214–1223. doi:10.1021/ci960039g

Pierre, M., Ueda, N., Akutsu, T., Perret, J.-L., & Vert, J.-P. (2005). Graph kernels for molecular structure-activity relationship analysis with support vector machines. Journal of Chemical Information and Modeling, 939–951.

Platt, J. R. (1947). Influence of Neighbor Bonds on Additive Bond Properties in Paraffins. The Journal of Chemical Physics, 15, 419–420. doi:10.1063/1.1746554

Polanski, J. (2006). Drug design using comparative molecular surface analysis. Expert Opinion on Drug Discovery, 1(7), 693–707.

Postarnakevich, N., & Singh, R. (2009). Global-to-local representation and visualization of molecular surfaces using deformable models. ACM Symposium on Applied Computing (pp. 782-787).New York: ACM Press.

Press, W. H., Teukolsky, S. A., Vetterling, W. T., & Flannery, B. P. (1994). Numerical Recipies in C (2nd ed.). Cambridge, UK: Cambridge University Press.

Proschak, E., Wegner, J. K., Schüller, A., Schneider, G., & Fechner, U. (2007). Molecular query language (mql)–a context-free grammar for substructure matching. Journal of Chemical Information and Modeling, 47(2), 295–301. Retrieved from http://dx.doi.org/10.1021/ci600305h. doi:10.1021/ci600305h

Qin, S., & McAvoy, T. (1992). Non-linear PLS modelling using neural networks. Computers & Chemical Engineering, 16(4), 379–391.

Quinlan, J. R. (1986). Induction of decision trees. Machine Learning, 1, 81–106. doi:10.1007/BF00116251

Quinlan, J. R. (1990). Learning logical definitions from relations. Machine Learning, 239–266. doi:10.1007/BF00117105

Quinlan, J. R. (1993). C4.5: Programs for machine learning. New York: Morgan Kaufmann.

Quinlan, J. R. (1996a). Boosting first-order learning. In S. Arikawa & A. Sharma (Eds.), Proceedings of the 7th international workshop on algorithmic learning theory (Vol. LNAI, pp. 143–155). New York: Springer.

Quinlan, J. R. (1996b). Learning first-order definitions of functions. Journal of Artificial Intelligence Research, 5.

Raedt, L. D., & Kramer, S. (2001). Feature construction with version spaces for biochemical application. In Proc. 18th int. conf. on machine learning (pp. 258 - 265).

Ralaivola, L., Swamidass, S. J., Saigo, H., & Baldi, P. (2005). Graph kernels for chemical informatics. Neural Networks, 18(8), 1093–1110. doi:10.1016/j.neunet.2005.07.009

Ramon, J., & Gärtner, T. (2003). Expressivity versus efficiency of graph kernels. In T. Washio & L. De Raedt (Eds.), Proceedings of the First International Workshop on Mining Graphs, Trees and Sequences (pp. 65-74).

Randic, M. (1975). On Characterization of Molecular Branching. Journal of the American Chemical Society, 97, 6609–6615. doi:10.1021/ja00856a001

Randic, M. (1993). Comparative Regression Analysis. Regressions Based on a Single Descriptor. Croatica Chemica Acta, 66, 289–312.

Rännar, S., Lindgren, F., Geladi, P., & Wold, S. (1994). A PLS kernel algorithm for data sets with many variables and fewer objects. Part 1: Theory and algorithm. Chemometrics and Intelligent Laboratory Systems, 8, 111–125.

Rao, K. V., Wilson, R. A., & Cummings, B. (1971). Alkaloids of tylophora. 3. New alkaloids of Tylophora indica (Burm) Merrill and Tylophora dalzellii Hook f. Journal of Pharmaceutical Sciences, 60, 1725–1726. doi:10.1002/jps.2600601133

Rarey, M., & Dixon, S. (1998). Feature trees: A new molecular similarity measure based on tree-matching. Journal of Computer-Aided Molecular Design, 12, 471–490. doi:10.1023/A:1008068904628

Rarey, M., Kramer, B., Lengauer, T., & Klebe, G. (1996). A fast flexible docking method using an incremental construction algorithm. Journal of Molecular Biology, 261, 470–489. doi:10.1006/jmbi.1996.0477

Rätsch, G., Mika, S., Schölkopf, B., & Müller, K.-R. (2002). Constructing boosting algorithms from SVMs: an application to one-class classification. IEEE Transactions on Pattern Analysis and Machine Intelligence, 24(9), 1184–1199. doi:10.1109/TPAMI.2002.1033211

Ravichandran, V., Kumar, B. R. P., Sankar, S., & Agrawal, R. K. (2008). Comparative molecular similarity indices analysis for predicting anti-HIV activity of phenyl ethyl thiourea (PET) derivatives. Medicinal Chemistry Research, 17(1), 1–11.

Raymond, J. W., & Willett, P. (2002). Maximum common subgraph isomorphism algorithms for the matching of chemical structures. Journal of Computer-Aided Molecular Design, 16(7), 521–533. doi:10.1023/A:1021271615909

Raymond, J. W., Gardiner, E. J., & Willett, P. (2002b). Heuristics for similarity searching of chemical graphs using a maximum common edge subgraph algorithm. Journal of Chemical Information and Computer Sciences, 42(2), 305–316. doi:10.1021/ci010381f

Raymond, J., Gardiner, E., Willett, P., & Rascal, P. (2002). Calculation of graph similarity using maximum common edge subgraphs. The Computer Journal, 45(6), 631–644. doi:10.1093/comjnl/45.6.631

Ren, Y., Chen, G., Hu, Z., Chen, X., & Yan, B. (2008). Applying novel Three-Dimensional Holographic Vector of Atomic Interaction Field to QSAR studies of artemisinin derivatives. QSAR & Combinatorial Science, 27(2), 198–207.

Rencher, A. C. (2002). Methods of Multivariate Analysis (2nd ed.). New York: John Wiley & Sons. doi:10.1002/0471271357

Ripley, B. D. (1996). Pattern recognition and neural networks. Cambridge, UK: Cambridge University Press.

Robertson, H. M., & Wanner, K. W. (2006, Nov). The chemoreceptor superfamily in the honey bee, apis mellifera: expansion of the odorant, but not gustatory, receptor family. Genome Research, 16(11), 1395–1403. Available from http://dx.doi.org/10.1101/gr.5057506. doi:10.1101/gr.5057506

Robertson, S. E., & Jones, K. S. (1976). Relevance weighting of search terms. Journal of the American Society for Information Science American Society for Information Science, 27, 129–146. doi:10.1002/asi.4630270302

Romeiro, N. C., Albuquerque, M. G., Bicca de Alencastro, R., Ravi, M., & Hopfinger, A. J. (2006). Free-energy force-field three-dimensional quantitative structure-activity relationship analysis of a set of p38-mitogen activated protein kinase inhibitors. Journal of Molecular Modeling, 12(6), 855–868.

Rosipal, R., & Krämer, N. (2006). Overview and recent advances in partial least squares . In Subspace, latent structure and feature selection techniques (pp. 34–51). New York: Springer. doi:10.1007/11752790_2

Rosipal, R., & Trejo, L. (2004). Kernel PLS Estimation of Single-trial Event-related Potentials. Psychophysiology, 41, S94. (Abstracts of The 44th Society for Psychophysiological Research Annual Meeting, Santa Fe, NM)

Rosipal, R., & Trejo, L. J. (2001). Kernel Partial Least Squares Regression in Reproducing Kernel Hilbert Space. Journal of Machine Learning Research, 2, 97–123.

Rosipal, R., Trejo, L., & Matthews, B. (2003). Kernel PLS-SVC for Linear and Nonlinear Classification. In Proceedings of the Twentieth International Conference on Machine Learning (pp. 640–647). Washington, DC.

Rosset, S., & Zhu, J. (2003). Piecewise linear regularized solution paths (Tech. Rep.). Stanford University. (Technical Report HAL:ccsd-00020066)

Rost, B. (2002). Enzyme function less conserved than anticipated. Journal of Molecular Biology, 318, 595–608.

Rouvray, D. H. (1983). Should We Have Designs on Topological Indices? In King, R. B. (Ed.), Chemical Applications of Topology and Graph Theory. Studies in Physical and Theoretical Chemistry (pp. 159–177). Amsterdam, The Netherlands: Elsevier.

Rucker, C., Rucker, G., & Meringer, M. (2007). Y-Randomization and its Variants in QSPR/QSAR. Journal of Chemical Information and Modeling, 47(6), 2345–2357. doi:10.1021/ci700157b

Ruckert, U., & Kramer, S. (2008). Margin-base first-order rule learning. Machine Learning, 70(2-3), 189–206. doi:10.1007/s10994-007-5034-6

Rupp, M., Proschak, E., & Schneider, G. (2007). Kernel approach to molecular similarity based on iterative graph similarity. Journal of Chemical Information and Modeling, 47(6), 2280–2286. Retrieved from http://dx.doi.org/10.1021/ci700274r. doi:10.1021/ci700274r

Sabet, R., & Fassihi, A. (2008). QSAR study of antimicrobial 3-hydroxypyridine-4-one and 3-hydroxypyran-4-one derivatives using different chemometric tools. International Journal of Molecular Sciences, 9(12), 2407–2423.

Sabidussi, G. (1966). The centrality index of a graph. Psychometrika, 31, 581–603.

Sadowski, J., Wagener, M., & Gasteiger, J. (1995). Assessing Similarity and Diversity of Combinatorial Libraries by Spatial Autocorrelation Functions and Neural Networks. Angewandte Chemie International Edition in English, 34, 2674–2677. doi:10.1002/anie.199526741

Sagrado, S., & Cronin, M. T. D. (2008). Application of the modelling power approach to variable subset selection for GA-PLS QSAR models. Analytica Chimica Acta, 609(2), 169–174.

Saigo, H., & Tsuda, K. (2008). Iterative subgraph mining for principal component analysis. In Proceedings of the 8th ieee international conference on data mining (icdm 2008) (pp. 1007–1012).

Saigo, H., Kadowaki, T., & Tsuda, K. (2006). A linear programming approach for molecular QSAR analysis. In Gärtner, T., Garriga, G., & Meinl, T. (Eds.), International workshop on mining and learning with graphs (mlg) (pp. 85–96).

Saigo, H., Krämer, N., & Tsuda, K. (2008). Partial least squares regression for graph mining. In Proceedings of the 14th acm sigkdd international conference on knowledge discovery and data mining (kdd2008) (pp. 578–586).

Saigo, H., Nowozin, S., Kadowaki, T., Kudo, T., & Tsuda, K. (2008). gboost: A mathematical programming approach to graph classification and regression. Machine Learning, 75(1), 69–89. doi:10.1007/s10994-008-5089-z

Saigo, H., Uno, T., & Tsuda, K. (2007). Mining complex genotypic features for predicting HIV-1 drug resistance. Bioinformatics (Oxford, England), 23(18), 2455–2462. doi:10.1093/bioinformatics/btm353

Saitoh, S. (1997). Integral Transforms, Reproducing Kernels and Their Applications. New York: Addison Wesley Longman.

Sakiyama, Y. (2009). The use of machine learning and nonlinear statistical tools for ADME prediction. Expert Opinion on Drug Metabolism & Toxicology, 5, 149–169. doi:10.1517/17425250902753261

Sanchez, C., Lachaize, C., Janody, F., Bellon, B., Röder, L., & Euzenat, J. (1999). Grasping at molecular interactions and genetic networks in Drosophila melanogaster using FlyNets, an internet database. Nucleic Acids Research, 27(1), 89–94. doi:10.1093/nar/27.1.89

Saunders, C., Hardoon, D., Shawe-Taylor, J., & Widmer, G. (2008). Using String Kernels to Identify Famous Performers from their Playing Style. Intelligent Data Analysis, 12(4), 425–440.

Schapire, R. E. (1999a). A brief introduction to boosting. In Proceedings of the sixteenth international conference on artificial intelligence (pp. 1401–1406).

Schapire, R. E. (1999b). Theoretical views of boosting. In European conference on computational learning theory (pp. 1–10).

Schapire, R. E. (1999c). Theoretical views of boosting and applications . In Tenth international conference on algorithmic learning theory (pp. 13–25). New York: Springer-Verlag. doi:10.1007/3-540-46769-6_2

Schapire, R. E., Freund, Y., Barlett, P., & Lee, W. S. (1998). Boosting the margin: A new explanation for the effectiveness of voting methods. Annals of Statistics, 5(26), 1651–1686.

Schemmel, J. Br¨uderle, D., Meier, K., & Ostendorf, B. (2007). Modeling synaptic plasticity within networks of highly accelerated I&F neurons. In Proceedings of the 2007 IEEE International Symposium on Circuits and Systems (iscas'07). Washington, DC: IEEE Press.

Schietgat, L., Ramon, J., Bruynooghe, M., & Blockeel, H. (2008). An Efficiently Computable Graph-Based Metric for the Classification of Small Molecules. In Discovery Science. 197-209.

Schmuker, M., & Schneider, G. (2007, Dec). Processing and classification of chemical data inspired by insect olfaction. Proceedings of the National Academy of Sciences of the United States of America, 104(51), 20285–20289. Available from http://dx.doi.org/10.1073/pnas.0705683104. doi:10.1073/pnas.0705683104

Schmuker, M., & Schwarte, F. Br¨uck, A., Proschak, E., Tanrikulu, Y., Givehchi, A., et al. (2007, January). SOMMER: self-organising maps for education and research. Journal of Molecular Modeling, 13(1), 225–228. Available from http://dx.doi.org/10.1007/s00894-006-0140-0

Schmuker, M., de Bruyne, M., H¨ahnel, M., & Schneider, G. (2007). Predicting olfactory receptor neuron responses from odorant structure. Chemistry Central Journal, 1, 11. Available from http://dx.doi.org/10.1186/1752-153X-1-11. doi:10.1186/1752-153X-1-11

Schneider, G., & Baringhaus, K.-H. (2008). Molecular design. Weinheim: Wiley-VCH.

Schneider, G., & Downs, G. (2003). Machine learning methods in QSAR modelling. QSAR & Combinatorial Science, 22, 485–486. doi:10.1002/qsar.200390046

Schneider, G., Neidhart, W., Giller, T., & Schmid, G. (1999). "Scaffold-Hopping" by Topological Pharmacophore Search: A Contribution to Virtual Screening. Angewandte Chemie International Edition in English, 38, 2894–2895. doi:10.1002/(SICI)1521-3773(19991004)38:19<2894::AID-ANIE2894>3.0.CO;2-F

Schneider, G., Tanrikulu, Y., & Schneider, P. (2009). (in press). Self-organizing molecular fingerprints: a ligand-based view on druglike chemical space and off-target prediction. Future Medicinal Chemistry. doi:10.4155/fmc.09.11

Schneider, P., & Schneider, G. (2003). Collection of bioactive reference compounds for focused library design. QSAR & Combinatorial Science, 22(7), 713–718. Available from http://dx.doi.org/10.1002/qsar.20033082514. doi:10.1002/qsar.200330825

Schneider, P., Tanrikulu, Y., & Schneider, G. (2009). Self-organizing maps in drug discovery: compound library design, scaffold-hopping, repurposing. Current Medicinal Chemistry, 16(3), 258–266. doi:10.2174/092986709787002655

Schölkopf, B., & Smola, A. J. (2002). Learning with kernels: Support vector machines, regularization, optimization, and beyond. Cambridge, MA: MIT Press.

Schölkopf, B., Smola, A., & Müller, K. (1998). Nonlinear Component Analysis as a Kernel Eigenvalue Problem. Neural Computation, 10, 1299–1319.

Scholkopf, B., Smola, A., & Muller, K.-R. (1998). Nonlinear component analysis as a kernel eigenvalue problem. Neural Computation, 10, 1299–1319. doi:10.1162/089976698300017467

Schölkopf, B., Tsuda, K., & Vert, J.-P. (Eds.). (2004). Kernel methods in computational biology. Cambridge, MA: The MIT Press.

Schomburg, I., Chang, A., Ebeling, C., Gremse, M., Heldt, C., & Huhn, G. (2004). Brenda, the enzyme database: updates and major new developments. Nucleic Acids Research, 32, D431–D433. doi:10.1093/nar/gkh081

Searson, D., Willis, M., & Montague, G. (2007). Co-evolution of non-linear PLS model components. Journal of Chemometrics, 21(12), 592–603.

Sebag, M., & Rouveirol, C. (1997). Tractable induction and classification in FOL via stochastic matching. In . Proceedings, IJCAI-97, 888–892.

Seber, G. A. F., & Lee, A. J. (2003). Linear Regression Analysis (2nd ed.). New York: Wiley-Interscience.

Shannon, C. E., & Weaver, W. (1963). The mathematical theory of communication. Urbana and Chicago, IL: University of Illinois Press.

Shawe-Taylor, J., & Cristianini, N. (2004). Kernel methods for pattern analysis. New York: Cambridge University Press.

Shen, M., Beguin, C., Golbraikh, A., Stables, J. P., Kohn, H., & Tropsha, A. (2004). Application of predictive QSAR models to database mining: Identification and experimental validation of novel anticonvulsant compounds. Journal of Medicinal Chemistry, 47, 2356–2364. doi:10.1021/jm030584q

Shen, Q., Jiang, J., Shen, G., & Yu, R. (2006). Ridge estimated orthogonal signal correction for data preprocessing prior to PLS modeling: QSAR studies of cyclooxygenase-2 inhibitors. Chemometrics and Intelligent Laboratory Systems, 82(1-2), 44–49.

Shen, Q., Jiang, J., Tao, J., Shen, G., & Yu, R. (2005). Modified Ant Colony Optimization Algorithm for Variable Selection in QSAR Modeling: QSAR Studies of Cyclooxygenase Inhibitors. Journal of Chemical Information and Modeling, 45(4), 1024–1029.

Sheng, Y. E., Xicheng, W., Jie, L., & Chunlian, L. (2003). A New Algorithm For Graph Isomorphism And Its Parallel Implementation. International Conference on Parallel Algorithms and Computing Environments ICPACE. Hong Kong, China.

Sheridan, R. P. (2007). Chemical similarity searches: when is complexity justified? Expert Opinion on Drug Discovery, 2, 423–430. doi:10.1517/17460441.2.4.423

Sheridan, R. P., & Kearsley, S. K. (2002). Why do we need so many chemical similarity search methods? Drug Discovery Today, 7, 903–911. doi:10.1016/S1359-6446(02)02411-X

Sheridan, R. P., Miller, M. D., Underwood, D. J., & Kearsley, S. K. (1996). Chemical similarity using geometric atom pair descriptors. Journal of Chemical Information and Computer Sciences, 36, 128–136. doi:10.1021/ci950275b

Sherman, D. B., Zhang, S., Pitner, J. B., & Tropsha, A. (2004). Evaluation of the relative stability of liganded versus ligand-free protein conformations using simplicial neighborhood analysis of protein packing (SNAPP) method. Proteins, 56, 828–838. doi:10.1002/prot.20131

Shigemizu, D., Araki, M., Okuda, S., Goto, S., & Kanehisa, M. (2009). Extraction and analysis of chemical modification patterns in drug development. Journal of Chemical Information and Modeling, 49(4), 1122–1129. doi:10.1021/ci8003804

Shimbel, A. (1953). Structural parameters of communication networks . Bulletin of Mathematical Biology, 15, 501–507.

Shindyalov, I. N., & Bourne, P. E. (1998). Protein structure alignment by incremental combinatorial extension (CE) of the optimal path. Protein Engineering, 11, 739–747.

Sigma Aldrich. (2004). Flavors and fragrances catalog. Milwaukee, WI.

Singh, R. (2007). Surface similarity-based molecular query-retrival. BMC Cell Biology, 8(supplement 1), S6. doi:10.1186/1471-2121-8-S1-S6

Singh, R. K., Tropsha, A., & Vaisman, I. I. (1996). Delaunay tessellation of proteins: four body nearest-neighbor propensities of amino acid residues. Journal of Computational Biology, 3, 213–221. doi:10.1089/cmb.1996.3.213

Sjöström, M., Rännar, S., & Wieslander, Å. (1995). Polypeptide sequence property relationships in Escherichia coli based on auto cross covariances. Chemometrics and Intelligent Laboratory Systems, 29, 295–305. doi:10.1016/0169-7439(95)00059-1

Smith, T., & Waterman, M. (1981). Identification of common molecular subsequences. Journal of Molecular Biology, 147, 195–197. doi:10.1016/0022-2836(81)90087-5

Smola, A., Schölkopf, B., & Müller, K. (1998). The connection between regularization operators and support vector kernels. Neural Networks, 11, 637–649.

Srinivasan, A., & King, R. (1999). Feature construction with inductive logic programming: A study of quantitative oredictions of biological activity aided by structural attributes. Knowledge Discovery and Data Mining Journal, 3(1), 37–57. doi:10.1023/A:1009815821645

Srinivasan, A., Muggleton, S., King, R. D., & Sternberg, M. (1996). Theories for mutagenicity: a study of first-order and feature based induction. Artificial Intelligence, 85(1-2), 277–299. doi:10.1016/0004-3702(95)00122-0

Stahura, F. L., & Bajorath, J. (2005). New methodologies for ligand-based virtual screening. Current Pharmaceutical Design, 11, 1189–1202. doi:10.2174/1381612053507549

Stanimirova, I., Zehl, K., Massart, D. L., Heyden, Y., & Einax, J. W. (2006). Chemometric analysis of soil pollution data using the Tucker N-way method. Analytical and Bioanalytical Chemistry, 385(4), 771–779.

Stark, A., & Russell, R. B. (2003). Annotation in three dimensions. PINTS: Patterns in Non-homologous Tertiary Structures. Nucleic Acids Research, 31, 3341–3344.

Stenberg, P., Luthman, K., & Artursson, P. (2000). Virtual screening of intestinal drug permeability. Journal of Controlled Release, 65(1-2), 231–243. doi:10.1016/S0168-3659(99)00239-4

Stockwell, B. (2000). Chemical genetics: ligand-based discovery of gene function. Nature Reviews. Genetics, 1, 116–125. doi:10.1038/35038557

Storm, C. E. V., & Sonnhammer, E. L. L. (2002). Automated ortholog inference from phylogenetic trees and calculation of orthology reliability. Bioinformatics (Oxford, England), 18, 92–99.

Strogatz, S. H. (2001). Exploring complex networks. Nature, 410, 268–276.

Suffness, M., & Cordell, G. A. (1985). The Alkaloids, Chemistry and Pharmacology (Vol. 25). New York: Academic Press.

Sutherland, J. J., O'Brien, L. A., & Weaver, D. F. (2003). Spline-fitting with a genetic algorithm: a method for developing classification structure-activity relationships. Journal of Chemical Information and Computer Sciences, 43(6), 1906–1915. doi:10.1021/ci034143r

Sutter, J. M., Peterson, T. A., & Jurs, P. C. (1997). Prediction of Gas Chromatographic Retention Indices of Alkylbenzene. Analytica Chimica Acta, 342, 113–122. doi:10.1016/S0003-2670(96)00578-8

Swain, M., & Ballard, D. (1991). Color indexing. International Journal of Computer Vision, 7(1), 11–32. doi:10.1007/BF00130487

Swamidass, S. J., Chen, J., Bruand, J., Phung, P., Ralaivola, L., & Baldi, P. (2005). Kernels for small molecules and the prediction of mutagenicity, toxicity and anti-cancer activity. Bioinformatics (Oxford, England), 21(Suppl. 1), i359–i368. doi:10.1093/bioinformatics/bti1055

Swets, J. A. (1988, June). Measuring the accuracy of diagnostic systems. Science, 240(4857), 1285–1293. doi:10.1126/science.3287615

Tabarakia, R., Khayamiana, T., & Ensafia, A. A. (2006, September). Wavelet neural network modeling in QSPR for prediction of solubility of 25 anthraquinone dyes at di®erent temperatures and pressures in supercritical carbon dioxide. Journal of Molecular Graphics & Modelling, 25(1), 46–54. doi:10.1016/j.jmgm.2005.10.012

Takahashi, Y., Maeda, S., & Sasaki, S. (1987). Automated recognition of common geometrical patterns among a variety of three-dimensional molecular structures. Analytica Chimica Acta, 200(11), 363–377. doi:10.1016/S0003-2670(00)83783-6

Talafous, J., Sayre, L. M., Mieyal, J. J., & Klopman, G. (1994). META.2. A dictionary model of mammalian xenobiotic metabolism. Journal of Chemical Information and Computer Sciences, 34(6), 1326–1333. doi:10.1021/ci00022a015

Tang, K., & Li, T. (2002). Combining PLS with GA-GP for QSAR. Chemometrics and Intelligent Laboratory Systems, 64(1), 55–64.

Taylor, P., Blackburn, E., Sheng, Y. G., Harding, S., Hsin, K.-Y., & Kan, D. (2008). Ligand discovery and virtual screening using the program LIDAEUS. British Journal of Pharmacology, 153, S55S67. doi:10.1038/sj.bjp.0707532

Taylor, T. J., & Vaisman, I. I. (2006). Graph theoretic properties of networks formed by the Delaunay tessellation of protein structures. Physical Review E: Statistical, Nonlinear, and Soft Matter Physics, 73, 041925–041913.

Tetko, I. V., Gasteiger, J., Todeschini, R., Mauri, A., Livingstone, D., & Ertl, P. (2005). Virtual computational chemistry laboratory--design and description. Journal of Computer-Aided Molecular Design, 19, 453–463. doi:10.1007/s10822-005-8694-y

Thornton, J. M., Todd, A. E., Milburn, D., Borkakoti, N., & Orengo, C. A. (2000). From structure to function: Approaches and limitations. Nature Structural & Molecular Biology, 7, 991–994.

Tibshrani, R. (1996). Regression shrinkage and selection via the LASSO. Journal of the Royal Statistical Society. Series B. Methodological, 58(1), 267–288.

Tikhonov, A. (1963). On solving ill-posed problem and method of regularization. Doklady Akademii Nauk USSR, 153, 501–504.

Tipton, K. F., & Boyce, S. (2005). Enzyme classification and nomenclature . In John Wiley & Sons, Inc., Encyclopedia of Life Sciences. Chichester, England: Wiley.

Todeschini, R. (1997). Data Correlation, Number of Significant Principal Components and Shape of Molecules. The K Correlation Index. Analytica Chimica Acta, 348, 419–430. doi:10.1016/S0003-2670(97)00290-0

Todeschini, R., & Consonni, V. (2002). Handbook of molecular descriptors. New York: Wiley-VCH.

Todeschini, R., & Consonni, V. (2003). Descriptors from Molecular Geometry . In Gasteiger, J. (Ed.), Handbook of Chemoinformatics (pp. 1004–1033). Weinheim, Germany: Wiley-VCH. doi:10.1002/9783527618279.ch37

Todeschini, R., & Consonni, V. (2009). Molecular Descriptors for Chemoinformatics. Weinheim, Germany: Wiley-VCH GmbH.

Todeschini, R., Ballabio, D., Consonni, V., Manganaro, A., & Mauri, A. Canonical Measure of Correlation (CMC) and Canonical Measure of Distance (CMD) between sets of data. Part 1. Theory and simple chemometric applications. Analytica Chimica Acta, 648, 45–51. doi:10.1016/j.aca.2009.06.032

Todeschini, R., Consonni, V., Mauri, A., & Pavan, M. (2004). Detecting "bad" regression models: multicriteria fitness functions in regression analysis. Analytica Chimica Acta, 515, 199–208. doi:10.1016/j.aca.2003.12.010

Todeschini, R., Gramatica, P., & Navas, N. (1998). 3D-modeling and predication by WHIM descriptors. Part 9. Chromatographic relative retentation time and physicochemical properties of polychlorinated biphenyls (PCBs). Chemometrics and Intelligent Laboratory Systems, 40(1), 53–63. doi:10.1016/S0169-7439(97)00079-8

Todeschini, R., Lasagni, M., & Marengo, E. (1994). New molecular descriptors for 2D and 3D structures. Theory. Journal of Chemometrics, 8(4), 263–272. doi:10.1002/cem.1180080405

Tong, J., & Liu, S. (2008). Three-Dimensional Holographic Vector of atomic interaction field applied in QSAR of anti-HIV HEPT analogues. QSAR & Combinatorial Science, 27(3), 330–337.

Tonga, J., Liu, S., Zhou, P., Wu, B., & Li, Z. (2008, July). A novel descriptor of amino acids and its application in peptide QSAR. Journal of Theoretical Biology, 253(1), 90–97. doi:10.1016/j.jtbi.2008.02.030

Topliss, J. G., & Edwards, R. P. (1979). Chance Factors in Studies of Quantitative Structure-Activity Relationships. Journal of Medicinal Chemistry, 22, 1238–1244. doi:10.1021/jm00196a017

Tralau-Stewart, C. J., Wyatt, C. A., Kleyn, D. E., & Ayad, A. (2009). Drug discovery: new models for industry-academic partnerships. Drug Discovery Today, 14, 95–101. doi:10.1016/j.drudis.2008.10.003

Trejo, L., Rosipal, R., & Matthews, B. (2006). Brain-Computer Interfaces for 1-D and 2-D Cursor Control: Designs using Volitional Control of the EEG Spectrum or Steady-State Visual Evoked Potentials. IEEE Transactions on Neural Systems and Rehabilitation Engineering, 14(2), 225–229.

Trinajstic, N. (Ed.). (1992). Chemical Graph Theory. Boca Raton, FL: CRC Press.

Tropsha, A. (2005). Application of Predictive QSAR Models to Database Mining . In Oprea, T. (Ed.), Cheminformatics in Drug Discovery (pp. 437–455). New York: Wiley-VCH. doi:10.1002/3527603743.ch16

Tropsha, A. (2006). Predictive QSAR (Quantitative Structure Activity Relationships) Modeling . In Martin, Y. C. (Ed.), Comprehensive Medicinal Chemistry II (pp. 113–126). New York: Elsevier.

Tropsha, A. (2006). Variable selection QSAR modeling, model validation, and virtual screening. Annual Reports in Computational Chemistry, 2, 113–126.

Tropsha, A., & Golbraikh, A. (2007). Predictive QSAR modeling workflow, model applicability domains, and virtual screening. Current Pharmaceutical Design, 13, 3494–3504. doi:10.2174/138161207782794257

Tropsha, A., & Zheng, W. (2001). Computer Assisted Drug Design . In MacKerell, O. B. B. R. M. W. A. Jr., (Ed.), Computational Biochemistry and Biophysics (pp. 351–369). New York: Marcel Dekker, Inc. doi:10.1201/9780203903827.ch16

Tropsha, A., & Zheng, W. (2001). Identification of the descriptor pharmacophores using variable selection QSAR: applications to database mining. Current Pharmaceutical Design, 7(7), 599–612.

Tropsha, A., Gramatica, P., & Gombar, V. K. (2003). The importance of being earnest: Validation is the absolute essential for successful application and interpretation of QSPR models. QSAR & Combinatorial Science, 22, 69–77. doi:10.1002/qsar.200390007

Tropsha, A., Singh, R. K., Vaisman, I. I., & Zheng, W. (1996). Statistical geometry analysis of proteins: implications for inverted structure prediction. Pacific Symposium on Biocomputing. Pacific Symposium on Biocomputing, 614–623.

Truchon, J.-F., & Bayly, C. I. (2007). Evaluating virtual screening methods: good and bad metrics for the "early recognition" problem. Journal of Chemical Information and Modeling, 47, 488–508. doi:10.1021/ci600426e

Trygg, J., & Wold, S. (2002). Orthogonal projections to latent structures (O-PLS). Journal of Chemometrics, 16(3), 119–128.

Trygg, J., & Wold, S. (2003). O2-PLS, a two-block (X-Y) latent variable regression (LVR) method with an integral OSC filter. Journal of Chemometrics, 17(1), 53–64.

Trygg, J., Holmes, E., & Lundstedt, T. (2007). Chemometrics in Metabonomics. Journal of Proteome Research, 6(2), 469–479.

Tsuda, K., & Kudo, T. (2006). Clustering graphs by weighted substructure mining. In Proceedings of the 23rd international conference on machine learning (p. 953-960). New York: ACM Press.

Tsuda, K., & Kurihara, K. (2008). Graph mining with variational dirichlet process mixture models. In Siam conference on data mining (sdm).

Tweeddale, H., Notley-McRobb, L., & Ferenci, T. (1998). Effect of slow growth on metabolism of Escherichia coli, as revealed by global metabolite pool ("metabolome") analysis. Journal of Bacteriology, 180(19), 5109–5116.

Uchida, N., Takahashi, Y. K., Tanifuji, M., & Mori, K. (2000). Odor maps in the mammalian olfactory bulb: domain organization and odorant structural features. Nature Neuroscience, 3, 1035–1043. doi:10.1038/79857

Ullmann, J. R. (1976). An Algorithm for Subgraph Isomorphism. [JACM]. Journal of the ACM, 23, 31–42.

Uno, T., Kiyomi, M., & Arimura, H. (2005). LCM ver.3: collaboration of array, bitmap and prefix tree for frequent itemset mining. In Osdm '05: Proceedings of the 1st international workshop on open source data mining (pp. 77–86).

van de Waterbeemd, H., & Gifford, E. (2003). ADMET In Silico Modelling: Towards Prediction Paradise? Nature Reviews. Drug Discovery, 2, 192–204. doi:10.1038/nrd1032

van de Waterbeemd, H., Testa, B., & Folkers, G. (Eds.). (1997). Computer-Assisted Lead Finding and Optimization. Weinheim, Germany: Wiley-VCH. doi:10.1002/9783906390406

Vapnik, V. (2000). The nature of statistical learning theory. New York: Springer.

Vapnik, V., & Chervonenkis, A. (1971). On the uniform convergence of relative frequencies of events to their probabilities. Theory of Probability and Its Applications, 16(2), 264–280.

Vapnik, V., & Chervonenkis, A. (1974). Theory of Pattern Recognition [in Russian]. Nauka, Moscow. ((German Translation: W.N. Vapnik and A.J. Cherwonenkis (1979). Theorie der Zeichenerkennung. Akademia-Verlag, Berlin))

Varmuza, K., Demuth, W., Karlovits, M., & Scsibrany, H. (2005). Binary Substructure Descriptors for Organic Compounds. Croatica Chemica Acta, 78, 141–149.

Vassura, M., Margara, L., Di Lena, P., Medri, F., Fariselli, P., & Casadio, R. (2008). FT-COMAR: fault tolerant three-dimensional structure reconstruction from protein contact maps. Bioinformatics (Oxford, England), 24, 1313–1315.

Veber, D., Johnson, S., Cheng, H.-Y., Smith, B., Ward, K., & Kopple, K. (2002). Molecular properties that influence the oral bioavailability of drug candidates. Journal of Medicinal Chemistry, 45(12), 2615–2623. doi:10.1021/jm020017n

Velculescu, V. E., Zhang, L., Zhou, W., Vogelstein, J., Basrai, M. A., & Bassett, D. E. Jr (1997). Characterization of the yeast transcriptome. Cell, 88(2), 243–251. doi:10.1016/S0092-8674(00)81845-0

Vendruscolo, M., Dokholyan, N. V., Paci, E., & Karplus, M. (2002). Small-world view of the amino acids that play a key role in protein folding. Physical Review E: Statistical, Nonlinear, and Soft Matter Physics, 65.

Vert, J. P. (2008). The optimal assignment kernel is not positive definite. Retrieved from http://www.citebase.org/abstract?id=oai:arXiv.org:0801.4061

Vert, J. P., & Jacob, L. (2008). Machine learning for in silico virtual screening and chemical genomics: new strategies. Combinatorial Chemistry & High Throughput Screening, 11, 677–685. doi:10.2174/138620708785739899

Vishwanathan, S. V. N., Borgwardt, K., & Schraudolph, N. (2007). Fast computation of graph kernels. [Cambridge, MA: MIT Press.]. Advances in Neural Information Processing Systems, 19, 1449–1456.

Vishwanathan, S., & Smola, A. (2003). Fast kernels for string and tree matching . In Becker, S., Thrun, S., & Obermayer, K. (Eds.), Advances in neural information processing systems 15 (pp. 569–576). Cambridge, MA: MIT Press.

Vishwanathan, S., & Smola, A. (2004). Fast Kernels for String and Tree Matching . In Schölkopf, B., Tsuda, K., & Vert, J. P. (Eds.), Kernel methods in computational biology (pp. 113–130). Cambridge, MA: MIT Press.

Vogt, M., & Bajorath, J. (2007). Introduction of an information-theoretic method to predict recovery rates of active compounds for Bayesian in silico screening: Theory and screening trials. Journal of Chemical Information and Modeling, 47, 337–341. doi:10.1021/ci600418u

Vogt, M., & Bajorath, J. (2008a). Bayesian screening for active compounds in high-dimensional chemical spaces combining property descriptors and molecular fingerprints. Chemical Biology & Drug Design, 71, 8–14.

Vogt, M., & Bajorath, J. (2008b). Bayesian similarity searching in high-dimensional descriptor spaces combined with Kullback-Leibler descriptor divergence analysis. Journal of Chemical Information and Modeling, 48, 247–255. doi:10.1021/ci700333t

Vogt, M., Godden, J., & Bajorath, J. (2007). Bayesian interpretation of a distance function for navigating high-dimensional descriptor spaces. Journal of Chemical Information and Modeling, 47, 39–46. doi:10.1021/ci600280b

Vosshall, L. B., Wong, A. M., & Axel, R. (2000, Jul). An olfactory sensory map in the fly brain. Cell, 102(2), 147–159. doi:10.1016/S0092-8674(00)00021-0

Wagener, M., Sadowski, J., & Gasteiger, J. (1995). Autocorrelation of Molecular Surface Properties for Modeling Corticosteroid Binding Globulin and Cytosolic Ah Receptor Activity by Neural Networks. Journal of the American Chemical Society, 117, 7769–7775. doi:10.1021/ja00134a023

Wahba, G. (1987). Spline models for observational data . In Cbms-nsf regional conference series in applied mathematics (Vol. 59). SIAM.

Wahba, G. (1990). Splines Models of Observational Data (Vol. 59). Philadelphia: SIAM.

Wale, N., & Karypis, G. (2006). Comparison of descriptor spaces for chemical compound retrieval and classification. In Proceedings of the 2006 ieee international conference on data mining (pp. 678–689).

Walters, W. P., Stahl, M. T., & Murcko, M. A. (1998). Virtual screening - an overview. Drug Discovery Today, 3, 160–178. doi:10.1016/S1359-6446(97)01163-X

Wanchana, S., Yamashita, F., & Hashida, M. (2003). QSAR Analysis of the Inhibition of Recombinant CYP 3A4 Activity by Structurally Diverse Compounds Using a Genetic Algorithm-Combined Partial Least Squares Method. Pharmaceutical Research, 20(9), 1401–1408.

Wang, G., & Dunbrack, R. L. Jr. (2003). PISCES: a protein sequence culling server. Bioinformatics (Oxford, England), 19, 1589–1591.

Wang, K., Fain, B., Levitt, M., & Samudrala, R. (2004). Improved protein structure selection using decoy-dependent discriminatory functions. Bioinformatics (Oxford, England), 4, 8.

Wang, R., Fang, X., Lu, Y., & Wang, S. (2004). The PDBbind database: collection of binding affinities for protein-ligand complexes with known three-dimensional structures. Journal of Medicinal Chemistry, 47, 2977–2980. doi:10.1021/jm0305801

Wang, Y., Li, Y., & Wang, B. (2007). An in silico method for screening nicotine derivatives as cytochrome P450 2A6 selective inhibitors based on kernel partial least squares. International Journal of Molecular Sciences, 8(2), 166–179.

Warmuth, M. K., Liao, J., Rätsch, G., Mathieson, M., Putta, S., & Lemmen, C. (2003). Active learning with support vector machines in the drug discovery process. Journal of Chemical Information and Computer Sciences, 43, 667–673. doi:10.1021/ci025620t

Washio, T., & Motoda, H. (2003). State of the art of graph-based data mining. SIGKDD Explorations, 5(1), 59–68. doi:10.1145/959242.959249

Watson, G. A. (1983). An algorithm for the single facility location problem using the Jaccard metric. SIAM Journal on Scientific and Statistical Computing, 4(4), 748–756. doi:10.1137/0904052

Wegelin, J. (2000). A survey of Partial Least Squares (PLS) methods, with emphasis on the two-block case (Tech. Rep.). Seattle: Department of Statistics, University of Washington.

Wegner, J. K. (2006). Data Mining und Graph Mining auf molekularen Graphen - Cheminformatik und molekulare Kodierungen fï¿œr ADME/Tox & QSAR-Analysen. Unpublished doctoral dissertation, Eberhard-Karls Universität Tübingen.

Wegner, J., Fröhlich, H., & Zell, A. (2003a). Feature selection for Descriptor based Classification Models: Part II - Human Intestinal Absorption (HIA). Journal of Chemical Information and Computer Sciences, 44, 931–939. doi:10.1021/ci034233w

Wegner, J., Fröhlich, H., & Zell, A. (2003b). Feature Selection for Descriptor based Classificiation Models: Part I - Theory and GA-SEC Algorithm. Journal of Chemical Information and Computer Sciences, 44, 921–930. doi:10.1021/ci0342324

Wei, L., Brossi, A., Kendall, R., Bastow, K. F., Morris-Natschke, S. L., & Shi, Q. (2006). Antitumor agents 251: synthesis, cytotoxic evaluation, and structure-activity relationship studies of phenanthrene-based tylophorine derivatives (PBTs) as a new class of antitumor agents. Bioorganic & Medicinal Chemistry, 14, 6560–6569. doi:10.1016/j.bmc.2006.06.009

Weinhold, N., Sander, O., Domingues, F. S., Lengauer, T., & Sommer, I. (2008). Local Function Conservation in Sequence and Structure Space. PLoS Computational Biology, 4, e1000105.

Weisel, M., Proschak, E., Kriegl, J. M., & Schneider, G. (2009, Jan). Form follows function: shape analysis of protein cavities for receptor-based drug design. Proteomics, 9(2), 451–459. Available from http://dx.doi.org/10.1002/pmic.200800092. doi:10.1002/pmic.200800092

Wessel, M. D., Jurs, P. C., Tolan, J. W., & Muskal, S. M. (1998). Prediction of Human Intestinal Absorption of Drug Compounds from Molecular Structure. Journal of Chemical Information and Computer Sciences, 38, 726–735. doi:10.1021/ci980029a

Westerhuis, J. A., Kourti, T., & Macgregor, J. F. (1998). Analysis of multiblock and hierarchical PCA and PLS models. Journal of Chemometrics, 12(5), 301–321.

Wettschereck, D., Aha, D. W., & Mohri, T. (1997). A review and empirical evaluation of feature weighting methods for a class of lazy learning algorithms. Artificial Intelligence Review, 11, 273–314. doi:10.1023/A:1006593614256

Wheeler, D., Barrett, T., Benson, D., Bryant, S., Canese, K., & Chetvernin, V. (2006). Database resources of the national center for biotechnology information. Nucleic Acids Research, 34, D173–D180. doi:10.1093/nar/gkj158

Whittle, M., Gillet, V. J., Willett, P., & Loesel, J. (2006a). Analysis of data fusion methods in virtual screening: theoretical model. Journal of Chemical Information and Modeling, 46, 2193–2205. doi:10.1021/ci049615w

Whittle, M., Gillet, V. J., Willett, P., & Loesel, J. (2006b). Analysis of data fusion methods in virtual screening: similarity and group fusion. Journal of Chemical Information and Modeling, 46, 2206–2219. doi:10.1021/ci0496144

Wiener, H. (1947). Influence of Interatomic Forces on Paraffin Properties. The Journal of Chemical Physics, 15, 766. doi:10.1063/1.1746328

Wiklund, S., Nilsson, D., Eriksson, L., Sjoestrom, M., Wold, S., & Faber, K. (2007). A randomization test for PLS component selection. Journal of Chemometrics, 21(10-11), 427–439.

Wildman, S. A., & Crippen, G. M. (1999). Prediction of Physicochemical Parameters by Atomic Contributions. Journal of Chemical Information and Computer Sciences, 39, 868–873. doi:10.1021/ci990307l

Wilkins, M. R., Sanchez, J. C., Gooley, A. A., Appel, R. D., Humphery-Smith, I., Hochstrasser, D. F., & Williams, K. L. (1996). Progress with proteome projects: why all proteins expressed by a genome should be identified and how to do it. Biotechnology & Genetic Engineering Reviews, 13, 19–50.

Willett, P. (1987). Similarity and clustering in chemical information systems. Letchworth: Research Studies Press.

Willett, P. (1998). Chemical similarity searching. Journal of Chemical Information and Computer Sciences, 38, 983–996. doi:10.1021/ci9800211

Willett, P. (2005). Searching techniques for databases of two- and three-dimensional chemical structures. Journal of Medicinal Chemistry, 48, 4183–4199. doi:10.1021/jm0582165

Willett, P. (2006a). Similarity-based virtual screening using 2D fingerprints. Drug Discovery Today, 11, 1046–1053. doi:10.1016/j.drudis.2006.10.005

Willett, P. (2006b). Data fusion in ligand-based virtual screening. QSAR & Combinatorial Science, 25, 1143–1152. doi:10.1002/qsar.200610084

Willett, P. (2009). Similarity methods in chemoinformatics. Annual Review of Information Science & Technology, 43, 3–71.

Wilson, C. A., Kreychman, J., & Gerstein, M. (2000). Assessing annotation transfer for genomics: quantifying the relations between protein sequence, structure and function through traditional and probabilistic scores. Journal of Molecular Biology, 297, 233–249.

Wilson, D., Irwin, G., & Lightbody, G. (1997). Nonlinear PLS modeling using radial basis functions. In American Control Conference. Albuquerque, New Mexico.

Wilton, D. J., Harrison, R. F., Willett, P., Delaney, J., Lawson, K., & Mullier, G. (2006). Virtual screening using binary kernel discrimination: Analysis of pesticide data. Journal of Chemical Information and Modeling, 46, 471–477. doi:10.1021/ci050397w

Wilton, D., Willett, P., Lawson, K., & Mullier, G. (2003). Comparison of ranking methods for virtual screening in lead-discovery programs. Journal of Chemical Information and Computer Sciences, 43, 469–474. doi:10.1021/ci025586i

Wishart, D., Knox, C., Guo, A., Cheng, D., Shrivastava, S., & Tzur, D. (2007). Drugbank: A knowledgebase for drugs, drug actions and drug targets. Nucleic Acids Res.

Witten, I. H., & Frank, E. (2005). Data mining: Practical machine learning tools and techniques (2nd ed.). San Francisco: Morgan Kaufmann.

Wold, H. (1966). Estimation of principal components and related models by iterative least squares. In P. R. Krishnaiaah (Ed.), Multivariate analysis (pp. 391–420). Maryland Heights, MO: Academic Press.

Wold, H. (1975). Path models with latent variables: The NIPALS approach. In H. B., et al. (Ed.), Quantitative Sociology: International perspectives on mathematical and statistical model building (pp. 307–357). Maryland Hieghts, MO: Academic Press.

Wold, H. (1985). Partial least squares. In Kotz, S., & Johnson, N. (Eds.), Encyclopedia of the Statistical Sciences (Vol. 6, pp. 581–591). New York: John Wiley & Sons.

Wold, S. (1992). Nonlinear partial least squares modeling II. Spline inner relation. Chemometrics and Intelligent Laboratory Systems, 14, 71–84.

Wold, S., Albano, C., Wold, H., Dunn, W. III, Edlund, U., & Esbensen, K. (1984). Multivariate data analysis in chemistry . In Kowalski, B. (Ed.), Chemometrics. Mathematics and Statistics in Chemistry. The Netherlands: Reidel. Dordrecht.

Wold, S., Antti, H., Lindgren, F. & Ohman, J. (1998). Orthogonal signal correction of near-infrared spectra. Chemometrics and Intelligent Laboratory Systems, 44(1,2), 175-185.

Wold, S., Jonsson, J., Sjöström, M., Sandberg, M., & Rännar, S. (1993). DNA and peptide sequences and chemical processes multivariately modelled by principal component analysis and partial least-squares projections to latent structures. Analytica Chimica Acta, 277, 239–253. doi:10.1016/0003-2670(93)80437-P

Wold, S., Kettaneh-Wold, N., & Skagerberg, B. (1989). Nonlinear PLS modeling. Chemometrics and Intelligent Laboratory Systems, 7(1-2), 53–65.

Wold, S., Ruhe, A., Wold, H., & Dunn, W. J. (1984). The Collinearity Problem in Linear-Regression - the Partial Least-Squares (Pls) Approach to Generalized Inverses. Siam Journal on Scientific and Statistical Computing, 5, 735–743. doi:10.1137/0905052

Wold, S., Ruhe, A., Wold, H., & Dunn, W. J. III. (1984). The collinearity problem in linear regression. the partial least squares (PLS) approach to generalized inverses. SIAM J. Sci. Stat. Comput., 5(3), 735–743. doi:10.1137/0905052

Wold, S., Sjöstöm, M., & Erikkson, L. (2001). PLS-regression: a basic tool of chemometrics. Chemometrics and Intelligent Laboratory Systems, 58, 109–130. doi:10.1016/S0169-7439(01)00155-1

Wold, S., Trygg, J., Berglund, A., & Antti, H. (2001). Some recent developments in PLS modeling. Chemometrics and Intelligent Laboratory Systems, 58(2), 131–150.

Woo, H. J., & Roux, B. (2005). Calculation of absolute proteinligand binding free energy from computer simulations. Proceedings of the National Academy of Sciences of the United States of America, 102(19), 6825–6830. doi:10.1073/pnas.0409005102

Worsley, K. (1997). An overview and some new developments in the statistical analysis of PET and fMRI data. Human Brain Mapping, 5, 254–258.

Wu, W., Massarat, D., & De Jong, S. (1997). The kernel PCA algorithms for wide data. Part I: theory and algorithms. Chemometrics and Intelligent Laboratory Systems, 36, 165–172.

Xu, L., Liang, G., Li, Z., Wang, J., & Zhou, P. (2008). Three-dimensional holographic vector of atomic interaction field for quantitative structure-activity relationship of Aza-bioisosteres of anthrapyrazoles (Aza-APs). Journal of Molecular Graphics & Modelling, 26(8), 1252–1258.

Xue, C. X., Zhang, R. S., Liu, H. X., Yao, X. J., Liu, M. C., & Hu, Z. D. (2004). An accurate QSPR study of O-H bond dissociation energy in substituted phenols based on support vector machines. Journal of Chemical Information and Computer Sciences, 44, 669–677. doi:10.1021/ci034248u

Xue, L., & Bajorath, J. (2000). Molecular descriptors for effective classification of biologically active compounds based on principal component analysis identified by a genetic algorithm. Journal of Chemical Information and Computer Sciences, 40, 801–809. doi:10.1021/ci000322m

Xue, L., & Bajorath, J. (2000). Molecular descriptors in chemoinformatics, computational combinatorial chemistry, and virtual screening. Combinatorial Chemistry & High Throughput Screening, 3(5), 363–372.

Xue, Y., Li, Z. R., Yap, C. W., Sun, L. Z., & Chen, X. (2004). Effect of molecular descriptor feature selection in support vector machine classification of pharmacokinetic and toxicological properties of chemical agents. Journal of Chemical Information and Computer Sciences, 44(5), 1630–1638. doi:10.1021/ci049869h

Yamanishi, Y. (2009). Supervised bipartite graph inference . In Koller, D., Schuurmans, D., Bengio, Y., & Bottou, L. (Eds.), Adv. Neural Inform. Process. Syst. 21 (pp. 1841–1848). Cambridge, MA: MIT Press.

Yamanishi, Y., Araki, M., Gutteridge, A., Honda, W., & Kanehisa, M. (2008). Prediction of drug-target interaction networks from the integration of chemical and genomic spaces. Bioinformatics (Oxford, England), 24, i232–i240. doi:10.1093/bioinformatics/btn162

Yamanishi, Y., Hattori, M., Kotera, M., Goto, S., & Kanehisa, M. (2009). E-zyme: predicting potential EC numbers from the chemical transformation pattern of substrate-product pairs. Bioinformatics (Oxford, England), 25(12), i179–i186. doi:10.1093/bioinformatics/btp223

Yamanishi, Y., Mihara, H., Osaki, M., Muramatsu, H., Esaki, N., & Sato, T. (2007). Prediction of missing enzyme genes in a bacterial metabolic network. The FEBS Journal, 2262–2273. doi:10.1111/j.1742-4658.2007.05763.x

Yamanishi, Y., Vert, J.-P., & Kanehisa, M. (2004). Heterogeneous data comparison and gene selection with kernel canonical correlation analysis . In Scholkopf, B., Tsuda, K., & Vert, J.-P. (Eds.), Kernel Methods in Computational Biology (pp. 209–230). Cambridge, MA: MIT Press.

Yamashita, F., Fujiwara, S., Wanchana, S., & Hashida, M. (2006). Quantitative structure/activity relationship modelling of pharmacokinetic properties using genetic algorithm-combined partial least squares method. Journal of Drug Targeting, 14(7), 496–504.

Yamashita, F., Wanchana, S., & Hashida, M. (2002). Quantitative structure/property relationship analysis of Caco-2 permeability using a genetic algorithm-based partial least squares method. Journal of Pharmaceutical Sciences, 91(10), 2230–2239.

Yan, X., & Han, J. (2002a). gSpan: graph-based substructure pattern mining. In Proceedings of the 2002 ieee international conference on data mining (p. 721-724). IEEE Computer Society.

Yan, X., & Han, J. (2003). CloseGraph: mining closed frequent graph patterns. ACM International Conference on Knowledge Discovery and Data Mining (pp. 286-295). New York: ACM Press.

Yan, X., Yu, P. S., & Han, J. (2004). Graph indexing: a frequent structure-based approach. In Proceedings of the acm sigmod international conference on management of data (p. 335-346).

Yao, X. J., Panaye, A., Doucet, J. P., Zhang, R. S., Chen, H. F., & Liu, M. C. (2004). Comparative study of QSAR/QSPR correlations using support vector machines, radial basis function neural networks, and multiple linear regression. Journal of Chemical Information and Computer Sciences, 44, 1257–1266. doi:10.1021/ci049965i

Yap, C. W., Cai, C. Z., Xue, Y., & Chen, Y. Z. (2004). Prediction of torsade-causing potential of drugs by support vector machine approach. Toxicological Sciences, 79, 170–177. doi:10.1093/toxsci/kfh082

Yap, C. W., Li, H., Ji, Z. L., & Chen, Y. Z. (2007). Regression methods for developing QSAR and QSPR models to predict compounds of specific pharmacodynamic, pharmacokinetic and toxicological properties. Mini Reviews in Medicinal Chemistry, 7(11), 1097–1107.

Yazdanian, M., Glynn, S., Wright, J., & Hawi, A. (1998). Correlating partitioning and caco-2 cell permeability of structurally diverse small molecular weight compounds. Pharmaceutical Research, 15, 1490–1494. doi:10.1023/A:1011930411574

Yee, S. (1997). In vitro permeability across caco-2 cells (colonic) can predict in vivo (small intestinal) absorption in man - fact or myth. Pharmaceutical Research, 14, 763–766. doi:10.1023/A:1012102522787

Yi, P., Fang, X., & Qiu, M. (2008). 3D-QSAR studies of Checkpoint Kinase Weel inhibitors based on molecular docking, CoMFA and CoMSIA. European Journal of Medicinal Chemistry, 43(5), 925–938.

Yildirim, M., Goh, K., Cusick, M., Barabasi, A., & Vidal, M. (2007). Drug-target network. Nature Biotechnology, 25, 1119–1126. doi:10.1038/nbt1338

Yoshida, F., & Topliss, J. (2000). QSAR model for drug human oral bioavailability. Journal of Medicinal Chemistry, 43, 2575–2585. doi:10.1021/jm0000564

Yoshida, H., & Funatsu, K. (1997). Optimization of the inner relation function of QPLS using genetic algorithm. Journal of Chemical Information and Computer Sciences, 37(6), 1115–1121.

Yu, H. Yang, J., Wang, W., & Han, J. (2003). Discovering compact and highly discriminative features or feature combinations of drug activities using support vector machines, IEEE Computer Society Bioinformatics Conference (pp. 220-228). Washington, DC: IEEE Press.

Yuan, C., & Casasent, D. (2003). A novel support vector classifier with better rejection performance. In Proceedings of 2003 ieee computer society conference on pattern recognition and computer vision (cvpr) (pp. 419–424).

Yuan, Y., Zhang, R., Hu, R., & Ruan, X. (2009). Prediction of CCR5 receptor binding affinity of substituted 1-(3,3-diphenylpropyl)-piperidinyl amides and ureas based on the heuristic method, support vector machine and projection pursuit regression. European Journal of Medicinal Chemistry, 44, 25–34. doi:10.1016/j.ejmech.2008.03.004

Yuehua, X., & Alan, F. (2007). On learning linear ranking functions for beam search. Proceedings of the 24th international conference on Machine learning. Corvalis, Oregon: ACM. Zhang, Y. & Skolnick, J. (2005). TM-align: a protein structure alignment algorithm based on the TM-score(pp. 2302-2309).

Zaki, M. J. (2005). Efficiently mining frequent trees in a forest: algorithms and applications. In Ieee transactions on knowledge and data engineering (pp. 1021–1035).

Zhang, H. (2004). The optimality of naive Bayes. In Proceedings of the seventeenth Florida artificial intelligence research society conference (pp. 562–567). Menlo Park, CA: The AAAI Press.

Zhang, H., & Su, J. (2004). Naive Bayesian classifiers for ranking. In Lecture notes in computer science [Berlin: Springer.]. Machine Learning, ECML-04, 501–512.

Zhang, S., Golbraikh, A., & Tropsha, A. (2006b). Development of quantitative structure-binding affinity relationship models based on novel geometrical chemical descriptors of the protein-ligand interfaces. Journal of Medicinal Chemistry, 49, 2713–2724. doi:10.1021/jm050260x

Zhang, S., Golbraikh, A., Oloff, S., Kohn, H., & Tropsha, A. (2006a). A Novel Automated Lazy Learning QSAR (ALL-QSAR) Approach: Method Development, Applications, and Virtual Screening of Chemical Databases Using Validated ALL-QSAR Models. Journal of Chemical Information and Modeling, 46, 1984–1995. doi:10.1021/ci060132x

Zhang, S., Kaplan, A. H., & Tropsha, A. (2008). HIV-1 protease function and structure studies with the simplicial neighborhood analysis of protein packing method. Proteins, 73, 742–753. doi:10.1002/prot.22094

Zhang, S., Wei, L., Bastow, K., Zheng, W., Brossi, A., & Lee, K. H. (2007). Antitumor agents 252. Application of validated QSAR models to database mining: discovery of novel tylophorine derivatives as potential anticancer agents. Journal of Computer-Aided Molecular Design, 21, 97–112. doi:10.1007/s10822-007-9102-6

Zhang, X., & Firestein, S. (2002, Feb). The olfactory receptor gene superfamily of the mouse. Nature Neuroscience, 5(2), 124–133. Available from http://dx.doi.org/10.1038/nn800.

Zheng, C. J., Han, L. Y., Yap, C. W., Ji, Z. L., Cao, Z. W., & Chen, Y. Z. (2006). Therapeutic targets: progress of their exploration and investigation of their characteristics. Pharmacological Reviews, 58, 259–279. doi:10.1124/pr.58.2.4

Zheng, W., & Tropsha, A. (2000). Novel Variable Selection Quantitative Structure-Property Relationship Approach Based on the k-Nearest-Neighbor Principle. Journal of Chemical Information and Computer Sciences, 40, 185–194. doi:10.1021/ci980033m

Zhu, S., Okuno, Y., Tsujimoto, G., & Mamitsuka, H. (2005). A probabilistic model for mining implicit 'chemical compound-gene' relations from literature. Bioinformatics (Oxford, England), 21(Suppl 2), ii245–ii251. doi:10.1093/bioinformatics/bti1141

Zou, H., & Hastie, T. (2005). Regularization and variable selection via the elastic net. Journal of the Royal Statistical Society. Series B. Methodological, 67(2), 301–320. doi:10.1111/j.1467-9868.2005.00503.x

Zozulya, S., Echeverri, F., & Nguyen, T. (2001). The human olfactory receptor repertoire. Genome Biology, 2(6), RESEARCH0018.

Zuegge, J., Fechner, U., Roche, O., Parrott, N. J., Engkvist, O., & Schneider, G. (2002). A fast virtual screening filter for cytochrome P450 3A4 inhibition liability of compound libraries. Quantitative Structure-Activity Relationships, 21(3), 249–256.

About the Contributors

Huma Lodhi obtained her Ph.D. in computer science from University of London. She is a researcher with the department of Computing, Imperial College London. She has published in leading international journals, books, conference proceedings and has edited a volume "Elements of Computational Systems Biology (Wiley Series in Bioinformatics)", (2010) by Huma M Lodhi and Stephen H Muggleton (Editors), Wiley. Her research interests are machine learning and data mining and their application to tasks in bioinformatics, chemoinformatics and computation systems biology.

Yoshihiro Yamanishi is a faculty member at Centre for Computational Biology, Mines ParisTech, France. He is also a researcher in the department of Bioinformatics and Computational Systems Biology of Cancer, Mines ParisTech - Institut Curie - INSERM U900. He is working on statistics and machine learning for bioinformatics, chemoinformatics, and genomic drug discovery. He obtained his Ph.D in 2005 from Kyoto University in Japan. He was a post-doctoral research fellow at Center for Geostatistics, Ecole des Mines de Paris from 2005 to 2006. He was an assistant professor at Institute for Chemical Research, Kyoto University from 2006 to 2007.

* * *

Nicos Angelopoulos Dr Angelopoulos is a senior researcher in computational Biology at the Institute for Animal Health, UK. He has worked as a researcher in a number of computer science projects on the interface with other disciplines. He has an Artificial Intelligence background and has published in a number of leading conferences in the area. His computing interests focus on the integration of logic programming and probability theory. Specially, machine learning applications of formalisms that use knowledge representation techniques to express existing knowledge within a scientific domain. He has authored a number of open source programs.

Jürgen Bajorath received a diploma and Ph.D. degree (1988) in biochemistry from the Free University Berlin and was a postdoctoral fellow at Biosym Technologies in San Diego (1989-1991). In 1991, he joined the Bristol-Myers Squibb Pharmaceutical Research Institute in Seattle where he became a Principal Scientist. From 1997-2004 he was Senior Director, Computer-Aided Drug Discovery, at MDS Panlabs that, through a series of acquisitions, became the AMRI Bothell Research Center. Since 1995 he has been an Affiliate Professor in the Department of Biological Structure at the University of Washington, Seattle. In 2004, he was appointed Professor and Chair of Life Science Informatics at the University of Bonn. His current research focuses on the development and application of chemoinformatics and computational medicinal chemistry methods.

Viviana Consonni received her PhD in chemical sciences from the University of Milano in 2000 and is now full researcher of chemometrics and chemoinformatics at the Department of Environmental Sciences of the University of Milano-Bicocca (Milano, Italy). She is a member of the Milano Chemometrics and QSAR Research Group and has 10 years experience in multivariate analysis, QSAR, molecular descriptors, multicriteria decision making, and software development. She is author of more than 40 publications in peer-reviewed journals and of the books "Handbook of Molecular Descriptors" (2000) and "Molecular Descriptors for Chemoinformatics" (2009) by R. Todeschini and V. Consonni, Wiley-VCH. In 2006, she obtained the International Academy of Mathematical Chemistry Award for distinguished young investigators and, in June 2009, was elected as the youngest Member of the Academy.

Andreas Hadjiprocopis obtained his PhD at the Computer Science Department of City University, London. He also holds a BEng in Electrical and Control Engineering from the same university. In 2001, Andreas joined, for three years, the Institute of Neurology, University College London as a Research Fellow, funded by Multiple Sclerosis Society, UK. In 2005, Andreas joined the Higher Technical Institute in Lefkosia, Cyprus, as a Lecturer of Computer Science. He teaches artificial intelligence, data communication systems and programming. In 2008, Andreas has been invited to participate as an expert evaluator and rapporteur for the FP7 Marie Curie Actions .Initial Training Networks (ITN). proposals (FP7-People-ITN-2008) / Mathematics and Engineering Panel. Andreas' main interests lie in the field of artificial intelligence (connectionism, neural networks, cellular automata), optimization (simulated annealing, markov chain monte carlo), evolutionary computer graphics and (medical) image processing. His research has yielded many journal and conference papers.

Alper Kucukural is a postdoc researcher in the Department of Molecular Biosciences, Center for Bioinformatics at University of Kansas. He graduated from Istanbul Technical University, Turkey with a B.S. degree in Mathematical Engineering in 2000. He then got his master degree in Systems Analysis in 2004 at the same university. Beside master studies, he worked as a Software Engineer in several companies. Then he got his PhD degree in Bioinformatics field of Biological Sciences and Bioengineering department at Sabanci University, Istanbul, Turkey in 2009. Mainly, he is working on protein structure and ligand binding site characterization using graph theoretical approaches.

Roman Rosipal is a research associate in the Department of Medical Cybernetics and Artificial Intelligence, Center for Brain Research, Medical University of Vienna. He obtained the MSc. degree in mathematics (1999) and the MSc. degree in electrical engineering (1993). After receiving his Ph.D. in computer science at the University of Paisley, UK, in 2001, he was a postdoctoral fellow at the NASA Ames Research Center (2001-2004) and at the Austrian Research Institute for Artificial Intelligence (2004-2007). Dr. Rosipal is an applied statistician with over fifteen years of experience in mathematical modeling, signal processing, pattern recognition, and statistical methods and algorithms development. He has worked on a variety of problems including brain-computer interface; human fatigue, workload and alertness monitoring and prediction; modeling of the human sleep process; automated monitoring of anesthesia; mass spectrometry based detection of primary lung cancer; and chemometric data analysis. He is a pioneer in the area of support vector machines and nonlinear kernel-based learning applied to neurophysiological data. He is the author or co-author of more than twenty refereed research papers and four book chapters. He has given several invited talks at universities, research institutions and conferences in the USA, Europe and Japan. Dr. Rosipal is a member of two editorial boards and he regularly reviews for several international journals in the field of machine learning, computational and applied statistics, and biomedical engineering. He also regularly serves as a program committee member of international conferences related to his research topics.

Michael Schmuker studied Biology and Computer Science in Freiburg, Germany and Montpellier, France and graduated in 2003 with a thesis on computational principles in the primate visual system. He got to know Gisbert Schneider during two interships at Hoffmann-La Roche in Basel, Switzerland. When Gisbert Schneider was appointed the head Beilstein-endowed chair for Chem- and Bioinformatics at Goethe-University in Frankfurt, Germany, Michael joined him to do his PhD. Michael now workings as a Postdoc in Randolf Menzel's lab at Freie Universität Berlin, Germany, where he is developing computational models of the olfactory system in order to investigate olfactory learning and memory. Michael Schmuker is associated to the Bernstein Center for Computational Neuroscience (BCCN) in Berlin and a member of the German Neuroscience Society.

Gisbert Schneider studied Biochemistry and Computer Science at the Free University of Berlin, Germany, where he received his Ph.D. in 1994. After several post-doctoral research studies he joined F.Hoffmann La-Roche Pharamceuticals in Basel, Switzerland, in 1997, where he was appointed head of the cheminformatics group. From 2002 to 2009 he held the Beilstein Endowed Chair for Chem- and Bioinformatics at Goethe-University in Frankfurt, Germany. Currently, he is full Professor for Computer-Assisted Drug Design at the Eidgenössische Technische Hochschule (ETH) in Zürich, Switzerland. Gisbert Schneider published several textbooks and over 150 research articles on molecular design, and is editor of the journal Molecular Informatics.

O. Uğur Sezerman graduated from Bogaziçi University, Istanbul, (B. Sc. Elect. Eng. 1985, M.Sc. Biomedical Eng. 1987) and received a Ph. D. in Biomedical Engineering (1993) from Boston University, MA, USA. Previously he worked at Boston University and Bogazici University as a researcher and an instructor. He has been at Sabanci University Biological Sciences and Bioengineering Program since 1999. He has established the Computational Biology Laboratory at Sabanci University. His current research interests are molecular modelling, synthetic vaccine and drug design, protein engineering, DNA chips, functional genomics, systems biology ,biomarker detection and other related problems in computational biology. He has over 50 peer reviewed conference and journal publications.

Rahul Singh is currently an associate professor in the department of Computer Science at San Francisco State University. His technical interests are in computational drug discovery and biology, multimedia systems, and multimedia search & retrieval. Prior to joining academia, he was in the industry at Scimagix Inc, where he worked on various problems related to image-based biological information management and at Exelixis Inc, where he founded and headed the computational drug discovery group, which worked on various problems across the genomics-drug discovery spectrum. Dr. Singh received his MSE degree in Computer Science with "excellence" from the Moscow Power Engineering Institute and the MS and PhD degrees in Computer Science from the University of Minnesota. His research has received several awards including the prestigious CAREER award of the National Science Foundation and the Frost & Sullivan Technology Innovation Award.

András Szilágyi obtained his MSc in physics and biophysics and his PhD in 1999 in biology from the Eötvös Loránd University in Budapest, Hungary. He conducted research in the Institute of Enzymology of the Hungarian Academy of Sciences in Budapest, Hungary from 1993 to 2001 and 2005 to this day. He was a postdoctoral fellow in the Center for Bioinformatics at the University at Buffalo in Buffalo, New York from 2002 to 2005. Currenty, he is a visiting professor in the Center for Bioinformatics at the University of Kansas in Lawrence, Kansas. He is mainly interested in computational approaches to study protein structure and folding.

Roberto Todeschini is full professor of chemometrics at the Department of Environmental Sciences of the University of Milano-Bicocca (Milano, Italy), where he constituted the Milano Chemometrics and QSAR Research Group. His main research activities concern chemometrics in all its aspects, QSAR, molecular descriptors, multicriteria decision making, and software development. President of the International Academy of Mathematical Chemistry, president of the Italian Chemometric Society, and "ad honorem" professor of the University of Azuay (Cuenca, Ecuador), he is author of more than 170 publications in international journals and of the books "The Data Analysis Handbook" by I.E. Frank and R. Todeschini, 1994, "Handbook of Molecular Descriptors" by R. Todeschini and V. Consonni, 2000, and "Molecular Descriptors for Chemoinformatics" by R. Todeschini and V. Consonni, 2009.

Martin Vogt Dr. Martin Vogt studied mathematics and computer science at the University of Bonn and holds a degree in computer science. He is currently a research associate at the department of Life Science Informatics at the University of Bonn where he also completed his doctoral thesis on Bayesian methods for virtual screening. Previously he was employed at the Fraunhofer Institut für Angewandte Informationstechnik where he worked on algorithms for image recognition for bioinformatic applications. His research interests focus on algorithmic method development in chemoinformatics especially focusing on data mining and machine learning methods.

Malcolm Walkinshaw obtained both his BSc (1973) and PhD (1976) degrees from the Chemistry Department at the University of Edinburgh. After leading a structure-based drug design group in Sandoz in Switzerland for ten years, he took up the Chair of Structural Biochemistry in 1995 at the University of Edinburgh. He has published over 200 papers on molecular recognition, protein structure and drug discovery. His lab currently consists of 20 research fellows, PhD students and support staff using crystallographic, biophysical and computational approaches to study protein-ligand interactions.

Martin Whittle received a B.Sc. in Chemistry in 1974 and a Ph.D. in 1979, both awarded by the University of Manchester. He is a researcher with an experimental background in laser light-scattering and infra-red photo-ionization. Since 1982 he has worked on the computer simulation of liquids, gels, emulsions and slurries using a variety of methods. He moved to the Department of Information Studies at the University of Sheffield in 2001 where he has worked on data mining of web-search databases and virtual screening in a drug-discovery context. He is currently studying methods of detecting tropical deforestation using satellite radar imagery.

Shuxing Zhang Dr. Shuxing Zhang received his Ph.D. from the School of Pharmacy at the University of North Carolina at Chapel Hill. He is currently Assistant Professor in the Department of Experimental Therapeutics at the University of Texas MD Anderson Cancer Center. With expertise in QSAR, molecular docking and *in silico* ADMET, Dr. Zhang's research is focused on ligand-based and structure-based drug discovery and development, cheminformatics, structural bioinformatics, and systems biology. Dr. Zhang is the recipient of ACS Hewlett-Packard Junior Faculty Award and Chemical Computing Group (CCG) Excellence Award, and his research is mainly supported by US NIH, DoD and other funding agencies. In 2009, Dr. Zhang co-founded Phusis Therapeutics Inc. based at Houston, Texas, involved in the development of pleckstrin homology (PH) domain inhibitors for the treatment of cancer and other diseases.

Yang Zhang received his PhD in Physics at the Central China Normal University in 1996. After spending two years as the Alexander von Humboldt research fellow at Free University of Berlin, Germany, he joined the group of Jeffrey Skolnick as a postdoctoral researcher at the Danforth Center and the University at Buffalo. He is now an associate professor at the Center for Bioinformatics, University of Kansas. The research interest of Zhang's Lab is in protein folding and protein structure prediction. The I-TASSER algorithm developed in his lab was ranked as the No 1 server for automated protein structure prediction in the worldwide CASP competitions in 2006 and 2008. Dr. Zhang is a recipient of the NSF Career Award and the Alfred P Sloan Award 2008.

Holger Fröhlich studied computer science at the University of Marburg (Germany) 1996 - 2002. He did his Master's thesis under the supervision of Prof. Bernhard Schölkopf at the Max-Planck-Institute for Biological Cybernetics in Tübingen, Germany. Afterwards he moved to Prof. Andreas Zell's group at the University of Tübingen, where he finished his PhD about "Kernel Methods in Chemo- and Bioinformatics" in 2006. From 2006 – 2009 he worked as a Postdoc in Bioinformatics at the German Cancer Research Center in Heidelberg (Germany). In February 2009 he joined the Biotech company "Cellzome" in Heidelberg as a Senior Scientist in Bioinformatics and Biostatistics. Holger Fröhlich's research interests cover various applications of machine learning in Bio- and Chemoinformatics, such as QSAR and ADME in silico prediction, estimation of cellular signalling networks from perturbation experiments, biomarker development, gene function prediction and modelling of treatment effects in signalling networks.

Kimito Funatsu Prof. Kimito Funatsu was born on December 3, 1955, Japan. He has got his Doctor degree (Dr. Sci.) from Kyushu University which is one of the old imperial universities (1983) and joined in Prof. Shin-ichi Sasaki's group, Toyohashi University of Technology (1984). During researching in Prof. Sasaki's group he has worked on chemoinformatics field, i.e., Ssructure elucidation system CHEMICS, organic synthesis design system AIPHOS, molecular design. In 2004 he moved to The University of Tokyo as full Professor to continue the same research work. Recently he is trying material design and soft sensor. He is a chair of division of chemical information and computer sciences, Chemical Society of Japan.

Val Gillet is Professor of Chemoinformatics at The University of Sheffield. Her research interests include applications of evolutionary algorithms to problems in chemoinformatics including de novo design, library design, structure-activity relationships and pharmacophore elucidation. She has authored over 90 research papers and has collaborated extensively with industry. She serves on the Editorial Advisory Board of Journal of Chemical Information and Modeling and regularly reviews manuscripts for this and related journals including Journal of Medicinal Chemistry. She is programme coordinator of the first masters programme in chemoinformatics worldwide, co-author of the textbook An Introduction to Chemoinformatics and organiser of the Sheffield triennial conference in Chemoinformatics and the annual short course to industry A Practical Introduction to Chemoinformatics, both of which attract delegates from around the globe.

Kiyoshi Hasegawa Dr. Kiyoshi Hasegawa was born on August 18, 1963 in Kanazawa, Japan. He has got his Ph.D. from Toyohashi University of Technology (1995) and worked at Biostructure group in Chugai Pharmaceutical Company in Japan. During working in Chugai, he has stayed in three months at Swiss Basel Research Center in Roche for developing new de novo design program (2003). He is now principal scientist in his group and is involved in many SBDD projects. He is also the Roche data mining working group member and the secretary general of Japan Chemical Information Science. His interests are de novo design, Chemogenomics, data mining and web application.

Masahiro Hattori is an assistant professor of Bioknowledge Systems Laboratory (Minoru Kanehisa Laboratory), which develops the KEGG database, in Kyoto University, Japan. He is working on developments of both database and algorithm and their applications to various

problems in bioinformatics, including chemical genomics, metabolomics, and chemoinformatics. He obtained his B.S. degree in 1995 from Kyoto University with a major in Physics. He started his bioinformatics researches under Professor Minoru Kanehisa to earn his M.S. degree in 1997 and Ph.D. of Science in 2004 from Kyoto University, Japan. He was an associate instructor at Institute for Chemical Research, Kyoto University from 2000 to 2003.

Hisashi Kashima received his B.E., M.E. and Ph.D. degrees from Kyoto University in Japan in 1997, 1999 and 2007, respectively. He was a researcher of Tokyo Research Laboratory of IBM Research from 1999 to 2009. Since 2009, he has been an Associate Professor at the University of Tokyo in Japan. His research interests include machine learning and its application to bioinformatics, autonomic computing and business intelligence.

Masaaki Kotera is an assistant professor of Bioknowledge Systems Laboratory (Minoru Kanehisa Laboratory), which develop the KEGG database, in Kyoto University of Japan. His main research topic is analyzing enzyme reactions in genomic aspect. He studied organic chemistry to earn his B.S. degree in 1999 at Ritsumeikan University, and studied enzymology to earn his M.S. degree in 2001 at Kyoto University. He earned his Ph.D Science degree in 2005 at Kyoto University for the research analyzing the relationships between genomic information and enzyme reactions. He was a post-doctral research fellow from 2005 to 2008 at Laboratory of Neurochemistry, Enzymology and Systems Biology (Keith Tipton Laboratory), which are responsible for curation of the IUBMB's Enzyme List, in Trinity College Dublin of Ireland.

Hiroto Saigo received the PhD degree in Informatics from Kyoto University, Japan, in 2006. While pursuing PhD, he studied one year at the Institute for Genomics and Biology of University of California, Irvine, USA. He worked as a research scientist for Department of Empirical Inference of Max Planck Institute for Biological Cybernetics, Tuebingen, Germany from 2006 to 2008. He is currently a research scientist at the Department of Computational Biology and Applied Algorithms of Max Planck Institute for Informatics, Saarbruecken, Germany. He is working on development of machine learning and data mining methods and their applications to bioinformatics and chemoinformatics.

Koji Tsuda is research scientist at Computational Biology Research Center, National Institute of Advanced Industrial Science and Technology (AIST), Tokyo, Japan. He was senior research scientist at Max Planck Institute for Biological Cybernetics, Tuebingen, Germany from 2006 to 2008. Also he was research scientist at the same institute in 2003 and 2004. In 2000, he was visiting researcher at GMD FIRST, Berlin, Germany. He received Doctor of Engineering from Kyoto University in 1998. His scientific interests are in the fields of bioinformatics, machine learning, and data mining.

Jean-Philippe Vert is director of the Centre for Computational Biology at Mines ParisTech, and adjunct director of a joint laboratory dedicated to computational and systems biology of cancer at Mines ParisTech, Institut Curie and Inserm, Paris, France. He graduated from Ecole Polytechnique, Palaiseau, France, in 1995, and Ecole des Mines de Paris, France, in 1998. He earned his M.Sc. in 1997 and his PhD in 2001, both in mathematics from Paris 6 University. His research interests include statistics, machine learning, in particular kernel methods, and their applications to chemistry and biology.

Peter Willett Following a first degree in Chemistry from Oxford, Peter Willett obtained MSc and PhD degrees in Information Science from the Department of Information Studies at the University of Sheffield. He joined the faculty of the Department in 1979, was awarded a Personal Chair in 1991 and a DSc in 1997. He was the recipient of the 1993 Skolnik Award of the American Chemical Society Division of Chemical Information, of the 1997 Distinguished Lecturer Award of the New Jersey Chapter of the American Society for Information Science, of the 2001 Kent Award of the Institute of Information Scientists, of the 2002 Lynch Award of the Chemical Structure Association Trust, and of the 2005 American Chemical Society Award for Computers in Chemical and Pharmaceutical Research. Professor Willett is included in Who's Who, is a member of the editorial boards of three international journals, and has over 480 publications describing novel computational techniques for the processing of chemical, biological and textual information. His current interests include: chemical applications of cluster analysis and graph theory; machine learning and similarity approaches for ligand-based virtual screening; and the use of citation data for the evaluation of academic research performance.

Index